AF863155

Topics in Current Chemistry Collections

Editor-in-Chief

Massimo Olivucci, Siena, Italy and Bowling Green, USA
Wai-Yeung Wong, Hong Kong, China

Series Editors

Hagan Bayley, Oxford, UK
Greg Hughes, Codexis Inc, USA
Christopher A. Hunter, Cambridge, UK
Seong-Ju Hwang, Seoul, Korea (Republic of)
Kazuaki Ishihara, Nagoya, Japan
Barbara Kirchner, Bonn, Germany
Michael J. Krische, Austin, USA
Delmar Larsen, Davis, USA
Jean-Marie Lehn, Strasbourg, France
Rafael Luque, Córdoba, Spain
Jay S. Siegel, Tianjin, China
Joachim Thiem, Hamburg, Germany
Margherita Venturi, Bologna, Italy
Chi-Huey Wong, Taipei, Taiwan
Henry N.C. Wong, Hong Kong, China
Vivian Wing-Wah Yam, Hong Kong, China
Chunhua Yan, Beijing, China
Shu-Li You, Shanghai, China

The series *Topics in Current Chemistry Collections* presents critical reviews from the journal *Topics in Current Chemistry* organized in topical volumes. The scope of coverage is all areas of chemical science including the interfaces with related disciplines such as biology, medicine and materials science. The goal of each thematic volume is to give the non-specialist reader, whether in academia or industry, a comprehensive insight into an area where new research is emerging which is of interest to a larger scientific audience. Each review within the volume critically surveys one aspect of that topic and places it within the context of the volume as a whole. The most significant developments of the last 5 to 10 years are presented using selected examples to illustrate the principles discussed. The coverage is not intended to be an exhaustive summary of the field or include large quantities of data, but should rather be conceptual, concentrating on the methodological thinking that will allow the non-specialist reader to understand the information presented. Contributions also offer an outlook on potential future developments in the field.

More information about this series at https://link.springer.com/bookseries/14181

Michael Filatov • Cheol H. Choi • Massimo Olivucci
Editors

New Horizons in Computational Chemistry Software

With contributions from

Mario Barbatti • Leonardo Barneschi • Filip Cernatic
Vsevolod D. Dergachev • Ilya D. Dergachev • Pavlo O. Dral
Emmanuel Fromager • Fuchun Ge • Yang Guo • Jong-Kwon Ha
Yi-Fan Hou • Jianxing Huang • Max Pinheiro Jr • Tae In Kim
Rika Kobayashi • Yibo Lei • Wenjian Liu • Aleksandr O. Lykhin
Robert C. Mauban • Seung Kyu Min • Massimo Olivucci
Yuriko Ono • Daniele Padula • Laura Pedraza-González
Vincent Robert • Mitra Rooein • Bruno Senjean • Yangyang Song
Tetsuya Taketsugu • Takuro Tsutsumi • Sergey A. Varganov
Luca De Vico • Bao-Xin Xue • Ning Zhang

Springer

gives a wide and deep overview of the existing and emerging collaborative tools encompassing the more traditional technologies, such as subversion and git, as well as the novel extended reality and virtual reality technologies, which have a potential to add the elements of "real" human interactions to remote collaboration and education.

One of the emerging technologies that has taken computational chemistry by storm is machine learning – a technology based on the use of the neural networks for organizing and interpolating massive arrays of heterogeneous data. Dral et al. describes a novel platform for atomistic machine learning simulations, the MLatom 2 suite, that, thanks to its modular design, enables an easy extension of its capabilities and simulation models. A novel approach to analyzing chemical reaction dynamics based on the reaction path network is presented by Taketsugu and coworkers. The Reaction Space Projector (ReSPer) technology enables reduction of the complexity of reaction paths on multiple potential energy surfaces by considerably reducing its dimensionality and projecting the dynamics trajectories onto the easily comprehensible low-dimensional reaction space. Along a similar venue, Varganov and coworkers propose a nonadiabatic statistical theory package NAST designed for analyzing and predicting kinetics of spin-dependent chemical processes. At a modest computational cost, NAST is capable of predicting the probabilities and rate constants of transitions between the electronic states of different spin multiplicities with the account for statistical and quantum effects.

One of the main hurdles in computational modeling of the reaction dynamics, especially the dynamics of the non-adiabatic reactions occurring in the excited electronic states, is setting up the appropriate and reproducible models with the inclusion of the effect of environment. Perdaza-González et al. describe the current status and future development perspectives of the Automatic Rhodopsin Modeling (ARM) protocol – a web based platform for designing multiscale computational models for the simulation of light induced processes occurring in a diverse family of light sensitive proteins. A special emphasis is put on standardization of the models and reproducibility of the computational results. A novel concept in the nonadiabatic dynamics simulations is presented by Min and coworkers, which is based on the exact factorization of the electron-nuclear wavefunction approach. The PyUNIxMD package implements a series of simulation methodologies derived from the exact factorization and enables transparent and efficient setup of the trajectory simulations. When simulating the nonadiabatic dynamics of photoexcited molecules, the computational chemistry methodologies are tasked with providing on-the-fly information on the energies, and gradients of the electronic states characterized by the strong electron correlation and multiconfigurational electronic structure. Song et al. present an overview of a series of novel approaches to the computation of such electronic states within the framework of multiconfigurational many-body perturbation theory. An emerging concept of ensemble density functional theory and its application to obtaining the excited states in a time-independent fashion is described by Fromager and coworkers. The emerging methodology holds a great promise to produce the accurate information on various excited states at a modest computational cost; what is precisely needed in the realm of modeling the excited state reaction dynamics.

We believe that this collection will prove useful and interesting for readers as an overview of the emerging computational tools and concepts in the realm of computational and theoretical chemistry. We would like to thank all the contributors to this collection. A special gratitude goes to the staff at Topics in Current Chemistry and the editorial board for offering us the opportunity to put together this edition.

Topics in Current Chemistry (2021) 379:41
https://doi.org/10.1007/s41061-021-00354-6

REVIEW

Check for updates

Technological Advances in Remote Collaborations

Rika Kobayashi[1]

Received: 16 August 2021 / Accepted: 23 September 2021 / Published online: 15 October 2021
© The Author(s), under exclusive licence to Springer Nature Switzerland AG 2021

Abstract
Sustainable scientific software needs a strong collaboration framework to ensure continuity by passing on the tools, skills and knowledge needed to the next generation. The COVID-19 pandemic triggered the unexpected effect of accelerating the development of remote platforms and tools to open up collaborations to a wider global community. In this article we outline the elements needed for such a framework, such as education, tools and community building, and discuss the current advances in technology with a nod to the future.

Keywords Remote collaboration · Computational chemistry · Software development · The Future of Meetings · XR

1 Introduction

Computational chemistry has been going strong since the 1950s when theoretical chemists first embraced the advent of digital computers to carry out ab initio molecular orbital calculations [1]. The earliest computational chemistry programs, such as POLYATOM [2], ATMOL [3] and Gaussian [4, 5], of which only Gaussian is still in use today, began to appear in the 1970s. Since then, the list of programs has proliferated, including a wider variety of methods, such as molecular mechanics, molecular dynamics and plane-wave codes. Some software packages are still being developed long after the people who started them have gone. Some have fallen by the wayside. This widely recognised problem of continuity in maintaining these packages—keeping up with new operating systems/architectures and support—has been coined software sustainability [6–8]. Software sustainability requires the inextricably linked resources of funding and people. Certainly, in computational chemistry,

Chapter 1 was originally published as Kobayashi, R. Topics in Current Chemistry (2021) 379: 41. https://doi.org/10.1007/s41061-021-00354-6.

✉ Rika Kobayashi
Rika.Kobayashi@anu.edu.au

1 ANU Supercomputer Facility, Leonard Huxley Building 56, Mills Rd, Acton, ACT 2601, Australia

programming skills do not lead to a secure career path, especially in academia. As Katz [7] recognises, Software Developers need to "keep themselves supported/employed". In recognition of this, in 2016 the United States government funded the Molecular Sciences Software Institute (MolSSI) [9] to serve "as a nexus for science, education and cooperation serving the worldwide community of computational molecular scientists". This is the only such initiative in existence to our knowledge. A similar initiative led by Peter Gill was proposed to the Australian Government in 2014 but failed to gain funding. However, recognition and funding are only the beginning. Sustainable scientific software needs a strong collaboration framework.

The 2020 COVID-19 pandemic had an unintended effect of bringing the world closer together. The halt to international travel disrupted scientific exchange and researchers scrambled to find ways of making working virtually more effective and software platforms reinvented themselves. One such endeavour was the Future of Meetings (TFOM) symposium, which was organised by a like-minded community with the aim of investigating best practice in how to work, educate, collaborate and interact virtually. It became a truly cross-disciplinary symposium, encouraging us to explore key themes and share ideas from a variety of people and perspectives we don't normally interact with. The symposium proved highly successful and our findings were made into a report [10] and an invited comment from Nature [11]. One theme of the symposium was "Technology to help us collaborate virtually" and this brought to the fore a range of online tools and initiatives that could provide the infrastructure for scientific software sustainability.

2 Education

Probably the foremost issue for the continued development of scientific software is the shortage of programmers. This is recognised worldwide, as reflected in the many articles decrying the shortage of software engineers, e.g. Lee [12, 13] estimates that by the year 2024 that number is expected to reach 1 million. This problem is perceived as being one of education, spawning many government initiatives. In Australia, the government introduced the Coding Across the Curriculum initiative with the aim to promote the teaching of digital technologies, including coding, across the different year levels in Australian schools [14]. In 2016, President Obama proposed a CS For All initiative with a US$4 billion dollar budget for computer science education in the United States that did not get approved [15]. The European Commission produced a digital education report for Europe in 2019 [16] and a general worldwide overview can be found in a 2019 UNESCO report [17]. Most of these initiatives concern the Tech and IT sector and target languages, such as JavaScript, Java and Python [18]. Few address the shortage of scientific programmers, except in the rapidly growing field of data science. However, when it comes to software in the applied sciences, certainly in high-performance computing (HPC), the majority of programs are written in Fortran and C/C++ [19]. From a survey in 2015, Rouson (Rouson, personal communication) reported on the programming languages used at NERSC (National Energy Research Scientific Computing); Fortran accounted for close to 60%, followed by C++ and C, at about 35% and 31%, respectively. Fortran was found to be the primary language for 23 of the 36 top codes, yet it is a language that is no longer widely taught. See also the entertaining talk by Roland Lindh in this series [20].

The technical skills shortage has been partly addressed by the rise in online learning [21]—a natural extension of distance education—suited particularly to programming and computer-related courses because of its basis in a digital environment. Worldwide government cuts in funding for education have made distance education more appealing as an easy source of revenue, leading to a highly competitive, and growing, education industry. At the turn of the century, the market for higher education through distance learning was estimated at US$ 300 billion. In 2019, pre-COVID times, Renub Research estimated that the online education market would reach US$ 350 billion by 2025 [22]. Renub's report highlighted online course providers Coursera [23] and Udacity [24] and indeed their most popular offerings are computer science related: programming, machine learning/artificial intelligence (ML/AI), data science. The majority of courses, as mentioned above, do not target scientific programmers. To date, the only online Fortran programming courses that could be found online were provided through Udemy [25] and Tutorialspoint [26], though there is further material available online in the form of university lecture notes and tutorial handouts. In fact, many online courses just provide lecture material, often in the form of videos and handouts. The more sophisticated online programming courses leverage the digital environment by embedding interactive exercises, as with the Tutorialspoint Fortran course, assessment and a discussion forum. At this point in time until this skills shortage is redressed, as has been done for the IT sector, passing on these advanced programming skills is in the hands of the scientific community.

3 Collaboration Tools

The global COVID-19 lockdowns found many technical platforms and tools coming into being or reinventing themselves to adapt to the remote working landscape as can be seen in Fig. 1 (a composite adapted and updated from several sources [27–29]). Video conferencing platforms became schools, fitness centers, places of worship; social apps, such as for chat and gaming, turned into tools for remote work.

The relevant tools from a programming perspective without a doubt would start with code hosting platforms or version control repositories. These have been in use for many years, starting with early version control systems [30], such as CVS [31] and Subversion [32], which were essentially a mechanism for tracking code revisions. However, as software development expanded and became more complex, often involving many authors working concurrently on different parts of code there was a need to have some form of coordination giving rise to distributed revision control, led by Git from Linus Torvalds in 2005 for development of the Linux kernel [33]. The majority of software packages are probably now hosted on such a code repository, the main ones in the scientific community being Github [34] and Gitlab [35]. This is often coupled with a Continuous Integration system. The idea behind the workflow is to keep a master copy of the code on the

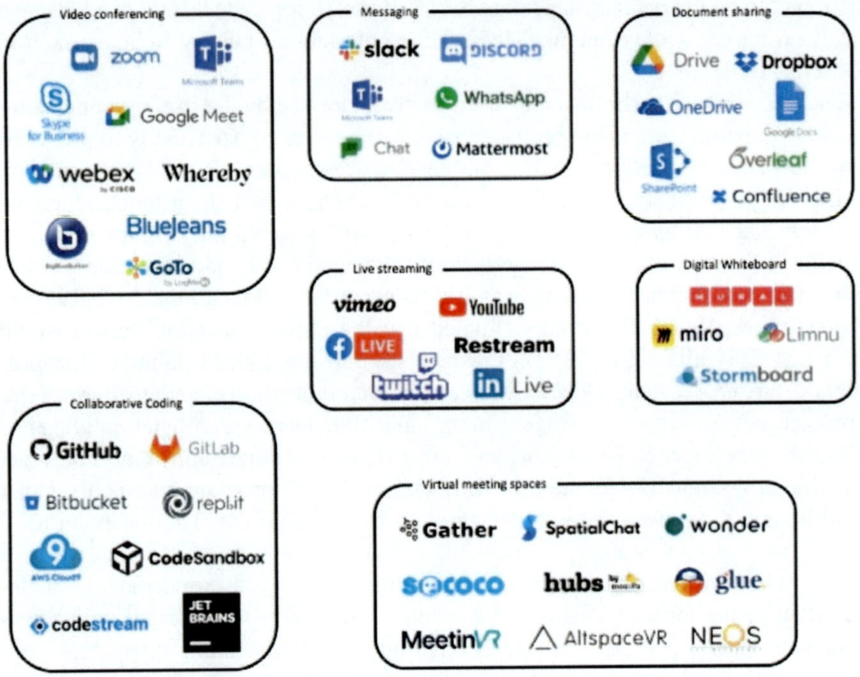

Fig. 1 Remote working tools for collaboration landscape as of 2021

repository which can be downloaded individually and worked on. Changes from the various developers' working copies are then merged back, with the platform providing a mechanism for tracking and resolving dependencies and conflicts. This is the most problematic part of collaborative software development, where independent developers can introduce changes incompatible with each other, e.g. reusing the same variable or changing the underlying structure. One way to mitigate this is to check out the code regularly to keep as close to the master copy as possible, the ideal being to spend less time merging the change than making the change itself—"integration hell". Continuous integration automates this merge to a frequent basis, at least daily, together with running a set of unit tests.

For the extended complicated software suites that make up the bulk of computational chemistry software, new developments can be a long time in the making, rendering "integration hell" unavoidable. Recent times, perhaps in mitigation of this or in conjunction with the popularity of "pair programming", have seen the springing up of real-time collaborative coding platforms where developers work on the same piece of code. This is already something that has been in place for a while with document sharing such as Google Docs [36] and Overleaf [37]. There will probably still be a need for a set of privileged developers to approve code changes but it is believed that the main advantage will be being able to see concurrent development as it happens, making potential conflicts more noticeable. Again, of the most popular collaborative coding tools [38], few target Fortran and it is still early days so not clear whether this approach can scale to a complicated software suite with a number of modules and developers.

4 Community

The biggest concern during the COVID-19 global shutdown was arguably what could be termed the human factor, described variously as lack of connectedness, the need for "real" interactions, and is the main argument for "return to the office" (see next section). There was already a movement towards the formation of global research communities through the rise of collaborative hubs, also known as virtual research environments or science gateways [39]. The idea behind these is to provide a technological infrastructure of shared computational resources: software, data, tools, workflows, HPC access, through a web portal or apps, thus enabling research to a broader scientific community. There now exist many well-established communities, as can be seen in the Special Issue on International Science Gateways 2017 [39], and these have proved successful in some disciplines, notably the Galaxy Project [40], started in 2005, which, according to their latest report [41], had "served hundreds of thousands of users, been used in >5700 scientific publications, and provided 500+ developers with a framework provisioning accessible, transparent and reproducible data analysis".

There is already some online community through the various code repository platforms but a true collaborative hub should have elements of:

- venue—a central website accessible to all and possibly distributed regional hubs to serve a more localized community;
- repositories for collecting and sharing software, tools, data, workflows;
- active community that can communicate synchronously and asynchronously for the exchange of ideas and training;
- (optionally) access to high-end HPC facilities as part of the workflow.

In the computational chemistry world there are existing initiatives, such as NanoHub [42], and nascent hubs, such as AiiDA [43] and Edison [44]. Nanohub, having been established in 2002, is the oldest of these and describes itself as a science gateway supporting a global Network for Computational Nanotechnology community within a cloud environment. It provides a wealth of resources, including training courses, discussion forums, simulation and modeling tools, and a computing environment in which to run them [45]. Its main usage and success appears, however, to be in delivering courses and enabling simulations—supporting an application community rather than a programming and development one.

On the other hand, AiiDA started from the Quantum Espresso [46] developers community, primarily for building an infrastructure providing tools for designing, deploying and analysing materials science simulations, integrated into HPC environments. However, simultaneous efforts since into education and collaboration have expanded its range into Materials Cloud—a web platform for computational materials scientists to "share their work and promote open science" [47].

Furthermore, the Materials Cloud community have begun taking steps towards defining standards, such as file formats and metadata, to facilitate interoperability. The concept of interoperability, the ability for different groups to exchange data consistently, is recognized as an important part of software sustainability, but as yet has been addressed comparatively little in computational chemistry.

A final quick mention should be given to the talk that initiated and possibly inspired this symposium series "A Web Platform for Scientific Collaborations", lectured by Cheol Ho Choi [48]. Leveraging the principles of modular environments, they have created a web platform based on sharing and running computational chemistry modules and workflows via a graphical pipeline and opened it up to the wider quantum chemistry community in the hopes of establishing a scientific software ecosystem.

However, these collaboration hubs are still rooted in two-dimensional screens and do not authentically fill the gap of the lamented missing "real" human interactions—the body language, corridor conversations and serendipitous interactions. Continued TFOM activities have allowed us to explore further these social aspects, especially whether extended reality (XR) and immersive technology can help add a social human factor into our virtual offices and conferences. To this end, we have held a variety of events in virtual reality (VR) platforms such as Altspace [49], NEOS [50] and Glue [51]. These were engaging and fully immersive, but the immaturity of the platforms and the technological requirements do not make them practical today. We were able to explore this aspect further by being given the opportunity to discuss "The Future of Meetings: Working in XR?" with the XR Developer Community through a Birds of a Feather Session at SIGGRAPH 2021 [52]. As part of the session, the attendees were polled informally on a variety of questions concerning the state of XR (for the complete set see Ref. [53]). The poll was not a rigorously conducted exercise so not too much can be read into it, but it was indicative that the albeit small sample felt XR was able to substitute "real" human interactions and that the industry believed that we will be seeing meaningful change in the near future (Fig. 2).

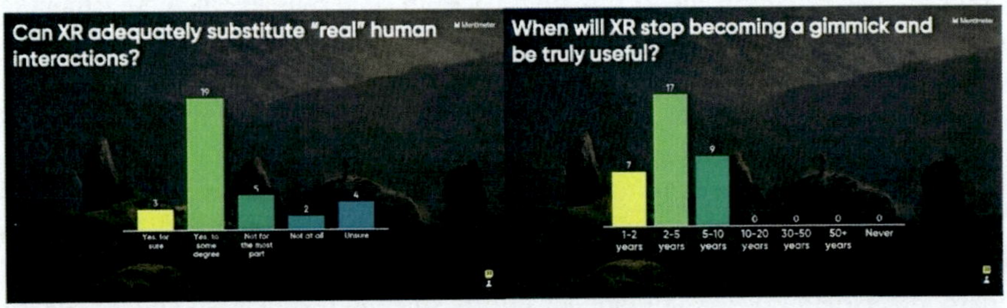

Fig. 2 Informal poll results on various aspects of working in XR from our Birds of a Feather Session at SIGGRAPH 2021

5 Virtual Conferences

In the early days of the COVID-19 pandemic an article appeared in Nature headed "A year without conferences?" [54] discussing the impact on researchers and raising the prospect of a need to rethink the concept of meetings. This article was very quickly countered by examples of successful virtual conferences from around the world, spanning many disciplines, notably one from the Virtual Winter School on Computational Chemistry, which has been running annually since 2015 [55]. Traditionally, conferences have been a means for scientific communities to meet and share knowledge, but these stemmed from a time when communication was slow and travel not so easy. However, technological advances as described here have blurred the need to travel to achieve these outcomes. The pros and cons of virtual conferences were explored in depth in the aforementioned Future of Meetings symposium [10, 11] and continue to be explored through various initiatives by the TFOM community [56]. The definite "pro" of virtual conferences has, without doubt, been their accessibility, inclusivity and sustainability. On the whole, virtual attendance figures have been much higher than their in-person equivalents as the cost of travel is no longer a barrier to attendance. Junior researchers have reporting increased confidence and feeling of safety in the virtual environment and there is significantly less harm to the environment. TFOM calculated the symposium produced 1420 kg of CO_2 equivalent compared to an in-person equivalent of 280,000 kg of CO_2. The biggest disadvantage, certainly for Australia, has been juggling timezones, coupled with the difficulty in separating conference and domestic duties as few international conferences overlap with normal working hours. For developing countries technological accessibility is the biggest problem and a recent a poll of 900 Nature readers [57] cited "poor networking opportunities" as the biggest drawback.

6 Future Outlook

With the effects of the COVID-19 pandemic still being felt around the world today, it is probable that there will be long-lasting changes to the way we work and meet. Major companies, especially the tech giants, are reducing their office space with some going fully remote [58] following the lead in May 2020 of Twitter and Facebook who announced that they would give staff the option to work remotely permanently [59]. Together with the rise in the globally competitive Distance Education industry, online learning is becoming more accepted. The COVID-19 pandemic provided momentum for overcoming the potential barrier to adoption of remote teaching practices that had been considered niche, such as flipped teaching. Such practices have become mainstream, especially as more educators are recognising their effectiveness in this digital age. The perception that online degrees were not "real" degrees is diminishing now that many students have been able to experience direct comparisons. Similarly, Virtual Winter

School, which had been motivated by the desire to make accessible to a wider audience experts in the field they would not normally be able to hear live or interact with, was attended regularly by participants from less advantaged countries. The 2021 School had noticeably more attendance, typically reaching about 200 for most sessions, more than double the usual attendance, and especially from the more established community. The level of engagement demonstrated that the Virtual School is a viable format for fruitful scientific exchange and hopefully this participation level will continue.

There is evidence that scientists want virtual meetings to stay after the COVID pandemic from the Nature poll [57] and reflected in our own survey from TFOM shown in Fig. 3. The concept of "hybrid" is gaining in popularity and it could be that the future will be some mixture of in-person and virtual. However, to be done well, virtual collaborations, whether conferences, meetings or teaching, need more effort, from planning to delivery. TFOM activities have shown how virtual conferences can be effective and subsequently we have been continually called on to advise on best practice for a variety of virtual initiatives. The obvious benefits of accessibility, inclusivity and sustainability, as highlighted in our conference experiences, are still competing with the drawbacks of time-zones, technological accessibility and the human factor, though XR may soon be able to provide a solution for that. TFOM is seeing disheartening signs of people wanting to take what they think is the easy option, i.e. go back to the way things were before. The future will be determined by the people who can see what virtual can do versus the people who see what it can't.

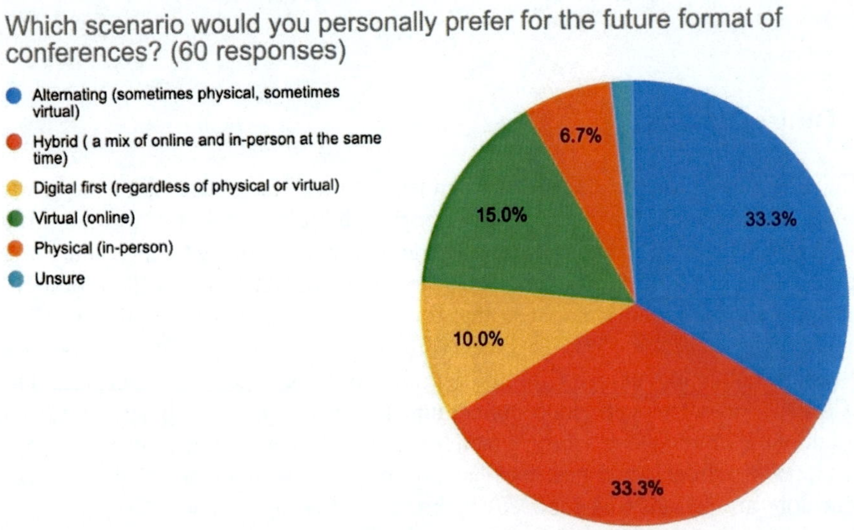

Fig. 3 Attendee preferences for the future format of conferences from The Future of Meetings symposium [10]

7 Conclusion

Sustainable scientific software needs a strong collaboration framework. With the lifetime of computational chemistry software packages exceeding that of the developers who began them, there needs to be a mechanism to pass down the generations the software, tools, skills and knowledge to maintain continuity. The COVID-19 pandemic opened up the world to a potentially global community of developers, whether through the imaginative creation of remote collaboration platforms and tools or just by forcing people to take the digital plunge. It is still early days to know what work in the post-pandemic world will look like, whether we just go back to how we used to do things or whether these remote innovations will be embraced and developed further. There is now a wealth of tools out there to help us meet and work, and possibly even develop software, better virtually. Through TFOM, we have demonstrated how these tools can be used to return to a better normal. Now is the time to keep the momentum going and make use of them to build the foundation of a solid and global software development community.

Acknowledgements Thanks to the TFOM community, especially Vanessa Moss, Glen Rees and Patrice Rey for accompanying me on this journey, and Aidan Hotan, Chenoa Tremblay, Claire Trenham, Roger Amos and "Earthmark" for cheering us on (and providing invaluable material). Thanks also to Damian Rouson of Berkeley Lab for his excellent "Why Fortran persists" presentation, and the Australian Government Department of Education, Skills and Employment for providing a copy of their Coding Across the Curriculum report.

Declarations

Conflict of interest The author has no conflicts of interest or relevant financial interests to declare.

References

1. Roothaan CCJ (1951) New developments in molecular orbital theory. Rev Mod Phys 23:69. https://doi.org/10.1103/RevModPhys.23.69
2. Moskowitz JW, Snyder LC (1977) Polyatom: a general computer program for ab initio calculations. In: Schaefer HF (ed) Methods of electronic structure theory. Modern theoretical chemistry, vol 3. Springer, Boston
3. ATMOL. http://www.chilton-computing.org.uk/acl/applications/qc/p003.htm. Accessed 23 July 2021
4. Gaussian.com. Expanding the limits of computational chemistry. https://gaussian.com/. Accessed 23 July 2021
5. Frisch MJ, Trucks GW, Schlegel HB, Scuseria GE, Robb MA, Cheeseman JR, Scalmani G, Barone V, Petersson GA, Nakatsuji H, Li X, Caricato M, Marenich AV, Bloino J, Janesko BG, Gomperts R, Mennucci B, Hratchian HP, Ortiz JV, Izmaylov AF, Sonnenberg JL, Williams-Young D, Ding F, Lipparini F, Egidi F, Goings J, Peng B, Petrone A, Henderson T, Ranasinghe D, Zakrzewski VG, Gao J, Rega N, Zheng G, Liang W, Hada M, Ehara M, Toyota K, Fukuda R, Hasegawa J, Ishida M, Nakajima T, Honda Y, Kitao O, Nakai H, Vreven T, Throssell K, Montgomery JA Jr, Peralta JE, Ogliaro F, Bearpark MJ, Heyd JJ, Brothers EN, Kudin KN, Staroverov VN, Keith TA, Kobayashi R, Normand J, Raghavachari K, Rendell AP, Burant JC, Iyengar SS, Tomasi J, Cossi M, Millam JM,

Klene M, Adamo C, Cammi R, Ochterski JW, Martin RL, Morokuma K, Farkas O, Foresman JB, Fox DJ (2016) Gaussian 16, Revision C.01. Gaussian Inc., Wallingford
6. Blanton B, Lenhardt C (2014) A scientist's perspective on sustainable scientific software. J Open Res Softw 2:17. https://doi.org/10.5334/jors.ba
7. Katz DS (2018) Fundamentals of software sustainability. https://danielskatzblog.wordpress.com/2018/09/26/fundamentals-of-software-sustainability/. Accessed 11 July 2021
8. Hannay J, MacLeod C, Singer J, Langtangen HP, Pfahl D, Wilson G (2009) How do scientists develop and use scientific software?" In: Proceedings of the 2009 ICSE Workshop on Software Engineering for Computational Science and Engineering, SECSE '09, Washington, DC, USA, IEEE Computer Society, pp 1–8
9. MolSSI–The Molecular Sciences Software Institute. https://molssi.org/. Accessed 23 July 2021
10. Moss VA, Hotan AW, Kobayashi R, Rees GA, Siegel C, Tremblay CD, Trenham CE, Engelke U, Gray A, Hurley-Walker N, Roos G (2020) The future of meetings: outcomes and recommendations. Zenodo. https://doi.org/10.5281/zenodo.4345562
11. Moss VA, Adcock M, Hotan AW, Kobayashi R, Rees GA, Siégel C, Tremblay CD, Trenham CE (2021) Forging a path to a better normal for conferences and collaboration. Nat Astron 5:213–216. https://doi.org/10.1038/s41550-021-01325-z
12. Lee T (2019) How to close the tech skills gap. Scientific American. https://blogs.scientificamerican.com/observations/how-to-close-the-tech-skills-gap/. Accessed 19 July 2021
13. Scarpelli B, Miller N, Stephens R (2017) State of the app economy, 5th edn. Act: The App Association
14. Australian Government Department of Education and Training (2018) Evaluation of the Coding Across the Curriculum program Australia: dandolopartners
15. Wills B (2016) The United States of coding. New America. https://www.newamerica.org/weekly/united-states-coding/. Accessed 19 July 2021
16. European Commission/EACEA/Eurydice (2019) Digital education at school in Europe. Eurydice Report. Publications Office of the European Union, Luxembourg
17. Storte D, Webb M, Bottino R, Passey D, Kalas I, Bescherer C, Smith J, Angeli C, Katz Y, Micheuz P, Røsvik S, Brinda T, Fluck A, Magenheim J, Anderson B, Fuschek G (2019) Coding, programming and the changing curriculum for computing in schools. Report of UNESCO/IFIP TC3 Meeting at OCCE—Wednesday 27th of June 2018, Linz, Austria
18. Hughes O (2021) These are the programming languages most in-demand with companies hiring. TechRepublic. https://www.techrepublic.com/article/these-are-the-programming-languages-most-in-demand-with-companies-hiring/. Accessed 19 July 2021
19. Partee S (2021) Why are climate models written in programming languages from 1950? PARTEE.IO. https://partee.io/2021/02/21/climate-model-response/. Accessed 15 July 2021
20. Lindh R (2021) Strategies for the OpenMolcas Legacy codes—fifty shades of Fortran [Video]. YouTube. https://youtu.be/LcDU_O4fmXg. Accessed 8 Mar 2021
21. Koksal I (2020) The rise of online learning. Forbes. https://www.forbes.com/sites/ilkerkoksal/2020/05/02/the-rise-of-online-learning/. Accessed 13 Nov 2020
22. Renub Research (2019) Online education market & global forecast, by end user, learning mode (self-paced, instructor led), technology, country, company. https://www.renub.com/online-education-market-p.php. Accessed 13 Nov 2020
23. Coursera. https://www.coursera.org/. Accessed 31 July 2021
24. Udacity. https://www.udacity.com/. Accessed 31 July 2021
25. Udemy Fortran Courses. https://www.udemy.com/topic/fortran/. Accessed 31 July 2021
26. Tutorialspoint —Learn Fortran. https://www.tutorialspoint.com/fortran/index.htm. Accessed 31 July 2021
27. Wengi F (2020) Remote work (WFH) tech landscape pauaventures. https://medium.com/paua-insights/remote-work-wfh-tech-landscape-85627ced3410. Accessed 23 Oct 2020
28. Out of office: 65+ startups helping you work from home (2020) CBINSIGHT. https://www.cbinsights.com/research/remote-work-from-home-market-map/. Accessed 23 Oct 2020
29. The ultimate remote work tools landscape. https://www.holloway.com/g/remote-work/sections/the-ultimate-remote-work-tools-landscape. Accessed 23 October 2020
30. Rochkind MJ (1975) The source code control system. IEEE Trans Softw Eng 1:364–370
31. Tichy WF (1985) Rcs — a system for version control. Softw Pract Exp 15:637–654
32. Collins-Sussman B, Fitzpatrick BW, Pilato CM (2004) Version control with subversion, 1st edn. O'Reilly Media, Sebastopol, CA

33. Chacon S, Straub B (2014) Pro git. Apress, Berkeley. https://doi.org/10.1007/978-1-4842-0076-6
34. GitHub. https://github.com/. Accessed 4 Aug 2021
35. GitLab. https://gitlab.com/. Accessed 4 Aug 2021
36. Google Docs. https://docs.google.com/. Accessed 4 Aug 2021
37. Overleaf. https://www.overleaf.com/. Accessed 4 Aug 2021
38. Jones B (2021) The top 5 online IDEs for in-browser development developer.com. https://www.developer.com/cloud/top-online-ide/. Accessed 31 July 2021
39. Dahan M, Pirzl R, Gesing S (2020) International science gateways 2017 special issue. Futur Gener Comput Syst 110:320–322. https://doi.org/10.1016/j.future.2020.04.045
40. Galaxy Community Hub. https://galaxyproject.org/. Accessed 4 Aug 2021
41. Afgan E, Baker D, Batut B, van den Beek M, Bouvier D, Čech M, Chilton J, Clements D, Coraor N, Grüning BA, Guerler A, Hillman-Jackson J, Hiltemann S, Jalili V, Rasche H, Soranzo N, Goecks J, Taylor J, Nekrutenko A, Blankenberg D (2018) The Galaxy platform for accessible, reproducible and collaborative biomedical analyses: 2018 update. Nucleic Acids Res 46:537–544. https://doi.org/10.1093/nar/gky379
42. nanoHUB. https://nanohub.org/. Accessed 4 Aug 2021
43. Huber SP, Zoupanos S, Uhrin M, Talirz L, Kahle L, Häuselmann R, Gresch D, Müller T, Yakutovich AV, Andersen CW, Ramirez FF, Adorf CS, Gargiulo F, Kumbhar S, Passaro E, Johnston C, Merkys A, Cepellotti A, Mounet N, Marzari N, Kozinsky B, Pizzi G (2020) AiiDA 1.0, a scalable computational infrastructure for automated reproducible workflows and data provenance. Sci Data 7:300. https://doi.org/10.1038/s41597-020-00638-4
44. Edison MQCP-Modular Quantum Chemistry Package https://mqcp.edison.re.kr/. Accessed 15 Aug 2021
45. Madhavan K, Zentner L, Farnsworth V, Shivarajapura S, Zentner M, Denny N, Klimeck G (2013) nanoHUB.org: cloud-based services for nanoscale modeling, simulation, and education. Nanotechnol Rev 2:107–117. https://doi.org/10.1515/ntrev-2012-0043
46. Giannozzi P, Baroni S, Bonini N, Calandra M, Car R, Cavazzoni C, Ceresoli D, Chiarotti GL, Cococcioni M, Dabo I, Dal Corso A, de Gironcoli S, Fabris S, Fratesi G, Gebauer R, Gerstmann U, Gougoussis C, Kokalj A, Lazzeri M, Martin-Samos L, Marzari N, Mauri F, Mazzarello R, Paolini S, Pasquarello A, Paulatto L, Sbraccia C, Scandolo S, Sclauzero G, Seitsonen AP, Smogunov A, Umari P, Wentzcovitch RM (2009) QUANTUM ESPRESSO: a modular and open-source software project for quantum simulations of materials. J Phys Condens Matter 21:395502. https://doi.org/10.1088/0953-8984/21/39/395502
47. Talirz L, Kumbhar S, Passaro E, Yakutovich AV, Granata V, Gargiulo F, Borelli M, Uhrin M, Huber SP, Zoupanos S, Adorf CS, Andersen CW, Schütt O, Pignedoli CA, Passerone D, Van de Vondele J, Schulthess TC, Smit B, Pizzi G, Marzari N (2020) Materials cloud, a platform for open computational science. Sci Data 7:299. https://doi.org/10.1038/s41597-020-00637-5
48. Choi C (2021) A web platform for scientific collaborations [Video]. YouTube. https://youtu.be/2OnZ0DUigVA. Accessed 10 Mar 2021
49. AltspaceVR https://altvr.com/ Accessed 15 Aug 2021
50. NEOS https://neos.com/ Accessed 15 Aug 2021
51. Glue https://glue.work/ Accessed 15 Aug 2021
52. Rees GA, Kobayashi R and Moss VA (2021) The future of meetings: working in XR? [Conference session]. SIGGRAPH 2021, United States. https://siggraph2021.hubb.me/fe/schedule-builder/schedule/sessions/868127. Accessed 12 Aug 2021
53. SIGGRAPH (2021) Birds of a Feather. https://thefutureofmeetings.wordpress.com/siggraph-2021-birds-of-a-feather/. Accessed 15 August 2021
54. Viglione G (2020) A year without conferences? How the coronavirus pandemic could change research. Nature 579:327–328. https://doi.org/10.1038/d41586-020-00786-y
55. Roos G, Oláh J, Ingle R, Kobayashi R, Feldt M (2020) Online conferences—towards a new (virtual) reality. Comput Theor Chem 1189:112975. https://doi.org/10.1016/j.comptc.2020.112975
56. The Future of Meetings. https://thefutureofmeetings.wordpress.com/. Accessed 22 September 2021
57. Remmel A (2021) Scientists want virtual meetings to stay after the COVID pandemic. Nature 591:185–186. https://doi.org/10.1038/d41586-021-00513-1

58. Stahl A (16 April 2021) The future of offices and workspaces, post-pandemic. Forbes. https://www.forbes.com/sites/ashleystahl/2021/04/16/the-future-of-offices-and-workspaces-post-pandemic/. Accessed 29 July 2021
59. Brownlee D (2020) Twitter, Square announce work from home forever option: what are the risks. Forbes. https://www.forbes.com/sites/danabrownlee/2020/05/18/twitter-square-announce-work-from-home-forever-optionwhat-are-the-risks/. Accessed 31 July 2021

Publisher's Note Springer Nature remains neutral with regard to jurisdictional claims in published maps and institutional affiliations.

Topics in Current Chemistry (2021) 379:27
https://doi.org/10.1007/s41061-021-00339-5

REVIEW

MLatom 2: An Integrative Platform for Atomistic Machine Learning

Pavlo O. Dral[1,2] · Fuchun Ge[2] · Bao-Xin Xue[1,2] · Yi-Fan Hou[1,2] · Max Pinheiro Jr[3] · Jianxing Huang[1,2] · Mario Barbatti[3]

Received: 22 February 2021 / Accepted: 7 May 2021 / Published online: 8 June 2021
© The Author(s) 2021

Abstract

Atomistic machine learning (AML) simulations are used in chemistry at an ever-increasing pace. A large number of AML models has been developed, but their implementations are scattered among different packages, each with its own conventions for input and output. Thus, here we give an overview of our MLatom 2 software package, which provides an integrative platform for a wide variety of AML simulations by implementing from scratch and interfacing existing software for a range of state-of-the-art models. These include kernel method-based model types such as KREG (native implementation), sGDML, and GAP-SOAP as well as neural-network-based model types such as ANI, DeepPot-SE, and PhysNet. The theoretical foundations behind these methods are overviewed too. The modular structure of MLatom allows for easy extension to more AML model types. MLatom 2 also has many other capabilities useful for AML simulations, such as the support of custom descriptors, farthest-point and structure-based sampling, hyperparameter optimization, model evaluation, and automatic learning curve generation. It can also be used for such multi-step tasks as Δ-learning, self-correction approaches, and absorption spectrum simulation within the machine-learning nuclear-ensemble approach. Several of these MLatom 2 capabilities are showcased in application examples.

Keywords Machine learning · Quantum chemistry · Kernel ridge regression · Neural networks · Gaussian process regression

Chapter 2 was originally published as Dral, P. O., Ge, F., Xue, B-X., Hou, Y-F., Pinheiro Jr, M., Huang, J. & Barbatti, M. Topics in Current Chemistry (2021) 379: 27. https://doi.org/10.1007/s41061-021-00339-5.

✉ Pavlo O. Dral
dral@xmu.edu.cn

[1] State Key Laboratory of Physical Chemistry of Solid Surfaces, Fujian Provincial Key Laboratory of Theoretical and Computational Chemistry, Xiamen 361005, China

[2] Department of Chemistry, and College of Chemistry and Chemical Engineering, Xiamen University, Xiamen 361005, China

[3] Aix Marseille University, CNRS, ICR, Marseille, France

1 Introduction

Machine learning (ML) has taken computational chemistry by storm [1–4]. It is often applied to find a relationship between given molecular geometry and quantum chemical (QC) properties. A particularly useful application of such atomistic ML (AML) models is mapping molecular potential energy surfaces (PESs) [5–8]. Creating AML models is, however, a complicated task and requires domain knowledge. Thus, much effort has been put into developing a mathematical foundation and writing specialized software for such simulations.

One of us (POD) started to develop the MLatom program package [9, 10] for atomistic simulations with ML already in 2013 when not many such packages were available. At first, it was written entirely in Fortran and parallelized with OpenMP as a self-contained black-box program for user-friendly calculations. Now, this part comprises the Fortran core of MLatom called MLatomF. Later, MLatomF added a Python wrapper called MLatomPy implementing multi-step tasks such as Δ-learning [11] and self-correction [12]. We have implemented these and other methods developed by ourselves, such as structure-based sampling [12], the KREG model [12], ML-nuclear ensemble approach (ML-NEA) for precise absorption spectrum simulations [13], as well as selected literature methods, such as those based on the Coulomb matrix descriptor [14, 15], in MLatom for development purposes, tighter integration, and higher efficiency (see "Native Implementations"). We have used these native implementations also for developing methods for improving QC Hamiltonian [16], accelerating ML nonadiabatic excited-state dynamics [17], and for PES construction with spectroscopic accuracy by introducing a hierarchical ML (hML) [18] approach.

In recent years, we have witnessed the rapid rise of many exciting new ML models [4, 5]. They are often designed for different applications ranging from very accurate ML PES trained on as few as a hundred molecular configurations of small- and medium-sized molecules [19, 20] to ML models trained on thousands or millions of points to be transferable to large molecules [21, 22]. Each has its own advantages and disadvantages. It is, therefore, highly desirable to be able to test different models before applying them to the problem at hand. This is, however, a formidable task because these models are scattered in many different software packages. Each has its own conventions for input and output.

We face the same problem in our research: when we want to test some promising ML model, there is often a high entry barrier for learning how to use the corresponding package. Sometimes the documentation is very poor, and only interaction with experienced users or developers enabled us to use some packages. Often, some critical functionality, such as hyperparameter optimization, is missing.

Thus, as a pragmatic solution, we have provided the community with an integrated platform based on MLatom that interfaces the selection of popular third-party software packages via MLatomPy written in Python 3.6+[23, 24]. This platform is released as MLatom 2 with all Python interfaces available as open-source, free software for non-commercial use. Importantly, the same input and output structure can now be used for many state-of-the-art, third-party ML models (see Interfaces). We have implemented the interfaces with sGDML [19, 25] (symmetrized gradient-domain ML), GAP and QUIP (providing GAP [26]-SOAP [27] method), TorchANI [28] (providing ANI [21] methods), DeePMD-kit [29] (providing the DPMD [30] and DeepPot-SE [31] methods), and PhysNet [22] programs. This selection of methods covers popular representatives of various types of methods, ranging from those based on kernel methods (KMs) to neural networks (NNs). Our implementation also supports hyperparameter optimization using Bayesian methods with Tree-structured Parzen Estimator (TPE) [32] via the hyperopt [33] package.

The modular structure of MLatom allows easy extension to other models in the future, as it requires only writing a separate independent module for converting MLatom input to the input of the third-party software and parsing the latter's output. A similar approach is also used for interfacing various QC software packages [34]. This differs, however, from an alternative approach where some packages offer only part of an ML model, e.g., only a descriptor of a molecule to be used as an input for ML, as in DScribe [35].

In the following, we provide an overview of MLatom 2 capabilities, and details of native implementations and interfaces. We also demonstrate the application of MLatom 2 to several typical AML simulation tasks (hyperparameter optimization and generation of learning curves), Δ-learning and structure-based sampling, and calculation of absorption spectra.

2 Overview

The philosophy behind MLatom is to provide the community with a black-box-like program that allows a variety of calculation tasks required for ML atomistic simulations to be performed (Fig. 1). The program provides only with user-friendly and intuitive input, and no scripting is required. Under the hood, MLatom is built of modules designed to be independent of each other as much as possible to the extent that many modules can be used as stand-alone programs entirely independent from the main program. As needed, the modules are combined to create a seamless workflow eliminating step-by-step manual calculations.

The calculation tasks in MLatom are ML tasks and data set tasks. ML tasks are calculations involving training and using ML models. Our implementation includes a wide range of such tasks from basic to multi-step procedures: using an existing ML model from a file, creating an ML model and saving it to a file for later use, estimating ML model accuracy (model evaluation to determine the generalization error), Δ-learning [11], self-correction [12], learning curve and nuclear-ensemble spectrum generation [13]. Data set tasks perform all the operations necessary for ML simulations, such as converting geometries given in XYZ format to the input vector $\mathbf{x}$ for ML, splitting the data set into the required subsets (e.g., training, validation, and test), sampling points into these subsets, and performing statistical analysis of ML estimated values (e.g., error concerning reference values). These tasks can be performed either independently from each other, e.g., creating an ML model from

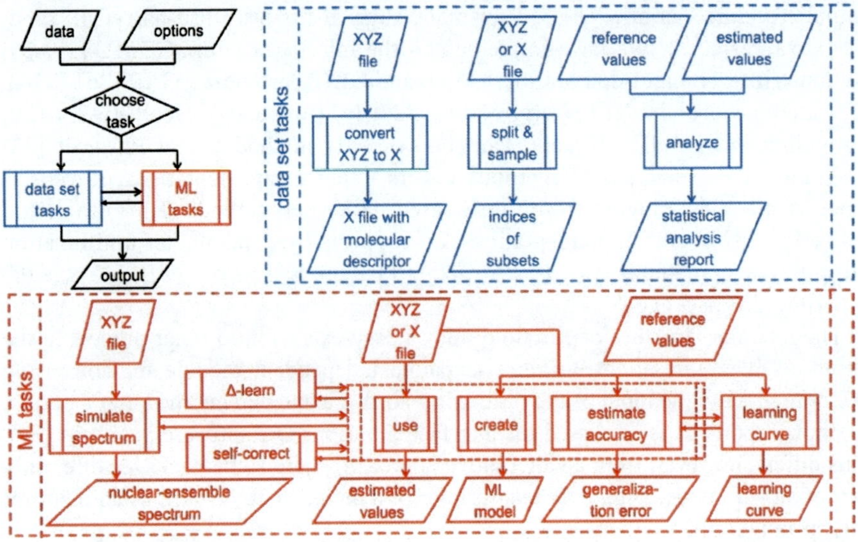

Fig. 1 Overview of tasks performed by MLatom

available input vectors **x** or combined, e.g., by first converting XYZ coordinates to **x** and then creating an ML model. The user just needs to modify several input lines to perform simulations using the first or second option. In the following, we describe each of these tasks and define the most important concepts in ML atomistic simulations.

2.1 ML Tasks

Currently, MLatom supports only supervised learning, which boils down to finding and using an approximating function $\hat{f}(\mathbf{x};\mathbf{h};\mathbf{p})$ that establishes a relationship between the reference values y and input vectors **x** in the training set based on statistically motivated assumptions [36] rather than on purely physical modelling. The approximating function typically has a large number of parameters **p** and so-called hyperparameters **h**.

2.1.1 Using ML Models

Using an existing ML model is conceptually simple as it requires information about the mathematical form of the approximating function and (hyper)parameters. It includes knowing how to transform a molecular geometry into an input vector **x**. One should pay attention, however, to many technical issues, such as ensuring consistent conventions for storing and retrieving this information from the file for long-term re-usability and reproducibility of scientific simulations. Another technical issue is related to performance and accuracy, as the information to be stored can be quite sizable, which can quickly lead to storage and input/output bottlenecks. MLatom saves ML model parameters and other information in a binary format file with a fixed structure for the core ML models and uses native formats of interfaced third-party software without converting them.

In atomistic simulations, we are also often interested in derivatives of properties. For example, in molecular dynamics simulations, we need to know derivatives of energy with respect to atomic coordinates (energy gradients = − forces). Thus, MLatom can estimate both property values and partial derivatives with respect to elements of the input vector or atomic XYZ coordinates.

2.1.2 Creating ML Models

Creating an ML model is already a much more complicated task as one needs to find the model (hyper)parameters in the right way (Fig. 2). This means that one needs to search for optimal values in the parameter space, leading to as low a generalization error as possible [36]. This is not the same as fitting parameters (training) that would give as low an error in the training set as possible. Modern ML methods can easily and exactly reproduce (an extreme case of overfitting) the reference values in the training set [2]. Thus, the standard practice is to set aside a validation set to ensure that training on the remaining data points will not lead to a large error in the validation set, i.e., to avoid overfitting [36]. The remaining data points are called either training or sub-training set in the literature, which adds to the confusion. While both conventions are valid, we prefer to call them sub-training points both here and in MLatom. All data points that are used in creating the ML model we call the training set. This set includes the sub-training and the validation sets in our nomenclature. This convention allows for a fairer and more straightforward comparison of ML models trained on the same number of training points as the validation set, which is used indirectly in training, to be accounted

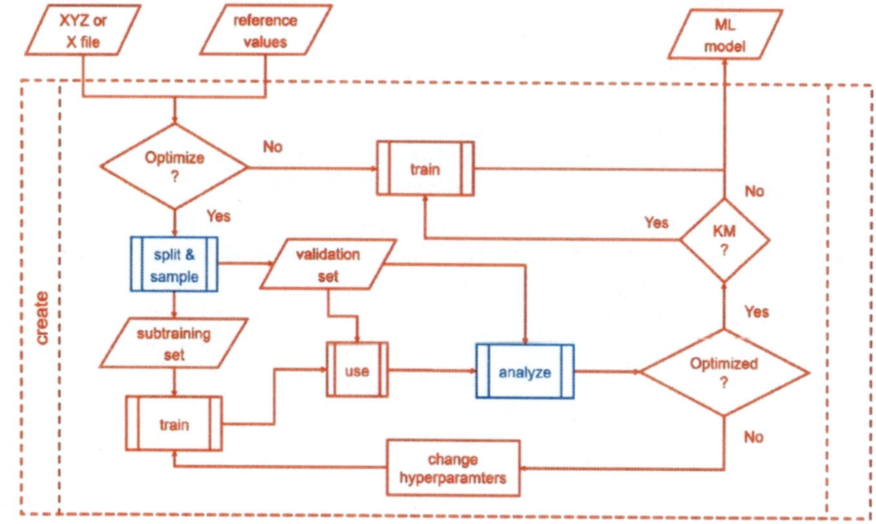

Fig. 2 Creating a machine learning (ML) model with MLatom can involve automatic model selection (hyperparameter tuning) using different types of the training set split into sub-training and validation sets and different sampling procedures

for. For example, if the model is trained on 1000 points, but used another 1000 points for validation, then the reference data is needed for all 2000 points, and such a model cannot be compared to another model trained on only 1000 points without using such a validation set.

When additional information such as derivatives of properties is available, it can be included into the training set too. It is common to train ML models for PESs simultaneously on energies and gradients (or only gradients), which is known to improve the quality of ML PESs significantly compared to fitting only on energies [7, 37, 38]. NNs simply fit parameters to the reference properties and their derivatives [38], while KMs can include the derivative information into their model explicitly [7, 37].

Many knobs exist and can be tuned in the process of finding suitable parameters for an ML model. One such knob concerns the ML model itself, and another deals with the splitting into sub-training and validation sets. As with the first type, while we do not need to touch the model parameters as this is the machine's task, we can influence the model by changing its hyperparameters manually [36, 39]. As a side note, the difference between parameters and hyperparameters is somewhat fussy as the latter can be found by a machine too. Some hyperparameters also enter the ML model, while others do not. The external hyperparameters that do not enter the ML model are clearly different from parameters, but influence the training process, e.g., the regularization hyperparameter in KRR [36].

In any case, (hyper)parameters can be fitted to attempt to reduce the generalization error of the ML model by minimizing the error in the validation set (Fig. 2). For so-called parametric models such as NNs, whose approximating function does not explicitly depend on the training points, finding (hyper)parameters reducing the validation error is usually the end of the story. However, nonparametric models such as kernel ridge regression (KRR) and Gaussian process regression (GPR) depend explicitly on the training points. In their case, not using the validation set for training the final ML model would lose valuable additional information available to the model and reduce its accuracy. Thus, after hyperparameters minimizing error in the validation set for models trained on the sub-training set are found (a procedure also known as model selection), MLatom uses them to train the final model on the whole training set.

During hyperparameter optimization in MLatom, by default, the root-mean-squared error (RMSE) is minimized, but the minimization of another type of error can be requested for native implementations. Alternatively, other defaults can be used by interfaces if they have their own hyperparameter optimization capabilities. When the ML model is trained on several different properties, the error (loss, L) should reflect the error for each of these properties. For example, for models trained on both property values and their derivatives, e.g., energies and energy gradients, the error in MLatom can be calculated as the sum of error for values (L_{val}) and weighted error for gradients in XYZ coordinates (L_{grxyz}) as typically done in the literature [37]:

$$L = L_{val} + w_{grxyz} L_{grxyz} \tag{1}$$

Although this approach gives the user additional flexibility, it has a drawback in that one has to choose an arbitrary parameter w_{grxyz}. To eliminate this parameter, we introduce in MLatom the geometric mean of errors of different properties, which is used by default:

$$L = \sqrt{L_{val} L_{grxyz}} \quad (2)$$

The final model's accuracy also depends on how the training set is split into sub-training and validation sets; this topic is overviewed below in "Splitting and Sampling".

2.1.3 Estimating Accuracy of ML Models

MLatom also provides the means to estimate the accuracy (generalization error) of the ML model (Fig. 3). Since the validation set has already been used to create the ML model, it is good to use another independent test set to assess the model's performance [36]. The entire data set can be split into training and test sets, and points can be sampled into these subsets using procedures similar to those used for hyperparameter tuning (model selection; see "Splitting and Sampling"). Naturally, hyperparameter tuning and model evaluation can be combined in a single calculation task with MLatom.

2.1.4 Multi-step Tasks

The tasks above can be considered as basic. We now turn to describe multi-step tasks built upon these basic tasks. One such task is a Δ-learning task that combines predictions at the low-level QC method with ML corrections to make estimations approaching high-level QC accuracy [11]. Another is a self-correction task that combines several layers of ML models, with each layer correcting the previous layer's residual errors, which is useful for KRR models created with a focus on some region of the molecular PES [12]. Other multi-step tasks are learning curve generation and ML spectrum simulations [13], covered in the next two sub-sections in more detail.

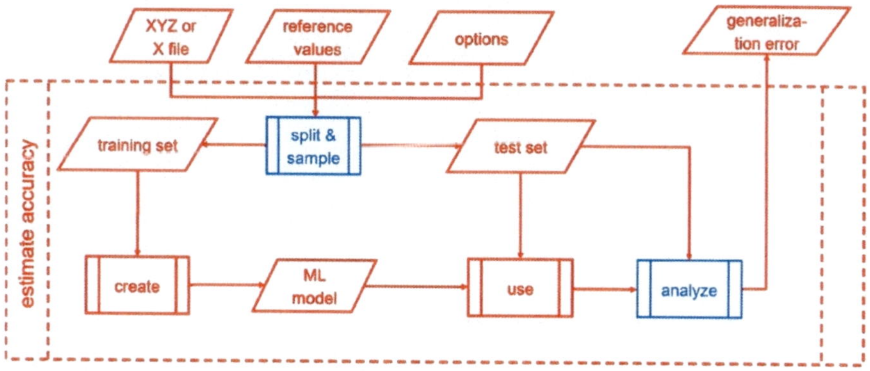

Fig. 3 Estimating the accuracy of the ML model

2.1.5 Learning Curves

Here, the concept of learning curves is used to investigate how ML generalization error depends on training set size. The relationship between ML error ε and training set size typically follows the power-law decay [40]:

$$\varepsilon = \varepsilon_{\mathrm{a}} + \frac{a}{N_{\mathrm{tr}}^{b}} \tag{3}$$

where ε_{a} is the asymptotic error in the limit of the infinitely large training set, a is a nonnegative amplitude, and b is the positive exponent telling us how fast the ML improves by training with more data. These three parameters define the learning curve, giving a more complete characterization of the performance of a given ML model type than a single generalization error estimated for one training set size.

Since the errors drop that fast, the learning curves are often plotted on a log–log scale. In this case, particularly for not too large training sets and small asymptotic errors, a linear curve is often observed [37]:

$$\log(\varepsilon) \approx \log(a) - b\log N_{\mathrm{tr}} \tag{4}$$

A learning curve cannot be drawn without the accuracy-estimating step (see "Estimating Accuracy of ML Models") being taken multiple times. Thus, we provide a dedicated task to automate this procedure (Fig. 4) in MLatom 2.

In the learning curve task, the accuracy of each training set size requested is examined in multiple repeats, where different training points are sampled. Testing with repeats helps to reduce the bias introduced by a specific combination of training and test sets, investigates the statistical variance, and reflects the robustness of an ML model. The results (RMSEs, wall-clock training and prediction times) from each test are recorded in the .csv database file.

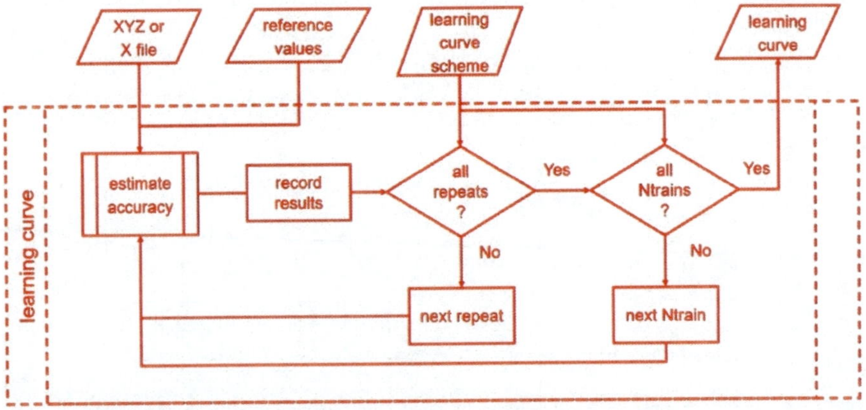

Fig. 4 Flowchart for the learning curve task

2.1.6 ML Nuclear Ensemble Spectra

Electronic spectrum simulation is yet another multi-step task that uses the ML-nuclear ensemble approach (ML-NEA) to automatically calculate quantities like absorption cross sections from as few QC calculations as possible [13]. This approach accelerates the traditional NEA, which usually requires thousands of points in a nuclear ensemble to generate a statistically converged spectrum [41]. Most nuclear ensemble points are relatively similar, making them suitable for using ML for efficient interpolation and replacing most QC calculations. Figure 5 shows this approach and its implementation in MLatom schematically.

The calculations require only an initial geometry, Gaussian 16 [42] input files for optimization and normal-mode calculations as well as for calculation of excited-state properties (excitation energies and oscillator strengths) with the QC method of choice. The user can also provide available pre-calculated data, such as output files with normal-mode calculations or nuclear ensemble geometries. Existing reference QC data can also be provided. MLatom has an interface to Gaussian 16, which automatically invokes and parses the QC calculations' output to get the equilibrium geometry and frequencies. Then, it passes the required data to the interface to Newton-X [43, 44] to generate a nuclear ensemble of 50k conformations sampled from a harmonic-oscillator Wigner distribution of the nuclear coordinates [45].

QC properties for training ML models are calculated iteratively. The number of training points is increased at each iteration to train $2N_{\text{fs}}$ models for excitation energies (ΔE_{0n}) and oscillator strengths (f_{0n}) of transitions from ground-state to each excited state n. Then, we evaluate the convergence of ML predictions. We

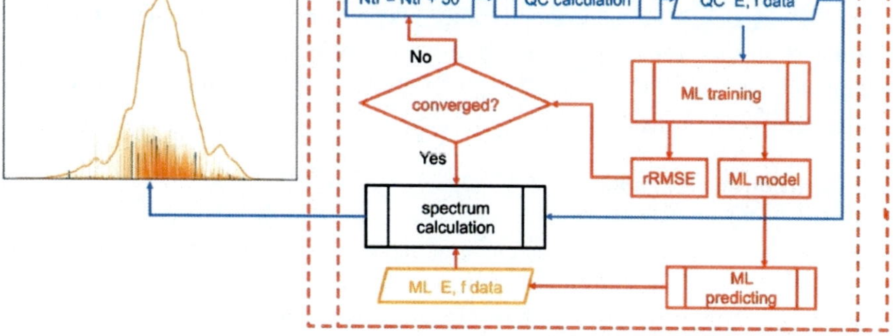

Fig. 5 *Left* Schematic representation of the machine learning–nuclear ensemble approach (ML-NEA). *Right* Implementation of ML-NEA for calculating absorption spectra. *Blue* quantum chemical (QC) data, *orange* ML

first calculate the geometric mean of the RMSE in the validation sets ($\text{RMSE}_{\text{geom}}$) for all ML models, then take the relative change to the previous step (which had N_{batch} = 50 fewer training points) to get the relative RMSE (rRMSE) as the convergence criterion (we consider it converged when rRMSE < 0.1):

$$\text{RMSE}_{\text{geom}}(N) = \sqrt[2N_{fs}]{\prod_{i=1}^{N_{fs}} \text{RMSE}_{\Delta E_i}(N) \bullet RMSE_{f_i}(N)} \tag{5}$$

$$\text{rRMSE} = \frac{\text{RMSE}_{\text{geom}}(N_{\text{tr}}) - \text{RMSE}_{\text{geom}}(N_{\text{tr}} - N_{batch})}{\text{RMSE}_{\text{geom}}(N_{\text{tr}})} \tag{6}$$

If the criterion is satisfied, we use current trained ML models to make predictions for the remaining nuclear ensemble points, substitute any negative ML oscillator strengths with zeros, and then calculate the absorption spectrum with the following equation [41]:

$$\sigma(E) = \frac{\pi e^2 \hbar}{2mc\varepsilon_0 E} \sum_{n}^{N_{fs}} \frac{1}{N_p} \sum_{i}^{N_p} \Delta E_{0n}(\mathbf{x}_i) f_{0n}(\mathbf{x}_i) \frac{1}{\sqrt{2\pi(\delta/2)^2}} \exp\left(-\frac{(E - \Delta E_{0n})^2}{2(\delta/2)^2}\right) \tag{7}$$

where σ is the absorption cross section, m and e are the electron mass and charge, $\hbar$ is the reduced Planck constant, c is the speed of light, ε_0 is the vacuum permittivity, N_p is the number of ensemble points (50k by default), and δ = 0.01 eV is the broadening factor [13]. This summation can become quite computationally intensive for a large number of ensemble points; thus, we implemented it in C++. Currently, only absorption cross sections are available, but the code can be trivially adapted for other electronic spectrum types, like steady-state fluorescence.

2.2 Data Set Tasks

The quality of ML models depends strongly on the descriptor, i.e., the chosen transformation of the molecular structure into the ML input vector **x**. While they are part of any ML model, MLatom currently only allows converting data sets with XYZ coordinates to the descriptors available in its native implementations (see "Native Implementations"). Other data set operations are discussed in the sub-sections below.

2.2.1 Splitting and Sampling

As we have seen, for tasks such as creating and estimating the accuracy of the ML model, the data set can be split into sub-sets: sub-training, validation, training, and test sets [36]. Data points can also be assigned to these sub-sets in different ways.

The simplest approach is to split the sets into sub-sets just once and sample the points randomly [10, 36]. When a data set to be explored with ML is known in advance, farthest-point (FPS) and structure-based sampling (SBS) [12] is possible and preferable to random sampling [10]. In both FPS and SBS, the data set points are sorted using an iterative procedure so that each next point is as far as possible from all the previous selected points [10]. The distance between points i and j is judged by the Euclidean distance $\|\mathbf{x}_i - \mathbf{x}_j\|_2$ between the corresponding descriptors, input vectors $\mathbf{x}_i$ and $\mathbf{x}_j$. In the case of SBS, the first point is the near-equlibrium geometry, in the case of FPS, the two most distant points are chosen as the first two points. Since these sampling procedures are based on an iterative greedy algorithm, they are implemented in MLatomF using Fortran and parallelized with OpenMP, which allows efficient sampling for data sets with tens of thousands of points [10]. SBS applied to sampling points from the PES of a single molecule would lead to underrepresentation of the near-equilibrium geometries as the most distorted geometries will be chosen. As a solution, the geometries are sorted by their Euclidean distance to the equilibrium geometry, sliced into regions corresponding to different degrees of deformation, and, finally, SBS can be performed from each of these regions to obtain a balanced set [2, 10, 12]. This slicing procedure is implemented in MLatomPy.

A more elaborate and slow technique for data set splitting is k-fold cross-validation, with leave-one-out cross-validation being the slowest [2, 10, 36]. In brief, the data set is split into k roughly equal parts, and then each of these parts is used for validation/testing, with the remaining parts used for training; after k-rotations of parts, the whole data set is effectively reused for validation/testing purposes. Thus, this procedure is useful for relatively small training sets.

All these techniques are available for native implementations of MLatom. User-defined sampling into these subsets can also be requested, and then the indices for each of the subsets should be provided to MLatom. By default, 80%:20% random splitting is used for native implementations, while interfaced third-party programs may use their default splitting and sampling for hyperparameter optimization and training (but not for model evaluation).

2.2.2 Analysis of Data

MLatom calculates several built-in statistical metrics for analyzing data, particularly for comparing ML estimations to the reference values. They are overviewed below for the sake of completeness. For N estimated values $\hat{y}$ and reference values y:

- Mean absolute error (MAE):

$$\mathrm{MAE} = \frac{1}{N}\sum_{i}^{N} |\hat{y}_i - y_i| \quad (8)$$

- Mean signed error (MSE, not to be confused with mean squared error):

$$\mathrm{MSE} = \frac{1}{N}\sum_{i}^{N}\left(\widehat{y}_i - y_i\right) \tag{9}$$

- Root-mean-squared error (RMSE):

$$\mathrm{RMSE} = \sqrt{\frac{1}{N}\sum_{i}^{N}\left(\widehat{y}_i - y_i\right)^2} \tag{10}$$

- Arithmetic means of estimated and reference values, respectively:

$$\mu_{\widehat{y}} = \frac{1}{N}\sum_{i}^{N}\widehat{y}_i \tag{11}$$

$$\mu_y = \frac{1}{N}\sum_{i}^{N}y_i \tag{12}$$

- Largest positive and negative outliers as judged by $\widehat{y} - y$.
- Linear regression coefficients a and b in $\hat{y}=a+by$, their standard errors (SEs) with the corresponding correlation coefficient R and its squared value R^2 found by least-squares fitting [46]:

$$b = \frac{\mathrm{ss}_{y\widehat{y}}}{\mathrm{ss}_{yy}} \tag{13}$$

$$a = \mu_{\widehat{y}} - b\mu_y \tag{14}$$

$$\mathrm{SE}(a) = s\sqrt{\frac{1}{N} + \frac{\mu_y^2}{\mathrm{ss}_{yy}}} \tag{15}$$

$$\mathrm{SE}(b) = \frac{s}{\sqrt{\mathrm{ss}_{yy}}} \tag{16}$$

$$R^2 = \frac{ss_{y\widehat{y}}^2}{\mathrm{ss}_{yy}\mathrm{ss}_{\widehat{y}\widehat{y}}} \tag{17}$$

where

$$\mathrm{ss}_{yy} = -N\mu_y^2 + \sum_{i}^{N}y_i^2 \tag{18}$$

$$\mathrm{ss}_{\widehat{y}\widehat{y}} = -N\mu_{\widehat{y}}^2 + \sum_i^N \widehat{y}_i^2 \tag{19}$$

$$\mathrm{ss}_{y\widehat{y}} = -N\mu_y\mu_{\widehat{y}} + \sum_i^N y_i\widehat{y}_i \tag{20}$$

$$s = \sqrt{\frac{\mathrm{ss}_{\widehat{y}\widehat{y}} - ss_{y\widehat{y}}^2/\mathrm{ss}_{yy}}{N-2}} \tag{21}$$

Analogous expressions are used for derived properties, such as partial derivatives; in the latter case, each partial derivative is treated as a data point, e.g. for RMSE of energy gradients $\frac{\partial E}{\partial M}$ in XYZ coordinates evaluated for the PES of a single molecule with N_{at} atoms:

$$\mathrm{RMSE}_{grxyz} = \sqrt{\frac{1}{N \bullet N_{at} \bullet 3} \sum_{i=1}^{N} \sum_{a=1}^{N_{\mathrm{at}}} \sum_{t=1}^{3} \left(\frac{\partial \widehat{E}_i}{\partial M_{i,at}} - \frac{\partial E_i}{\partial M_{i,at}} \right)^2} \tag{22}$$

3 Native Implementations

This section overviews the theory and provides technical details behind the native implementations available in MLatom. All native ML models are currently based on KRR, so this approach is described first, and then we describe the details behind the KREG model and approaches based on the Coulomb matrix. The code for the KREG model, hyperparameter grid search, farthest-point, and structure-based sampling was optimized for efficient computing, while no such efforts were necessarily made for other implementations.

3.1 Kernel Ridge Regression

The approximating function $\widehat{f}(\mathbf{x};\mathbf{h};\mathbf{p})$ in KRR is the sum over all training points N_{tr}:[36]

$$\widehat{f}(\mathbf{x};\mathbf{h};\mathbf{p}) = \sum_{j=1}^{N_{\mathrm{tr}}} \alpha_j k(\mathbf{x}, \mathbf{x}_j;\mathbf{h}), \tag{23}$$

where k is the kernel function, $\mathbf{p}$ are model parameters that include the set of the regression coefficients α and $\mathbf{h}$ are parameters present in the kernel function.

The regression coefficients are found by solving the linear system of equations regularized by adding a small, nonnegative constant value λ to the diagonal elements [36]:

$$\begin{pmatrix} k(\boldsymbol{x}_1,\boldsymbol{x}_1)+\lambda & \cdots & k(\boldsymbol{x}_1,\boldsymbol{x}_{N_{tr}}) \\ \vdots & \ddots & \vdots \\ k(\boldsymbol{x}_{N_{tr}},\boldsymbol{x}_1) & \cdots & k(\boldsymbol{x}_{N_{tr}},\boldsymbol{x}_{N_{tr}})+\lambda \end{pmatrix} \begin{pmatrix} \alpha_1 \\ \vdots \\ \alpha_{N_{tr}} \end{pmatrix} = \begin{pmatrix} y_1 \\ \vdots \\ y_{N_{tr}} \end{pmatrix} \tag{24}$$

or in matrix form:

$$(\mathbf{K}+\lambda\mathbf{I})\boldsymbol{\alpha}=\mathbf{y} \tag{25}$$

where **I** is the identity matrix, **K** is the kernel matrix that evaluates the kernel function for each pair of the training points, and **y** is the vector with reference values. λ is called the regularization parameter and is an external hyperparameter not entering the approximating function itself but used for model selection.

This system of equations has an analytical solution that makes it very attractive. The solution is, however, computationally costly for large data sets as it involves some kind of matrix decomposition that scales as $O(N_{tr}^{3})$, followed by solving the system of equations, which scales as $O(N_{tr}^{2})$. MLatom uses the very computationally efficient Cholesky decomposition by default. Bunch–Kaufman and LU decomposition approaches are also available, which are sometimes necessary when Cholesky decomposition fails [47]. The solution of the above system of equations also requires calculation of the kernel matrix of size N_{tr}^{2}, which can become very large and no longer fit in the available computer memory. Note that, by default, MLatom does not invert the matrix to solve the system of equations using the common expression [36]:

$$\boldsymbol{\alpha}=(\mathbf{K}+\lambda\mathbf{I})^{-1}\mathbf{y} \tag{26}$$

as it is much less computationally efficient and numerically less precise [39].

The kernel functions supported by MLatom are [2, 10, 36, 39]:

- Gaussian:

$$k(\mathbf{x},\mathbf{x}_j)=\exp\left(-\frac{1}{2\sigma^2}\sum_{s}^{N_x}(x_s-x_{j,s})^2\right) \tag{27}$$

- Laplacian:

$$k(\mathbf{x},\mathbf{x}_j)=\exp\left(-\frac{1}{\sigma}\sum_{s}^{N_x}\left|x_s-x_{j,s}\right|\right) \tag{28}$$

- Exponential:

$$k(\mathbf{x},\mathbf{x}_j)=\exp\left(-\frac{1}{\sigma}\left[\sum_{s}^{N_x}(x_s-x_{j,s})^2\right]^{1/2}\right) \tag{29}$$

- Matérn:

$$k(\mathbf{x},\mathbf{x}_j) = \exp\left(-\frac{1}{\sigma}\left[\sum_s^{N_x}(x_s - x_{j,s})^2\right]^{1/2}\right)\sum_{k=0}^{n}\frac{(n+k)!}{(2n)!}\binom{n}{k}\left(\frac{2}{\sigma}\left[\sum_s^{N_x}(x_s - x_{j,s})^2\right]^{1/2}\right)^{n-k} \quad (30)$$

where N_x is the dimensionality of the input vector **x**, and the symbol for hyperparameters **h** entering the kernel function was dropped. All of them have the length-scale parameter σ, which is an internal hyperparameter. The Matérn kernel function has an additional integer hyperparameter *n*. The choice of the kernel function depends on the application, and it can be considered a hyperparameter itself, although MLatom does not automatically make this choice. MLatom performs automatic optimization of the hyperparameters λ and σ on the nested logarithmic grid [10]. Alternatively, the hyperparameters can be optimized using a third-party hyperopt package (see "Interfaces" and its subsection "Hyperopt" below).

As discussed above, we are often interested in the derivatives of properties. Once the KRR approximating function is trained on reference values of properties, it can be differentiated to obtain the required derivatives with respect to the *d* dimension of the input vector **x**:

$$\frac{\partial\hat{f}(\mathbf{x})}{\partial x_d} = \sum_{j=1}^{N_{tr}}\alpha_j\frac{\partial k(\mathbf{x},\mathbf{x}_j)}{\partial x_d}. \quad (31)$$

Thus, calculating the approximating function derivatives requires calculation of kernel function derivatives. The expressions for analytical first-order derivatives of the kernel functions are:

- Gaussian:

$$\frac{\partial k(\mathbf{x},\mathbf{x}_j)}{\partial x_d} = \frac{1}{\sigma^2}(x_{j,d} - x_d)k(\mathbf{x},\mathbf{x}_j) \quad (32)$$

- Matérn with $n>0$:

$$\frac{\partial k(\mathbf{x},\mathbf{x}_j)}{\partial x_d} = \exp\left(-\frac{\|\mathbf{x}-\mathbf{x}_j\|_2}{\sigma}\right)\sum_{k=0}^{n-1}\frac{(n+k-1)!}{(2n)!}\binom{n}{k}\times(n-k)\frac{2}{\sigma^2}\left(\frac{2\|\mathbf{x}-\mathbf{x}_j\|_2}{\sigma}\right)^{n-k-1}(x_{j,d}-x_d) \quad (33)$$

where $\left\|\mathbf{x}-\mathbf{x}_j\right\|_2 = \left[\sum_x^{N_s}(x_s - x_{j,s})^2\right]^{1/2}$ is the Euclidean distance. The exponential and Laplacian kernel functions, as well as the Matérn kernel function with $n=0$ are not differentiable.

Often, we need to know the derivatives in XYZ coordinates. Let us define the molecular XYZ coordinates as **M** that can be transformed into the input vector x via descriptor function *x*(**M**). Then, the partial derivatives in XYZ coordinates for atom *a* and dimension *t* can be obtained using the chain rule as:

$$\frac{\partial \hat{f}(x(\mathbf{M}))}{\partial M_{at}} = \sum_{j=1}^{N_{tr}} \alpha_j \sum_{d=1}^{N_d} \frac{\partial k(\mathbf{x}, \mathbf{x}_j)}{\partial x(\mathbf{M})_d} \frac{\partial x(\mathbf{M})_d}{\partial M_{at}} \tag{34}$$

where $x(\mathbf{M})_d$ are N_d elements of the input vector $\mathbf{x}$ depending on M_{at}. The expression for the first-order derivatives of the descriptors $\partial x(\mathbf{M})_d / \partial M_{at}$ available in MLatom are given in the following sections.

Note that the derivative information can also be included in the training data for KRR, which usually requires implementing higher-order derivatives of the kernel functions [26, 37, 39, 47–50]. Such implementations in MLatom are currently underway and will be released in the near future.

All KRR calculations are implemented in Fortran and OpenMP, and we use the efficient Intel® Math Kernel Library (MKL) LAPACK[51] routines for calculating regression coefficients.

3.2 KREG

KREG is the ML model type designed for constructing accurate PESs of a single molecule [12]. Its name is strictly speaking not an acronym but can be somewhat loosely derived from the first letters of the underlying components: KRR with the RE descriptor and the Gaussian kernel function. The RE descriptor is a vector of all inverse distances $r_{a,b\neq a}$ between atoms a and $b \neq a$ in a molecule normalized relative to the corresponding distances in some reference structure of the same molecule, usually its equilibrium geometry:

$$\mathbf{x}^{\mathrm{T}} = \left[\cdots \; \frac{r^{\mathrm{ref}}_{a,b\neq a}}{r_{a,b\neq a}} \; \cdots \right] \tag{35}$$

The RE descriptor is a global descriptor, meaning that it describes the molecule as a whole, and the KREG model learns the QC property directly without partitioning it into, e.g., atomic contributions as done by many models discussed below. This descriptor is also a complete descriptor, meaning that it uniquely represents the molecular geometry as the latter can always be derived from the descriptor up to rotation, translation, and such symmetry operations as reflection [10]. Rotational and translation invariance is a feature of the RE descriptor that ensures that scalar properties such as total energy are invariant to these operations according to the laws of physics [10].

The RE descriptor does not, however, ensure homonuclear permutational invariance, e.g., that interchange of hydrogen atoms in the methyl group CH_3 does not change the total energy [10, 12]. Thus, several variants of the KREG model are possible depending on how this issue is dealt with. The simplest approximation is to neglect permutational invariance, i.e., by using the unsorted RE descriptor as obtained after transforming XYZ coordinates [10]. Another approximation is to sort homonuclear atoms using some criteria. In MLatom, this is done by sorting homonuclear atoms in descending order with respect to the sum of nuclear repulsions to other atoms [12]. This may help ensure the same sorting of atoms while doing structure-based sampling but may lead to discontinuities in the approximating function, which is problematic, e.g., in molecular dynamics [12]. Finally, permutational invariance can be ensured using the permutational invariant kernel that takes as input the permuted RE descriptor (see "Permutationally invariant kernel") [12]. The latter is the most accurate but also most computationally expensive [10].

If the derivatives of the KREG model are necessary, they can be easily obtained using the expressions discussed in the previous sub-section and the first-order derivative of the RE descriptor:

$$\frac{\partial x(\mathbf{M})_d}{\partial M_{at}} = x_d \frac{1}{r_{a,b}^2} \left(M_{bt} - M_{at} \right), \tag{36}$$

where

$$x_d = \frac{r_{a,b}^{\mathrm{ref}}}{r_{a,b}} = \frac{r_{a,b}^{\mathrm{ref}}}{\sqrt{\left(\sum_{s=1}^{3} \left(M_{as} - M_{bs} \right)^2 \right)}} \tag{37}$$

The first-order derivatives of the KREG models are implemented for both the unsorted and permuted RE descriptor variants.

3.3 Coulomb Matrix

The Coulomb matrix (CM) descriptor is popular in ML studies, where its vectorized form is used as the input vector [14]:

$$\mathbf{x} = \mathbf{vec} \begin{pmatrix} 0.5Z_1^{2.4} & \cdots & \frac{Z_1 Z_{N_{\mathrm{at}}}}{r_{1N_{\mathrm{at}}}} \\ \vdots & \ddots & \vdots \\ \frac{Z_{N_{\mathrm{at}}} Z_1}{r_{N_{\mathrm{at}}1}} & \cdots & 0.5Z_{N_{\mathrm{at}}}^{2.4} \end{pmatrix}$$

Like the RE descriptor, it is a global, complete descriptor based on the internuclear distances (in Bohr), ensuring rotational and translational invariance. However, it can also differentiate between molecules of different compositions by including nuclear charges Z (in a.u.) essentially to calculate internuclear repulsions in its off-diagonal elements. Its dimensionality, however, is limited by the largest number of N_{at} atoms among molecules in the training set, and, for a smaller number of atoms, the CM descriptor elements are padded with zeros. Better ML models nowadays exist for treating molecules with a different number of atoms [4], and some of them are interfaced to MLatom, as discussed below. The CM matrix can also be used for constructing the PES of a single molecule though.

Like the RE descriptor, three variants of the CM matrix are available in MLatom: unsorted, sorted, and permuted. In the sorted variant, the atoms are sorted so that the Euclidean norms of columns and rows of the matrix are in descending order [15]. In the case of using the CM descriptor for a single molecule PES, the unsorted CM matrix has many redundant elements (as it is symmetric, only one-half of it is needed without diagonal elements), while the sorting can lead to large discontinuities in the approximating function [2]. Despite this, the unsorted Coulomb matrix was used in the molecular dynamics studies [52, 53], where its first-order derivative is also needed. It is implemented in MLatom for the unsorted variant as:

$$\frac{\partial x(\mathbf{M})_d}{\partial M_{at}} = -\frac{Z_a Z_b}{r_{ab}^3}\left(M_{at} - M_{bt}\right) \tag{39}$$

3.4 Permutationally Invariant Kernel

Permutationally invariant (symmetrized) kernel is employed to take into account the permutational invariance of the homonuclear atoms [7, 10, 12, 18]:

$$\bar{k}\left(x(\mathbf{M}), x(\mathbf{M}_j)\right) = \frac{\sum_{\hat{P}}^{N_{perm}} k\left(x(\mathbf{M}), x\left(\hat{P}\mathbf{M}_j\right)\right)}{\sqrt{\sum_{\hat{P}}^{N_{perm}} k\left(x(\mathbf{M}), x\left(\hat{P}\mathbf{M}\right)\right)}\sqrt{\sum_{\hat{P}}^{N_{perm}} k\left(x(\mathbf{M}_j), x\left(\hat{P}\mathbf{M}_j\right)\right)}}, \tag{40}$$

where k is one of the kernel functions mentioned above and $\hat{P}$ permutes the order of atoms that are selected by the user. The descriptor is calculated for each such permutation, and hence it is called the permuted descriptor. The denominator normalizes the kernel function [39]. Unnormalized variants of this kernel symmetrization approach were used to extend the original GDML model type[48] to be permutationally invariant (creating the sGDML model type [19] interfaced to MLatom 2 and discussed below) and to create conceptually related RKHS+F [20] (reproducing kernel Hilbert space using energies and forces) model type.

The first-order derivative of the permutationally invariant kernel defined in Eq. 40 is given by:

$$\begin{aligned}\frac{\partial \bar{k}\left(x(\mathbf{M}), x(\mathbf{M}_j)\right)}{\partial M_{at}} &= \left[\sum_{\hat{P}}^{N_{\text{perm}}} k\left(x(\mathbf{M}), x\left(\hat{P}\mathbf{M}\right)\right) \sum_{\hat{P}}^{N_{\text{perm}}} k\left(x(\mathbf{M}_j), x\left(\hat{P}\mathbf{M}_j\right)\right)\right]^{-1/2} \\ &\left\{\left[\sum_{\hat{P}}^{N_{\text{perm}}} \frac{\partial k\left(x(\mathbf{M}), x\left(\hat{P}\mathbf{M}_j\right)\right)}{\partial M_{at}}\right] - \frac{1}{2}\sum_{\hat{P}}^{N_{\text{perm}}} \frac{\partial k\left(x(\mathbf{M}), x\left(\hat{P}\mathbf{M}\right)\right)}{\partial M_{at}} \frac{\sum_{\hat{P}}^{N_{\text{perm}}} k\left(x(\mathbf{M}), x\left(\hat{P}\mathbf{M}_j\right)\right)}{\sum_{\hat{P}}^{N_{\text{perm}}} k\left(x(\mathbf{M}), x\left(\hat{P}\mathbf{M}_i\right)\right)}\right\}\end{aligned} \tag{41}$$

The derivative $\frac{\partial k\left(x(\mathbf{M}), x\left(\hat{P}\mathbf{M}_j\right)\right)}{\partial M_{at}}$ is analogous to the derivatives $\frac{\partial k(x(\mathbf{M}), x(\mathbf{M}_j))}{\partial M_{at}}$ shown above with the difference that elements $x_{j,d}$ change with each permutation as $x\left(\hat{P}\mathbf{M}_j\right)_d$. The derivatives $\frac{\partial k\left(x(\mathbf{M}), x\left(\hat{P}\mathbf{M}\right)\right)}{\partial M_{at}}$ require additional derivation because M_{at} enters both $x(\mathbf{M})$ and $x\left(\hat{P}\mathbf{M}\right)$. The dth element stemming from atoms a and b in the original atom order corresponds to $x(\mathbf{M})_d$ from which $x\left(\hat{P}\mathbf{M}\right)_d$ is subtracted in both the Gaussian and Matérn kernel functions. On the other hand, the element stemming from atoms a and b in the permuted atom order will be $x\left(\widehat{P}\mathbf{M}\right)_{\widehat{P}d} = x(\mathbf{M})_d$, from which $x(\mathbf{M})_{\widehat{P}d}$ is subtracted in both the Gaussian and Matérn kernel functions. Thus, the expressions for this term are for these kernel functions:

- Gaussian:

$$\frac{\partial k\left(x(\mathbf{M}), x\left(\widehat{P}\mathbf{M}\right)\right)}{\partial M_{at}} = \frac{1}{\sigma^2}\left[\left(x(\mathbf{M})_d - x\left(\widehat{P}\mathbf{M}\right)_d\right) + \left(x(\mathbf{M})_d - x(\mathbf{M})_{\widehat{P}d}\right)\right] k\left(x(\mathbf{M}), x\left(\widehat{P}\mathbf{M}\right)\right) \quad (42)$$

- Matérn:

$$\begin{aligned}\frac{\partial k\left(x(\mathbf{M}), x\left(\widehat{P}\mathbf{M}\right)\right)}{\partial M_{at}} = & \frac{2}{\sigma^2}\left[\left(x(\mathbf{M})_d - x\left(\widehat{P}\mathbf{M}\right)_d\right) + \left(x(\mathbf{M})_d - x(\mathbf{M})_{\widehat{P}d}\right)\right] \\ & \times \exp\left(-\frac{\|x(\mathbf{M}) - x\left(\widehat{P}\mathbf{M}\right)\|_2}{\sigma}\right) \sum_{k=0}^{n-1} \frac{(n+k-1)!}{(2n)!} \\ & \times \binom{n}{k}(n-k)\left(\frac{2\|x(\mathbf{M}) - x\left(\widehat{P}\mathbf{M}\right)\|_2}{\sigma}\right)^{n-k-1}\end{aligned} \quad (43)$$

4 Interfaces

As mentioned in the "Introduction", it is not that easy to start using a new ML model, especially for novices who did not get their feet wet in this field, even sometimes for experienced researchers who have some preconceptions from their familiar frameworks. Such difficulties could be alleviated by just implementing all models we want into a single all-in-one software. However, this approach is labor-intensive and unsustainable, considering the fast-growing numbers of ML models. A better solution to tackle this problem is to make interfaces to third-party software, which is easier to implement and modularize. The drawbacks of such interface-based solutions compared to all-in-one software are the need to install multiple third-party software packages and the decreased computational efficiency due to converting data and communication bottlenecks between programs. The interface-based approach has, however, the considerable benefit of the rapid development of one uniform workflow, which eliminates the problem at its origin: the lack of standardization. This allows researchers to quickly test multiple ML model types, an advantage that outweighs the drawbacks in many cases.

Thus, we introduced the interfaces to third-party ML software in MLatom 2. Each interface should have four main functions inside, which are shown in Fig. 6. *Arguments parsing* translates MLatom-style input arguments to third-party equivalent. *Data conversion* takes in MLatom-format data then converts them into the corresponding format required by third-party software. *Model training* communicates

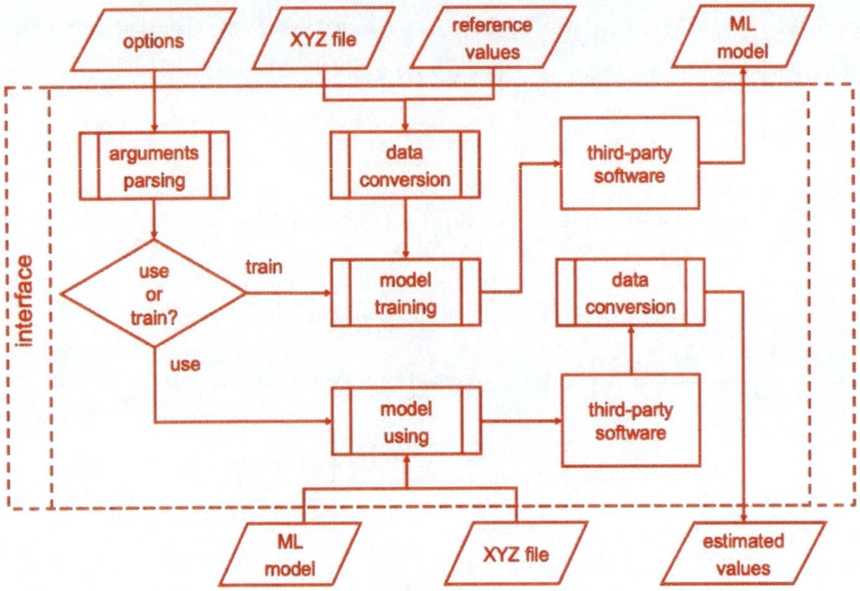

Fig. 6 Flowchart for interfaces

with third-party software to get output models by sending prepared data and arguments. *Model using* uses third-party software to get estimated values from the trained model. The interfaces are modularized Python 3 scripts stored in the subfolder *interfaces* in the MLatom program directory.

The currently interfaced third-party software packages are listed in Table 1 and will be introduced in the sub-sections below. Note that some programs provide several ML model types other than the listed ones. Using them through the interfaces might already be supported in MLatom or needs some slight modifications in the interface modules. However, only the listed model types are tested and will be discussed in this paper. Generally, we follow the strategy of making no modifications in the interfaced programs so that they can be taken "as is" and only the path to their executable should be known to MLatom. The only exception is the PhysNet, which uses TensorFlow v1. To make it compatible with the newer version of Python and TensorFlow [54], names of some legacy functions need to be changed to their equivalent in TensorFlow v2. Also, it is important to note that the main hyperparameters of each interfaced model can be modified easily through the standard input file of MLatom such that no modifications to the third-party software are required.

Besides the third-party ML software, we also implemented the interface to a hyperparameter optimization library called hyperopt [33], to provide a solution for hyperparameter optimization, which is unavailable in most of third-party software packages.

Table 1 Interfaced third-party software with their versions and ML model types tested here

Program (version)	ML model types	Developers	Language	URL
sGDML [25] (0.4.4)	sGDML [19]	Chmiela et al	Python	www.sgdml.org
GAP (1598976566) QUIP (5c61598e4)	GAP [26]-SOAP [27]	Csányi, Bartók, Kermode et al	Fortran	www.libatoms.org github.com/libAtoms/QUIP
TorchANI [28] (2.2)	ANI [21]	Gao, Ramezanghorbani, Smith, Isayev, Roitberg, et al	Python	github.com/aiqm/torchani
DeePMD-kit [29] (v1.2.2)	DPMD [30] DeepPot-SE [31]	Wang, Zhang, Han, E et al	C++ Python	www.deepmd.org/ github.com/deepmodeling/deepmd-kit
PhysNet [22]	PhysNet [22]	Unke, Meuwly	Python	github.com/MMunibas/PhysNet

Main program developers, programming languages of the majority of the code, and URL addresses to access the program are provided. References to ML model types (and third-party packages themselves where available) are also given

4.1 Hyperopt

Hyperparameters—the parameters that tune the shape of the ML model and stay unchanged in the training—have a huge impact on the performance of an ML model. To unleash the full potential of an ML model, the hyperparameters need to be well optimized, and here comes the hyperparameter optimization problem. Solving this problem manually is too cumbersome and would rarely lead to the optimal solution. Hence, many packages that are capable of automatic hyperparameter optimization are available; hyperopt [33] is one of them. Hyperopt is an open-source Python library that uses a Bayesian method with tree-structured Parzen estimator (TPE) [32] to find the optima in hyperparameter spaces.

In MLatom 2, we added the interface that uses the hyperopt library as a convenient solution to the hyperparameter optimization problem. By simply substituting hyperparameters that need to be tuned with keywords for hyperopt search space, the interface will be activated to perform the automatic optimizing process (see Fig. 2). Optimization splits the training set into the sub-training and validation sets. The trained model's error on the validation set will be sent to hyperopt to get the next searching point in hyperparameter space. If KMs are used, the final model will be generated by training on the entire training set with optimized hyperparameters after the optimization, while for NNs, the best model trained during the hyperparameter optimization will be the final ML model.

4.2 sGDML

sGDML [19] is a symmetrized variant of the gradient-domain machine learning (GDML [48]) model type and is interfaced to MLatom through the sGDML program [25]. The sGDML method is a KRR model with the descriptor of unnormalized inverse distances (the ID descriptor)

$$\mathbf{x}^{\mathrm{T}} = \left[\cdots \ \frac{1}{r_{a,b}} \ \cdots \right] \tag{44}$$

Thus, it has the same basic properties as the RE descriptor, i.e., the ID descriptor is also a global and complete descriptor ensuring rotational and translational invariance, but not permutational invariance of homonuclear atoms.

In contrast to the KREG model, the GDML model learns only from the energy gradients. It also uses the Matérn kernel function, whose expression differs slightly from that implemented in MLatom. For learning from gradients, the common KRR approximating function for scalar properties is modified by using covariance between derivatives to predict the XYZ components of energy gradients:

$$\frac{\partial f(\mathbf{x})}{\partial M_{\mathrm{at}}} = \sum_{j=1}^{N_{\mathrm{tr}}} \sum_{b=1}^{N_{\mathrm{at}}} \sum_{u=1}^{3} \alpha_{j,bu} \frac{\partial^2 k(\mathbf{x}, \mathbf{x}_j)}{\partial M_{\mathrm{at}} \partial M_{j,bu}}. \tag{45}$$

As usual, the linear system of KRR equations is solved with the kernel matrix **K**, now of a size $3N_{\mathrm{at}}N_{\mathrm{tr}} \times 3N_{\mathrm{at}}N_{\mathrm{tr}}$:

$$\begin{pmatrix} \frac{\partial^2 k(\mathbf{x}_1,\mathbf{x}_1)}{\partial M_{1,11}\partial M_{1,11}} + \lambda & \cdots & \frac{\partial^2 k(\mathbf{x}_1,\mathbf{x}_{N_{tr}})}{\partial M_{1,11}\partial M_{N_{tr},N_{at}3}} \\ \vdots & \ddots & \vdots \\ \frac{\partial^2 k(\mathbf{x}_{N_{tr}},\mathbf{x}_1)}{\partial M_{N_{tr},N_{at}3}\partial M_{1,11}} & \cdots & \frac{\partial^2 k(\mathbf{x}_{N_{tr}},\mathbf{x}_{N_{tr}})}{\partial M_{N_{tr},N_{at}3}\partial M_{N_{tr},N_{at}3}} + \lambda \end{pmatrix} \begin{pmatrix} \alpha_{1,11} \\ \vdots \\ \alpha_{N_{tr},N_{at}3} \end{pmatrix} = \begin{pmatrix} \frac{\partial y_1}{\partial M_{1,11}} \\ \vdots \\ \frac{\partial y_{N_{tr}}}{\partial M_{N_{tr},N_{at}3}} \end{pmatrix}. \tag{46}$$

When the energies are needed, they are obtained by integrating the energy gradients, and the integration constant is found by fitting to the reference energy values:

$$f(\mathbf{x}) = const + \sum_{j=1}^{N_{tr}} \sum_{b=1}^{N_{at}} \sum_{u=1}^{3} \alpha_{j,bu} \frac{\partial k(\mathbf{x}, \mathbf{x}_j)}{\partial M_{j,bu}}. \tag{47}$$

The sGDML tackles the permutational invariance by using a modified, unnormalized permutational invariant kernel with permutations chosen automatically to minimize the matching cost between pairs of training points. The sGDML program also supports automatic hyperparameter optimization via cross-validation.

The sGDML method achieves remarkable accuracy given only a very small number of training points, as was shown for molecular PESs visited during dynamics [19]. Because of the large size of the kernel matrix, the method is practically applicable only for hundreds, up to several thousand training points. Note that when we refer to a single "training point," we mean all the associated information for one geometry, and the real number of reference values available for sGDML is $3N_{at}$ times larger (the number of Cartesian energy gradient components). sGDML efficiently utilizes all this available information, which explains its accuracy.

The sGDML program requires a proprietary data format that uses NumPy's [55] npz file as the container. Scripts to convert from other data formats (e.g., extended XYZ) are included in the program. Like MLatom, sGDML has a built-in hyperparameter optimization function for the hyperparameter σ using a grid search, which is enabled by default; λ is not optimized. The users can also specify σ or a list of σ values for the grid search, but only integers are acceptable.

4.3 GAP-QUIP

The Gaussian approximation potential [26] (GAP) model is interfaced to MLatom through QUIP and GAP suite programs. Like native implementations in MLatom, GAP is also based on a kernel method, although it was developed within Gaussian process regression (GPR) formalism rather than KRR. In the GAP model, the total energy of a system is represented as the sum of atomic energies:

$$E = \sum_{i}^{\text{atoms}} \varepsilon_i \tag{48}$$

As a result, local descriptors, rather than global descriptors as used in MLatom's native models and sGDML, describing atomic environments for every single atom are used. The GAP suite provides a smooth overlap of atomic positions [27] (SOAP) descriptor for this purpose. The construction of a SOAP descriptor is quite involved as it has to respect all the required symmetries (rotational, translational, permutational), and its derivations are given in the literature [7, 27, 56]. Alternative versions of SOAP descriptor also exist [35].

Here, we describe the main points behind this descriptor. The local environment of an atom is represented by the atomic neighborhood density $\rho_i(\mathbf{r})$ that is constructed using Gaussians for the ith atom as:

$$\rho_i(\mathbf{r}) = \sum_j \exp\left(-\frac{\left|\mathbf{r} - \mathbf{r}_{ij}\right|^2}{2\sigma_{atom}^2} \right) f_{\text{cut}}\left(\left|\mathbf{r}_{ij}\right|\right), \tag{49}$$

where $\mathbf{r}_{ij}$ is a vector pointing from atom i to the neighboring atom j, σ_{atom} reflects the atom "size", and f_{cut} is the cutoff function with the width of the cutoff region r_Δ approaching the limits of r_{cut}

$$f_{\text{cut}}(r) = \begin{cases} 1, & r \le r_{\text{cut}} - r_\Delta, \\ \frac{1}{2}\left(\cos\left(\pi \frac{r - r_{\text{cut}} + r_\Delta}{r_\Delta}\right) + 1\right), & r_{\text{cut}} - r_\Delta < r \le r_{\text{cut}}, \\ 0, & r > r_{\text{cut}}, \end{cases} \tag{50}$$

The Gaussians are expanded with a set of orthonormal radial basis functions g_n [57]:

$$\rho_i(\mathbf{r}) = \sum_{\substack{n < n_{\max} \\ l < l_{\max} \\ |m| \le l}} c^i_{nlm} g_n(r) Y_{lm}(\hat{\mathbf{r}}). \tag{51}$$

where $\hat{\mathbf{r}}$ projects the direction of the vector $\mathbf{r}$ on the unit sphere and Y_{lm} are spherical harmonics. For better efficiency, the choices of n and l are limited by $n_{\max}$ and $l_{\max}$, respectively. The orthonormal radial basis functions are constructed from

$$\phi_n(r) = exp\left(-\frac{\left(r - \frac{r_{\text{cut}} n}{n_{\max}}\right)^2}{2\sigma_{atom}^2} \right) \tag{52}$$

and the overlap matrix $\mathbf{S} = \mathbf{U}^T\mathbf{U}$ with elements $S_{nn'} = \int_0^{r_{\text{cut}}} \phi_n(r)\phi_{n'}(r) r^2 dr$ as

$$g_n(r) = \sum\nolimits_{n'} (\mathbf{U}^{-1})_{nn'} \phi_{n'}(r). \tag{53}$$

The coefficients c_{nlm} are obtained as

$$c^i_{nlm} = \langle g_n Y_{lm} | \rho_i \rangle. \tag{54}$$

The SOAP descriptor $\mathbf{p}_{\mathrm{i}}$ consists of $\sum_m \left(c^i_{nlm}\right)^* c^i_{n'lm}$, which correspond to the power spectrum elements.

The kernel matrix elements are calculated using the dot-product kernel function

$$K_{ij} = \left|\mathbf{p}_i \bullet \mathbf{p}_j\right|, \tag{55}$$

which are subsequently normalized, raised to a power of ζ (a positive integer) to tune its sensitivity, and scaled by σ_ω:

$$\overline{K}_{ij} = \sigma_\omega \left(\frac{K_{ij}}{\sqrt{K_{ii}}\sqrt{K_{jj}}} \right)^\zeta. \tag{56}$$

Then the SOAP descriptor and kernel are used in estimating what the values of atomic energies are most likely to be by performing GPR that uses the same expression for making estimations as KRR:

$$\varepsilon(\mathbf{p}) = \sum_j \alpha_j K\left(\mathbf{p}, \mathbf{p}_j\right) \tag{57}$$

The problem with this expression is that the α coefficients cannot be obtained directly using the similar expression as described in "Kernel ridge regression", because there are (usually) no reference N_{tr} atomic energies $\boldsymbol{\varepsilon}$ and only N_{obs} total energies $\mathbf{E}$ are available. In the GAP approach, this is solved by introducing a linear transformation using the $N_{\mathrm{tr}} \times N_{\mathrm{obs}}$ matrix $\mathbf{L}$ with elements 1 or 0 so that

$$\mathbf{E} = \mathbf{L}^T \boldsymbol{\varepsilon}, \tag{58}$$

Then, the kernel matrix becomes

$$\mathbf{K}_{N_{\mathrm{obs}}N_{\mathrm{obs}}} = \mathbf{L}^T \mathbf{K}_{N_{\mathrm{tr}}N_{\mathrm{tr}}} \mathbf{L}, \tag{59}$$

Using this kernel matrix, the regression coefficients α can be calculated in the usual manner as in the KRR approach. In the GAP–SOAP notation, the regularization hyperparameter is denoted σ_E^2 for energies. The GAP–QUIP implementation also allows for using sparsification to reduce the size of the kernel matrix and, in this case, additional parameters defining the size N_{sparse} of the sparse kernel matrix and its regularization parameter σ_{jitter} added to its diagonal elements can be set by the user.

GAP–SOAP implementation allows the inclusion of energy gradient information to the kernel matrix. In this case, the transformation matrix $\mathbf{L}$ has additional elements with the differentiation operators $\partial/\partial M_{\mathrm{at}}$, which results in calculating covariance between energies and their partial derivatives and also between derivatives [7, 26, 56].

The GAP software suite from the libatoms.org website contains a gap_fit program that trains the GAP model with the SOAP descriptor. The QUIP program is used to get predictions from the model trained by gap_fit. To use this combination

Table 2 Main tunable hyperparameters in the Gaussian approximation potential (GAP) model type and their corresponding keywords in the gap_fit program

Hyperparameter	Keyword	Description	Default values in MLatom[a]
σ	default_sigma	List of regularization parameters for energy, force, viral and hessian	$\{0.0005, 0.001, 0.1, 0.1\}$
ζ	zeta	Power of kernel	4
δ	delta	Scaling of kernel	1
r_{cut}	cutoff	Cutoff radius	6
r_{Δ}	cutoff_transition_width	Cutoff transition width	0.5
n_{max}	n_max	Number of radial basis functions	6
l_{max}	l_max	Number of angular basis functions	6
σ_{atom}	atom_sigma	Gaussian smearing width of atom density	0.5

[a]Values chosen to provide reasonable accuracy for a small molecule (ethanol) by manual testing on the MD17 data set [48]

for PES training and prediction, the data need to be formatted to extended XYZ format.

The gap_fit program provides tons of options that enable users to make fine adjustments to the training process, including settings for atomic energies' offsets, sparsification, etc. (Table 2). However, the regularization hyperparameter and hyperparameters in the SOAP descriptor need to be set by the user manually, which makes it harder to realize the model's full potential.

4.4 TorchANI

TorchANI [28] is a Python library that implements the ANI model type [21], with PyTorch [58] as its backend for NN construction.

ANI is the abbreviation for ANAKIN-ME, which was back-engineered to Accurate NeurAl networK engINe for Molecular Energies. The ANI atomic environmental vector used in this model is a local descriptor, and is derived from the descriptor in Behler and Parrinello's work [59].

The ANI descriptor $\overrightarrow{G_i^X}$ for element X's ith atom contains a radial and an angular part, and both parts are further subdivided into subAEVs, in which the atoms taken into consideration will be limited by the same element.

For each element, the radial subAEV consists of input vector elements $G_k^R \in \mathbf{x}$ for different values of radial shift hyperparameters $R_s^{(k)}$:

$$G_k^R = \sum_{j \neq i} e^{-\eta\left(R_{ij} - R_s^{(k)}\right)^2} f_c\left(R_{ij}\right). \tag{60}$$

Similarly, for each pair of elements, the angular subAEV consists of input vector elements $G_{p,q}^A \in \mathbf{x}$ for different combinations of angular shift hyperparameter $\theta_s^{(q)}$ and another set of radial shift hyperparameters $R_s^{(p)}$:

$$G_{p,q}^A = 2^{1-\zeta} \sum_{j,k \neq i} \left(1 + \cos\left(\theta_{ijk} - \theta_s^{(q)}\right)\right)^{\zeta} e^{-\eta\left(\frac{R_{ij}+R_{ik}}{2} - R_s^{(p)}\right)^2} f_C\left(R_{ij}\right) f_C\left(R_{ik}\right). \tag{61}$$

In the equations above, f_c is the cutoff function used in Behler–Parrinello NN potentials [59] and similar to that used in GAP-SOAP, η, $R_s^{(k)}$,$R_s^{(p)}$, $\theta_s^{(q)}$ and ζ are predefined hyperparameters. Parameters η are defined separately for radial part and angular part similarly to R_s.

After being computed, each atom's AEV will be plugged into its own NN as the input vector to predict atomic energy, and the atoms of the same element share the same NN structure to ensure the permutational invariance of the trained model.

The total energies are obtained by summing all atomic energies, while atomic forces are generated by differentiating atomic energies using PyTorch's automatic differentiation engine. TorchANI reads HDF5 files, where training data are organized and stored.

As shown in Table 3, many hyperparameters can be tuned in ANI descriptors, not to mention the hyperparameters of NNs. However, as a Python library, TorchANI provides neither default values nor optimization method for hyperparameters, but only the basic core functions of ANI model type as building blocks. The final training scripts need to be written by the users themselves.

Table 3 Table of the main tunable hyperparameters in ANI model type related to the local AEV descriptor and their corresponding keywords in the TorchANI program

Hyperparameter	Keyword	Description	Default values in MLatom[a]
R_C (radial)	Rcr	radial cutoff radius	5.3
R_C (angular)	Rca	angular cutoff radius	3.5
η (radial)	EtaR	radial smoothness in radial part	{16}
R_s (radial)	ShfR	list of radial shifts in radial part	{0.90, 1.17, 1.44, 1.71, 1.98, 2.24, 2.51, 2.78, 3.05, 3.32, 3.59, 3.86, 4.12, 4.39, 4.66, 4.93}
η (angular)	EtaA	radial smoothness in angular part	{8}
R_s (angular)	ShfA	list of radial shifts in angular part	{0.90, 1.55, 2.20, 2.85}
θ_s	ShfZ	list of angular shifts	{0.19, 0.59, 0.98, 1.37, 1.77, 2.16, 2.55, 2.95}
ζ	Zeta	angular smoothness	{32}

Hyperparameters for neural networks are not listed

[a]Taken from the example script on the website of the program (https://aiqm.github.io/torchani-test-docs/examples/nnp_training.html)

4.5 DeePMD-kit

DeePMD-kit [29] is a software with the Deep Potential Molecular Dynamics (DPMD) [30] ML model type and its successor Deep Potential—Smooth Edition (DeepPot-SE) [31] built-in. Like ANI, DPMD and DeepPot-SE are also based on NNs with local descriptors. Nevertheless, DeepPot-SE switched to a learned local descriptor rather than the fixed one in its predecessor.

In DeepPot-SE, the generalized local environment matrix $\widetilde{\mathbf{R}}^i$ (which is the descriptor of original DPMD) and the local embedding matrix $\mathbf{G}^i$ are used in representing the local environment of atom i with N_i neighboring atoms. The matrix $\widetilde{\mathbf{R}}^i$ has N_i rows and each row are defined from relative coordinates and distances as:

$$\left\{ s\left(r_{ij}\right), \frac{x_{ij}}{r_{ij}} s\left(r_{ij}\right), \frac{y_{ij}}{r_{ij}} s\left(r_{ij}\right), \frac{z_{ij}}{r_{ij}} s\left(r_{ij}\right) \right\}, \tag{62}$$

where:

$$s\left(r_{ij}\right) = \begin{cases} \frac{1}{r_{ij}}, & r_{ij} < r_{cs} \\ \frac{1}{r_{ij}}\left(\frac{1}{2}\cos\left(\pi\frac{r_{ij}-r_{cs}}{r_c-r_{cs}}\right)+\frac{1}{2}\right), & r_{cs} < r_{ij} < r_c \\ 0, & r_{ij} > r_c \end{cases} \tag{63}$$

The $N_i \times M$ matrix $\mathbf{G}^i$ is generated from the local embedding network (also called filter network), which outputs a M-dimensional vector for each neighboring atom j:

$$\mathbf{G}^i_{jk} = g_k\left(s\left(r_{ij}\right)\right) \tag{64}$$

where g_k is the kth output of local embedding network applied to $s\left(r_{ij}\right)$.

The final descriptor, or the feature matrix $\mathbf{D}^i$ of atom i is defined by

$$\mathbf{D}^i = \left(\mathbf{G}^{i1}\right)^* \widetilde{\mathbf{R}}^i \left(\widetilde{\mathbf{R}}^i\right)^* \mathbf{G}^{i2}, \tag{65}$$

where $\mathbf{G}^{i1}$ and $\mathbf{G}^{i2}$ are first M_1 and M_2 columns of $\mathbf{G}^i$. The translational, rotational and permutational invariance is preserved in such expressions.

The feature matrices are then passed to NNs that generate atomic energies as the ANI model does.

DeePMD-kit program comes with its Python 3 environment, including TensorFlow and LAMMPS interface for MD simulations. The training data need to be saved in plain text in a specified style and then be transformed to what the program can read by the scripts they provide.

Training with the DeePMD-kit needs to be initialized with json input script, in which options and parameters are defined. The main tunable hyperparameters of DeepPot-SE are listed in Table 4, while hyperparameters in NNs (e.g., hyperparameters for network architecture, learning rate schedule, etc.) are not listed. However, this package cannot optimize those hyperparameters. Also, DeePMD-kit does not include in its native implementation the regularization scheme called *early stopping* often required in NN models to control the number of iterations performed during

Table 4 Main hyperparameters in DeepPot-SE model type, and their corresponding keywords in the DeePMD-kit program. Hyperparameters for neural networks are not listed

Hyperparameter	Keyword	Description	Default values in MLatom[a]
filter_neuron	filter_neuron	List of numbers of neurons in filter network	{30, 60}
M_2	n_axis_neuron	Number of columns in $\mathbf{G}^{i2}$	6
n_neuron	n_neuron	List of numbers of neurons in fitting net	{80, 80, 80}
r_c	rcut	Cutoff radius	6.5
r_{cs}	rcut_smth	Radius cutoff transition starts	6.3
sel_a	sel_a	Maximum numbers of neighboring atoms	10 for each element

[a]Taken from [31]

training, to stop the simulation before the model can reach an overfitting stage. Thus, we provide an external early stopping function as part of the interface module that monitors the training progress (based on the loss for the validation set) in the MLatom/DeePMD-kit output to stop the simulation when the criterion defined in the input has been reached.

4.6 PhysNet

PhysNet [22] is another ML model type based on learned local descriptor but using a message-passing NN architecture as the underlying model.

In PhysNet, the embedding vectors $\mathbf{e}_z$ are used as the input vectors:

$$\mathbf{x}_i^0 = \mathbf{e}_{z_i}, \tag{66}$$

where the superscript l over a vector denotes the layer number ($l = 0$ stands for the input vector), and z_i is the nuclear charge of atom i. Moreover, the number of features is defined by hyperparameter F.

Coordinates are transformed to $\mathbf{g}$ by applying K radial basis functions and cutoff functions to internuclear distances:

$$g_k(r_{ij}) = f_c(r_{ij}) \cdot e^{-\beta_k\left(e^{-r_{ij}} - \mu_k\right)^2}, \tag{67}$$

where β and μ_k are parameters of radial basis functions and r_{ij} denotes pairwise Euclidean distance between atoms i and j.

Then $\mathbf{x}^0$ is passed through a stack of N_{module} modules which are connected in series, and $\mathbf{g}$ is passed to each module.

There is a building block that will be repeatedly used in modules called a residual block. The residual block is a 2-layer mini residual neural network (ResNet), in which input vectors will also directly contribute to output vectors by skipping over the layers in between:

$$\mathbf{x}_i^{l+2} = \mathbf{x}_i^l + \mathbf{W}^{l+1}\sigma\left(\mathbf{W}^l\sigma\left(\mathbf{x}_i^l\right) + \mathbf{b}^l\right) + \mathbf{b}^{l+1}, \tag{68}$$

where $\mathbf{W}$ and $\mathbf{b}$ consist of learnable weights and biases, σ is the activation function.

Inside each module, a prototype vector $\tilde{\mathbf{v}}$ will be generated by a NN first:

$$\tilde{\mathbf{v}}_i^l = \sigma\left(\mathbf{W}_\mathbf{I}^l\sigma\left(\mathbf{x}_i^l\right) + \mathbf{b}_\mathbf{I}^l\right) + \sum_{j\neq i} \mathbf{G}^l\mathbf{g}\left(r_{ij}\right)\circ\sigma\left(\mathbf{W}_\mathbf{J}^l\sigma\left(\mathbf{x}_j^l\right) + \mathbf{b}_\mathbf{J}^{'}\right), \tag{69}$$

where the elements of matrix $\mathbf{G}$ are learnable coefficients for $g_k\left(r_{ij}\right)$, and $\circ$ is the Hadamard product operator.

This prototype $\tilde{\mathbf{v}}$ then will be tuned by $N_{\text{residual}}^{\text{interaction}}$ residual blocks to get the optimized vector $\mathbf{v}$ which will then 'interact' with $\mathbf{x}$:

$$\mathbf{x}_i^{l+1} = \mathbf{u}^l\circ\mathbf{x}_i^l + \mathbf{W}^l\sigma\left(\mathbf{v}_i^l\right) + \mathbf{b}^l, \tag{70}$$

where $\mathbf{u}$ is also a learnable parameter.

After going through another $N_{\text{residual}}^{\text{atomic}}$ residual blocks, input $\mathbf{x}$ will be passed separately to the next module (if it exists) and the output block which turns $\mathbf{x}$ to module's output $\mathbf{y}^m$, which contributes to the final output $\mathbf{y}$ after a linear transformation whose parameters are also learnable.

Unlike previously described models trained with local descriptors (GAP–SOAP, ANI, DeepPot-SE), PhysNet may also take long-range interactions (e.g., electrostatic and dispersion) into account. By default, dispersion corrections are enabled in the MLatom interface, while electrostatic corrections are disabled because their calculations require additional input (reference dipole moments).

The official implementation of the PhysNet model is programmed in Python 3.6 with TensorFlow v1, and the data need to be stored in the Numpy's npz format of a specific structure. Similar to TorchANI, using the PhysNet program also needs much manual work on script-writing and hyperparameter-tuning (see Table 5 for the list of the main hyperparameters).

Table 5 Main tunable hyperparameters in PhysNet, and their corresponding keywords

Hyperparameter	Keyword	Description	Default values in MLatom[a]
F	num_features	Number of input features	128
K	num_basis	Number of radial basis functions	64
N_{module}	num_blocks	Number of modules	5
$N_{\text{residual}}^{\text{atomic}}$	num_residual_atomic	Number of residual blocks after interaction	2
$N_{\text{residual}}^{\text{interaction}}$	num_residual_interaction	Number of residual blocks in interaction	3
$N_{\text{residual}}^{\text{output}}$	num_residual_output	Number of residual blocks in output block	1
r_{cut}	cutoff	Cutoff radius	10

[a]Taken from [22]

5 Applications

In this section, we present several case studies demonstrating the capabilities of MLatom 2.

5.1 Case Study 1: Hyperparameter Optimization

As mentioned in section "Hyperopt" in "Interfaces", a solution for hyperparameter optimization problem is given in MLatom 2 by introducing the interface to the hyperopt package. Here, we demonstrate a case using hyperopt interface to optimize the hyperparameters of the learning rate schedule in DeepPot-SE model type (start_lr and decay_rate, Table 6). For this, we used PES data (including energy gradients) of ethanol from MD17 data set [48]. A total of 1 k training points and 20 k test points were chosen randomly from the data set without overlapping. Other technical details can be found in Fig. 7 and Table 6.

For comparison, we also took two sets of start_lr and decay_rate from original literature on the DeepPot-SE [31] and DPMD models [30] (see Table 6). These values were used for the same data set but with much larger training set sizes (50 and 95 k).

As shown in Table 6, the optimization process achieved significantly better accuracies in both energies and gradients prediction, despite being used for a related task as reported in the literature. This indicates that hyperparameter optimization is highly recommended for each new case, even for cases similar to previously published ones.

Table 6 Root-mean-squares errors (RMSEs) in energies and energy gradients for DeepPot-SE potential of ethanol potential energy surface trained on 1 k random training points for the independent test set of 20 k randomly chosen test points for hyperparameters start_lr and decay_rate taken from the literature (Sets A[a] and B[b]) and optimized using MLatom's interfaces.[c]

	Set A from [31][a]	Set B from [30][b]	Optimized
start_lr (starting learning rate)	0.005	0.001	0.005675
decay_rate (decay rate)	0.96	0.95	0.9688
RMSE in energies (kcal/mol)	0.96	3.20	0.74
RMSE in gradients (kcal/mol/Å)	2.53	6.36	1.77

[a]Hyperparameters are taken for the DeepPot-SE model used for MD17 data set

[b]Hyperparameters are taken for the DPMD model used for MD17 data set

[c]In DeePMD-kit, a step decay schedule is used for learning rate decay. The related hyperparameter starting learning rate (start_lr) and the decay rate (decay_rate) were optimized, while the decay steps (decay_steps) were fixed to 200 with a stopping batch (stop_batch) set to 40,000. The search space was set to be from 0.0001 to 0.01 for starting learning rate and from 0.9 to 0.99 for the decay rate. Both spaces were set to be linear for 10 attempts of searching. The geometric mean of RMSE in energies and its gradients was used as the validation error

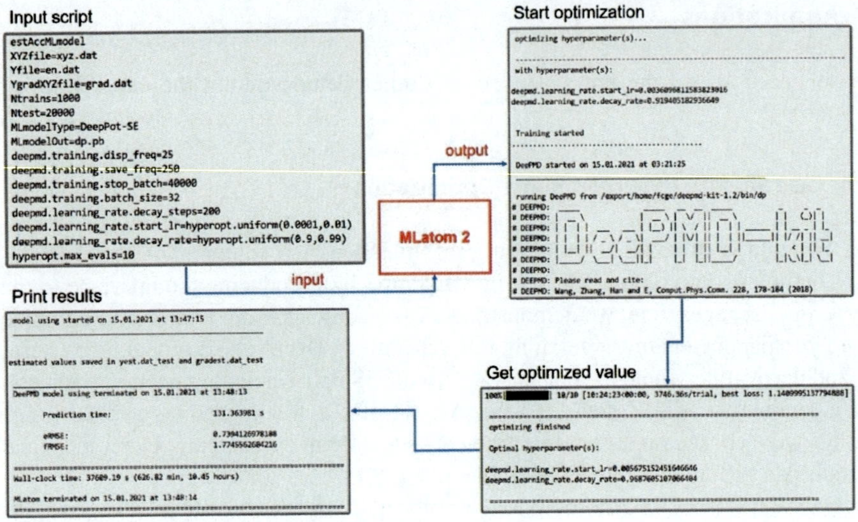

Fig. 7 Part of input and output of MLatom for hyperparameter optimization of DeepPot-SE model using the interfaces to the hyperopt and DeePMD-kit packages

5.2 Case Study 2: Learning Curves

In this part, we provide cases of MLatom's learning curve task (see "Learning curves") to show how KRR performance varies with the different molecular descriptors being used. Unsorted, sorted, and permuted RE descriptors, unsorted and sorted Coulomb matrix, and unsorted inverse distances descriptor were examined on ethanol, with energy data from MD17 data set [48]. The descriptors are denoted as uRE, sRE, pRE, uCM, sCM, and uID, respectively, and details for RE and CM can be found in subsections "KREG" and "Coulomb matrix" in section "Native implementations". The uID descriptor is not among native implementations. Thus, it is provided to MLatom, which also demonstrates support for external, user-defined descriptors. We used the Gaussian kernel throughout, i.e., KRR with the uRE, sRE, and pRE descriptors are equivalent to the corresponding KREG model variants. All these descriptors were tested with seven training set sizes roughly evenly spaced on the log scale from 100 to 10 k. Other training and testing details can be found in Fig. 8a.

The results (Fig. 8b) show the big impact of molecular descriptor choice on ML performance. First of all, let us look at the unsorted descriptors. The RE descriptor and CM are both related to the unnormalized inverse distances (ID) (see subsections "Native implementations" and "sGDML"). The uRE descriptor is a normalized version of the uID descriptor, with its advantages and disadvantages manifested in differences in the corresponding learning curves. Normalization gives equal importance to both close-range and long-range interactions, which is detrimental to accuracy for scarce training data (up to 1 k training points in case of this data set) but is advantageous when more training data are available. uCM has a product of nuclear charges in the nominator, which is different for each pair of elements and may

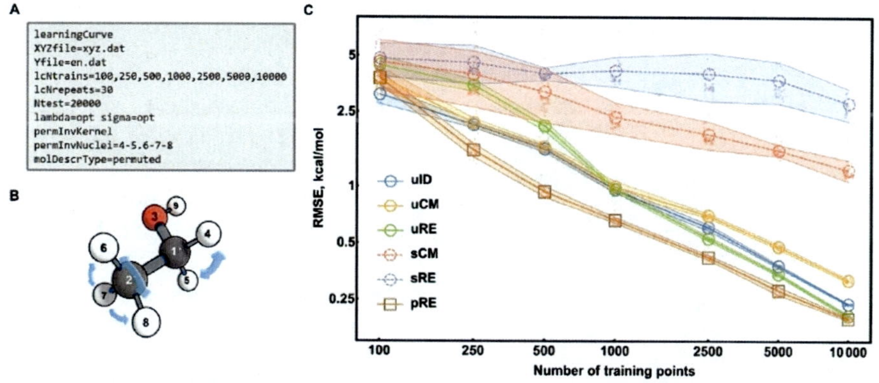

Fig. 8 **a** Input file for learning curve task using the permuted RE descriptor with kernel ridge regression (KRR) (used with the Gaussian kernel, i.e., the ML model type is a permutationally invariant KREG). The scheme for the learning curve is defined with keywords *lcNtrains* and *lcNrepeats*. **b** A three-dimensional (3D) representation of an ethanol molecule. Atoms are numbered by their order in the MD17 data set [41]. Hydrogen atoms in methyl and methylene groups are permuted separately, as defined in the input using the option *permInvNuclei=4–5.6–7–8*. **c** Model performances with different descriptors and training set sizes. Hyperparameter optimization was performed throughout. *Markers* and *error bars* show the mean and standard deviation values of RMSEs in predictions for 20 k independent test points. All data sets were randomly sampled

introduce sub-optimal weighting of the descriptor elements, leading to increased error relative to the uID descriptor [2].

The importance of properly taking into account the permutational invariance is demonstrated by using the sorted and permuted descriptors. Sorting is the simplest approach, but it causes discontinuities in the interpolant and leads to much worse results even compared to the unsorted descriptors. One of the more solid approaches, the permutationally invariant kernel using the permuted RE descriptors, can preserve permutational invariance without malfunctioning and achieves much better performance than uRE and better than uID (except for a very small number of training points).

5.3 Case Study 3: Δ-Learning and Structure-Based Sampling

Training with Δ-learning [9] and choosing training points using structure-based sampling [5] can offer ML models with better accuracy. To showcase the superiority of these approaches, we compared them to the direct ML models of the target property and random sampling. Here, we provide test results with four combinations of learning (direct vs. Δ) and sampling (random vs. structure-based) approaches using the KREG model type. The data set from reference [9] containing the PES of CH_3Cl calculated at several QM levels was used.

For the Δ-learning in this case study, CCSD(T)/CBS energy $E_{\mathrm{CCSD(T)}}$ served as the target energy, while MP2/aug-cc-pVQZ energy E_{MP2} was considered as the baseline energy. Thus, the Δ-learning model $\Delta_{\mathrm{MP2}}^{\mathrm{CCSD(T)}}$ is defined by:

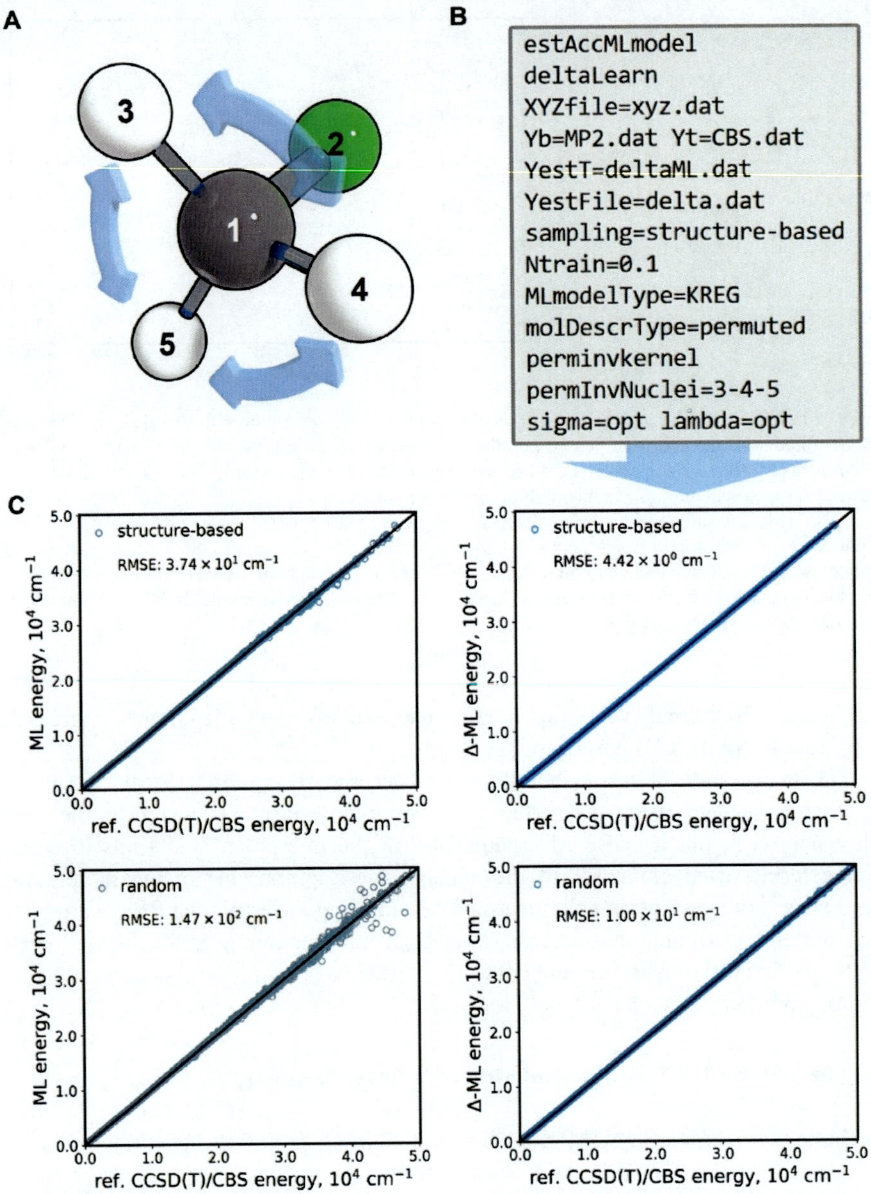

Fig. 9 **a** A 3D representation of the CH_3Cl molecule. Atoms are numbered by their order in the data set from reference [9]. Inside CH_3Cl, hydrogen atoms 3, 4, and 5 are indistinguishable, and thus their permutations should result in no difference in molecular properties. **b** Sample input script for training ML models using the Δ-learning and structure-based sampling [5] for the selection of the training set. **c** ML energies vs. reference CCSD(T)/CBS energies. ML models were trained with the 10% points of the whole data set and were tested with the remaining 90% points. R^2 is approaching 1 in all cases, with slightly larger values for more accurate models and thus are not shown for clarity. *Right column* Δ-learning models with MP2/aug-cc-pVQZ energies as a baseline. *Left column* ML model trained with reference CCSD(T)/CBS energies directly. *Bottom row* Data sets were split by random sampling. *Top row* Data sets were split by structure-based sampling

$$\Delta_{\mathrm{MP2}}^{\mathrm{CCSD(T)}}(\mathbf{x}) = E_{\mathrm{MP2}}(\mathbf{x}) + \hat{f}(\mathbf{x}) \tag{71}$$

where the ML model giving predictions $\hat{f}(\mathbf{x})$ is trained on the differences between target and baseline methods $E_{\mathrm{CCSD(T)}} - E_{\mathrm{MP2}}$. Thus, the cost of the Δ-learning model is determined by the cost of the baseline QC method, and the user should provide MLatom with the values calculated using the baseline QC method for both training and prediction.

For sampling, we used 10% points of the whole data set as the training set and the remaining 90% as the test set. Also, as illustrated in Fig. 9a, the CH_3Cl molecule has three indistinguishable hydrogen atoms, so the permutational invariant kernel was used (see "Permutationally invariant kernel").

Figure 9c shows a sample input file, and the scatter plots with RMSEs for all four combinations. Both Δ-learning and structure-based sampling led to much more accurate predictions for CCSD(T)/CBS energy than the simplest combination of random sampling and direct learning, and even more so when these two approaches were combined.

5.4 Case Study 4: Absorption Spectrum

In this case study, we will calculate the absorption cross section for an acridophosphine derivative [60] (Fig. 10) using ML-NEA implementation in MLatom, and discuss the effect of the number of points in the training set and nuclear ensemble. MLatom allows refining cross sections using existing data. Therefore, we used this feature to perform all the simulations using QC data at the ωB97XD[61]/def2-SVP [62–64] level of theory from our previous publication (see [60] for computational details; energies and oscillator strengths for 30 excited states are available at https://doi.org/10.6084/m9.figshare.13635353).

By default, MLatom determines the optimal training point number iteratively by adding 50 points at each step, and the cross section is calculated using 50 k in the nuclear ensemble. In our example, the ML-NEA procedure converged after 200 points. For comparison, the cross section obtained without ML using these 200 points in the nuclear ensemble (QC-NEA spectrum, Fig. 10a) curve has many peaks, and it is hard to judge what are the actual peaks and what are the artifacts of insufficient statistical sampling, while the ML-NEA spectrum is much smoother and ML calculations incur only small additional computational cost. The popular single-point convolution (SPC) approach gives a blue-shifted spectrum with an arbitrary bandwidth [65].

Although MLatom determines the training set size automatically, one can always request calculations of additional training points to check whether the spectrum has "truly converged" by visual inspection. For example, by comparing ML-NEA spectra with the ones calculated using 200, 250, and 300 points, one can see that they are very close to each other with some minor deviations (Fig. 10b). The ML-NEA spectrum obtained with 300 points is nearly the same as the spectrum calculated with 2 k points. One should remember, however, that the accuracy of ML-NEA

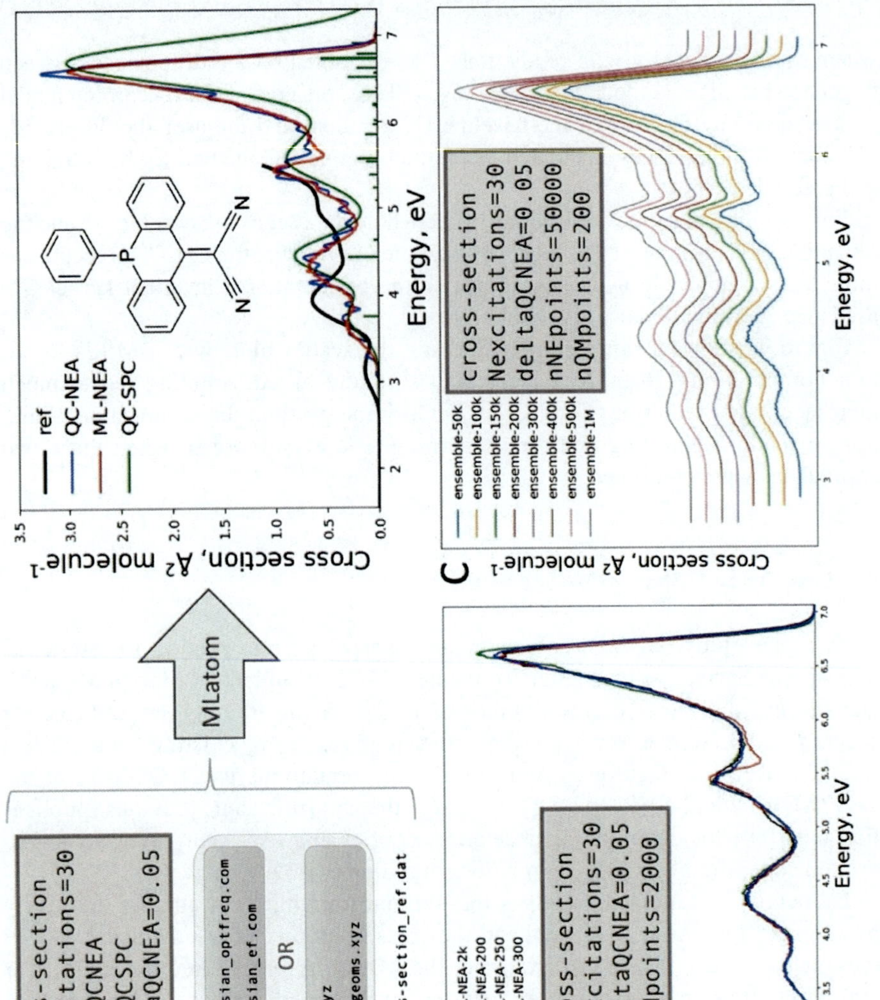

Fig. 10 **a** Structure of the acridophosphine derivative molecule investigated here. MLatom input file and the list of additional required files for the ML-NEA calculations and the resulting spectrum. QC calculation details are defined in the Gaussian input files. Alternatively, the user can provide pre-calculated results (useful to refine spectra). The number of training points (200) was determined automatically by MLatom, and the resulting cross section ML-NEA is compared to the cross section obtained using traditional QC-NEA with the same points and single-point convolution approach (QC-SPC) based on broadening lines only for the equilibrium geometry. The broadening factor for QC-NEA is 0.05 eV and for QC-SPC 0.3 eV. The reference (ref) spectrum is the experimental cross section from [60]. **b** ML-NEA spectra with sample input file for 200, 250, 300, and 2 k training points. **c** Sample input file and spectra calculated with 50 k, 100 k, 150 k, 200 k, 300 k, 400 k, 500 k, and 1 M points in the nuclear ensemble. The spectra are shifted vertically for clarity

depends on the accuracy of the underlying QC method, and the difference between the experimental spectrum and ML-NEA, even with the largest 2 k number of training points, is bigger than the difference between ML-NEA spectra with 200, 250, and 300 points (compare Fig. 10a, b). One minor aspect is that, although the default 50 k points is a rather large ensemble, it still leads to slightly rugged curves. Perfectly smooth curves can be obtained by simply increasing the number of ensemble points (Fig. 10c), and MLatom is a computationally efficient tool for this task; one can calculate very smooth curves, e.g., with 1 M points. One should keep in mind though, that such a large number is not necessary, and it will take about 10 GB of disk space to store the file with the nuclear ensemble geometries and take more time to make ML prediction and cross section calculations.

6 Conclusions

In this review article, we have described the MLatom 2 software package, which provides an integrative platform for AML simulations. Unlike other specialized AML packages, MLatom has been developed with the aim of facilitating the application of ML models to the wide variety of tasks often required in computational chemistry research.

Its capabilities range from native implementations such as the KREG model and other KRR model types (with the Coulomb matrix or any other user-defined descriptors as well as the Gaussian, Matérn, Laplacian, and exponential kernel functions) to interfaces to the third-party packages with popular models. The latter models are overviewed here for the sake of completeness and include sGDML, GAP–SOAP, ANI, DPMD, DeepPot-SE, and PhysNet. Other AML model types can be implemented easily by using the modular approach adopted in MLatom for third-party interfaces.

Other important features of MLatom for AML simulations such as model evaluation, hyperparameter optimization, sampling procedures (including farthest-point and structure-based sampling), Δ-learning, self-correction, and automatic learning curve generation are overviewed too. We also discussed how all steps required for the absorption spectrum simulation with the machine learning-nuclear ensemble approach (ML-NEA) are integrated in MLatom. Finally, we provided examples of how MLatom can be used for selected applications: hyperparameter optimization, learning curve generation, Δ-learning and structure-based sampling, and absorption spectrum simulation.

MLatom provides a user-friendly, integrated platform for researchers who want to use a wide variety of AML approaches and related techniques. It is also a useful package for educational purposes as it is used for teaching the basic and advanced concepts behind ML use in quantum chemistry (see, e.g., the book chapter [2], and online tutorial at http://MLatom.com). We are continually developing this platform based on the needs for practical AML computations such as dynamics, calculation of excited-state properties, and rovibrational spectrum simulations, improvement of QC methods, and materials design.

Authors' contributions P.O.D. is MLatom project supervisor and lead developer. P.O.D. and Y.F.H. derived and implemented analytical gradients in MLatomF and made other improvements to the KREG model. F.G. implemented learning curves and interfaces to hyperopt, PhysNet, DeePMD-kit, TorchANI, and GAP-SOAP packages. M.P.J. assisted in implementation of PhysNet, DeePMD-kit, and TorchANI interfaces. B.X.X. implemented ML-NEA approach. J.H. assisted in implementation of interfaces to GAP-SOAP and DeePMD-kit. P.O.D. has done all other implementations in MLatom. P.O.D. and M.B. conceived the project of testing third-party software. P.O.D., F.G., B.X.X., and Y.F.H. wrote the manuscript and prepared all illustrations; all other authors revised, discussed, and commented on the manuscript.

Funding P.O.D. acknowledges funding by the National Natural Science Foundation of China (No. 22003051) and via the Lab project of the State Key Laboratory of Physical Chemistry of Solid Surfaces. M.B. and M.P.J. acknowledges the support of the European Research Council (ERC) Advanced grant SubNano (Grant agreement 832237).

Availability of Data and Material All the data used in this manuscript are available from literature and online databases as cited in the article. No new data was generated in this study.

Code Availability MLatom 2 is available from http://MLatom.com free of charge for non-commercial and non-profit uses, such as academic research and education. All interfaced third-party software should be obtained and installed by the user separately.

Declarations

Conflict of Interest The authors declare that they have no conflict of interest.

Open Access This article is licensed under a Creative Commons Attribution 4.0 International License, which permits use, sharing, adaptation, distribution and reproduction in any medium or format, as long as you give appropriate credit to the original author(s) and the source, provide a link to the Creative Commons licence, and indicate if changes were made. The images or other third party material in this article are included in the article's Creative Commons licence, unless indicated otherwise in a credit line to the material. If material is not included in the article's Creative Commons licence and your intended use is not permitted by statutory regulation or exceeds the permitted use, you will need to obtain permission directly from the copyright holder. To view a copy of this licence, visit http://creativecommons.org/licenses/by/4.0/.

References

1. Dral PO (2020) Quantum chemistry in the age of machine learning. J Phys Chem Lett 11(6):2336–2347. https://doi.org/10.1021/acs.jpclett.9b03664
2. Dral PO (2020) Quantum chemistry assisted by machine learning. In: Ruud K, Brändas EJ (eds) Advances in quantum chemistry. Chemical physics and quantum chemistry, vol 81. Elsevier, Amdsterdam, pp 291–324. https://doi.org/10.1016/bs.aiq.2020.05.002
3. Butler KT, Davies DW, Cartwright H, Isayev O, Walsh A (2018) Machine learning for molecular and materials science. Nature 559(7715):547–555. https://doi.org/10.1038/s41586-018-0337-2
4. von Lilienfeld OA, Müller K-R, Tkatchenko A (2020) Exploring chemical compound space with quantum-based machine learning. Nat Rev Chem 4(7):347–358. https://doi.org/10.1038/s41570-020-0189-9
5. Manzhos S, Carrington T Jr (2020) Neural network potential energy surfaces for small molecules and reactions. Chem Rev. https://doi.org/10.1021/acs.chemrev.0c00665
6. Mueller T, Hernandez A, Wang C (2020) Machine learning for interatomic potential models. J Chem Phys 152(5):050902. https://doi.org/10.1063/1.5126336
7. Bartók AP, Csányi G (2015) Gaussian approximation potentials: a brief tutorial introduction. Int J Quantum Chem 115(16):1051–1057. https://doi.org/10.1002/qua.24927

8. Behler J (2016) Perspective: machine learning potentials for atomistic simulations. J Chem Phys 145(17):170901. https://doi.org/10.1063/1.4966192
9. Dral PO, Xue B-X, Ge F, Hou Y-F, Pinheiro Jr M (2013–2021) *MLatom*: A Package for Atomistic Simulations with Machine Learning. Xiamen University, Xiamen, China, http://MLatom.com Accessed 23 Feb 2021
10. Dral PO (2019) *MLatom*: a program package for quantum chemical research assisted by machine learning. J Comput Chem 40(26):2339–2347. https://doi.org/10.1002/jcc.26004
11. Ramakrishnan R, Dral PO, Rupp M, von Lilienfeld OA (2015) Big data meets quantum chemistry approximations: the Δ-machine learning approach. J Chem Theory Comput 11(5):2087–2096. https://doi.org/10.1021/acs.jctc.5b00099
12. Dral PO, Owens A, Yurchenko SN, Thiel W (2017) Structure-based sampling and self-correcting machine learning for accurate calculations of potential energy surfaces and vibrational levels. J Chem Phys 146(24):244108. https://doi.org/10.1063/1.4989536
13. Xue B-X, Barbatti M, Dral PO (2020) Machine learning for absorption cross sections. J Phys Chem A 124(35):7199–7210. https://doi.org/10.1021/acs.jpca.0c05310
14. Rupp M, Tkatchenko A, Müller K-R, von Lilienfeld OA (2012) Fast and accurate modeling of molecular atomization energies with machine learning. Phys Rev Lett 108(5):058301. https://doi.org/10.1103/Physrevlett.108.058301
15. Hansen K, Montavon G, Biegler F, Fazli S, Rupp M, Scheffler M, von Lilienfeld OA, Tkatchenko A, Müller K-R (2013) Assessment and validation of machine learning methods for predicting molecular atomization energies. J Chem Theory Comput 9(8):3404–3419. https://doi.org/10.1021/ct400195d
16. Dral PO, von Lilienfeld OA, Thiel W (2015) Machine learning of parameters for accurate semiempirical quantum chemical calculations. J Chem Theory Comput 11(5):2120–2125. https://doi.org/10.1021/acs.jctc.5b00141
17. Dral PO, Barbatti M, Thiel W (2018) Nonadiabatic excited-state dynamics with machine learning. J Phys Chem Lett 9:5660–5663. https://doi.org/10.1021/acs.jpclett.8b02469
18. Dral PO, Owens A, Dral A, Csányi G (2020) Hierarchical machine learning of potential energy surfaces. J Chem Phys 152(20):204110. https://doi.org/10.1063/5.0006498
19. Chmiela S, Sauceda HE, Müller K-R, Tkatchenko A (2018) Towards exact molecular dynamics simulations with machine-learned force fields. Nat Commun 9(1):3887. https://doi.org/10.1038/s41467-018-06169-2
20. Koner D, Meuwly M (2020) Permutationally invariant, reproducing kernel-based potential energy surfaces for polyatomic molecules: from formaldehyde to acetone. J Chem Theory Comput 16(9):5474–5484. https://doi.org/10.1021/acs.jctc.0c00535
21. Smith JS, Isayev O, Roitberg AE (2017) ANI-1: an extensible neural network potential with DFT accuracy at force field computational cost. Chem Sci 8(4):3192–3203. https://doi.org/10.1039/c6sc05720a
22. Unke OT, Meuwly M (2019) PhysNet: a neural network for predicting energies, forces, dipole moments, and partial charges. J Chem Theory Comput 15(6):3678–3693. https://doi.org/10.1021/acs.jctc.9b00181
23. Gv R (1995) Python tutorial, Technical Report CS-R9526. Centrum voor Wiskunde en Informatica (CWI), Amsterdam
24. Rossum GV, Drake FL (2009) Python 3 Reference Manual. CreateSpace, 100 Enterprise Way, Suite A200, Scotts Valley, CA
25. Chmiela S, Sauceda HE, Poltavsky I, Müller K-R, Tkatchenko A (2019) sGDML: constructing accurate and data efficient molecular force fields using machine learning. Comput Phys Commun 240:38–45. https://doi.org/10.1016/j.cpc.2019.02.007
26. Bartók AP, Payne MC, Kondor R, Csányi G (2010) Gaussian approximation potentials: the accuracy of quantum mechanics, without the electrons. Phys Rev Lett 104(13):136403. https://doi.org/10.1103/Physrevlett.104.136403
27. Bartók AP, Kondor R, Csányi G (2013) On representing chemical environments. Phys Rev B 87(18):187115. https://doi.org/10.1103/physrevb.87.184115
28. Gao X, Ramezanghorbani F, Isayev O, Smith JS, Roitberg AE (2020) TorchANI: a free and open source PyTorch-based deep learning implementation of the ANI neural network potentials. J Chem Inf Model 60(7):3408–3415. https://doi.org/10.1021/acs.jcim.0c00451
29. Wang H, Zhang L, Han J, Weinan E (2018) DeePMD-kit: a deep learning package for many-body potential energy representation and molecular dynamics. Comput Phys Commun 228:178–184. https://doi.org/10.1016/j.cpc.2018.03.016

30. Zhang L, Han J, Wang H, Car R, Weinan E (2018) Deep potential molecular dynamics: a scalable model with the accuracy of quantum mechanics. Phys Rev Lett 120(14):143001. https://doi.org/10.1103/PhysRevLett.120.143001
31. Zhang LF, Han JQ, Wang H, Saidi WA, Car R (2018) End-to-end symmetry preserving inter-atomic potential energy model for finite and extended systems. Adv Neural Inf Process Syst 31:4436–4446
32. Bergstra J, Bardenet R, Bengio Y, Kégl B (2011) Algorithms for hyper-parameter optimization. In: Shawe-Taylor J, Zemel R, Bartlett P, Pereira F, Weinberger KQ (eds) Advances in neural information processing systems, vol 24. Curran Associates, Red Hook, NY
33. Bergstra J, Yamins D, Cox DD Making a Science of Model Search: Hyperparameter Optimization in Hundreds of Dimensions for Vision Architectures. In: Proceedings of the 30th International Conference on International Conference on Machine Learning, Atlanta, GA, 2013. ICML'13. JMLR.org, pp I–115–I–123. https://doi.org/10.5555/3042817.3042832
34. Rezac J (2016) Cuby: an integrative framework for computational chemistry. J Comput Chem 37(13):1230–1237. https://doi.org/10.1002/jcc.24312
35. Himanen L, Jäger MOJ, Morooka EV, Federici Canova F, Ranawat YS, Gao DZ, Rinke P, Foster AS (2020) DScribe: library of descriptors for machine learning in materials science. Comput Phys Commun 247:106949. https://doi.org/10.1016/j.cpc.2019.106949
36. Hastie T, Tibshirani R, Friedman J (2009) The elements of statistical learning: data mining, inference, and prediction, 2nd edn. Springer, New York
37. Christensen AS, von Lilienfeld OA (2020) On the role of gradients for machine learning of molecular energies and forces. Mach Learn Sci Technol 1(4):045018. https://doi.org/10.1088/2632-2153/abba6f
38. Behler J (2011) Neural network potential-energy surfaces in chemistry: a tool for large-scale simulations. Phys Chem Chem Phys 13(40):17930–17955. https://doi.org/10.1039/C1cp21668f
39. Rasmussen CE, Williams CKI (2006) Gaussian processes for machine learning. MIT Press, Boston
40. Cortes C, Jackel LD, Solla SA, Vapnik V, Denker JS (1994) Learning curves: asymptotic values and rate of convergence. Advances in neural information processing systems. Morgan Kaufmann, San Mateo, CA, pp 327–334
41. Crespo-Otero R, Barbatti M (2012) Spectrum simulation and decomposition with nuclear ensemble: formal derivation and application to benzene, furan and 2-phenylfuran. Theor Chem Acc 131(6):1237. https://doi.org/10.1007/s00214-012-1237-4
42. Frisch MJ, Trucks GW, Schlegel HB, Scuseria GE, Robb MA, Cheeseman JR, Scalmani G, Barone V, Petersson GA, Nakatsuji H, Li X, Caricato M, Marenich AV, Bloino J, Janesko BG, Gomperts R, Mennucci B, Hratchian HP, Ortiz JV, Izmaylov AF, Sonnenberg JL, Williams, Ding F, Lipparini F, Egidi F, Goings J, Peng B, Petrone A, Henderson T, Ranasinghe D, Zakrzewski VG, Gao J, Rega N, Zheng G, Liang W, Hada M, Ehara M, Toyota K, Fukuda R, Hasegawa J, Ishida M, Nakajima T, Honda Y, Kitao O, Nakai H, Vreven T, Throssell K, Montgomery Jr. JA, Peralta JE, Ogliaro F, Bearpark MJ, Heyd JJ, Brothers EN, Kudin KN, Staroverov VN, Keith TA, Kobayashi R, Normand J, Raghavachari K, Rendell AP, Burant JC, Iyengar SS, Tomasi J, Cossi M, Millam JM, Klene M, Adamo C, Cammi R, Ochterski JW, Martin RL, Morokuma K, Farkas O, Foresman JB, Fox DJ (2016) Gaussian 16 Rev. C.01. Wallingford, CT
43. Barbatti M, Granucci G, Ruckenbauer M, Plasser F, Crespo-Otero R, Pittner J, Persico M, Lischka H (2013) NEWTON-X: a package for Newtonian dynamics close to the crossing seam. http://www.newtonx.org. Accessed 23 Feb 2021
44. Barbatti M, Ruckenbauer M, Plasser F, Pittner J, Granucci G, Persico M, Lischka H (2014) Newton-X: a surface-hopping program for nonadiabatic molecular dynamics. WIREs Comp Mol Sci 4(1):26–33. https://doi.org/10.1002/wcms.1158
45. Schinke R (1995) Photodissociation dynamics: spectroscopy and fragmentation of small polyatomic molecules. Cambridge University Press, Cambridge
46. Weisstein EW (2020) "Least Squares Fitting." From MathWorld—A Wolfram Web Resource. https://mathworld.wolfram.com/LeastSquaresFitting.html. Accessed 25 Dec 2020
47. Schmitz G, Klinting EL, Christiansen O (2020) A Gaussian process regression adaptive density guided approach for potential energy surface construction. J Chem Phys 153(6):064105. https://doi.org/10.1063/5.0015344
48. Chmiela S, Tkatchenko A, Sauceda HE, Poltavsky I, Schütt KT, Müller K-R (2017) Machine learning of accurate energy-conserving molecular force fields. Sci Adv 3(5):e1603015. https://doi.org/10.1126/sciadv.1603015
49. Denzel A, Kästner J (2018) Gaussian process regression for geometry optimization. J Chem Phys 148(9):094114. https://doi.org/10.1063/1.5017103

50. Fdez Galván I, Raggi G, Lindh R (2021) Restricted-variance constrained, reaction path, and transition state molecular optimizations using gradient-enhanced kriging. J Chem Theory Comput 17(1):571–582. https://doi.org/10.1021/acs.jctc.0c01163
51. Anderson E, Bai Z, Bischof C, Blackford S, Demmel J, Dongarra J, Du Croz J, Greenbaum A, Hammarling S, McKenney A, Sorensen D (1999) LAPACK users' guide, 3rd edn. Society for Industrial and Applied Mathematics, Philadelphia, PA
52. Hu D, Xie Y, Li X, Li L, Lan Z (2018) Inclusion of machine learning kernel ridge regression potential energy surfaces in on-the-fly nonadiabatic molecular dynamics simulation. J Phys Chem Lett 9:2725–2732. https://doi.org/10.1021/acs.jpclett.8b00684
53. Krämer M, Dohmen PM, Xie W, Holub D, Christensen AS, Elstner M (2020) Charge and exciton transfer simulations using machine-learned hamiltonians. J Chem Theory Comput 16(7):4061–4070. https://doi.org/10.1021/acs.jctc.0c00246
54. Abadi M, Barham P, Chen J, Chen Z, Davis A, Dean J, Devin M, Ghemawat S, Irving G, Isard M, Kudlur M, Levenberg J, Monga R, Moore S, Murray DG, Steiner B, Tucker P, Vasudevan V, Warden P, Wicke M, Yu Y, Zheng X TensorFlow: A system for large-scale machine learning. In: 12th USENIX Symposium on Operating Systems, Savannah, GA, USA, 2016. USENIX Association. https://doi.org/10.5555/3026877.3026899
55. Harris CR, Millman KJ, van der Walt SJ, Gommers R, Virtanen P, Cournapeau D, Wieser E, Taylor J, Berg S, Smith NJ, Kern R, Picus M, Hoyer S, van Kerkwijk MH, Brett M, Haldane A, Del Rio JF, Wiebe M, Peterson P, Gerard-Marchant P, Sheppard K, Reddy T, Weckesser W, Abbasi H, Gohlke C, Oliphant TE (2020) Array programming with NumPy. Nature 585(7825):357–362. https://doi.org/10.1038/s41586-020-2649-2
56. Szlachta WJ, Bartók AP, Csányi G (2014) Accuracy and transferability of Gaussian approximation potential models for tungsten. Phys Rev B 90(10):104108. https://doi.org/10.1103/PhysRevB.90.104108
57. Taylor CD (2009) Connections between the energy functional and interaction potentials for materials simulations. Phys Rev B 80(2):024104. https://doi.org/10.1103/PhysRevB.80.024104
58. Paszke A, Gross S, Massa F, Lerer A, Bradbury J, Chanan G, Killeen T, Lin Z, Gimelshein N, Antiga L, Desmaison A, Kopf A, Yang E, DeVito Z, Raison M, Tejani A, Chilamkurthy S, Steiner B, Fang L, Bai J, Chintala S (2019) PyTorch: an imperative style, high-performance deep learning library. In: Wallach H, Larochelle H, Beygelzimer A, d'Alché-Buc F, Fox E, Garnett R (eds) Advances in neural information processing systems, vol 32. Curran Associates, Red Hook, NY, pp 8026–8037
59. Behler J, Parrinello M (2007) Generalized neural-network representation of high-dimensional potential-energy surfaces. Phys Rev Lett 98(14):146401. https://doi.org/10.1103/Physrevlett.98.146401
60. Schaub TA, Brülls SM, Dral PO, Hampel F, Maid H, Kivala M (2017) Organic electron acceptors comprising a dicyanomethylene-bridged acridophosphine scaffold: the impact of the heteroatom. Chem Eur J 23(29):6988–6992. https://doi.org/10.1002/chem.201701412
61. Chai J-D, Head-Gordon M (2008) Long-range corrected hybrid density functionals with damped atom-atom dispersion corrections. Phys Chem Chem Phys 10(44):6615–6620. https://doi.org/10.1039/b810189b
62. Weigend F, Ahlrichs R (2005) Balanced basis sets of split valence, triple zeta valence and quadruple zeta valence quality for H to Rn: design and assessment of accuracy. Phys Chem Chem Phys 7(18):3297–3305. https://doi.org/10.1039/B508541a
63. Schäfer A, Huber C, Ahlrichs R (1994) Fully optimized contracted Gaussian-basis sets of triple zeta valence quality for atoms Li to Kr. J Chem Phys 100(8):5829–5835. https://doi.org/10.1063/1.467146
64. Schäfer A, Horn H, Ahlrichs R (1992) Fully optimized contracted Gaussian-basis sets for atoms Li to Kr. J Chem Phys 97(4):2571–2577
65. Bai S, Mansour R, Stojanovic L, Toldo JM, Barbatti M (2020) On the origin of the shift between vertical excitation and band maximum in molecular photoabsorption. J Mol Model 26(5):107. https://doi.org/10.1007/s00894-020-04355-y

Publisher's Note Springer Nature remains neutral with regard to jurisdictional claims in published maps and institutional affiliations.

Topics in Current Chemistry (2022) 380:19
https://doi.org/10.1007/s41061-022-00377-7

REVIEW

Check for updates

Reaction Space Projector (ReSPer) for Visualizing Dynamic Reaction Routes Based on Reduced-Dimension Space

Takuro Tsutsumi[1] · Yuriko Ono[2] · Tetsuya Taketsugu[1,2]

Received: 31 October 2021 / Accepted: 21 February 2022 / Published online: 10 March 2022
© The Author(s), under exclusive licence to Springer Nature Switzerland AG 2022

Abstract

To analyze chemical reaction dynamics based on a reaction path network, we have developed the "Reaction Space Projector" (ReSPer) method with the aid of the dimensionality reduction method. This program has two functions: the construction of a reduced-dimensionality reaction space from a molecular structure dataset, and the projection of dynamic trajectories into the low-dimensional reaction space. In this paper, we apply ReSPer to isomerization and bifurcation reactions of the Au_5 cluster and succeed in analyzing dynamic reaction routes involved in multiple elementary reaction processes, constructing complicated networks (called "closed islands") of nuclear permutation-inversion (NPI) isomerization reactions, and elucidating dynamic behaviors in bifurcation reactions with reference to bundles of trajectories. Interestingly, in the second application, we find a correspondence between the contribution ratios in the ability to visualize and the symmetry of the morphology of closed islands. In addition, the third application suggests the existence of boundaries that determine the selectivity in bifurcation reactions, which was discussed in the phase space. The ReSPer program is a versatile and robust tool to clarify dynamic reaction mechanisms based on the reduced-dimensionality reaction space without prior knowledge of target reactions.

Keywords Reaction space projector · Dimensionality reduction · Reaction path network · On-the-fly molecular dynamics

---

Chapter 3 was originally published as Tsutsumi, T., Ono, Y. & Taketsugu, T. Topics in Current Chemistry (2022) 380: 19. https://doi.org/10.1007/s41061-022-00377-7.

✉ Tetsuya Taketsugu
take@sci.hokudai.ac.jp

1 Department of Chemistry, Faculty of Science, Hokkaido University, Sapporo 060-0810, Japan

2 Institute for Chemical Reaction Design and Discovery (WPI-ICReDD), Hokkaido University, Sapporo 001-0021, Japan

1 Introduction

In theoretical chemistry, the concept of reaction paths is used to understand the mechanism of chemical reactions. The intrinsic reaction coordinate (IRC) [1] is a representative reaction path that connects the reactant and product minima on the potential energy surface (PES) via the transition state (TS) structure (first-order saddle point). Once the TS structure is determined, the IRC can be calculated straightforwardly in a descending way on the PES from the TS to the reactant and product minima; thus, the bottleneck in IRC calculations is determination of the TS structure. A great deal of effort has been expended in developing efficient TS search methods [2]. In this context, the anharmonic downward distortion following (ADDF) [3, 4] and artificial force induced reaction (AFIR) [5–7] methods have been developed to find TS structures efficiently by moving up the PES from potential minima. Recently, the mathematical aspects of ADDF paths have been discussed by classifying several patterns of path bifurcation [8]. Combining the calculation of ADDF or AFIR with IRC provides the concept of a reaction path network consisting of multiple IRCs. Reaction path networks provide the basis for kinetic simulations by setting up a reaction rate equation based on TS theory for each elementary process and evolving the equation in a coupled manner over time [9, 10]. In the last decade, the concept of reaction path networks on PES has attracted much attention, and several groups have developed theoretical methodologies for constructing reaction path networks [11–15].

The reaction path itself is determined based on the geometrical features of the PES and does not take into account the dynamic effects caused by the momentum of the molecular system. There are two factors to be considered in reaction path-based dynamics: the first is the curvature of the reaction path, which induces vibrational excitation of orthogonal vibrational modes, leading to large deviations from the reaction path [16, 17]; the second is the instability of the reaction path, where the shape of the reaction path on the PES changes from a valley path to a ridge path, causing a bifurcation of the reaction path [18, 19]. To investigate dynamic routes in full-dimensional coordinate space, on-the-fly molecular dynamics (MD) [20, 21]—a classical trajectory method that determines forces acting on atoms by electronic structure calculations—is an effective approach. Analysis of how on-the-fly trajectories move in the vicinity of the IRC has shown that the molecular system deviates significantly from the reaction path due to centrifugal forces acting in the negative direction of the curvature coordinate [22, 23]. Kraka et al. [24] proposed the unified reaction valley approach, which relates changes in the components of reaction path curvature to changes in electronic structure along the reaction coordinate. The post-TS bifurcation has also attracted much attention, and several approaches have been proposed to analyze the bifurcation paths [25] and the branching ratio [26] based on the morphology of the PES. These dynamics effects have been reported for various types of chemical reactions in organic chemistry [27–30], biochemistry [31], and catalytic chemistry [32].

To investigate the dynamics aspects of chemical reaction processes, on-the-fly MD is a desirable method because it allows an unbiased exploration of the full-dimensional PES based on the atomic force determined by electronic structure calculations, but it comes at a high computational cost. Recently, much attention has been devoted to the artificial neural network approach to generate multi-dimensional PESs for real molecules, and machine learning-assisted dynamics simulation is becoming feasible to investigate polyatomic reactions [33–35]. MD simulations with chemical reactions considering all degrees of freedom provide dynamic reaction routes with various structural changes, but the trajectories involve complex time-series data containing many degrees of freedom, and prior knowledge of the reaction channels is required to elucidate the reaction mechanism. The reaction path network should be the basis for understanding the behavior of dynamic trajectories. Much effort has been devoted to the analysis of dynamic reaction processes based on static reaction path networks [36, 37]. We have recently developed a trajectory mapping method to track dynamic reaction routes based on the distance function between reference structures on the reaction path network and molecular structures along the trajectory, and revealed dynamic behaviors such as "IRC jump" [38]. However, since the reaction path network and trajectory lie on a manifold of PES in a multi-dimensional coordinate space, it is difficult to visualize the trajectory running on PES in a low-dimensional subspace that well represents the characteristics of chemical reactions.

To overcome this problem, we have developed the Reaction Space Projector (ReSPer) method, which automatically generates principal coordinates (PCos) that reproduce the structural similarity of multi-dimensional data and projects trajectories onto a reduced-dimensionality subspace determined from reference structures [39–41]. The ReSPer method successfully visualizes dynamic reaction routes on a two- or three-dimensional reaction space spanned by PCos without arbitrary selection of reaction coordinates. In this paper, we present the functions of the ReSPer program as currently implemented. In the following section, we give an overview of the ReSPer program and the classical multidimensional scaling method, which is a dimensionality reduction method available in ReSPer. Applications to Au_5 Cluster highlights the advantages of ReSPer analysis with three applications: isomerization reactions based on dimension-reduced reaction space, closed islands of nuclear permutation-inversion isomers of a single IRC, and bifurcation reactions based on partial reaction space.

2 Reaction Space Projector

Dimensionality reduction has been used widely to classify huge and complex data in various fields. In the field of theoretical chemistry, linear dimensionality reduction methods such as principal component analysis (PCA) [37, 42–46] and (classical) multidimensional scaling (MDS) [39–41, 47–49], as well as non-linear methods such as locally linear embedding (LLE) [50] and isometric feature mapping (ISOMAP) [48, 50], have been applied to reproduce the chemical and structural features of data with minimal degrees of freedom. The disconnectivity graph can visualize the topography of high-dimensional PESs and has been applied in particular to visualize the free energy landscape of proteins, where the representation of reaction barriers is maintained [42, 51, 52].

Classical MDS (CMDS), one of the linear dimensionality reduction methods, provides a reduced-dimensional subspace determined by the PCos and maintains the dissimilarity of multidimensional data as much as possible [53, 54]. The projection of out-of-sample data that are not included in the original data, can be done using the out-of-sample extended CMDS method [55]. In applications to molecular chemistry, dissimilarity is usually defined as the Euclidean distance between pairwise molecules. CMDS using Euclidean distance is mathematically equivalent to PCA, but CMDS has a very significant advantage in optimizing the distance between a pair of molecules, which is a measure of dimensionality reduction. Therefore, we developed a Python-based program, ReSPer, to visualize the structural datasets of N atomic molecular systems in a $3N$–6 dimensional configurational space and to investigate the dynamic reaction behavior in a reduced-dimensional reaction subspace [39–41].

The ReSPer scheme is simple: (1) pre-processing, (2) dimensionality reduction of the original data, and (3) (if necessary) projection of the out-of-sample data onto the dimensionality reduction reaction subspace. Note that the dimensionality reduction methods available in the ReSPer scheme are not limited to CMDS. We are currently implementing new capabilities to perform other dimensionality reduction methods. It should also be noted that ReSPer is an analysis tool, not a quantum chemical calculation program, and thus requires the prior preparation of molecular structure data at the same calculation level. The details of the ReSPer flowchart are summarized in Fig. 1.

2.1 Pre-Processing for CMDS

Dimensionality reduction by CMDS requires the calculation of a dissimilarity matrix for all pairs of molecular structures in the dataset. Although practical and sophisticated structural dissimilarities, such as the Tanimoto coefficient [56] and the generalized root-mean-squared derivation (G-RMSD) [57], have been proposed in various situations, we adopted the simplest and uniquely defined Euclidean distance (linear distance) in a mass-weighted Cartesian coordinate system. Mass-weighted coordinate systems are suitable for describing changes in molecular structures composed of atoms of different masses.

Let $\mathbf{x}^{(i)} = \left(x_1^{(i)}, y_1^{(i)}, z_1^{(i)}, \ldots, z_N^{(i)}\right)$ and $\boldsymbol{\xi}^{(i)} = \left(\xi_1^{(i)}, \ldots, \xi_{3N}^{(i)}\right) = \left(\sqrt{m_1}x_1^{(i)}, \ldots, \sqrt{m_N}z_N^{(i)}\right)$ be the $3N$-dimensional Cartesian coordinates and mass-weighted Cartesian coordinates of an N atomic molecule, where m_a is the mass of the ath atom. From these notations, the linear distance between two molecules i and j is defined as follows:

$$d_{ij} = \sqrt{\sum\nolimits_k^{3N} \left(\xi_k^{(i)} - \xi_k^{(j)}\right)^2} = \left|\boldsymbol{\xi}^{(i)} - \boldsymbol{\xi}^{(j)}\right|. \tag{1}$$

To obtain a linear distance between two molecular structures, both the superposition of the respective x–y–z coordinate axes and the numbering of the identical

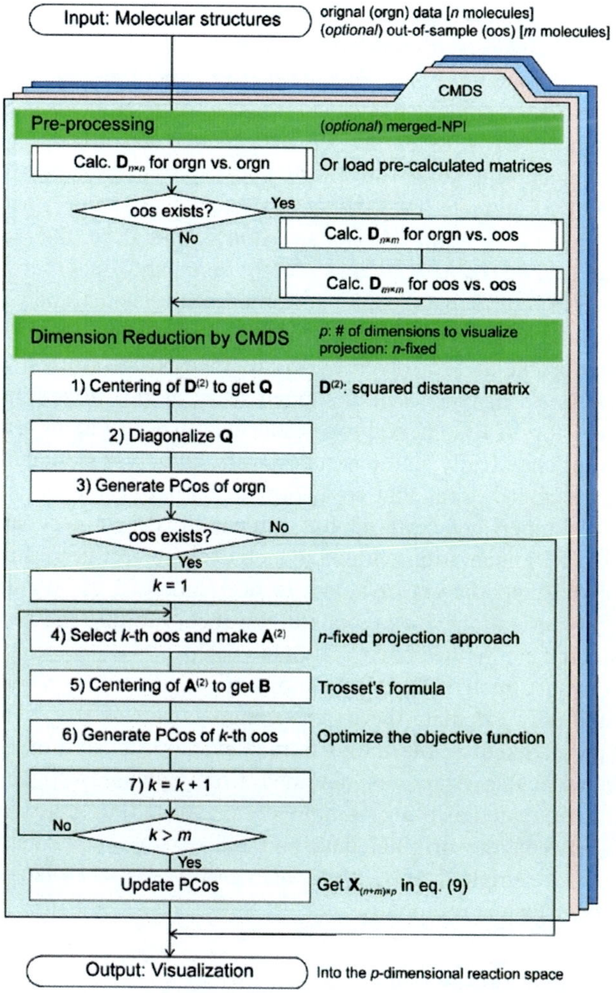

Fig. 1 Scheme of the Reaction Space Projector (ReSPer) method. The only dimensionality reduction method implemented in ReSPer at present is the classical multidimensional scaling (CMDS) method, but we plan to implement other dimensionality reduction methods in the future

atoms need to be handled properly; ReSPer pre-processing addresses these issues technically by considering the orientation alignment of the coordinate axes and the nuclear permutation-inversion (NPI) isomers.

In general, the Cartesian coordinates of a prepared molecular structure dataset depend on the origin of the coordinate system and the direction of the coordinate axes, which are altered by the translational and rotational degrees of freedom of the molecular structure. The displacement of the translational degrees of freedom can be removed completely by superimposing the center of mass of the molecular structure on the origin of the coordinate system. For the rotational degrees of freedom, the Kabsch algorithm determines the coordinate axis that minimizes d_{ij}

in Eq. (1) [58, 59]. Note that one should be careful when applying the Kabsch algorithm in a mass-weighted coordinate system. In the current ReSPer scheme, the molecular coordinates are mass-weighted after structural alignment.

The linear distances defined in Eq. (1) may be overestimated if one does not take into account the NPI isomers, which are generated by both the permutations of identical atoms and the spatial inversion operation of the molecular structure. In other words, NPI isomers have the same molecular structures as the original one, differing only in the numbering of the atoms. Therefore, the linear distance between a pair of NPI isomers is not exactly zero, and, in order to obtain an appropriate interstructural distance, all possible linear distances between NPI isomers must be compared for molecular structures i and j. In the ReSPer program, such NPI-based problems are solved by the merged-NPI option, which searches for the smallest linear distance within a given set of atomic numbering. Since the reaction barriers for NPI isomerization reactions related to atom exchange reactions in distant or chemically stable parts are very high, it is sufficient to consider only those NPI isomerizations that are energetically easy to generate. By default, all possible NPI isomers are explored, but a subset of NPI isomers can be selected by the restricted NPI generation options, which exclude NPI generation of certain elements or give a specific set of atoms to be referenced in NPI generation. If possible NPI isomerizations are obvious in advance, limiting the number of NPI isomers considered can reduce computational cost.

Since the distance matrix $\mathbf{D}_{n \times n}$ for n molecular structures is a real symmetric matrix, we need only calculate the upper triangular part of $\mathbf{D}_{n \times n}$. These calculations can be parallelized by the joblib library available in Python. If one wants to project m molecular structures obtained from dynamics trajectories, these structural data are treated as out-of-sample data, and ReSPer generates a distance matrix $\mathbf{D}_{n \times m}$ between the original data and the out-of-sample data, and $\mathbf{D}_{m \times m}$ between the out-of-sample data. One can also easily reuse the distance matrix in pre-calculated csv format optionally.

2.2 Classical Multidimensional Scaling

CMDS has been described in the literature [53–55, 60] and in our previous publications [39–41]. Here, we briefly present its implementation in the ReSPer program. CMDS analysis, which visualizes the original dataset with n reference structures in a p-dimensional space, usually set to $p = 2$ or 3, is performed in the following three steps:

(1) Center the squared distance matrix $\left(\mathbf{D}^{(2)}\right)_{ij} = \left(d_{ij}\right)^2$ to generate the inner-product matrix $\mathbf{Q}$:

$$\mathbf{Q} = -\frac{1}{2}\left(\mathbf{E} - \frac{1}{n}\mathbf{1}\right)\mathbf{D}^{(2)}\left(\mathbf{E} - \frac{1}{n}\mathbf{1}\right)^T, \tag{2}$$

where **E** and **1** are the unit matrix and a square matrix with all elements set to 1, respectively. This process is called the Young-Householder transformation [61].

(2) The inner-product matrix **Q** is diagonalized to obtain the eigenvalues $\{\lambda_1, \ldots, \lambda_n\}$ and the corresponding eigenvectors $\{\mathbf{L}_1, \ldots, \mathbf{L}_n\}$. The eigenvalues and eigenvectors are sorted in descending order, $\lambda_1 \geq \lambda_2 \cdots \geq \lambda_n$. The inner-product matrix **Q** and the configuration matrix **Y** of the full-dimensional space are associated as follows

$$\mathbf{Q} = \mathbf{L}\Lambda\mathbf{L}^T = \mathbf{Y}\mathbf{Y}^T. \tag{3}$$

(3) The PCos are generated by $\mathbf{Y}_a = \mathbf{L}_a\sqrt{\lambda_a}(1 \leq a \leq p)$. Here, PCos obtained from the original dataset are denoted as $\mathbf{Y}_{n\times p}$.

The normalized eigenvalues indicate the contribution of each PCo, and the proportion of variances Λ_a and the cumulated proportion Λ are useful for estimating the visibility of a PCo in a p-dimensional subspace [54]:

$$\Lambda_a = \frac{\lambda_a}{\sum_c^b \lambda_c}, \tag{4}$$

$$\Lambda = \frac{\sum_c^p \lambda_c}{\sum_c^b \lambda_c}, \tag{5}$$

where b is the number of non-negative eigenvalues of λ_c.

If the user wants only to visualize a reduced-dimensionality reaction space determined from the original dataset, ReSPer usually exits after recording $\mathbf{Y}_{n\times p}$ and these contribution rates in a log file. If the user wants to project an out-of-sample dataset onto the constructed reaction space, ReSPer will proceed to the out-of-sample extension scheme [55]. The current ReSPer program has three options for out-of-sample projection: m out-of-samples are (1) projected at once; (2) projected one by one into the p-dimensional space of $\mathbf{Y}_{n\times p}$ (n-fixed); or (3) projected one by one into the p-dimensional space of $\mathbf{Y}_{n\times p} \leftarrow [\mathbf{Y}_{n\times p}, \mathbf{Z}_{1\times p}]$ ($i.e.\, n \leftarrow n+1$), which is updated for each out-of-sample projection (n-expanded). In the (1) at-once approach, the dissimilarity of the molecular structure of the out-of-sample data affects the projection, and in the (3) n-expanded approach, the p-dimensional reaction space consisting of $\mathbf{Y}_{n\times p}$ is updated with the trajectory data, so we use the (2) n-fixed approach for the out-of-sample projection in this study. In Fig. 1, we show the outline of the ReSPer method using the (2) n-fixed approach. The ReSPer procedure for out-of-sample projection can be written as follows:

(4) Make the squared distance matrix $\mathbf{A}^{(2)}$, defined as

$$\mathbf{A}^{(2)} = \begin{pmatrix} \mathbf{D}^{(2)}_{n\times n} & \mathbf{D}^{(2)}_{n\times m} \\ \mathbf{D}^{(2)}_{m\times n} & \mathbf{D}^{(2)}_{m\times m} \end{pmatrix}, \tag{6}$$

where $\mathbf{D}^{(2)}_{n\times n}$, $\mathbf{D}^{(2)}_{n\times m}$, and $\mathbf{D}^{(2)}_{m\times m}$ are the squared distance matrices of $\mathbf{D}_{n\times n}$, $\mathbf{D}_{n\times m}$, and $\mathbf{D}_{m\times m}$, and $\mathbf{D}^{(2)}_{m\times n}$ is the transpose matrix of $\mathbf{D}^{(2)}_{n\times m}$. It should be noted that m has different meanings depending on the projection method, such as the total number of out-of-sample data in (1) and the index of the component of the distance matrix corresponding to the kth out-of-sample data in (2) and (3) $(1 \leq k \leq m)$. Thus, for example, $\mathbf{D}^{(2)}_{m\times m}$ is always a scalar of zero in (2) and (3).

(5) The centering of $\mathbf{A}^{(2)}$ is done according to Trosset's formula [55]:

$$\mathbf{B} = \tau_w(\mathbf{A}) = -\frac{1}{2}\left(\mathbf{E} - \frac{\mathbf{e}\cdot\mathbf{w}^T}{n}\right)\mathbf{A}^{(2)}\left(\mathbf{E} - \frac{\mathbf{e}\cdot\mathbf{w}^T}{n}\right)^T, \tag{7}$$

where $\mathbf{E}$, $\mathbf{e}$, and $\mathbf{w}$ are the unit matrix, the $(n + m)$ dimensional column vector with all elements set to 1, and the $(n + m)$-dimensional column vector with 1 (the first n components) and 0 (the remaining m components).

(6) Optimize the following objective function [55] with respect to $\mathbf{Z}_{m\times p}$.

$$\min\|\mathbf{B} - \mathbf{X}\mathbf{X}^T\|^2 = \min\left\|\begin{pmatrix} \mathbf{B}_{n\times n} - \mathbf{Y}\mathbf{Y}^T & \mathbf{B}_{n\times m} - \mathbf{Y}\mathbf{Z}^T \\ \mathbf{B}_{m\times n} - \mathbf{Z}\mathbf{Y}^T & \mathbf{B}_{m\times m} - \mathbf{Z}\mathbf{Z}^T \end{pmatrix}\right\|^2, \tag{8}$$

$$\mathbf{X}_{(n+m)\times p} \equiv \begin{pmatrix} \mathbf{Y}_{n\times p} \\ \mathbf{Z}_{m\times p} \end{pmatrix}, \tag{9}$$

where $\mathbf{B}_{n\times n}$, $\mathbf{B}_{n\times m}$, $\mathbf{B}_{m\times n}$, and $\mathbf{B}_{m\times m}$ are part of the $\mathbf{B}$ matrix given in step (5). In this ReSPer program, the initial p-dimensional coordinate $\mathbf{Z}_{m\times p}$ of the out-of-sample data is set as the origin. At present, the optimization of Eq. (8) is done by the modified Powell algorithm implemented in SciPy software.

(7) In case (2) or (3), increase k by 1, and if $1 \leq k \leq m$ is satisfied, return to step (4). Otherwise, ReSPer will exit normally after recording $\mathbf{X}_{(n+m)\times p}$.

2.3 Recent Applications of ReSPer

We have so far applied ReSPer to several chemical reaction systems: the intramolecular proton transfer of malonaldehyde [39], the collision reaction of $OH^- + CH_3F \rightarrow CH_3OH + F^-$ [39–41], and the isomerization and bifurcation reactions of a small gold cluster [39–41]. Recent applications of ReSPer are summarized in Table 1. These studies show that ReSPer analysis can construct a low-dimensional space that depends only on the linear distance between given reference structures, and that the projection of dynamic trajectories by the out-of-sample extended method is useful for understanding complex dynamic reaction processes. In the next section, we will show what has been revealed by applying ReSPer to the small gold cluster Au_5. Note that the mass-weighting procedure does not affect the linear distance of Au_5, but we have confirmed that it works well when applied to organic molecules composed of different atoms. Currently, the ReSPer program is being extended to analyze more complex reaction systems: exhaustive reaction path

Table 1 Recent applications of Reaction Space Projector (ReSPer). *IRC* intrinsic reaction coordinate, *NPI* nuclear permutation-inversion

Molecular system	Reference structures	Contents
Malonaldehyde	Single IRC	Two-dimensional reaction space [39]
$OH^- + CH_3F \rightarrow CH_3OH + F^-$	Single IRC	Two-dimensional reaction space [39], energy landscape [38], and dynamic trajectories [40, 41]
Au_5 cluster	Multiple IRCs	Two- and three-dimensional reaction space [39, 40], energy landscape [41], and dynamic trajectories related to isomerization and bifurcation reactions [40, 41],[a] connectivity networks of NPI isomers[a]

[a]This paper

networks including a huge number of IRCs, organic chemical reactions with C–C bond formation, and excited-state reaction dynamics.

3 Applications to Au_5 Cluster

It is known that the catalytic activity of gold nanoparticles increases dramatically with decreasing particle size [62–64], and gold nanoparticles and gold nanoclusters have attracted much attention due to their catalytic [62, 63] and optical [65, 66] properties. In our previous work, we explored the reaction pathways of structural changes in Au_5 cluster by the ADDF method to obtain five stable structures and 14 TSs connecting them (14 IRCs), revealing a global reaction path network on the full-dimensional PES [67]. We searched for valley-ridge transition (VRT) points along 14 IRCs and found five VRTs where the nature of the IRC valley changes to a ridge and the reaction path shows a bifurcation. Furthermore, the dynamic bifurcation mechanism was analyzed by on-the-fly MD simulation [38, 67]. In this paper, to demonstrate the capabilities of the ReSPer program, we perform the analysis on the previously calculated global reaction path network of Au_5 and on-the-fly MD trajectories [67]. These calculations were performed by density functional theory (DFT) using the Perdew–Burke–Ernzerhof (PBE) functional [68, 69] and the LanL2DZ [70] basis set with Gaussian09 [71]. Reaction path search and on-the-fly MD simulations were performed using GRRM11 [72] and SPPR [73], respectively.

3.1 Reduced-Dimensionality Reaction Space and Potential Energy Landscape

First, to visualize the global reaction pathway map of the Au_5 cluster, we apply ReSPer analysis with the merged-NPI option [40, 41]. Data of 336 molecular structures along 14 IRCs, including five minima and 14 TSs, are input into the ReSPer program to construct a reduced-dimensionality reaction space. In principle, there is no upper limit to the number of structures ReSPer can handle, but since it is time-consuming and the reaction space becomes cumbersome, we have reduced the number of structures along each IRC to about 20. Figure 2a shows the molecular structures at the stationary point and one VRT point associated with the bifurcation reaction described below. The five minima are denoted as MIN*i* (i=1–5), the seven TSs connecting MIN*i* and MIN*j* as TS*i*-*j*, and the TS connecting MIN*i* and the NPI isomer of MIN*i* as TS*i*-*i*. Since there are multiple structures in TS*i*-*i*, we distinguish them by adding letters (a, b, c, d) in order of decreasing energy.

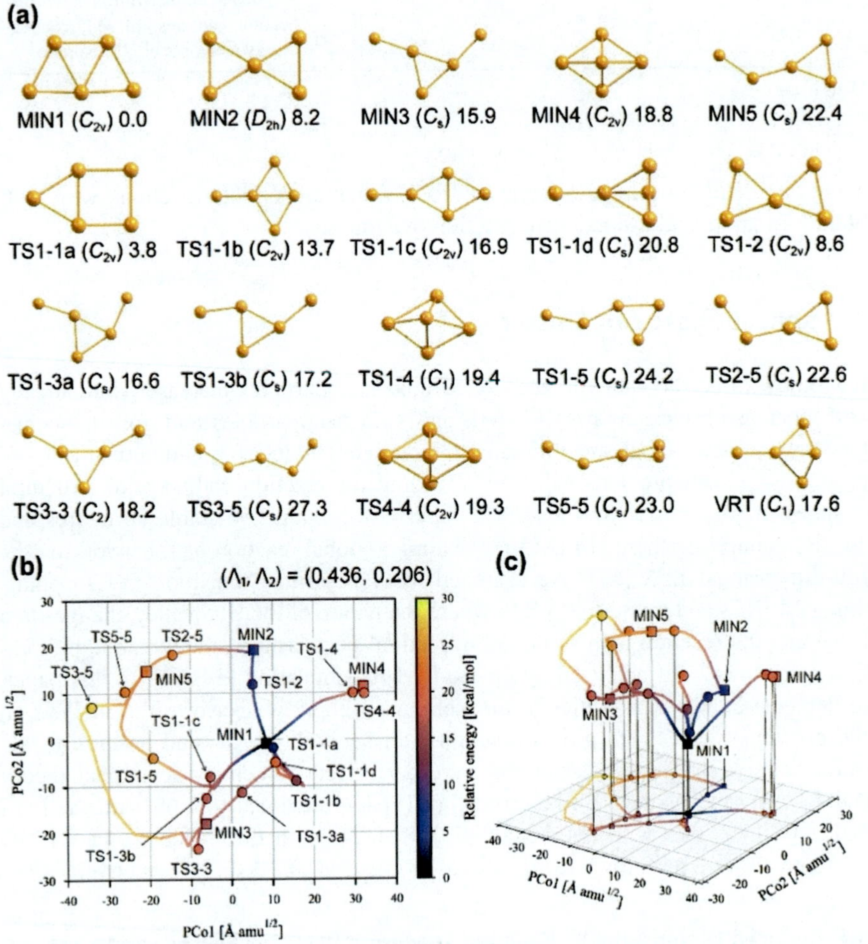

Fig. 2 **a** Molecular structures of the five minima, 14 transitions states (TSs), and one valley-ridge transition (VRT) on the intrinsic reaction coordinate (IRC) connecting TS1-1d and MIN1 in the global reaction path network of the Au_5 cluster. The point group of each structure and the relative energy (in kcal mol^{-1}) to MIN1 are also shown. **b** Two-dimensional reaction space determined by 336 reference structures along all IRCs. **c** Potential energy landscape as determined by the ReSPer program. For simplicity of visualization, the data points of the molecular structure along the IRC were omitted. *Squares* minima, *circles* TS, *color map* potential energy relative to MIN1

Figure 2b shows the two-dimensional reaction space determined by the 336 reference structures. The contribution ratio is $(\Lambda_1, \Lambda_2) = (0.436, 0.206)$, and the cumulative percentage is $\Lambda = 0.642$. This result shows that about 64% of the network structure of the reaction pathway in nine dimensions (=3*N*-6 dimensions) is reproduced by the two PCos extracted automatically by ReSPer. The global minimum of MIN1 is located near the origin of the coordinate system, and the other four minima are distributed around MIN1. Interestingly, TS*i*-*i* is located near MIN*i*, and TS*i*-*j* is located between MIN*i* and MIN*j*. These mutual locations clearly reflect the connectivity of each reaction pathway, even though ReSPer knows the structural dissimilarities only of the given molecular structures. Furthermore, due to the merged-NPI option, the IRC connecting MIN*i* and TS*i*-*i* almost overlaps the outgoing and returning pathways. The physical meaning of each PCos and the three-dimensional reaction space has been mentioned in previous papers [40, 41].

Figure 2c shows the potential energy landscape constructed by the two PCos, with the potential energy of each molecular structure displayed as the third axis. Similar to disconnectivity graphs [52], ReSPer can visualize the high-dimensional PES that contains information about reaction barriers. It should be emphasized that the energy landscape constructed by ReSPer can also reproduce mutual structural dissimilarities on the full-dimensional configurational space and can reproduce the morphology of the PES in which chemical reactions take place. Although beyond the scope of this paper, the analysis of reaction dynamics based on the energy landscape has provided deep insights into the picture of inherent reaction processes [40, 41]. We are currently working on the visualization of the multi-state energy landscape, including the potential energy surfaces of the ground and excited states.

3.2 Reaction Dynamics of Isomerization Reaction on Two-Dimensional Reaction Space

We have analyzed trajectories on the basis of the reduced-dimensionality reaction space using the ReSPer program and the trajectory mapping method [40, 41]. In the case of Au_5, we performed on-the-fly MD simulations from TS1-1d and analyzed the dynamic reaction behavior mainly from TS1-1d to MIN1. In this section, we present the various dynamic reaction processes based on the two-dimensional global reaction path network shown in Fig. 2b. In the previous MD simulations, the initial velocity was generated by randomly assigning a total of 5 kcal mol^{-1} to the nine vibrational modes of TS1-1d. The initial coordinates were fixed to one of the NPI isomers of TS1-1d. The time step was set to 5 fs, and 200 trajectories up to 3 ps were run. For the same reason as described in the section above, only the points where the potential energy shows a minimum along the trajectory were added as out-of-sample data.

Figure 3 shows the projection of the four trajectories onto the reduced-dimensionality reaction space of Fig. 2b. Overall, as shown in Fig. 2a, none of the 200 trajectories proceeded to the higher energy region of PCo1 < − 15 Å amu$^{1/2}$, even though the total energy of the simulation (20.8 + 5 = 25.8 kcal mol^{-1}) is higher than

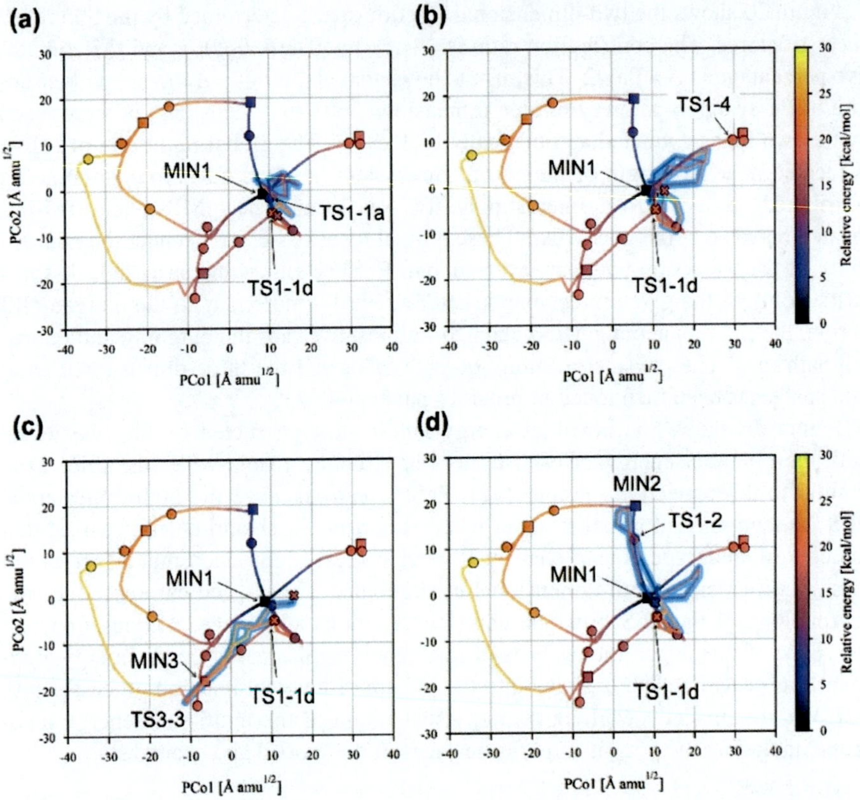

Fig. 3 Four trajectories projected into the two-dimensional reaction space shown in Fig. 2b: **a** a trajectory that travels in a narrow region between MIN1 and TS1-1a after reaching MIN1, **b** a trajectory that travels several times between MIN1 and TS1-4, which has a non-planar structure, **c** a trajectory that proceeds to MIN1 after reaching MIN3 from the initial structure of TS1-1d, and **d** a trajectory that proceeds to MIN2 via TS1-2 before returning to MIN1. *Blue lines* trajectories, *cross marks* initial and final points of the trajectories, *color map* potential energy relative to MIN1

TS except for TS3-5. These results are attributed to the fact that the energy is rarely concentrated so as to cross the TS with a relatively high reaction barrier.

Figure 3a shows a trajectory that starts with TS1-1d and then wanders between MIN1 and TS1-1a, which has the lowest reaction barrier. This process is confirmed to be a sequential isomerization from one NPI isomer of MIN1 to another NPI isomer via a corresponding NPI isomer of TS1-1a, as explained in the next section. Figure 3b shows the trajectory of Au_5 continuing to change from the planar MIN1 to the non-planar TS1-4 due to the momentum induced by the structural change along the IRC connecting TS1-1d and MIN1. Actually, the atomic motions of these two trajectories are so similar that it was difficult to distinguish them without prior knowledge. However, as discussed, we have succeeded in classifying these different reaction processes by projecting the dynamic reaction routes with ReSPer.

Figure 3c shows an unusual trajectory, approaching MIN3 directly from TS1-1d and then reaching MIN1. Figure 3d shows a typical trajectory that crosses TS1-2 and proceeds to MIN2, the second-lowest MIN. Conventional analysis of reaction dynamics requires the selection of internal coordinates that are suitable for discussing the assumed reaction mechanism. In the case of reactions involving bond cleavage and formation, it is not easy to analyze the reaction dynamics for multiple elementary reaction processes because a suitable set of internal coordinates cannot be uniquely defined. ReSPer overcomes these challenges and enables the analysis of chemical reactions involving multiple bond cleavages and formations on a uniquely defined low-dimensional reaction space.

3.3 Closed Islands: Linkage of NPI Isomers of MIN1

The permutation-inversion group has been studied in the areas of spectroscopy and nuclear physics. The concept of NPI isomers, which are generated by substitution of identical atoms and spatial inversion operations, has been used in the analysis of rearrangement reactions in water clusters [74], the construction of quantum graph models in quantum dynamics [75], and the improvement of PESs generated by artificial neural networks [76]. So far, the connectivity of NPI isomers has been illustrated as a conceptual diagram because it is a seriously complex network on a high-dimensional manifold. Here, a simple question emerges: is it possible to visualize the connectivity of NPI isomers in low-dimensional space?

Since there are five identical Au atoms in the Au_5 cluster, the maximum number of NPI isomers is 240 (=5!×2). The number of NPI isomers decreases with the symmetry of the molecule, with 120 in the C_s and C_2 structures, 60 in the C_{2v} structure, and 30 in the D_{2h} structure, respectively [38]. The 60 NPI isomers of MIN1 are connected by TS1-1a, TS1-1b, TS1-1c, and TS1-1d, with activation barriers of 3.8, 13.7, 16.9, and 20.8 kcal mol^{-1}, respectively. We have previously reported the concept of "closed islands" in which different NPI isomers are linked by structural transformation via NPI isomers of a particular TS; in the case of the Au_5 cluster, the 60 NPI isomers of MIN1 were found to be partitioned into 12, 15, 10, 30 closed islands by TS1-1a, TS1-1b, TS1-1c, and TS1-1d, respectively [38, 41]. In this paper, we performed ReSPer analysis without using the merged-NPI option to visualize the connection between the closed island of TS1-1a, which has the lowest energy barrier, and the closed islands of TS1-1a and TS1-1b.

Figure 4a shows one of the closed islands of TS1-1a constructed by ReSPer. For simplicity, the molecular structures along the five IRCs have been omitted, but, actually, this map contains 500 structures as reference structures. These structures were generated by the NPI operation for one IRC related to the two NPI isomers of MIN1 linked by TS1-1a. The contribution ratio is $(\Lambda_1, \Lambda_2) = (0.484, 0.484)$, and the cumulated proportion is $\Lambda = 0.968$. Furthermore, other proportions of $\Lambda_i (1 \leq i \leq 15)$ are displayed in Fig. 4d. The first and second ratios are degenerate, reflecting the E representation of the D_5 point group to which the regular pentagon of the closed-island linkage belongs. Figure 4b shows the visualization of 12 independent closed islands of TS1-1a in a three-dimensional subspace containing 600 structures constructed by

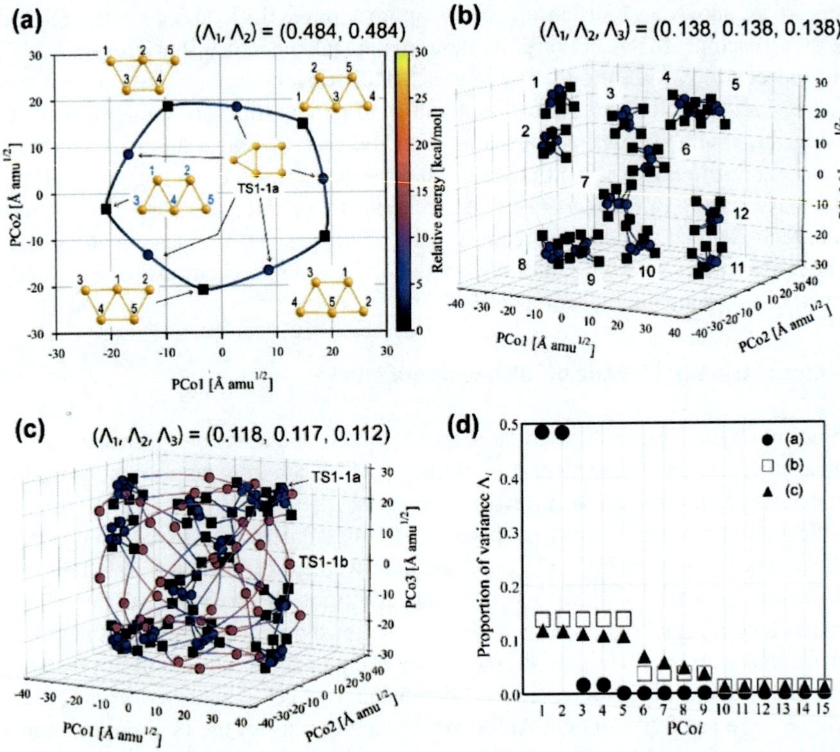

Fig. 4 Connectivity network of nuclear permutation-inversion (NPI) isomers of MIN1 in the reduced-dimensionality reaction space: **a** closed island of TS1-1a, **b** 12 independent closed islands of TS1-1a, and **c** complete closed island of TS1-1a and TS1-1b. **b**, **c** Visualized in the three-dimensional reaction space spanned by PCo1, PCo2, and PCo3. The structure dataset used for the analysis of **b**, **c** was generated by thinning out the molecular structure by 10 points along the original IRC. Data points along the IRC are omitted. *Squares* minima, *circles* TS, *color map* potential energy relative to MIN1. The variance proportions of the ReSPer analysis of **a–c** are shown in **d**

ReSPer. The results confirm that TS1-1a divides the 60 NPI isomers of MIN1 into 12 closed islands, each of which has five different MIN1s, and that TS1-1a alone does not link all NPI isomers of MIN1.

In order to fully connect all MIN1, at least TS1-1b is considered necessary, as shown in Fig. 4c. Note that Fig. 4c was constructed using a dataset of 1800 structures, which we call the "complete closed island" of TS1-1a and TS1-1b. We also verified the previously reported complete closed islands of (TS1-1a and TS1-1c), (TS1-1a and TS1-1d), and (TS1-1b and TS1-1d) with ReSPer [38, 41]. Interestingly, the contribution in Fig. 4b consists of fivefold and fourfold degeneracies, while that of the degeneracy in Fig. 4c slightly broken down. This is due to the fact that the dodecahedron with 12 isolated closed islands as vertices in Fig. 4b belongs to the I_h point group, while the symmetry in Fig. 4c is reduced by considering the TS1-1b linkage. ReSPer has succeeded in automatically constructing complex linkages of NPI isomers based on their structural dissimilarities, and has also found a correspondence between the contribution ratios involved in visibility and symmetry in closed island morphology.

3.4 A Bundle of Trajectories Related to Bifurcation Reaction

Finally, we move on to the discussion of the bifurcation reaction starting from TS1-1d. In previous studies, ReSPer has visualized two-dimensional bifurcation maps determined from carefully collected NPI isomers of TS1-1d, TS1-b, and MIN1, and their corresponding IRCs, revealing dynamic reaction behaviors such as "IRC-jump" [40, 41] that induce dynamic bifurcation [77]. In this section, we discuss a bundle of trajectories that can be regarded as dynamic reaction paths based on the partial branching reaction space with restricted NPI isomers. Among the 200 pre-calculated trajectories, we classify 94 trajectories that proceed into the partial reaction space. These trajectories are thinned out as in "Reaction Dynamics of Isomerization Reaction on Two-Dimensional Reaction Space" and terminated when the linear distance from the NPI isomers of MIN1 (called MIN1 and MIN1'') associated with the bifurcation reaction of interest is 20 Å amu^{½}. The merged-NPI option is turned off to distinguish NPI isomers.

As a result of the ReSPer analysis, four typical trajectories bundles appear in the two-dimensional reaction space, with a cumulated proportion of 0.961. Figure 5a shows a bundle of 28 trajectories that reach MIN1 directly, suggesting a dynamic route that climbs a shallow PES orthogonal to the original IRC and is attracted to the product region by the potential gradient in the MIN1 direction. Figure 5b contains 12 trajectories that reach MIN1 after bouncing off in the TS1-1b region. From the motion of the trajectory bundle, we can see the dynamic effect of the highly curved nature of the original IRC. The molecular system cannot resist the centrifugal force and goes straight ahead, encountering an energy barrier near TS1-1b. Figure 5c shows the dynamic bifurcation process caused by the IRC-jump that occurs in the region between TS1-1d and VRT. After the dynamic bifurcation, some trajectories shown in Fig. 5d bounce off the energy barrier near TS1-1b and proceed to the bifurcation product, MIN1''. These results indicate that the trajectory bundles projected by ReSPer depict realistic reaction paths where dynamic effects are explicitly taken into account in the low-dimensional reaction space, suggesting that they can be used as dynamic reaction paths instead of static ones. Conventionally, such statistical trajectory analysis requires scrutinizing a large number of trajectories by tracking temporal changes in molecular structure variables, potential energies, and molecular orbitals. ReSPer allows us to investigate reaction mechanisms using only the molecular structure dataset relevant to the chemical reaction of interest, making it a robust tool that does not require prior knowledge and complicated operations.

Finally, two very interesting dynamic processes are highlighted. Figure 6a shows a bundle of four trajectories that experience two dynamic bifurcations. First, Au_5 moves toward MIN1'' on the branching product side by IRC-jump, then back to the MIN1 region by a second IRC-jump. Figure 6c shows the trajectory bundle in Fig. 6a based on the potential energy landscape. Figure 6b shows a bundle of 10 trajectories that revisit the initial TS1-1d region after bouncing off the energy barrier

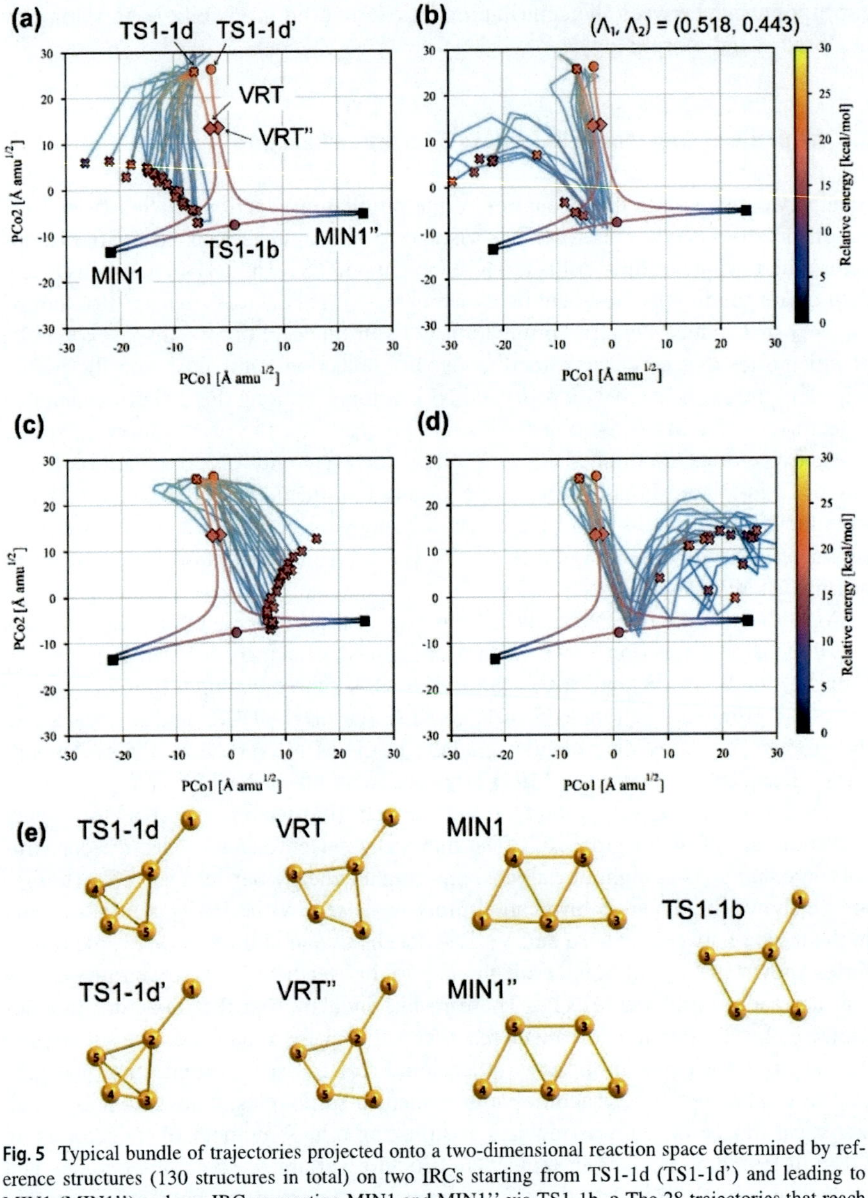

Fig. 5 Typical bundle of trajectories projected onto a two-dimensional reaction space determined by reference structures (130 structures in total) on two IRCs starting from TS1-1d (TS1-1d') and leading to MIN1 (MIN1'') and one IRC connecting MIN1 and MIN1'' via TS1-1b. **a** The 28 trajectories that reach MIN1 along the original IRC, **b** 12 trajectories that bounce off TS1-1b and reach MIN1, **c** 25 trajectories that reach MIN1'' by dynamic bifurcation, and **d** 15 trajectories that bounce off TS1-1b and reach MIN1''. Data points along the IRC and trajectories are omitted. *Squares* minima, *circles* TS, *diamonds* VRT, *cross marks* initial and final points of the trajectories, *color map* potential energy relative to MIN1. **e** Representative NPI isomers of TS1-1d, TS1-1d', MIN1, MIN1'', and TS1-1b

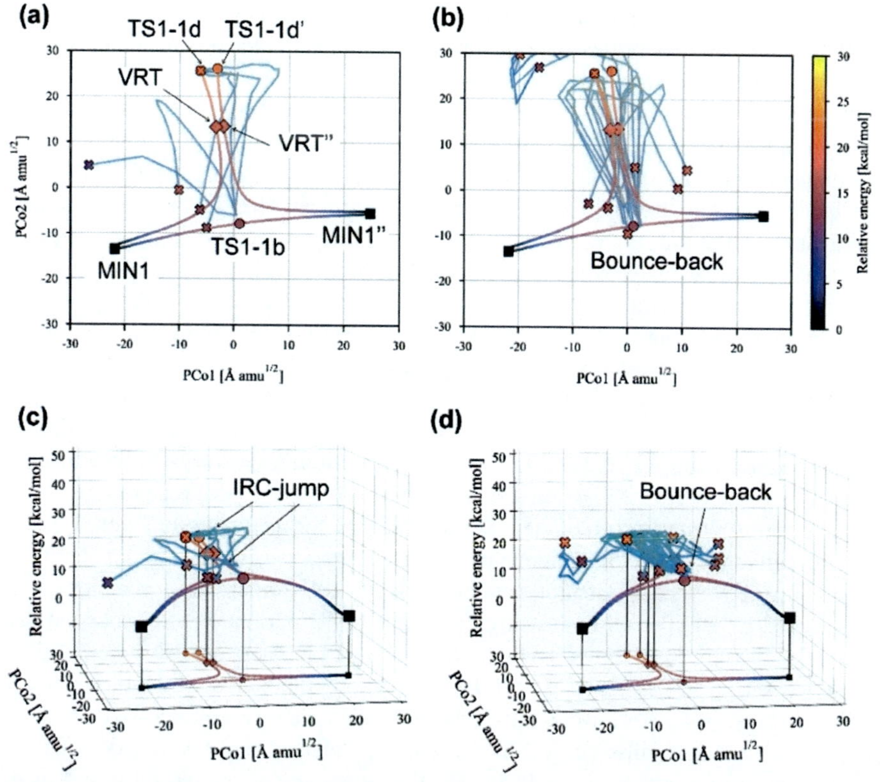

Fig. 6 Bundle of trajectories projected in two-dimensional space shown in Fig. 5: **a** four trajectories that experienced two dynamic bifurcations, **b** ten trajectories that bounced back to the region of TS1-1b and then revisited the region of TS1-1d. **c**, **d** The potential energy landscapes in which the bundles of trajectories shown in **a**, **b** are projected, respectively

near TS1-1b, similar to Fig. 5b,d, while Fig. 6d shows the corresponding figure based on the potential energy landscape. Wiggins et al. [78] studied the selectivity of the post-transition-state bifurcation in phase space based on a two-dimensional symmetric model surface, and discussed boundary that separates trajectories that proceed to one well from those that recross to the initial TS. This elaborated result is consistent with the bouncing behavior obtained by ReSPer and supports the existence of a boundary that determines the selectivity of the bifurcation reaction in the 9-dimensional asymmetric PES of the Au_5 cluster.

4 Conclusion

In this paper, we introduced the ReSPer method, which constructs a reduced-dimensionality reaction space using dimensionality reduction methods and projects trajectories into the reaction space determined from the molecular structure dataset involved in chemical reactions of interest. The current ReSPer program allows only CMDS dimensionality reduction, but we plan to implement parsers for other dimensionality reduction methods in future developments.

In the CMDS method for visualizing chemical reaction space, the distance matrix between a pair of molecular structures is transformed into PCos to properly evaluate their mutual structural dissimilarities in the full-dimensional space. Therefore, the quality of interstructural distances has a significant impact on the visualization results. To obtain the optimal Euclidean distances in mass-weighted Cartesian coordinate, several features are implemented in the ReSPer program: the pairwise structure alignment with the Kabsch algorithm [58, 59], the merged-NPI option, and the restricted NPI generation option. In addition, ReSPer can project out-of-sample data, such as trajectories, into a pre-constructed, low-dimensional reaction space, which has been implemented with reference to Trosset's formulation [55].

To demonstrate the ReSPer analysis, we revisited the isomerization and bifurcation reactions of the Au_5 cluster. In the first application, trajectories were projected into the two-dimensional reaction space determined by the global reaction path network, demonstrating that ReSPer is a powerful tool for analyzing reaction dynamics with reference to multiple elementary reaction processes with different chemical events. In the second application, the network of NPI isomerization reactions was constructed by ReSPer without using the merged-NPI option. ReSPer successfully visualized the closed island of TS1-1a, where the NPI isomers of MIN1 were linked through a set of NPI isomers of TS1-1a, and the complete closed island, where all 60 NPI isomers of MIN1 were fully linked through TS1-1a and TS1-1b. Furthermore, the ReSPer results suggested the correspondence between the degeneracy in the contribution ratios of the PCos and the symmetry of the closed island morphology. In the last application, we displayed trajectory bundles projected onto the reduced-dimensionality reaction space associated with the bifurcation reaction without using the merged-NPI option. The trajectory bundles described realistic reaction paths where dynamic effects were explicitly taken into account, and supported the existence of boundaries that determine selectivity in bifurcation reactions, as discussed in previous work [78].

ReSPer is a robust and versatile tool for reaction dynamics analysis based on a reduced-dimensionality reaction space constructed from a dataset of molecular structures relevant to the chemical reaction of interest. Once some development is complete, we plan to release ReSPer as a Python program in the near future. We have applied ReSPer to several simple organic chemical reactions and to the isomerization reaction of the small gold cluster. These studies have validated the advantages of ReSPer analysis, which does not rely on prior knowledge of chemical reactions, and established the basis for a methodology to track complex reaction dynamics by referring to a reaction path network in reduced dimension. Currently, the developmental goals of ReSPer are directed toward applications to more general organic chemical reactions, comprehensive reaction path networks for organic compounds, and comprehensive photoreaction analysis, including excited-state reaction dynamics and relaxation processes to the ground state. Reaction mechanisms in high-energy chemistry, such as photochemistry and combustion chemistry, are strongly dominated by reaction dynamics, and static reaction paths can no longer serve as reference paths on the PES. The reduced-dimensionality reaction space and energy landscape served by ReSPer will provide a platform to capture the characteristics of complex reaction dynamics even under high-energy conditions. We hope that ReSPer will reveal the close relationship between the morphology of the PES and reaction dynamics, and provide guidelines for controlling reaction dynamics.

Acknowledgements The authors are sincerely grateful to Z. Arai (Chubu Univ.) for his collaborated works on the Reaction Space Projector method. T. Tsutsumi was financially supported by Wakeshima Makoto Research Fellowship. This work was partly supported by the Elements Strategy Initiative of MEXT (JPMXP0112101003), the Photo-excitonix Project in Hokkaido University, and JST CREST Grant Number JPMJCR1902, Japan. A part of calculations was performed using the Research Center for Computational Science, Okazaki, Japan.

Author Contributions T. Tsutsumi was involved in conceptualization, data curation, methodology, investigation, software, visualization, writing (original draft, writing), review and editing. Y.O. was involved in conceptualization, methodology, investigation, software, visualization, writing (original draft, writing), review and editing. T. Taketsugu was involved in conceptualization, methodology, investigation, software, visualization, writing (original draft, writing), review and editing.

Funding This article was funded by Ministry of Education, Culture, Sports, Science and Technology (Grant no. JPMXP0112101003), Core Research for Evolutional Science and Technology (Grant no. JPMJCR1902).

Declarations

Competing interests The authors have declared no conflicts of interest for this paper.

References

1. Fukui K (1970) Formulation of the reaction coordinate. J Phys Chem 74:4161–4163. https://doi.org/10.1021/j100717a029
2. Schlegel HB (2011) Geometry optimization. Wiley Interdiscip Rev Comput Mol Sci 1:790–809. https://doi.org/10.1002/wcms.34
3. Ohno K, Maeda S (2004) A scaled hypersphere search method for the topography of reaction pathways on the potential energy surface. Chem Phys Lett 384:277–282. https://doi.org/10.1016/j.cplett.2003.12.030
4. Maeda S, Taketsugu T, Morokuma K, Ohno K (2014) Anharmonic downward distortion following for automated exploration of quantum chemical potential energy surfaces. Bull Chem Soc Jpn 87:1315–1334. https://doi.org/10.1246/bcsj.20140189
5. Maeda S, Morokuma K (2011) Finding reaction pathways of type A + B → X: toward systematic prediction of reaction mechanisms. J Chem Theory Comput 7:2335–2345. https://doi.org/10.1021/ct200290m
6. Maeda S, Taketsugu T, Morokuma K (2014) Exploring transition state structures for intramolecular pathways by the artificial force induced reaction method. J Comput Chem 35:166–173. https://doi.org/10.1002/jcc.23481
7. Maeda S, Harabuchi Y (2021) Exploring paths of chemical transformations in molecular and periodic systems: an approach utilizing force. WIREs Comput Mol Sci. https://doi.org/10.1002/wcms.1538
8. Ebisawa S, Tsutsumi T, Taketsugu T (2021) Geometric analysis of anharmonic downward distortion following paths. J Comput Chem 42:27–39. https://doi.org/10.1002/jcc.26430
9. Sumiya Y, Nagahata Y, Komatsuzaki T, Taketsugu T, Maeda S (2015) Kinetic analysis for the multistep profiles of organic reactions: significance of the conformational entropy on the rate constants

of the Claisen rearrangement. J Phys Chem A 119:11641–11649. https://doi.org/10.1021/acs.jpca.5b09447
10. Sumiya Y, Maeda S (2020) Rate constant matrix contraction method for systematic analysis of reaction path networks. Chem Lett 49:553–564. https://doi.org/10.1246/cl.200092
11. Martínez-Núñez E (2015) An automated method to find transition states using chemical dynamics simulations. J Comput Chem 36:222–234. https://doi.org/10.1002/jcc.23790
12. Dewyer AL, Argüelles AJ, Zimmerman PM (2018) Methods for exploring reaction space in molecular systems. Wiley Interdiscip Rev Comput Mol Sci 8:1–20. https://doi.org/10.1002/wcms.1354
13. Mitsuta Y, Shigeta Y (2020) Analytical method using a scaled hypersphere search for high-dimensional metadynamics simulations. J Chem Theory Comput 16:3869–3878. https://doi.org/10.1021/acs.jctc.0c00010
14. Garay-Ruiz D, Álvarez-Moreno M, Bo C, Martínez-Núñez E (2022) New tools for taming complex reaction networks: the unimolecular decomposition of indole revisited. ACS Phys Chem Au. https://doi.org/10.1021/acsphyschemau.1c00051
15. Field-Theodore TE, Taylor PR (2020) ALTRUISM: a higher calling. J Chem Theory Comput 16:4388–4398. https://doi.org/10.1021/acs.jctc.0c00388
16. Miller WH, Handy NC, Adams JE (1980) Reaction path Hamiltonian for polyatomic molecules. J Chem Phys 72:99–112. https://doi.org/10.1063/1.438959
17. Kato S, Morokuma K (1980) Potential energy characteristics and energy partitioning in chemical reactions: ab initio MO study of $H_2CCH_2F \rightarrow H_2CCHF+H$ reaction. J Chem Phys 72:206–217. https://doi.org/10.1063/1.438877
18. Taketsugu T, Tajima N, Hirao K (1996) Approaches to bifurcating reaction path. J Chem Phys 105:1933–1939. https://doi.org/10.1063/1.472063
19. Maeda S, Harabuchi Y, Ono Y, Taketsugu T, Morokuma K (2015) Intrinsic reaction coordinate: calculation, bifurcation, and automated search. Int J Quantum Chem 115:258–269. https://doi.org/10.1002/qua.24757
20. Gordon MS, Chaban G, Taketsugu T (1996) Interfacing electronic structure theory with dynamics. J Phys Chem 100:11512–11525. https://doi.org/10.1021/jp953371o
21. Pratihar S, Ma X, Homayoon Z, Barnes GL, Hase WL (2017) Direct chemical dynamics simulations. J Am Chem Soc 139:3570–3590. https://doi.org/10.1021/jacs.6b12017
22. Taketsugu T, Gordon MS (1995) Dynamic reaction path analysis based on an intrinsic reaction coordinate. J Chem Phys 103:10042–10049. https://doi.org/10.1063/1.470704
23. Taketsugu T, Gordon MS (1996) Reaction path Hamiltonian based on a reaction coordinate and a curvature coordinate. J Chem Phys 104:2834–2840. https://doi.org/10.1063/1.471019
24. Zou W, Sexton T, Kraka E, Freindorf M, Cremer D (2016) A new method for describing the mechanism of a chemical reaction based on the unified reaction valley approach. J Chem Theory Comput 12:650–663. https://doi.org/10.1021/acs.jctc.5b01098
25. Ito T, Harabuchi Y, Maeda S (2020) AFIR explorations of transition states of extended unsaturated systems: automatic location of ambimodal transition states. Phys Chem Chem Phys 22:13942–13950. https://doi.org/10.1039/D0CP02379E
26. Lee S, Goodman JM (2020) Rapid route-finding for bifurcating organic reactions. J Am Chem Soc 142:9210–9219. https://doi.org/10.1021/jacs.9b13449
27. Sun L, Song K, Hase WL (2002) A SN2 reaction that avoids its deep potential energy minimum. Science 296:875–878. https://doi.org/10.1126/science.1068053
28. Hare SR, Li A, Tantillo DJ (2018) Post-transition state bifurcations induce dynamical detours in Pummerer-like reactions. Chem Sci 9:8937–8945. https://doi.org/10.1039/C8SC02653J
29. Campos RB, Tantillo DJ (2019) Designing reactions with post-transition-state bifurcations: asynchronous nitrene insertions into C-C σ bonds. Chem 5:227–236. https://doi.org/10.1016/j.chempr.2018.10.019
30. Kraka E, Cremer D (2010) Computational analysis of the mechanism of chemical reactions in terms of reaction phases: hidden intermediates and hidden transition states. Acc Chem Res 43:591–601. https://doi.org/10.1021/ar900013p
31. Hong YJ, Tantillo DJ (2014) Biosynthetic consequences of multiple sequential post-transition-state bifurcations. Nat Chem 6:104–111. https://doi.org/10.1038/nchem.1843
32. Kraka E, Zou W, Tao Y, Freindorf M (2020) Exploring the mechanism of catalysis with the unified reaction valley approach (URVA)—a review. Catalysts 10:691. https://doi.org/10.3390/catal10060691

33. Jiang B, Li J, Guo H (2016) Potential energy surfaces from high fidelity fitting of ab initio points: the permutation invariant polynomial - neural network approach. Int Rev Phys Chem 35:479–506. https://doi.org/10.1080/0144235X.2016.1200347
34. Liu Y, Song H, Xie D, Li J, Guo H (2020) Mode specificity in the OH + $HO_2 \rightarrow H_2O + O_2$ reaction: enhancement of reactivity by exciting a spectator mode. J Am Chem Soc 142:3331–3335. https://doi.org/10.1021/jacs.9b12467
35. Lin S, Peng D, Yang W, Gu FL, Lan Z (2021) Theoretical studies on triplet-state driven dissociation of formaldehyde by quasi-classical molecular dynamics simulation on machine-learning potential energy surface. J Chem Phys 155:214105. https://doi.org/10.1063/5.0067176
36. Olasz B, Czakó G (2019) Uncovering the role of the stationary points in the dynamics of the F − + CH 3 I reaction. Phys Chem Chem Phys 21:1578–1586. https://doi.org/10.1039/C8CP06207B
37. Sharma N, Biswas R, Lourderaj U (2020) Dynamics of a gas-phase S_NAr reaction: non-concerted mechanism despite the Meisenheimer complex being a transition state. Phys Chem Chem Phys 22:26562–26567. https://doi.org/10.1039/D0CP05567K
38. Tsutsumi T, Harabuchi Y, Ono Y, Maeda S, Taketsugu T (2018) Analyses of trajectory on-the-fly based on the global reaction route map. Phys Chem Chem Phys 20:1364–1372. https://doi.org/10.1039/C7CP06528K
39. Tsutsumi T, Ono Y, Arai Z, Taketsugu T (2018) Visualization of the intrinsic reaction coordinate and global reaction route map by classical multidimensional scaling. J Chem Theory Comput 14:4263–4270. https://doi.org/10.1021/acs.jctc.8b00176
40. Tsutsumi T, Ono Y, Arai Z, Taketsugu T (2020) Visualization of the dynamics effect: projection of on-the-fly trajectories to the subspace spanned by the static reaction path network. J Chem Theory Comput 16:4029–4037. https://doi.org/10.1021/acs.jctc.0c00018
41. Tsutsumi T, Ono Y, Taketsugu T (2021) Visualization of reaction route map and dynamical trajectory in reduced dimension. Chem Commun 57:11734–11750. https://doi.org/10.1039/D1CC04667E
42. Komatsuzaki T, Hoshino K, Matsunaga Y, Rylance GJ, Johnston RL, Wales DJ (2005) How many dimensions are required to approximate the potential energy landscape of a model protein? J Chem Phys 122:084714. https://doi.org/10.1063/1.1854123
43. Hare SR, Bratholm LA, Glowacki DR, Carpenter BK (2019) Low dimensional representations along intrinsic reaction coordinates and molecular dynamics trajectories using interatomic distance matrices. Chem Sci 10:9954–9968. https://doi.org/10.1039/C9SC02742D
44. Peng J, Xie Y, Hu D, Lan Z (2021) Analysis of bath motion in MM-SQC dynamics via dimensionality reduction approach: principal component analysis. J Chem Phys 154:094122. https://doi.org/10.1063/5.0039743
45. Rashmi R, Yadav K, Lourderaj U, Paranjothy M (2021) Second-order saddle dynamics in isomerization reaction. Regul Chaotic Dyn 26:119–130. https://doi.org/10.1134/S1560354721020027
46. Casier B, Carniato S, Miteva T, Capron N, Sisourat N (2020) Using principal component analysis for neural network high-dimensional potential energy surface. J Chem Phys 152:234103. https://doi.org/10.1063/5.0009264
47. Pisani P, Caporuscio F, Carlino L, Rastelli G (2016) Molecular dynamics simulations and classical multidimensional scaling unveil new metastable states in the conformational landscape of CDK2. PLoS One 11:e0154066. https://doi.org/10.1371/journal.pone.0154066
48. Li X, Xie Y, Hu D, Lan Z (2017) Analysis of the geometrical evolution in on-the-fly surface-hopping nonadiabatic dynamics with machine learning dimensionality reduction approaches: classical multidimensional scaling and isometric feature mapping. J Chem Theory Comput 13:4611–4623. https://doi.org/10.1021/acs.jctc.7b00394
49. Oliveira AB, Yang H, Whitford PC, Leite VBPP (2019) Distinguishing biomolecular pathways and metastable states. J Chem Theory Comput 15:6482–6490. https://doi.org/10.1021/acs.jctc.9b00704
50. Shi W, Jia T, Li A (2020) Quasi-classical trajectory analysis with isometric feature mapping and locally linear embedding: deep insights into the multichannel reaction on an $NH3^+$ (^4A) potential energy surface. Phys Chem Chem Phys 22:17460–17471. https://doi.org/10.1039/D0CP01941K
51. Evans DA, Wales DJ (2003) Free energy landscapes of model peptides and proteins. J Chem Phys 118:3891–3897. https://doi.org/10.1063/1.1540099
52. Becker OM, Karplus M (1997) The topology of multidimensional potential energy surfaces: theory and application to peptide structure and kinetics. J Chem Phys 106:1495–1517. https://doi.org/10.1063/1.473299
53. Torgerson WS (1952) Multidimensional scaling: I. Theory and method. Psychometrika 17:401–419. https://doi.org/10.1007/BF02288916

54. Härdle WK, Simar L (2015) Applied Multivariate Statistical Analysis, 3rd edn. Springer, Berlin
55. Trosset MW, Priebe CE (2008) The out-of-sample problem for classical multidimensional scaling. Comput Stat Data Anal 52:4635–4642. https://doi.org/10.1016/j.csda.2008.02.031
56. Rogers DJ, Tanimoto TT (1960) A computer program for classifying plants. Science 132:1115–1118. https://doi.org/10.1126/science.132.3434.1115
57. Fukutani T, Miyazawa K, Iwata S, Satoh H (2021) G-RMSD: root mean square deviation based method for three-dimensional molecular similarity determination. Bull Chem Soc Jpn 94:655–665. https://doi.org/10.1246/bcsj.20200258
58. Kabsch W (1976) A solution for the best rotation to relate two sets of vectors. Acta Crystallogr Sect A 32:922–923. https://doi.org/10.1107/S0567739476001873
59. GitHub Calculate Root-mean-square deviation (RMSD) of two molecules using rotation. In: GitHub. http://github.com/charnley/rmsd
60. Borg I, Groenen P (2005) Modern Multidimensional Scaling: Theory and Applications. Springer Series in Statistics. Springer, Berlin
61. Young G, Householder AS (1938) Discussion of a set of points in terms of their mutual distances. Psychometrika 3:19–22. https://doi.org/10.1007/BF02287916
62. Haruta M (2005) Gold rush. Nature 437:1098–1099. https://doi.org/10.1038/4371098a
63. Gao M, Lyalin A, Takagi M, Maeda S, Taketsugu T (2015) Reactivity of gold clusters in the regime of structural fluxionality. J Phys Chem C 119:11120–11130. https://doi.org/10.1021/jp511913t
64. Muramatsu S, Koyasu K, Tsukuda T (2018) Abstraction of the I atom from CH3I by gas-phase Au_n^- (n = 1–4) via reductive activation of the C-I bond. ACS Omega 3:16874–16881. https://doi.org/10.1021/acsomega.8b02809
65. Sugiuchi M, Maeba J, Okubo N, Iwamura M, Nozaki K, Konishi K (2017) Aggregation-induced fluorescence-to-phosphorescence switching of molecular gold clusters. J Am Chem Soc 139:17731–17734. https://doi.org/10.1021/jacs.7b10201
66. Miyamoto M, Taketsugu T, Iwasa T (2021) A comparative study of structural, electronic, and optical properties of thiolated gold clusters with icosahedral vs face-centered cubic cores. J Chem Phys 155:094304. https://doi.org/10.1063/5.0057566
67. Harabuchi Y, Ono Y, Maeda S, Taketsugu T (2015) Analyses of bifurcation of reaction pathways on a global reaction route map: a case study of gold cluster Au5. J Chem Phys 143:014301. https://doi.org/10.1063/1.4923163
68. Perdew JP, Burke K, Ernzerhof M (1996) Generalized gradient approximation made simple. Phys Rev Lett 77:3865–3868. https://doi.org/10.1103/PhysRevLett.77.3865
69. Perdew JP, Burke K, Ernzerhof M (1997) Generalized gradient approximation made simple [Phys. Rev. Lett. 77, 3865 (1996)]. Phys Rev Lett 78:1396–1396. https://doi.org/10.1103/PhysRevLett.78.1396
70. Hay PJ, Wadt WR (1985) Ab initio effective core potentials for molecular calculations. Potentials for K to Au including the outermost core orbitale. J Chem Phys 82:299–310. https://doi.org/10.1063/1.448975
71. Frisch MJ, Trucks GW, Schlegel HB, Scuseria GE, Robb MA, Cheeseman JR, Scalmani G, Barone V, Petersson GA, Nakatsuji H, Li X, Caricato M, Marenich A, Bloino J, Janesko BG, Gomperts R, Mennucci B, Hratchian HP, Ortiz J V, Izmaylov AF, Sonnenberg JL, Williams-Young D, Ding F, Lipparini F, Egidi F, Goings J, Peng B, Petrone A, Henderson T, Ranasinghe D, Zakrzewski VG, Gao J, Rega N, Zheng G, Liang W, Hada M, Ehara M, Toyota K, Fukuda R, Hasegawa J, Ishida M, Nakajima T, Honda Y, Kitao O, Nakai H, Vreven T, Throssell K, Montgomery, J. A. J, Peralta JE, Ogliaro F, Bearpark M, Heyd JJ, Brothers E, Kudin KN, Staroverov VN, Keith T, Kobayashi R, Normand J, Raghavachari K, Rendell A, Burant JC, Iyengar SS, Tomasi J, Cossi M, Millam JM, Klene M, Adamo C, Cammi R, Ochterski JW, Martin RL, Morokuma K, Farkas O, Foresman JB, Fox DJ (2013) Gaussian 09. Gaussian, Inc., Wallingford CT
72. Maeda S, Osada Y, Morokuma K, Ohno K (2011) GRRM11,version 11.01.
73. Harabuchi Y, Okai M, Yamamoto R, Tsutsumi T, Ono Y, Taketsugu T (2020) SPPR (a developmental version). Hokkaido University: Sapporo, Japan
74. Taketsugu T, Wales DJ (2002) Theoretical study of rearrangements in water dimer and trimer. Mol Phys 100:2793–2806. https://doi.org/10.1080/00268970210142648
75. Rawlinson JI, Fábri C, Császár AG (2021) Exactly solvable 1D model explains the low-energy vibrational level structure of protonated methane. Chem Commun 57:4827–4830. https://doi.org/10.1039/D1CC01214B

76. Williams DMG, Eisfeld W (2020) Complete nuclear permutation inversion invariant artificial neural network (CNPI-ANN) diabatization for the accurate treatment of vibronic coupling problems. J Phys Chem A 124:7608–7621. https://doi.org/10.1021/acs.jpca.0c05991
77. Thomas JB, Waas JR, Harmata M, Singleton DA (2008) Control elements in dynamically determined selectivity on a bifurcating surface. J Am Chem Soc 130:14544–14555. https://doi.org/10.1021/ja802577v
78. Agaoglou M, García-Garrido VJ, Katsanikas M, Wiggins S (2020) The phase space mechanism for selectivity in a symmetric potential energy surface with a post-transition-state bifurcation. Chem Phys Lett 754:137610. https://doi.org/10.1016/j.cplett.2020.137610

Publisher's Note Springer Nature remains neutral with regard to jurisdictional claims in published maps and institutional affiliations.

Topics in Current Chemistry (2022) 380:15
https://doi.org/10.1007/s41061-022-00366-w

REVIEW

NAST: Nonadiabatic Statistical Theory Package for Predicting Kinetics of Spin-Dependent Processes

Vsevolod D. Dergachev[1] · Mitra Rooein[1] · Ilya D. Dergachev[1] · Aleksandr O. Lykhin[1,2] · Robert C. Mauban[1] · Sergey A. Varganov[1]

Received: 3 September 2021 / Accepted: 15 January 2022 / Published online: 24 February 2022
© The Author(s), under exclusive licence to Springer Nature Switzerland AG 2022

Abstract
We present a nonadiabatic statistical theory (NAST) package for predicting kinetics of spin-dependent processes, such as intersystem crossings, spin-forbidden unimolecular reactions, and spin crossovers. The NAST package can calculate the probabilities and rates of transitions between the electronic states of different spin multiplicities. Both the microcanonical (energy-dependent) and canonical (temperature-dependent) rate constants can be obtained. Quantum effects, including tunneling, zero-point vibrational energy, and reaction path interference, can be accounted for. In the limit of an adiabatic unimolecular reaction proceeding on a single electronic state, NAST reduces to the traditional transition state theory. Because NAST requires molecular properties at only a few points on potential energy surfaces, it can be applied to large molecular systems, used with accurate high-level electronic structure methods, and employed to study slow nonadiabatic processes. The essential NAST input data include the nuclear Hessian at the reactant minimum, as well as the nuclear Hessians, energy gradients, and spin–orbit coupling at the minimum energy crossing point (MECP) between two states. The additional computational tools included in the NAST package can be used to extract the required input data from the output files of electronic structure packages, calculate the effective Hessian at the MECP, and fit the reaction coordinate for more advanced NAST calculations. We describe the theory, its implementation, and three examples of application to different molecular systems.

Keywords Intersystem crossings · Spin-forbidden reactions · Spin–orbit coupling · Reaction rates

---

Chapter 4 was originally published as Dergachev, V. D., Rooein, M., Dergachev, I. D., Lykhin, A. O., Mauban, R. C. & Varganov, S. A. Topics in Current Chemistry (2022) 380: 15. https://doi.org/10.1007/s41061-022-00366-w.

✉ Sergey A. Varganov
svarganov@unr.edu

Extended author information available on the last page of the article

1 Introduction

Spin-dependent processes, including transitions between electronic states characterized by different values of total electron spin and magnetic quantum numbers, play an important role in many areas of atomic and molecular science. These include multi-state reactivity in transition-metal-based catalysis [1–6] and on semiconductor surfaces [7], design of photosensitizers for various applications [8–10], and development of single-molecule magnets for applications in quantum sensing, quantum computing, and spintronics [11–15]. For example, in photodynamic therapy, intersystem crossing (ISC) populates a manifold of low-lying triplet states of a photosensitizer. The following spin-allowed triplet–triplet reaction with molecular oxygen produces highly reactive oxygen species, which destroy cancer cells [8–10]. Delayed fluorescence in organic light-emitting diodes (OLEDs) is due to a thermally activated reverse ISC between close-lying excited triplet and singlet electronic states [16–20]. A nitrogen-vacancy (NV) center in diamond—one of the most promising candidates for a spin-based qubit—is initialized through the ISC between optically excited triplet and singlet states [13]. Photolysis of axial ligands from active sites of heme proteins proceeds via several ISCs on a femtosecond time scale [21–25]. Spin-forbidden low-energy reaction pathways have been proposed for the catalytic mechanisms of molecular hydrogen activation on the NiFe-hydrogenase metalloprotein [6] and the C–H bond activation on Fe (II) [3]. Many ligand–metal binding and dissociation reactions, such as the CO binding to $Fe(CO)_4$ [26, 27] and the diatomic molecules binding to the active site of heme proteins [28–30], are formally spin-forbidden. Spin crossovers and spin–spin magnetic exchange interactions in d^4–d^7 transition metal complexes play a fundamental role in the design of magnetic bistable materials [31–39].

Predicting kinetics of spin-dependent processes is important for understanding the mechanisms of thermally activated spin-forbidden reactions, ISCs in photochemistry, and spin crossovers in transition metal-based systems. There are two main approaches to calculate the rate constants and lifetimes of electronic states in spin-dependent processes. In ab initio nonadiabatic molecular dynamics, the classical or quantum nuclei are time propagated on the coupled potential energy surfaces (PESs) of multiple spin states [40–49], and the time evolution of nuclear trajectories is used to statistically describe the population transfer between the interacting electronic states. Such molecular dynamics simulations can be very accurate; however, they often require thousands of electronic structure calculations and can be computationally expensive if not prohibitive. An alternative nonadiabatic statistical theory (NAST) approach, also called nonadiabatic transition state theory [4, 26, 30, 50–59], largely eliminates the computational burden by exploring only the critical points on PESs. This allows NAST to be used with the high-level electronic structure methods and model even slow nonadiabatic processes intractable for molecular dynamics. NAST can be viewed as an extension of traditional transition state theory (TST) [60–65] to the nonadiabatic processes. Similarly to TST, NAST assumes that (1) intramolecular energy is statistically

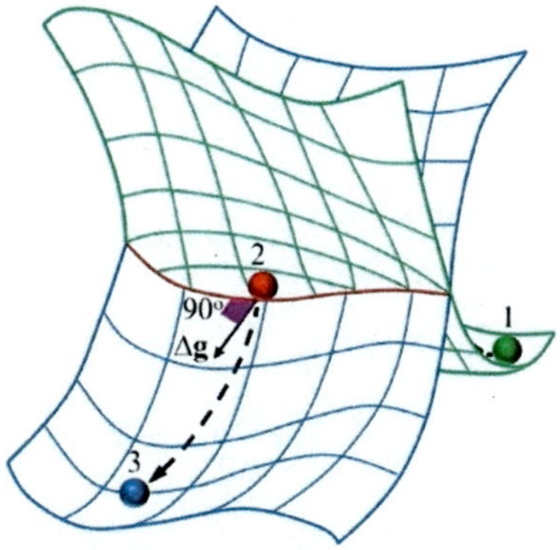

Fig. 1 Intersection of potential energy surfaces of two electronic states with different spin multiplicities. Point 1 is the reactant minimum, point 2 is the minimum energy crossing point (MECP), point 3 is the product minimum, and $\Delta\mathbf{g} = |\mathbf{g}_1 - \mathbf{g}_2|$ is the gradient vector orthogonal to the crossing seam and aligned with the reaction coordinate at the MECP. The dashed curve shows the minimum energy path connecting the MECP with the reactant and product minima

distributed among the molecular degrees of freedom (DOF) and (2) a one-dimensional reaction coordinate can be separated from the remaining spectator DOF. Both microcanonical (energy-dependent) and canonical (temperature-dependent) ensembles can be used to describe internal states leading to two formulations of NAST. Transitions between PESs of two electronic states with different spin multiplicities are driven by various spin-dependent interactions, among which the spin–orbit coupling (SOC) is often the strongest [66–69]. The crossing seam between such two PESs is $3N$–7 dimensional, where N is the number of atoms in a molecule, and the seven omitted dimensions include three translational, three rotational, and a reaction coordinate DOF. It is also assumed that nonadiabatic transitions between two crossing PESs can be described by effective transitions at a minimum energy crossing point (MECP) on the seam [4], which plays a similar role to that of a saddle point in TST (Fig. 1). At the MECP, the reaction coordinate is orthogonal to the rest of the (spectator) DOF. To evaluate the rate constant using NAST, the nuclear Hessian at the reactant minimum, as well as the energy gradients, nuclear Hessians, and the SOC at MECP, are needed. A more advanced treatment requires the knowledge of one-dimensional minimum energy paths from the MECP to the reactant and product minima [59, 70]. The statistical nature of NAST makes it relatively simple to account for the quantum effects, such as tunneling and zero-point vibrational energy (ZPE) [59].

In this article, we present the NAST package for investigating the mechanisms and predicting the rates of spin-dependent processes. The package includes an implementation of NAST and additional computational tools for processing the output of electronic structure calculations. In Sect. 2, we describe the fundamentals of NAST. In Sect. 3, we discuss the NAST package capabilities and implementation. In Sect. 4, we present several examples of NAST application. In conclusion, we summarize the main strengths and discuss future extensions of NAST.

2 Nonadiabatic Statistical Theory

2.1 Microcanonical Rate Constants

The microcanonical rate constant $k(E)$ of a unimolecular reaction is calculated as a function of the internal (rovibrational) energy E,

$$k(E) = \sigma \frac{N_\mathrm{X}(E)}{h\rho_\mathrm{R}(E)}, \tag{1}$$

$$N_\mathrm{X}(E) = \int_0^E \rho_\mathrm{X}(E - \varepsilon_\perp) P(\varepsilon_\perp) \mathrm{d}\varepsilon_\perp, \tag{2}$$

$$\sigma = \frac{\sigma_\mathrm{R}}{\sigma_\mathrm{X}} \gamma_\mathrm{X}, \tag{3}$$

where $N_\mathrm{X}(E)$ is the effective number of states at the MECP, ρ_R and ρ_X are the densities of rovibrational states at the reactant and MECP, respectively, and h is the Planck constant. Calculations of the densities of rovibrational states are described in the Supplementary Information (SI). The interstate transition probability $P(\varepsilon_\perp)$ is a function of the energy $\varepsilon_\perp$ partitioned in the reaction coordinate orthogonal to the crossing seam. The reaction path degeneracy σ is defined in terms of the symmetry numbers of reactant (σ_R) and MECP (σ_X) [71], and the number of chiral MECP isomers (γ_X) [72].

2.2 Microcanonical Transition Probabilities

The most popular ways to calculate the interstate transition probability $P(\varepsilon_\perp)$ in Eq. (2) are the double-passage Landau–Zener (LZ) and weak coupling (WC) formulas [4, 55, 73, 74]. These formulas yield a cumulative probability of transition at the MECP for the forward (primary passage) and backward (secondary passages) motions along the reaction coordinate,

$$P_\mathrm{LZ}(\varepsilon_\perp) = p_\mathrm{LZ} + (1 - p_\mathrm{LZ})p_\mathrm{LZ} = 2p_\mathrm{LZ} - p_\mathrm{LZ}^2, \tag{4}$$

$$p_\mathrm{LZ}(\varepsilon_\perp) = 1 - \exp\left(-\frac{2\pi H_\mathrm{SO}^2}{\hbar|\Delta \mathbf{g}|} \sqrt{\frac{\mu_\perp}{2(\varepsilon_\perp - E_X)}} \right), \tag{5}$$

$$P_\mathrm{WC}(\varepsilon_\perp) = 4\pi^2 H_\mathrm{SO}^2 \left(\frac{2\mu_\perp}{\hbar^2 \overline{\mathbf{g}} |\Delta \mathbf{g}|} \right)^{\frac{2}{3}} \mathrm{Ai}^2 \left(-(\varepsilon_\perp - E_X) \left(\frac{2\mu_\perp |\Delta \mathbf{g}|^2}{\hbar^2 \overline{\mathbf{g}}^4} \right)^{\frac{1}{3}} \right), \tag{6}$$

where p_{LZ} is the single-passage LZ probability, H_{SO} is the spin–orbit coupling constant, and $\hbar$ is the reduced Planck constant. The norm of the gradient parallel to the reaction coordinate $|\Delta \mathbf{g}| = |\mathbf{g}_1 - \mathbf{g}_2|$ and the mean gradient $\overline{\mathbf{g}} = \left(|\mathbf{g}_1||\mathbf{g}_2|\right)^{1/2}$ are defined in terms of the gradients of two crossing PESs at the MECP, $\mathbf{g}_1$ and $\mathbf{g}_2$. In Eqs. (5 and 6), $\mu_\perp$ is the reduced mass for the motion along the reaction coordinate, and E_X is the MECP energy barrier with respect to the reactant minimum. In Eq. (6), Ai is the Airy function.

To account for ZPE, the MECP energy can be redefined as $E_X \to E_X + \mathrm{ZPE}_X - \mathrm{ZPE}_R$. For a nonlinear molecule, the zero-point energies at the reactant minimum and MECP are defined as:

$$\mathrm{ZPE_R} = \frac{1}{2} \sum_{i=1}^{3N-6} \hbar\omega_i^{\mathrm{R}}, \tag{7}$$

$$\mathrm{ZPE_X} = \frac{1}{2} \sum_{i=1}^{3N-7} \hbar\omega_i^{\mathrm{X}}, \tag{8}$$

where ω_i^{R} and ω_i^{X} are the fundamental transition frequencies of the reactant and MECP, respectively. In Eq. (7), index i runs over $3N-6$ vibrational DOF, while only $3N-7$ DOF orthogonal to the reaction coordinate contribute to ZPE at the MECP. Note that using $\mathrm{ZPE_X}$ at the classical turning point along the minimum energy reaction path is equivalent to the zero-curvature tunneling approximation in TST, where the density of the rovibrational states is approximated by the density at TS, and the effective TS barrier is reduced by the difference between $\mathrm{ZPE_R}$ and $\mathrm{ZPE_{TS}}$.

The LZ probability is defined only at the reaction energy $\varepsilon_\perp$ above the MECP, and therefore does not account for quantum tunneling through the MECP barrier. In addition, the LZ probability formula does not describe the quantum interference between primary and secondary passages at the MECP [4]. These two quantum effects are included in the WC probability formula. However, both the LZ and WC formulas are only valid within the limited region of the parameters $\varepsilon_\perp$, $\mu_\perp$, H_{SO} and the energy gradients. For example, the WC formula can predict a greater than unit probability of transition if the interacting states are strongly coupled, as is often the case in the complexes containing second- and third-row transition metals [4]. In addition, the LZ and WC formulas assume a linear behavior of the two crossing potentials, which is often a reasonable approximation in the vicinity of the MECP but not in the regions closer to the reactant and product minima. A more sophisticated approach to predict transition probabilities, which does not suffer from these limitations, has been introduced by Zhu and Nakamura (ZN) [70, 75–79]. The ZN probability expressions require the knowledge of the entire one-dimensional minimum energy path connecting the MECP to the reactant and product minima. This path can be obtained in either spin-diabatic or spin-adiabatic representations [49]. The ZN theory distinguishes two intersection types between PESs: a sloped intersection $\left(\mathbf{g}_1 \cdot \mathbf{g}_2 > 0\right)$ and a peaked intersection $\left(\mathbf{g}_1 \cdot \mathbf{g}_2 < 0\right)$ (Fig. S1). Currently, in the NAST package, the ZN probability is implemented only for a sloped intersection in the spin-adiabatic representation. Because most of the electronic structure

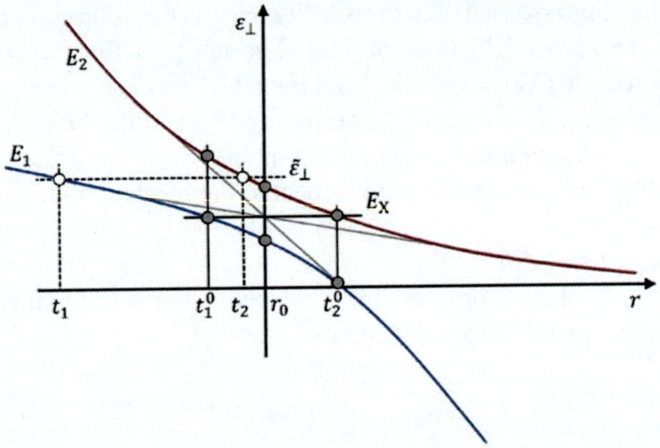

Fig. 2 Sloped intersection of two spin-adiabatic potentials with the energies E_1 (blue) and E_2 (red) along the reaction coordinate r. Points r_0, t_1^0, t_2^0, t_1, and t_2 are defined in the text

calculations produce spin-diabatic PESs, the adiabatic potentials are obtained by diagonalizing the following matrix:

$$\begin{pmatrix} E_1^d(r) & H_{SO} \\ H_{SO} & E_2^d(r) \end{pmatrix}, \tag{9}$$

where E_1^d and E_2^d are the spin-diabatic energies of two electronic states with arbitrary spin multiplicities, r is the arc length along the minimum energy reaction path in mass-scaled coordinates [80] with the reduced mass set to 1 amu, and H_{SO} is the spin–orbit coupling constant at the MECP. The diagonalization of the matrix defined by Eq. (9) produces two eigenvalues corresponding to the adiabatic (spin-mixed) state energies, E_1 and E_2 (Fig. 2). The ZN double-passage transition probability is given by

$$P_{ZN}(\varepsilon_\perp) = 4p_{ZN}(1 - p_{ZN})\sin^2(\psi), \tag{10}$$

where ψ is the overall transition phase. The single passage probability p_{ZN} defined in the SI depends on several parameters that are functions of the energies at the smallest energy gap point r_0, the turning points at the MECP energy t_1^0 and t_2^0, and the turning points t_1 and t_2 corresponding to the specific value of the energy $\varepsilon_\perp$ (Fig. 2).

2.3 Canonical Rate Constants

A canonical, temperature-dependent rate constant can be obtained by averaging the microcanonical rate constant (Eq. 1) over the internal energy Boltzmann distribution, leading to the following expression:

$$k(T) = \frac{\sigma}{hQ_{\mathrm{R}}(T)} \int_{0}^{\infty} N_{\mathrm{X}}(E) e^{-E/k_{\mathrm{B}}T} \mathrm{d}E, \tag{11}$$

$$Q_{\mathrm{R}}(T) = \int_{0}^{\infty} \rho_{\mathrm{R}}(E) e^{-E/k_{\mathrm{B}}T} \mathrm{d}E, \tag{12}$$

where Q_{R} is the partition function of the reactant, T is the temperature, and k_{B} is the Boltzmann constant.

2.4 Velocity-Averaged Probabilities

Velocity-averaged probabilities provide a simple estimate for the likelihood of spin-forbidden reaction mechanism as a function of temperature. The velocity-averaged single passage LZ probabilities (Eqs. 13 and 14) are derived by averaging the microcanonical LZ probability (Eq. 5) using the Maxwell–Boltzmann (MB) and normalized Kuki (K) distributions of the mass-weighted velocity $\upsilon^2 = 2(\varepsilon_{\perp} - E_{\mathrm{X}})$ [6, 81]. The corresponding double-passage LZ probabilities can be obtained using Eq. (4). The velocity-averaged WC probabilities (Eqs. 15 and 16) can also be derived using the Maxwell–Boltzmann and normalized Kuki distributions. It is important to note that, in contrast to the microcanonical WC probability, the velocity-averaged WC probabilities do not account for quantum tunneling.

$$\langle p_{\mathrm{LZ}}(T) \rangle_{\mathrm{MB}} = 1 - \left(\frac{2}{\pi k_{\mathrm{B}} T} \right)^{\frac{1}{2}} \int_{0}^{\infty} \exp\left(-\frac{2\pi H_{\mathrm{SO}}^2 \mu_{\perp}^{1/2}}{\hbar |\Delta \mathbf{g}| \upsilon} \right) \exp\left(-\frac{\upsilon^2}{2k_{\mathrm{B}}T} \right) \mathrm{d}\upsilon, \tag{13}$$

$$\langle p_{\mathrm{LZ}}(T) \rangle_{\mathrm{K}} = 1 - \frac{1}{k_{\mathrm{B}}T} \int_{0}^{\infty} \upsilon \exp\left(-\frac{2\pi H_{\mathrm{SO}}^2 \mu_{\perp}^{1/2}}{\hbar |\Delta \mathbf{g}| \upsilon} \right) \exp\left(-\frac{\upsilon^2}{2k_{\mathrm{B}}T} \right) \mathrm{d}\upsilon, \tag{14}$$

$$\langle P_{\mathrm{WC}}(T) \rangle_{\mathrm{MB}} = \alpha \left(\frac{2}{\pi k_{\mathrm{B}} T} \right)^{\frac{1}{2}} \int_{0}^{\infty} \mathrm{Ai}^2\left(-\frac{1}{2} \upsilon^2 \mu_{\perp} \gamma \right) \exp\left(-\frac{\upsilon^2}{2k_{\mathrm{B}}T} \right) \mathrm{d}\upsilon, \tag{15}$$

$$\langle P_{\mathrm{WC}}(T) \rangle_{\mathrm{K}} = \frac{\alpha}{k_{\mathrm{B}}T} \int_{0}^{\infty} \upsilon \mathrm{Ai}^2\left(-\frac{1}{2} \upsilon^2 \mu_{\perp} \gamma \right) \exp\left(-\frac{\upsilon^2}{2k_{\mathrm{B}}T} \right) \mathrm{d}\upsilon, \tag{16}$$

$$\alpha = 4\pi^2 H_{\mathrm{SO}}^2 \left(\frac{2\mu_{\perp}}{\hbar^2 \overline{\mathbf{g}} |\Delta \mathbf{g}|} \right)^{\frac{2}{3}}, \tag{17}$$

$$\gamma = \left(\frac{2\mu_{\perp}|\Delta \mathbf{g}|^2}{\hbar^2 \overline{\mathbf{g}}^4} \right)^{\frac{1}{3}}. \tag{18}$$

2.5 Rate Constants and Transition Probabilities Between Individual M_S Components of Spin States

A simple approach to model transitions between electronic states with the different spin quantum numbers S and S' is to calculate the effective probabilities and rate constants accounting for all M_S components of the spin multiplets. In this approach, the effective SOC, also called the SOC constant, is obtained as the RMS of the couplings between individual M_S components,

$$H_{SO} = \left(\sum_{M_s=-S}^{S} \sum_{M_{S'}=-S'}^{S'} \left| \left\langle S, M_S \middle| \hat{H}_{SO} \middle| S', M_{S'} \right\rangle \right|^2 \right)^{1/2}. \tag{19}$$

In Eq. (19), $\hat{H}_{SO}$ is the spin–orbit operator; for example, from the Breit–Pauli Hamiltonian [67]. This approach is easy to justify for a singlet–triplet crossing with the MECP energy gap between the two spin-adiabatic states equal to $2H_{SO}$ [4]. However, for the states with higher spin multiplicities, for example a triplet–quintet crossing, there are multiple energy gaps between the adiabatic states. Therefore, employing a single effective SOC to calculate the transition probability and rate constant is not well justified.

This issue can be addressed by calculating the rate constants and transition probabilities between individual M_S components of the spin multiplets. As an example, for a singlet–triplet crossing, the non-zero spin–orbit coupling matrix elements are

$$z = \langle 0, 0 | \hat{H}_{SO} | 1, -1 \rangle, \quad ib = \langle 0, 0 | \hat{H}_{SO} | 1, 0 \rangle, \quad z^* = \langle 0, 0 | \hat{H}_{SO} | 1, +1 \rangle, \tag{20}$$

where z and z^* are complex conjugate to each other, and b is real. The single-passage LZ probabilities $p_{LZ}^{M_S, M_{S'}}(\varepsilon_\perp)$ between the components M_S and $M_{S'}$ of the spin states $S = 0$ and $S' = 1$ read

$$p_{LZ}^{0,-1}(\varepsilon_\perp) = p_{LZ}^{0,+1}(\varepsilon_\perp) = 1 - \exp\left(-\frac{2\pi z z^*}{\hbar|\Delta \mathbf{g}|} \sqrt{\frac{\mu_\perp}{2(\varepsilon_\perp - E_X)}} \right), \tag{21}$$

$$p_{LZ}^{0,0}(\varepsilon_\perp) = 1 - \exp\left(-\frac{2\pi b^2}{\hbar|\Delta \mathbf{g}|} \sqrt{\frac{\mu_\perp}{2(\varepsilon_\perp - E_X)}} \right). \tag{22}$$

The double-passage probabilities $P_{LZ}^{0,\pm1}$ and $P_{LZ}^{0,0}$ can be obtained from the single-passage probabilities (Eq. 4) and employed to calculate the microcanonical rate constants between individual M_S components,

$$k_{0,\pm 1}(E) = \frac{\sigma}{h\rho_R(E)} \int_0^E \rho_X(E - \varepsilon_\perp) P_{LZ}^{0,\pm 1}(\varepsilon_\perp) d\varepsilon_\perp, \tag{23}$$

$$k_{0,0}(E) = \frac{\sigma}{h\rho_R(E)} \int_0^E \rho_X(E - \varepsilon_\perp) P_{LZ}^{0,0}(\varepsilon_\perp) d\varepsilon_\perp. \tag{24}$$

The probabilities and rate constants between individual M_S components can be calculated for any pair of spin states with $|S - S'| = 1$, as shown in the SI for a triplet-quintet crossing.

2.6 Transition State Theory Rate Constants

For adiabatic reactions, both microcanonical and canonical NAST rate constants can be reduced to the traditional TST rate constants by replacing MECP with a transition state (TS) and the transition probability in Eq. (2) with the Heaviside step function [82]. The canonical TST rate constant, obtained by averaging the microcanonical constant over the Boltzmann internal energy distribution, is equivalent to the traditional analytical TST expression [4]:

$$k(T) = \sigma \frac{k_B T}{h} \frac{Q_{TS}}{Q_R} e^{-E_{TS}/k_B T}, \tag{25}$$

where Q_{TS} and Q_R are partition functions of TS and reactant.

2.7 Effective Hessian

Calculation of the effective number of states $N_X(E)$ defined in Eq. 2 requires the density of vibrational states at the MECP, which can be obtained using the harmonic vibrational analysis [83, 84]. Since MECP is not a stationary point on either of the two crossing PESs, a conventional vibrational analysis is not valid. The vibrational frequencies at the MECP can be obtained from the effective Hessian matrix calculated using the state-specific Hessians $\mathbf{H}_1$ and $\mathbf{H}_2$, and gradients at MECP [85]:

$$\mathbf{H}_{eff} = \frac{|\mathbf{g}_1|\mathbf{H}_2 \pm |\mathbf{g}_2|\mathbf{H}_1}{|\mathbf{g}_1| \pm |\mathbf{g}_2|}. \tag{26}$$

The details of calculating the effective Hessian are given in the SI.

3 NAST Package Capabilities and Implementation

3.1 Forward and Reverse Rate Constants

By default, the NAST package calculates only forward rate constants with both the LZ and WC transition probabilities. A forward direction is defined as the transition from the higher energy spin state (reactant) to the lower energy spin state (product). Such calculations require molecular properties only at the reactant minimum and MECP. However, calculation of the reverse rate constants can be requested in the same run, if molecular properties at the product minimum are provided. For such reverse rate calculations, the above definition of reactant and product prevents the reverse unphysical tunneling to the region below the reactant minimum where the density of states is zero (Fig. 3). If only a forward rate constant with tunneling contribution is calculated, it is important to ensure that the reactant has a higher energy than the product to prevent contribution from unphysical tunneling to the rate constants. This is not required if the rate constants are calculated with the LZ probability or traditional TST, which do not account for tunneling. In the present implementation, the ZN transition probability can only be used to calculate a forward rate constant. For canonical rate constant calculations, the temperature range (default 290–300 K) and step (default 1 K) can be specified in the input file.

3.2 Transition Probabilities

The LZ, WC, and ZN probabilities (Eqs. 4–6 and 10) are used to calculate the microcanonical and canonical rate constants. The velocity-averaged LZ and WC probabilities (Eq. 13–16), which are calculated by default, can be used to obtain a qualitative understanding of the spin-dependent reaction kinetics. While calculations of the LZ and WC probabilities require molecular properties at reactants and MECP only, the ZN probability calculations require additional input data discussed together with the intrinsic reaction coordinate (IRC) fit code (Sect. 3.7).

Fig. 3 Definition of forward and reverse directions. The region of reverse unphysical tunneling is shown by the red box

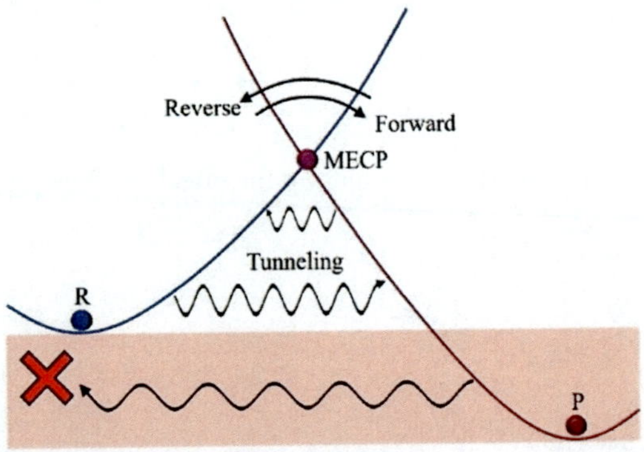

3.3 Rate Constants and Transition Probabilities Between Individual M_S Components of Spin States

In addition to effective transition probabilities and rate constants between the spin manifolds with the different values of quantum number S, the NAST package can calculate the transition probabilities and rate constants between individual M_S components of different spin manifolds. In the current implementation, the M_S-specific kinetics can be modeled only using the LZ formula. Such calculations can provide insight into the role of individual M_S components in the overall spin-dependent kinetics. Moreover, working in the basis of individual M_S components is necessary to study the effect of an external magnetic field on spin-dependent processes.

3.4 Rate Constants in Solution

Transition state theories for reactions in solution have been discussed extensively in previous works [60, 86–91]. In the current NAST implementation, the solution phase effects can be modeled by simply assuming that all molecular rotations are frozen. In such calculations, only vibrational states contribute to the total density of states, and contributions from rotational states are ignored.

3.5 Transition State Theory Rate Constants

The NAST package can calculate the traditional TST rate constants for single-state adiabatic reactions by replacing the MECP molecular properties with the TS properties (energy, vibrational frequencies, moments of inertia). Such TST calculations do not require spin–orbit couplings, energy gradients, or reduced mass, and can be invoked with a separate keyword.

3.6 Effective Hessian Tool *effhess*

The effective Hessian tool *effhess*, distributed as a part of the NAST package, calculates and diagonalizes the effective Hessian (Eq. 26) to generate the vibrational frequencies at the MECP, the reduced mass $\mu_\perp$, $|\Delta \mathbf{g}|$, and $\bar{\mathbf{g}}$. These quantities are needed for the following rate constant calculations. Currently, the tool can read the MECP energy gradient vectors and Hessian matrices from the output files generated by the GAMESS [92] and Molpro [93] electronic structure packages. The *effhess* tool generates a template of an input file for the main NAST code.

3.7 IRC Fitting Tool *ircfit*

The fitting tool *ircfit* produces the IRC potentials of two crossing spin states. These potentials are required for calculating the ZN probabilities. Currently, *ircfit* can read only the GAMESS IRC output files. Here we summarize the *ircfit* algorithm; the details are included in the SI.

1. Perform two IRC calculations from MECP to the minima of reactant and product, generating the two sets of geometries $\{Q_i, E_i\}_{X\to R}$ and $\{Q_i, E_i\}_{X\to P}$, where Q_i and E_i are the coordinates and energy of molecular geometry i. Index i runs from 1 (MECP) to n for the reactant side of IRC and m for the product side.
2. Define $n+m$ points along the reaction coordinate as

$$r_j = \left(\sum_{p=1}^{3N} \left(dQ_{j,p}\right)^2\right)^{1/2}, \tag{27}$$

where $dQ_{j,p}$ is the difference between the coordinates of the reactant and the geometry $j \in [1, n+m]$.
3. Build the crossing potentials by fitting the two sets of points $\{r_i, E_i\}_{X\to R}$ and $\{r_i, E_i\}_{X\to P}$ with the fourth-order polynomials:

$$f(\mathbf{c}, r) = \sum_{q=0}^{4} c_q r^q. \tag{28}$$

The vector of coefficients $\mathbf{c}$ is obtained by minimizing the linear least-squares function

$$F = \sum_{i=1}^{k} \left(E_i - f\left(\mathbf{c}, r_i\right)\right)^2, \tag{29}$$

where $k=n$ for the reactant side of the fit, and $k=m$ for the product side. The coefficients of the polynomials, which represent spin-diabatic states, are used by the main NAST code to generate spin-adiabatic states (Eq. 9) required for the ZN probability calculations.

3.8 Modular Structure of the NAST Package

The NAST package consists of a collection of modules written in modern Fortran and runs under Linux (Fig. 4). The package manual contains a full list of the input file keywords that control the type of rate calculations. Most of the keywords have default values appropriate for the common calculation types. The effective Hessian tool *effhess* can be used to generate an input file template containing MECP

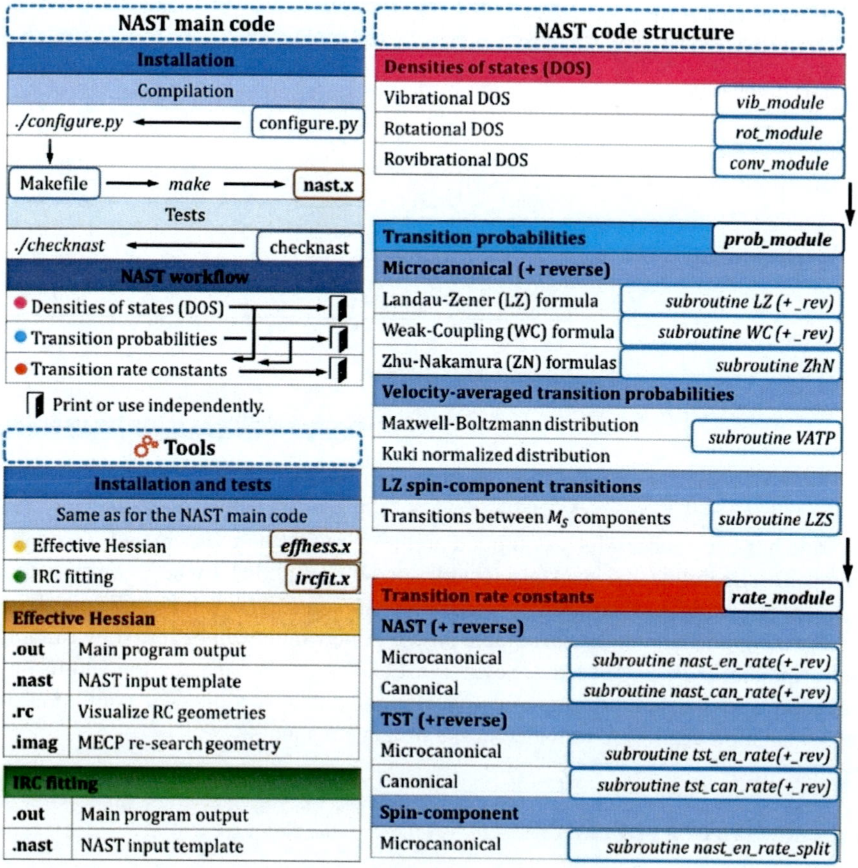

Fig. 4 Modular structure of the NAST package. The top left panel illustrates the main NAST code installation, testing, and workflow. The bottom left panel lists the output files of the effective Hessian (*effhess*) and IRC fitting (*ircfit*) tools. The right panel shows the main modules and subroutines of the NAST package

properties, further reducing the efforts required to set up a NAST calculation. The rest of the input data, including vibrational frequencies, moments of inertia, electronic energies, and SOC, must be extracted from the output of electronic structure calculations. The calculated canonical rate constants and velocity-averaged transition probabilities are saved to the main output file *nast.out*. Additional information, including microcanonical rate constants, transition probabilities, and density of states, are written to separate output files. The amount of output information can be controlled by the user.

Table 1 Canonical TST rate constants for isomerization of propylene oxide to acetone and propanal at T = 1000 K

Product	Source	E_{TS}, kcal·mol^{-1}	k_{1000}, s^{-1}
Acetone	NAST	53.2	67.8
	Analytical, Eq. (30)[a]	54.2	50.2
	Analytical, Eq. (25)	53.2	101.0
	Experiment[b]	–	30.0
Propanal	NAST	54.2	43.0
	Analytical, Eq. (30)[a]	54.4	69.6
	Analytical, Eq. (25)	54.2	88.6
	Experiment[b]	-	90.0

[a]Ref. [94]

[b]Experimental values are estimated from the log(k) vs. T plots of Ref. [95]

The NAST package requires an external math library. It has been tested with the Intel Math Kernel Library (MKL) but can be linked with other math libraries. The package has been successfully compiled and tested with the *gfortran* and *ifort* compilers. To streamline the package installation, the provided Python configuration script *configure.py* can be used to generate a *Makefile*. The Linux utility *make* is used to compile and link the executable *nast.x*. A new build can be tested by running the script *checknast* that executes several test examples. The installation and testing procedures are the same for the effective Hessian and IRC tools. The configuration and installation details are provided in the package manual.

4 Examples of Applications

4.1 Isomerization of Propylene Oxide to Acetone and Propanal

This example demonstrates how the NAST package can be used to calculate the reaction rates for single-state adiabatic reactions using the traditional TST. We consider the isomerization of propylene oxide to acetone and propanal, following the original theoretical work of Dubnikova and Lifshitz [94]. For each isomerization reaction, four canonical TST rate constants are compared: (1) predicted by the NAST package, (2) calculated analytically using Eq. (25), (3) calculated analytically using Eq. (30) below [94], and (4) obtained from an experiment [95]. In Ref. [94], the following equation is used:

$$k(T) = \sigma \frac{k_B T}{h} e^{-\Delta S^{\#}/R} e^{-\Delta H^{\#}/RT}, \tag{30}$$

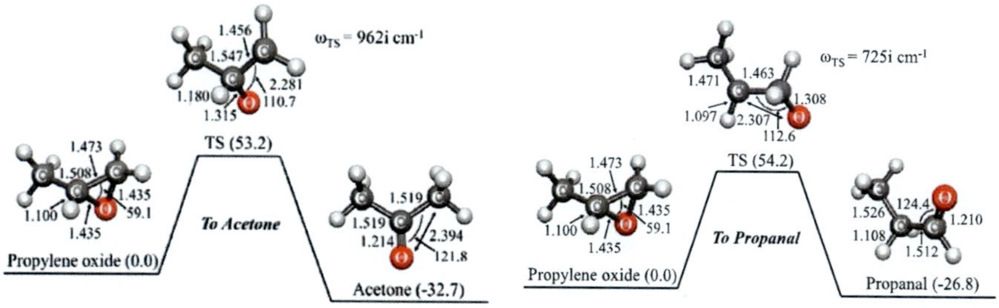

Fig. 5 Energies and structural parameters for the isomerization reactions of propylene oxide to acetone (left) and propanal (right). The relative CCSD(T) energies (kcal mol^{-1}) listed in parenthesis are corrected with $\Delta ZPE = ZPE_X - ZPE_R$ calculated with B3LYP. Bond lengths and angles are in Å and degrees, respectively

where R is the universal gas constant; $\Delta S^{\#}$ and $\Delta H^{\#}$ are the entropy and enthalpy of activation, respectively. Because for a unimolecular reaction, $\Delta H^{\#} = E_{TS} + \Delta ZPE$, the activation enthalpy is equal to the ZPE-corrected electronic barrier between transition state and reactant. The input data for NAST were obtained from the B3LYP geometry optimization and Hessian calculations in GAMESS, and the single-point CCSD(T) energy calculations in Molpro. The cc-pVDZ basis set was used in all calculations. The partition functions Q_{TS} and Q_R used in Eq. (25) were taken from the GAMESS Hessian calculations. In Eq. (30), the original data from Table 1 of Dubnikova and Lifshitz [94] were used. The energy profiles for two isomerization reactions are shown in Fig. 5.

The NAST input file and parts of the output file are shown below.

```
&keys zpe = 1 tst = .true. &end          ! TST calculation with ZPE correction

&inputdata
freX = 263, 294, 374, 580, 729, 871, 886, 958, 999, 1081, 1118, 1241, 1283, 1357, 1434, 1452, 1471, 2296, 3026,
3092, 3118, ,3142, 3238        ! Vibrational frequencies (cm-1) of TS/MECP
freR = 240, 370, 416, 773, 848, 903, 974, 102, 1117, 1136, 1153, 1173, 1277, 1381, 1423, 1451, 1471, 1515, 3025,
3059, 3076, 3098, 3117, 3153 ! Vibrational frequencies (cm-1) of reactant
inertX = 170.850, 217.264, 351.144       ! Moments of inertia for TS/MECP
inertR = 100.305, 272.514, 305.514       ! Moments of inertia for reactant
enX = 0.09129754                         ! Energy of TS in hartree (program will add ΔZPE)
enR = 0.0                                ! Energy of reactant in hartree
maxn = 30000                             ! Maximum energy bin - integration limit in cm-1
T1 = 1000                                ! Initial temperature in K for canonical rate constant
&end
```

```
*************************************************************
~~~~~~~ NAST: Nonadiabatic Statistical Theory ~~~~~~~~
        ~~~~~ v. 2021.1 ~~~~~
*************************************************************
------------------------------------------------------
NAST control parameters and related data

zpe = 1   sp = F   zn = F   solution = F
tst = T   printmore = F   rev = F   extern = F
------------------------------------------------------

zpe = 1: ZPE correction scheme I (eliminates turning points below ZPE).
Electronic barrier from reactant to MECP is  20037 cm-1
ZPE of reactants =  18586 cm-1
ZPE of MECP =  17150 cm-1
ZPE-corrected MECP energy bin =  18602 cm

 ----------------------------------------------------------
          Start NAST calculation
 ----------------------------------------------------------

  1. Calculating densities of states

......vibrational.
.......rotational.
........rovibrational.

  2. Calculating microcanonical TST rate constant.
.................................. Done.

(The forward rate constant k(E) is multiplied by reaction path degeneracy equal to 1 )

          Canonical rate constant, k(T)

   T(K)    TST rate constant

  1000.0        6.78E+01

 Total CPU time =    10.35
```

Table 1 shows the calculated and experimental rate constants for the isomerization of propylene oxide to acetone and propanal. The factor of two difference between the acetone rate constants calculated with two analytical expressions can be

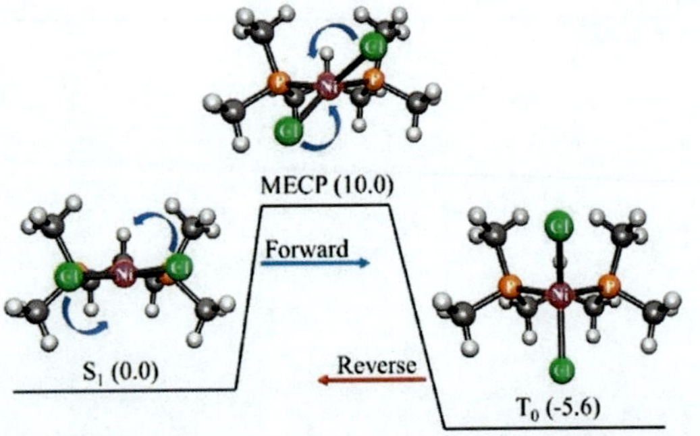

Fig. 6 Reaction path for the singlet–triplet isomerization of the Ni(dpp)Cl_2 model from the square-planar to tetrahedral geometry. The twist angle between the Cl–Ni–Cl and P–Ni–P planes is 6°, 42°, and 90° for the singlet, MECP, and triplet geometries, respectively. The phenyl groups of Ni(dpp)Cl_2 are replaced with the methyl groups

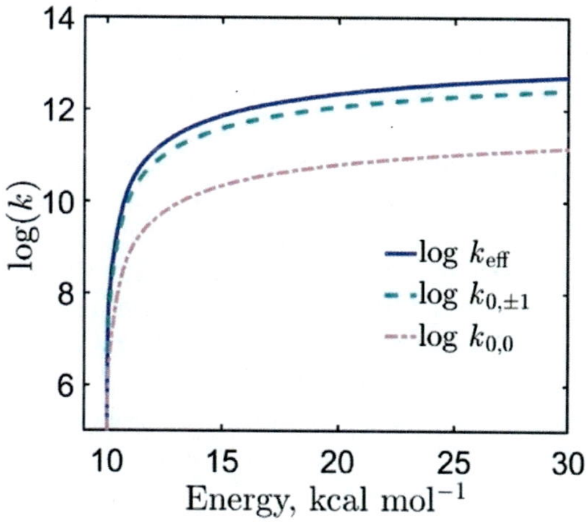

Fig. 7 The microcanonical rate constants $k_{0,0}$ and $k_{0,+1} = k_{0,-1}$ for transitions between the M_S components of the singlet and triplet states calculated using the LZ probabilities. The effective rate constant k_{eff} corresponds to the overall transition between the singlet state and all three components of the triplet state

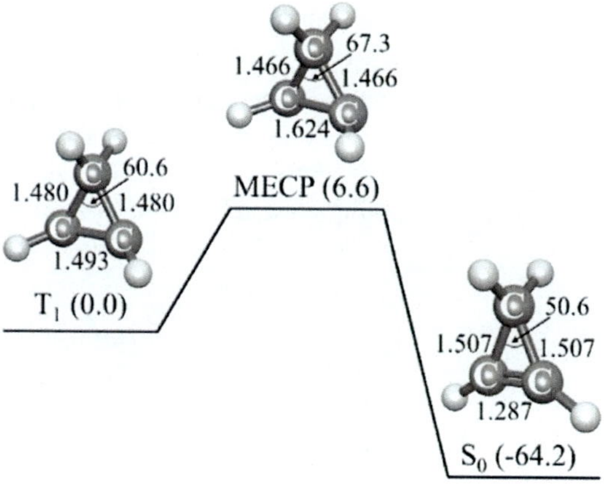

Fig. 8 The $T_1 \rightarrow S_0$ relaxation path in cyclopropene. The relative energies of the T_1 minimum, the S_0 minimum and MECP (kcal mol^{-1}) are in parentheses. The bond lengths and angles calculated at the B3LYP/cc-pVTZ level of theory are in Å and degrees

explained by the 1.0 kcal mol^{-1} discrepancy in the TS barrier (using the same barrier height of 53.2 kcal mol^{-1} in both analytical expressions reduces this difference to 18%). The difference between the NAST rate constant (67.8 s^{-1}) and the analytical constant obtained with Eq. (30) (101.0 s^{-1}) could be due to a finite accuracy of the numerical integration in NAST. For the isomerization to propanal, similar differences in the reaction rate constants calculated with different methods are observed. All calculated rate constants are in reasonable agreement with the experimental values for both isomerization reactions.

4.2 Spin-Forbidden Isomerization of Ni(dpp)Cl_2

This example demonstrates the forward and reverse rate constants calculation in a single NAST run. In addition to the overall rate constants, it predicts the rate constants between individual M_S components of two spin states. Spin-forbidden kinetics of interconversion between the singlet (planar) and triplet (tetrahedral) isomers of Ni(dpp)Cl_2 complex (dpp = 1,3-bis(diphenylphosphino)propane) has been studied in acetonitrile [96]. To reduce the computational cost, the phenyl groups were replaced by methyls (Fig. 6). The optimized geometries and Hessians for singlet, triplet, and MECP structures were obtained with the M11-L density functional, implicit solvation model, and def2-TZVP basis set, as implemented in GAMESS. The SOC value of 135 cm^{-1} was calculated with the second-order multiconfigurational quasi-degenerate perturbation theory (MCQDPT2) using the CASSCF(2e, 4o) wave function averaged over the lowest energy singlet and triplet electronic states. The forward ($S_1 \rightarrow T_0$) rate constant calculated using the WC transition probability at 296 K (4.96×10^6 s^{-1}) agrees almost within one order of magnitude with the experimental value of 4.5×10^5 s^{-1}. However, the reverse ($T_0 \rightarrow S_1$) rate constant predicted to be 3.50×10^2 s^{-1} is significantly smaller than the experimental value of 6.0×10^5 s^{-1}, which could indicate that the barrier for the reverse reaction is overestimated due to the low level of electronic structure calculations and the reduced model size.

Calculations of the transition probabilities and rate constants between individual M_S components of the singlet and triplet states were carried out using the spin–orbit matrix elements ($z = -69.0 - 63.8i$ cm^{-1} and $b = 21.3$ cm^{-1}, as defined in Eq. 20) obtained from the same MCQDPT2 calculations. The M_S-specific rate constants $k_{M_S,M_{S'}}$ for the transitions between the $M_S = 0$ component of singlet state and three $M_{S'} = \pm 1$ and $M_{S'} = 0$ components of the triplet state, calculated using the LZ transition probability, are plotted in Fig. 7. The stronger couplings between the $M_S = 0$ component of the singlet state and the $M_{S'} = \pm 1$ components of the triplet result

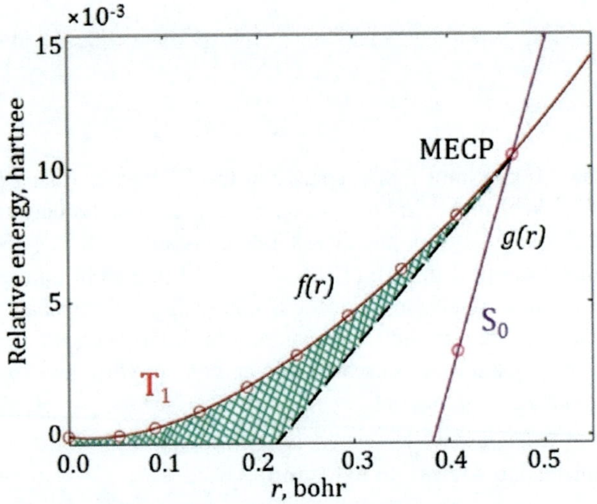

Fig. 9 Fitted $T_1 \rightarrow$ MECP $\rightarrow S_0$ reaction path. The circles mark the energies of the IRC geometries. The green area shows an increase in the tunneling barrier width compared to the linear model (dashed black line). For the rate constant calculations, the coefficients of the fitting polynomials ($f(r) = 0.092\, r^4 - 0.121\, r^3 + 0.093\, r^2 - 0.004\, r - 1.86 \times 10^{-6}$ and $g(r) = 0.127r - 0.048$) are included in the NAST input file

Table 2 The $T_1 \rightarrow S_0$ canonical rate constant (s^{-1}) at different temperatures

T, K	NAST	MESMER	Relative difference
100	1.52×10^{-16}	2.37×10^{-16}	−0.36
200	6.71×10^{-4}	7.88×10^{-4}	−0.15
300	1.19×10^{1}	1.29×10^{1}	−0.08
400	1.62×10^{3}	1.68×10^{3}	−0.03
500	3.16×10^{4}	3.09×10^{4}	0.02
600	2.33×10^{5}	2.13×10^{5}	0.09
700	9.81×10^{5}	8.43×10^{5}	0.16
800	2.91×10^{6}	2.35×10^{6}	0.24
900	6.85×10^{6}	5.18×10^{6}	0.32
1000	1.36×10^{7}	9.71×10^{6}	0.40
1100	2.41×10^{7}	1.62×10^{7}	0.49
1200	3.89×10^{7}	2.47×10^{7}	0.58
1300	5.85×10^{7}	3.52×10^{7}	0.66
1400	8.32×10^{7}	4.76×10^{7}	0.75
1500	1.13×10^{8}	6.17×10^{7}	0.83
1600	1.48×10^{8}	7.73×10^{7}	0.91
1700	1.88×10^{8}	9.42×10^{7}	1.00
1800	2.33×10^{8}	1.12×10^{8}	1.08
1900	2.82×10^{8}	1.31×10^{8}	1.15
2000	3.36×10^{8}	1.50×10^{8}	1.24

The relative difference between the NAST and MESMER values is defined as $(k_{NAST} - k_{MESMER})/k_{MESMER}$

in the faster population transfer, as evident from the $k_{0,\pm1}$ rate constant being larger than $k_{0,0}$.

4.3 $T_1 \rightarrow S_0$ Relaxation in Cyclopropene

This example demonstrates calculation of the rate constant using the ZN transition probability. Miller and Klippenstein published a detailed kinetics study of different reactions of C_3H_4, including those proceeding on multiple PESs with different spin multiplicities [97]. Here we consider the $T_1 \rightarrow S_0$ relaxation in cyclopropene (Fig. 8). The equilibrium geometries and Hessians of the S_0, T_1 states and MECP were obtained at the B3LYP/cc-pVTZ level of theory. The SOC constant of 4.0 cm^{-1} was calculated with the MCQDPT2 method based on the CASSCF(2,2) wave function averaged over the S_0 and T_1 states, using the same basis set. All calculations were performed in GAMESS. To calculate the $T_1 \rightarrow S_0$ rate constant with the ZN transition probability, the $T_1 \rightarrow$ MECP $\rightarrow S_0$ minimum energy reaction path was fitted with the *ircifit* tool using the geometries and energies generated by two IRC calculations starting from MECP and following to the T_1 and S_0 minima. The MECP $\rightarrow T_1$ path was fitted with the quartic polynomial, while the MECP $\rightarrow S_0$ path was approximated by a linear fit (Fig. 9).

The $T_1 \rightarrow S_0$ canonical rate constant calculated at 1000 K with the ZN transition probability (3.07×10^7 s^{-1}) is almost an order of magnitude smaller than the value obtained with the WC probability (1.24×10^8 s^{-1}). This difference can be explained by the fact that the linear-crossing WC model significantly underestimates the width of the tunneling barrier (Fig. 9). In contrast, the ZN model uses more realistic potential energy curves and predicts a wider barrier, leading to a significant reduction in the tunneling contribution to the rate constant. Table 2 compares the same rate constant calculated using the simple LZ transition probability with the values obtained by the MESMER 6.0 package [98]. There is reasonable agreement between the values predicted by the two packages. The larger discrepancies at high temperatures are attributed to the numerical integration errors in Eq. (2). It is important to note that these discrepancies in the rate constants are significantly smaller than the errors expected due to the limited accuracy of the MECP barrier calculated by commonly used electronic structure methods.

5 Conclusions

We introduced the NAST software package for predicting kinetics of spin-dependent processes, including intersystem crossings, spin-forbidden reactions, and spin crossovers. The package calculates both the microcanonical and canonical rate constants and can account for quantum tunneling, zero-point vibrational energy, and reaction path interference. Traditional single-state adiabatic TST calculations are also possible. The main strengths of NAST stem from its ability to (1) model nonadiabatic kinetics in large systems, (2) use high-level electronic structure methods for predicting molecular properties, which is critical for accurate kinetics calculations, and (3) study slow spin-dependent processes that present a great challenge for nonadiabatic molecular dynamics. Additional computational tools included in the NAST package simplify setting up kinetics calculations using molecular properties obtained with different electronic structure programs. The three presented examples demonstrate various capabilities of the NAST package. Future development will focus on multidimensional tunneling effects, accounting for MECP barrier recrossing, modeling spin-dependent processes in solution, and using an external magnetic field to control the kinetics of spin-dependent processes.

Supplementary Information The online version contains supplementary material available at https://doi.org/10.1007/s41061-022-00366-w.

Acknowledgements This work was supported by the National Science Foundation through a CAREER Award (CHE-1654547). We acknowledge Dr. Ryan Zaari for his contributions to the earlier version of the NAST package and thank Drs. Danil Kaliakin and Saikat Mukherjee for fruitful discussions.

Author contributions S.A.V. supervised the NAST package development. V.D.D., M.R., I.D.D., A.O.L. and R.C.M. wrote different parts of the code. V.D.D, M.R. and I.D.D. performed the electronic structure and kinetics calculations. All authors contributed to the preparation of the manuscript.

Funding This work was supported by the National Science Foundation through a CAREER Award (CHE 1654547).

Availability of data and material The input files for the electronic structure, NAST and MESMER calculations discussed in the main text are available in the SI.

Code availability The source code of the NAST package is available free of charge at https://github.com/svarganov/NAST.

Declarations

Conflict of interest The authors declare no conflict of interest.

References

1. Schröder D, Shaik S, Schwarz H (2000) Two-state reactivity as a new concept in organometallic chemistry. Acc Chem Res 33:139–145. https://doi.org/10.1021/ar990028j
2. Liao P, Carter EA (2013) New concepts and modeling strategies to design and evaluate photo-electro-catalysts based on transition metal oxides. Chem Soc Rev 42:2401–2422. https://doi.org/10.1039/c2cs35267b
3. Sun Y, Tang H, Chen K et al (2016) Two-state reactivity in low-valent iron-mediated C-H activation and the implications for other first-row transition metals. J Am Chem Soc 138:3715–3730. https://doi.org/10.1021/jacs.5b12150
4. Lykhin AO, Kaliakin DS, DePolo GE et al (2016) Nonadiabatic transition state theory: application to intersystem crossings in the active sites of metal-sulfur proteins. Int J Quantum Chem 116:750–761. https://doi.org/10.1002/qua.25124
5. Goodrow A, Bell AT, Head-Gordon M (2009) Are spin-forbidden crossings a bottleneck in methanol oxidation? J Phys Chem C 113:19361–19364. https://doi.org/10.1021/jp906603r
6. Kaliakin DS, Zaari RR, Varganov SA (2015) Effect of H_2 binding on the nonadiabatic transition probability between singlet and triplet states of the [NiFe]-hydrogenase active site. J Phys Chem A 119:1066–1073. https://doi.org/10.1021/jp510522z
7. Pohlman AJ, Kaliakin DS, Varganov SA, Casey SM (2020) Spin controlled surface chemistry: alkyl desorption from Si(100)-2×1 by nonadiabatic hydrogen elimination. Phys Chem Chem Phys 22:16641–16647. https://doi.org/10.1039/d0cp01913e
8. Zhao J, Wu W, Sun J, Guo S (2013) Triplet photosensitizers: from molecular design to applications. Chem Soc Rev 42:5323–5351. https://doi.org/10.1039/c3cs35531d
9. Zhao J, Chen K, Hou Y et al (2018) Recent progress in heavy atom-free organic compounds showing unexpected intersystem crossing (ISC) ability. Org Biomol Chem 16:3692–3701. https://doi.org/10.1039/c8ob00421h
10. Pordel S, Pickens RN, White JK (2021) Release of CO and production of $^1O_2$ from a Mn-BODIPY Photoactivated CO releasing molecule with visible light. Organometallics 40:2983–2994. https://doi.org/10.1021/acs.organomet.1c00331
11. Bogani L, Wernsdorfer W (2008) Molecular spintronics using single-molecule magnets. Nat Mater 7:179–186. https://doi.org/10.1038/nmat2133
12. Gaita-Ariño A, Luis F, Hill S, Coronado E (2019) Molecular spins for quantum computation. Nat Chem 11:301–309. https://doi.org/10.1038/s41557-019-0232-y
13. Goldman ML, Doherty MW, Sipahigil A et al (2015) State-selective intersystem crossing in nitrogen-vacancy centers. Phys Rev B 91:1–11. https://doi.org/10.1103/PhysRevB.91.165201
14. Bayliss SL, Laorenza DW, Mintun PJ et al (2020) Optically addressable molecular spins for quantum information processing. Science 370:1309–1312. https://doi.org/10.1126/science.abb9352
15. Ullah A, Cerdá J, Baldoví JJ et al (2019) In silico molecular engineering of dysprosocenium-based complexes to decouple spin energy levels from molecular vibrations. J Phys Chem Lett 10:7678–7683. https://doi.org/10.1021/acs.jpclett.9b02982

16. Mitschke U, Bäuerle P (2000) The electroluminescence of organic materials. J Mater Chem 10:1471–1507. https://doi.org/10.1039/a908713c
17. Goushi K, Yoshida K, Sato K, Adachi C (2012) Organic light-emitting diodes employing efficient reverse intersystem crossing for triplet-to-singlet state conversion. Nat Photonics 6:253–258. https://doi.org/10.1038/nphoton.2012.31
18. Endo A, Ogasawara M, Takahashi A et al (2009) Thermally activated delayed fluorescence from Sn^{4+}-porphyrin complexes and their application to organic light-emitting diodes—a novel mechanism for electroluminescence. Adv Mater 21:4802–4806. https://doi.org/10.1002/adma.200900983
19. Bergmann L, Hedley GJ, Baumann T et al (2016) Direct observation of intersystem crossing in a thermally activated delayed fluorescence copper complex in the solid state. Sci Adv 2:1–7. https://doi.org/10.1126/sciadv.1500889
20. Wada Y, Nakagawa H, Matsumoto S et al (2020) Organic light emitters exhibiting very fast reverse intersystem crossing. Nat Photonics 14:643–649. https://doi.org/10.1038/s41566-020-0667-0
21. Franzen S, Kiger L, Poyart C, Martin JL (2001) Heme photolysis occurs by ultrafast excited state metal-to-ring charge transfer. Biophys J 80:2372–2385. https://doi.org/10.1016/S0006-3495(01)76207-8
22. Wang W, Ye X, Demidov AA et al (2000) Femtosecond multicolor pump-probe spectroscopy of ferrous cytochrome C. J Phys Chem B 104:10789–10801. https://doi.org/10.1021/jp0008602
23. Dunietz BD, Dreuw A, Head-Gordon M (2003) Initial steps of the photodissociation of the CO ligated heme group. J Phys Chem B 107:5623–5629. https://doi.org/10.1021/jp0226376
24. Mara MW, Hadt RG, Reinhard ME et al (2017) Metalloprotein entatic control of ligand-metal bonds quantified by ultrafast x-ray spectroscopy. Science 356:1276–1280. https://doi.org/10.1126/science.aam6203
25. Falahati K, Tamura H, Burghardt I, Huix-Rotllant M (2018) Ultrafast carbon monoxide photolysis and heme spin-crossover in myoglobin via nonadiabatic quantum dynamics. Nat Commun 9:1–8. https://doi.org/10.1038/s41467-018-06615-1
26. Harvey JN, Aschi M (2003) Modelling spin-forbidden reactions: recombination of carbon monoxide with iron tetracarbonyl. Faraday Discuss 124:129–143. https://doi.org/10.1039/b211871h
27. Yang B, Gagliardi L, Truhlar DG (2018) Transition states of spin-forbidden reactions. Phys Chem Chem Phys 20:4129–4136. https://doi.org/10.1039/c7cp07227a
28. Jensen KP, Ryde U (2004) How O_2 binds to heme. Reasons for rapid binding and spin inversion. J Biol Chem 279:14561–14569. https://doi.org/10.1074/jbc.M314007200
29. Strickland N, Harvey JN (2007) Spin-forbidden ligand binding to the ferrous-heme group: Ab initio and DFT studies. J Phys Chem B 111:841–852. https://doi.org/10.1021/jp064091j
30. Harvey JN (2004) Spin-forbidden CO ligand recombination in myoglobin. Faraday Discuss 127:165–177. https://doi.org/10.1039/b314768a
31. Kahn O, Martinez CJ (1998) Spin-transition polymers: from molecular materials toward memory devices. Science 279:44–48. https://doi.org/10.1126/science.279.5347.44
32. Halder GJ, Kepert CJ, Moubaraki B et al (2002) Guest-dependent spin crossover in a nanoporous molecular framework material. Science 298:1762–1765. https://doi.org/10.1126/science.1075948
33. Gütlich PGH (2004) Spin crossover in transition metal compounds I-III. Top Curr Chem 23(234):235
34. Halcrom MA (2013) Spin-crossover materials, properities and applications. Wiley, Hoboken
35. Senthil Kumar K, Ruben M (2017) Emerging trends in spin crossover (SCO) based functional materials and devices. Coord Chem Rev 346:176–205. https://doi.org/10.1016/j.ccr.2017.03.024
36. Real JA, Gaspar AB, Carmen Muñoz M (2005) Thermal, pressure and light switchable spin-crossover materials. Dalt Trans 12:2062–2079. https://doi.org/10.1039/b501491c
37. Bonhommeau S, Molnár G, Galet A et al (2005) One shot laser pulse induced reversible spin transition in the spin-crossover complex $[Fe(C_4H_4N_2)\{Pt(CN)_4\}]$ at room temperature. Angew Chemie Int Ed 44:4069–4073. https://doi.org/10.1002/anie.200500717
38. Gaspar AB, Ksenofontov V, Seredyuk M, Gütlich P (2005) Multifunctionality in spin crossover materials. Coord Chem Rev 249:2661–2676. https://doi.org/10.1016/j.ccr.2005.04.028
39. Muller RN, Vander EL, Laurent S (2003) Spin transition molecular materials: Intelligent contrast agents for magnetic resonance imaging. J Am Chem Soc 125:8405–8407. https://doi.org/10.1021/ja0349599
40. Beck M (2000) The multiconfiguration time-dependent Hartree (MCTDH) method: a highly efficient algorithm for propagating wavepackets. Phys Rep 324:1–105. https://doi.org/10.1016/S0370-1573(99)00047-2

41. Barbatti M (2011) Nonadiabatic dynamics with trajectory surface hopping method. Wiley Interdiscip Rev Comput Mol Sci 1:620–633. https://doi.org/10.1002/wcms.64
42. Curchod BFE, Martinez TJ (2018) Ab initio nonadiabatic quantum molecular dynamics. Chem Rev 118:3305–3336. https://doi.org/10.1021/acs.chemrev.7b00423
43. Mai S, Marquetand P, González L (2018) Nonadiabatic dynamics: the SHARC approach. Wiley Interdiscip Rev Comput Mol Sci 8:1–23. https://doi.org/10.1002/wcms.1370
44. Hammes-Schiffer S, Tully JC (1995) Nonadiabatic transition state theory and multiple potential energy surface molecular dynamics of infrequent events. J Chem Phys 103:8513–8527. https://doi.org/10.1063/1.470162
45. Shenvi N, Subotnik JE, Yang W (2011) Simultaneous-trajectory surface hopping: a parameter-free algorithm for implementing decoherence in nonadiabatic dynamics. J Chem Phys 134:144102. https://doi.org/10.1063/1.3575588
46. Curchod BFE, Rauer C, Marquetand P et al (2016) Communication: GAIMS—generalized ab initio multiple spawning for both internal conversion and intersystem crossing processes. J Chem Phys 144:101102. https://doi.org/10.1063/1.4943571
47. Fedorov DA, Pruitt SR, Keipert K et al (2016) Ab initio multiple spawning method for intersystem crossing dynamics: spin-forbidden transitions between $^3B_1$ and $^1A_1$ states of GeH_2. J Phys Chem A 120:2911–2919. https://doi.org/10.1021/acs.jpca.6b01406
48. Fedorov DA, Lykhin AO, Varganov SA (2018) Predicting intersystem crossing rates with AIMS-DFT molecular dynamics. J Phys Chem A 122:3480–3488
49. Mukherjee S, Fedorov DA, Varganov SA (2021) Modeling spin-crossover dynamics. Annu Rev Phys Chem 72:515–540. https://doi.org/10.1146/annurev-physchem-101419-012625
50. Harvey JN (2007) Understanding the kinetics of spin-forbidden chemical reactions. Phys Chem Chem Phys 9:331–343. https://doi.org/10.1039/b614390c
51. Harvey JN, Aschi M (1999) Spin-forbidden dehydrogenation of methoxy cation: a statistical view. Phys Chem Chem Phys 1:5555–5563. https://doi.org/10.1039/a907723e
52. Harvey JN (2014) Spin-forbidden reactions: Computational insight into mechanisms and kinetics. Wiley Interdiscip Rev Comput Mol Sci 4:1–14. https://doi.org/10.1002/wcms.1154
53. Cui Q, Morokuma K, Bowman JM, Klippenstein SJ (1999) The spin-forbidden reaction $CH(^2\Pi)+N_2 \rightarrow HCN+N(^4S)$ revisited. II. Nonadiabatic transition state theory and application. J Chem Phys 110:9469–9482. https://doi.org/10.1063/1.478949
54. Lorquet JC, Leyh-Nihant B (1988) Nonadiabatic unimolecular reactions. 1. A statistical formulation for the rate constants. J Phys Chem 92:4778–4783. https://doi.org/10.1021/j100327a043
55. Nikitin EE, Umanskii SY (1984) Theory of slow atomic collisions. Springer, Berlin
56. Zhao Y, Mil'nikov G, Nakamura H (2004) Evaluation of canonical and microcanonical nonadiabatic reaction rate constants by using the Zhu-Nakamura formulas. J Chem Phys 121:8854–8860. https://doi.org/10.1063/1.1801971
57. Marks AJ (2001) Nonadiabatic transition-state theory: a Monte Carlo study of competing bond fission processes in bromoacetyl chloride. J Chem Phys 114:1700–1708. https://doi.org/10.1063/1.1333702
58. McLafferty FJ, George TF (1976) On nonadiabatic transition state theory. Chem Phys Lett 37:67–71. https://doi.org/10.1016/0009-2614(76)80163-7
59. Lykhin AO, Varganov SA (2020) Intersystem crossing in tunneling regime: $T_1 \rightarrow S_0$ relaxation in thiophosgene. Phys Chem Chem Phys 22:5500–5508. https://doi.org/10.1039/c9cp06956a
60. Truhlar DG, Garrett BC, Klippenstein SJ (1996) Current status of transition-state theory. J Phys Chem 100:12771–12800. https://doi.org/10.1021/jp953748q
61. Truhlar DG, Garrett BC (1984) Variational transition state theory. Ann Rev Phys Chem 35:159–189. https://doi.org/10.1146/annurev.pc.35.100184.001111
62. Laidler KJ, King MC (1983) The development of transition-state theory. J Phys Chem 87:2657–2664. https://doi.org/10.1021/j100238a002
63. Garrett BC, Truhlar DG (1979) Generalized transition state theory. Classical mechanical theory and applications to collinear reactions of hydrogen molecules. J Phys Chem 83:1052–1079. https://doi.org/10.1021/j100471a031
64. Miller WH (1998) "Direct" and "correct" calculation of canonical and microcanonical rate constants for chemical reactions. J Phys Chem A 102:793–806. https://doi.org/10.1021/jp973208o
65. Bao JL, Truhlar DG (2017) Variational transition state theory: Theoretical framework and recent developments. Chem Soc Rev 46:7548–7596. https://doi.org/10.1039/c7cs00602k

66. Marian CM (2012) Spin-orbit coupling and intersystem crossing in molecules. Wiley Interdiscip Rev Comput Mol Sci 2:187–203. https://doi.org/10.1002/wcms.83
67. Fedorov DG, Koseki S, Schmidt MW, Gordon MS (2003) Spin-orbit coupling in molecules: chemistry beyond the adiabatic approximation. Int Rev Phys Chem 22:551–592. https://doi.org/10.1080/0144235032000101743
68. Marian CM (2001) Spin-orbit coupling in molecules. In: Lipkowitz KB, Boyd DB (eds) Reviews in computational chemistry. Wiley, Hoboken, pp 99–204
69. Marian CM (2021) Understanding and controlling intersystem crossing in molecules. Annu Rev Phys Chem 72:617–640. https://doi.org/10.1146/annurev-physchem-061020-053433
70. Zhu C, Nakamura H (1994) Theory of nonadiabatic transition for general two-state curve crossing problems. I. Nonadiabatic tunneling case. J Chem Phys 101:10630–10647. https://doi.org/10.1063/1.467877
71. Fernández-Ramos A, Ellingson BA, Meana-Pañeda R et al (2007) Symmetry numbers and chemical reaction rates. Theor Chem Acc 118:813–826. https://doi.org/10.1007/s00214-007-0328-0
72. Lykhin AO (2019) Predicting kinetics of spin-forbidden unimolecular reactions with nonadiabatic transition state theory. Ph.D. Dissertation, University of Nevada, Reno
73. Delos JB (1973) On the reactions of N_2 with O. J Chem Phys 59:2365–2369. https://doi.org/10.1063/1.1680345
74. Dashevskaya EI, Nikitin EE (2017) Uniform airy approximation for nonadiabatic transitions in a curve-crossing weak-coupling. Z Phys Chem 232:311–323. https://doi.org/10.1515/zpch-2017-1025
75. Nakamura H (2012) Nonadiabatic transition: concepts, basic theories and applications, 2nd edn. World Scientific, Berlin
76. Zhu C, Teranishi Y, Nakamura H (2001) Nonadiabatic transitions due to curve crossings: complete solutions of the landau-zener-stueckelberg problems and their applications. In: Prigogine I, Rice SA (eds) Advances in chemical physics. Wiley Online Librabry, pp 127–133
77. Zhu C, Nakamura H, Re N, Aquilanti V (1992) The two-state linear curve crossing problems revisited. I. Analysis of Stokes phenomenon and expressions for scattering matrices. J Chem Phys 97:1892–1904. https://doi.org/10.1063/1.463178
78. Zhu C, Nakamura H (1992) The two-state linear curve crossing problems revisited. II. Analytical approximations for the Stokes constant and scattering matrix: the Landau-Zener case. J Chem Phys 97:8497–8514. https://doi.org/10.1063/1.464814
79. Zhu C, Nakamura H (1994) Two-state linear curve crossing problems revisited. IV. The best analytical formulas for scattering matrices. J Chem Phys 101:4855–4866. https://doi.org/10.1063/1.468505
80. Melissas VS, Truhlar DG, Garrett BC (1992) Optimized calculations of reaction paths and reaction-path functions for chemical reactions. J Chem Phys 96:5758–5772. https://doi.org/10.1063/1.462674
81. Kuki A (1993) Adiabaticity factor for electron transfer in the multimode case: an energy velocity perspective. J Phys Chem 97:13107–13116. https://doi.org/10.1021/j100152a013
82. Bracewell RN (2000) The fourier transform and its applications, 3d edn. McGraw-Hill, New York
83. Baer T, Hase WL (1996) Unimolecular reaction dynamics: theory and experiments. Oxford University Press, Oxford
84. Green NJB (2003) Unimolecular kinetics part 1. The reaction step. Elsevier, Hoboken
85. Lykhin AO, Truhlar DG, Gagliardi L (2021) Role of triplet states in the photodynamics of aniline. J Am Chem Soc 143:5878–5889. https://doi.org/10.1021/jacs.1c00989
86. Galano A, Alvarez-Idaboy JR (2013) A computational methodology for accurate predictions of rate constants in solution: application to the assessment of primary antioxidant activity. J Comput Chem 34:2430–2445. https://doi.org/10.1002/jcc.23409
87. Dzib E, Cabellos JL, Ortíz-Chi F et al (2019) Eyringpy: a program for computing rate constants in the gas phase and in solution. Int J Quant Chem 119:11–13. https://doi.org/10.1002/qua.25686
88. Garrett BC, Schenter GK (1994) Variational transition state theory for activated chemical reactions in solution. Int Rev Phys Chem 13:263–289. https://doi.org/10.1080/01442359409353296
89. Truhlar D, Pliego JR (2008) Transition state theory and chemical reaction dynamics in solution. In: Mennucci B, Cammi R (eds) Continuum solvation models in chemical physics: theory and application. Wiley, Hoboken
90. Hall DG (1986) The status of transition-state theory in non-ideal solutions and application of Kirkwood-Buff theory to the transition state. J Chem Soc Faraday Trans 2: Mol Chem Phys 82:1297–1303. https://doi.org/10.1039/F29868201297

91. Henriksen NE, Hansen FY (2018) Static solvent effects, transition-state theory. Theories of molecular reaction dynamics: the microscopic foundation of chemical kinetics. Oxford University Press, Oxford
92. Barca GMJ, Bertoni C, Carrington L et al (2020) Recent developments in the general atomic and molecular electronic structure system. J Chem Phys 152:154102. https://doi.org/10.1063/5.0005188
93. Werner HJ, Knowles PJ, Manby FR et al (2020) The Molpro quantum chemistry package. J Chem Phys 152:144107. https://doi.org/10.1063/5.0005081
94. Dubnikova F, Lifshitz A (2000) Isomerization of propylene oxide. Quantum chemical calculations and kinetic modeling. J Phys Chem A 104:4489–4496. https://doi.org/10.1021/jp004038+
95. Lifshitz A, Tamburu C (1994) Isomerization and decomposition of propylene oxide. Studies with a single-pulse shock tube. J Phys Chem 98:1161–1170. https://doi.org/10.1021/j100055a020
96. McGarvey JJ, Wilson J (1975) Photochemical perturbation and chemical relaxation of the planar—tetrahedral equilibrium in a di(tertiary phosphine) complex of nickel(II). J Am Chem Soc 97:2531–2532. https://doi.org/10.1021/ja00842a034
97. Miller JA, Klippenstein SJ (2003) From the multiple-well master equation to phenomenological rate coefficients: reactions on a C_3H_4 potential energy surface. J Phys Chem A 107:2680–2692. https://doi.org/10.1021/jp0221082
98. Glowacki DR, Liang CH, Morley C et al (2012) MESMER: an open-source master equation solver for multi-energy well reactions. J Phys Chem A 116:9545–9560. https://doi.org/10.1021/jp3051033

Publisher's Note Springer Nature remains neutral with regard to jurisdictional claims in published maps and institutional affiliations.

Authors and Affiliations

Vsevolod D. Dergachev[1] · Mitra Rooein[1] · Ilya D. Dergachev[1] · Aleksandr O. Lykhin[1,2] · Robert C. Mauban[1] · Sergey A. Varganov[1]

[1] Department of Chemistry, University of Nevada, Reno, 1664 N. Virginia Street, Reno, NV 89557-0216, USA

[2] Present Address: Department of Chemistry, The University of Chicago, Chicago, IL 60637, USA

Topics in Current Chemistry (2022) 380:21
https://doi.org/10.1007/s41061-022-00374-w

REVIEW

Evolution of the Automatic Rhodopsin Modeling (ARM) Protocol

Laura Pedraza-González[1,3] · Leonardo Barneschi[1] · Daniele Padula[1] · Luca De Vico[1] · Massimo Olivucci[1,2]

Received: 3 December 2021 / Accepted: 29 January 2022 / Published online: 15 March 2022
© The Author(s) 2022

Abstract
In recent years, photoactive proteins such as rhodopsins have become a common target for cutting-edge research in the field of optogenetics. Alongside wet-lab research, computational methods are also developing rapidly to provide the necessary tools to analyze and rationalize experimental results and, most of all, drive the design of novel systems. The Automatic Rhodopsin Modeling (**ARM**) protocol is focused on providing exactly the necessary computational tools to study rhodopsins, those being either natural or resulting from mutations. The code has evolved along the years to finally provide results that are **reproducible** by any user, **accurate** and **reliable** so as to replicate experimental trends. Furthermore, the code is **efficient** in terms of necessary computing resources and time, and **scalable** in terms of both number of concurrent calculations as well as features. In this review, we will show how the code underlying ARM achieved each of these properties.

Keywords QM/MM · Rhodopsins · Photochemistry · Photobiology · Python package

Laura Pedraza-González, Luca De Vico and Massimo Olivucci contributed equally to this work.

Chapter 5 was originally published as Pedraza-Gonzalez, L., Barneschi, L., Padula, D., Vico, L. D. & Olivucci, M. Topics in Current Chemistry (2022) 380: 21. https://doi.org/10.1007/s41061-022-00374-w.

✉ Laura Pedraza-González
laura.pedraza@unisi.it

✉ Luca De Vico
Luca.DeVico@unisi.it

✉ Massimo Olivucci
olivucci@unisi.it

Extended author information available on the last page of the article

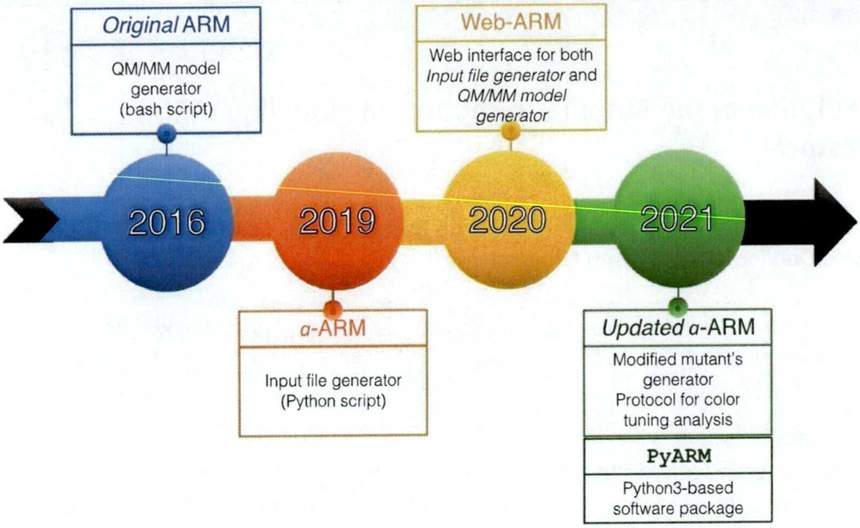

Fig. 1 Timeline for Automatic Rhodopsin Modeling (ARM) development detailing various parts of the ARM protocol and the years of the corresponding publications

1 Introduction: Contents and Scope

This review deals with the development of the Automatic Rhodopsin Modeling protocol (ARM), as seen in the last 5 years (Fig. 1). ARM represents a protocol capable of reliably building and analyzing computational models of rhodopsins, a superfamily of photoactive proteins. Given their major role in many aspects of life, from the basic act of ion-gating and ion-pumping in Archaea and Eubacteria to vision in animals, rhodopsins are currently studied extensively. Furthermore, rhodopsins and their engineered mutants represent powerful tools for potential biotechnological applications, the most prominent of which is optogenetics. Thus, researchers are constantly looking for new rhodopsins with particular photochemical properties, those being, among others, a specific absorption wavelength, a long or short excited state lifetime, and a strong fluorescence. Section 2 gives a panoramic view of rhodopsins and discusses current and future technological applications.

ARM was developed to provide a basic quantum-mechanics molecular mechanics (QM/MM) computational model, but sufficiently accurate so that differences in the model would reflect actual changes in the behavior of a different/mutated rhodopsin, and vice versa. This first iteration of ARM, called $ARM_{original}$, was created to generate models capable of accurately reproducing the spectral trends observed for a limited set of rhodopsins, and will be described in Sect. 3. In particular, $ARM_{original}$ models have been shown to be capable of reproducing trends in light absorbance maximum values in rhodopsin of different origins, and provide effective tools to discern the causes of effects such as blue- or red-shifting of the absorbed wavelength.

The natural extension of the ARM code has been to extend the general accuracy and applicability of the models and, most importantly, the level of automation in building protocol. Section 4 presents the *advanced* Automatic Rhodopsin Model building protocol (*a*-ARM), which meant completely rewriting the previous ARM code, and incorporating the possibility to automatically prepare inputs for the protocol itself. The completely new input preparation phase removed the need for user files manipulation and possible source of errors, hence achieving **reliability** and complete **reproducibility** of the results. The code behind *a*-ARM has also been used to power a web-interface, which allows, in principle, any user to obtain rhodopsin models of the same quality (Web-ARM, Sect. 5). This meant the extension of the benchmarking pool to more rhodopsins, thus increasing the applicability of the protocol, which, in turn, made it possible to use ARM models to guide the rational design of rhodopsins. More specifically, *a*-ARM can now be used to suggest which residue to mutate to obtain, for instance, a desired spectroscopic effect.

Finally, the ARM code was encapsulated inside a python-based package, called PyARM and described in Sect. 6. While *a*-ARM was already efficient in producing a single rhodopsin model, PyARM allows hundreds of concurrent rhodopsin models to be obtained automatically. This high level of efficiency is obtained by allowing the code to completely take care of all necessary calculations (complete automation), through the clever use of a highly modular code structure. Different types of analyses are now possible, all through the use of automatic, user-friendly, command-line Python drivers. Finally, the modular nature of PyARM allows the easy implementation of additional features, thus scaling-up the usability of the code with additional features.

2 Rhodopsins: a Family of Biological Photoreceptors

2.1 Structure and Diversity

Rhodopsins constitute a vast family of photoreceptive, membrane-embedded, seven-helix proteins, structurally formed by an opsin apoprotein and a retinal that serves as a chromophore[1] to absorb photons for either energy conversion or the initiation of intra- or intercellular signaling. As illustrated in Fig. 2, the opsin features an internal pocket hosting the retinal chromophore, which is covalently attached by a Schiff base linkage to the ε-amino group of a lysine side-chain in the middle of helix VII; the resulting retinal Schiff base (*r*SB) is protonated in most cases, constituting the so-called retinal protonated Schiff base (*r*PSB), illustrated in Fig. 3. Changes in protonation state of the *r*SB are crucial for both signaling and transport activity of rhodopsins [5–7].

Recent genomic advances have revealed that tens of thousands of rhodopsin genes are distributed widely in all domains of life (i.e., *Eukaryotes*, *Bacteria*, and *Archaea*) [5–10], and reside in many diverse organisms such as animals (e.g., vertebrates and invertebrates), microorganisms, and even viruses [11, 12]. Based on their

[1] Part of a molecular entity consisting of an atom or moiety in which the electronic transition responsible for a given spectral band above 200 nm is approximately localized [4].

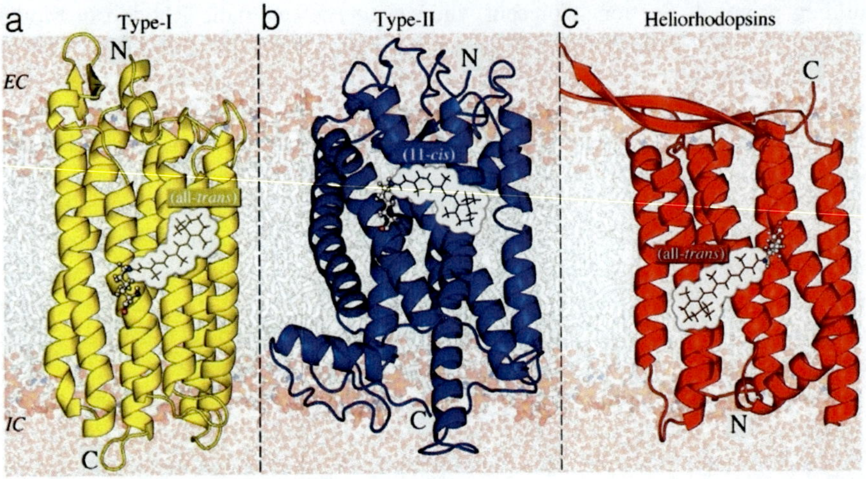

Fig. 2 **a–c** Rhodopsin types: structural similarities and differences. Rhodopsin proteins are classified into three types: yellow cartoon microbial or type-I, blue cartoon animal or type-II, red cartoon heliorhodopsins. For each type, the retinal chromophore in the dark adapted state (DA) is presented as lines and the covalently linked lysine is shown as ball-and-sticks. The orientation of the N terminus and C terminus residues with respect to the extracellular (EC) and intracellular (IS) surfaces of the membrane is specified. The structures correspond to **a** KR2 (type-I) [6REW [1]], **b** Rh (Type-II) [1U19 [2]] and **c** TaHeR (heliorhodopsin) [6IS6 [3]]

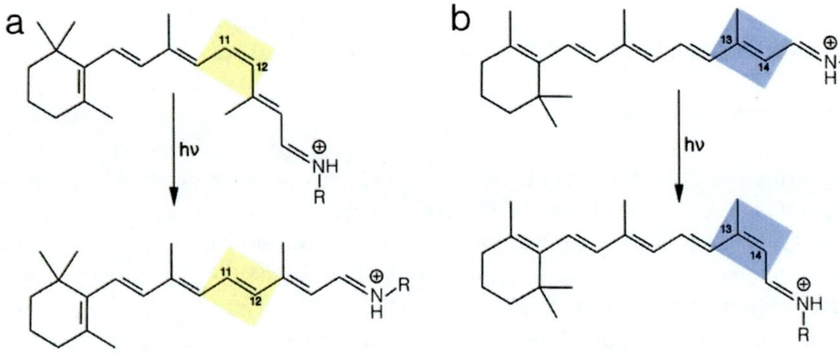

Fig. 3 Primary photoreaction in animal, microbial and helio-rhodopsins. Retinal isomerization from **a** the 11 – *cis* to the all-*trans* form and **b** from the all-*trans* to the 13 – *cis* form is the primary reaction in animal and microbial/helio-rhodopsins, respectively

host organism, rhodopsins are divided into different types (see Fig. 2): animal (type II) rhodopsins, a specialized subset of G-protein-coupled receptor (GPCR); microbial (type I) rhodopsins [6]; and heliorhodopsins—a recently discovered new type of light-sensing microbial rhodopsins [13].

Although members of the three types display a remarkably constant general architecture, they exhibit large differences in amino acid sequence (i.e., they have almost no sequence homology), as reflected in different chromophore cavities as well as in certain structural features [5, 7, 8, 13, 14]. In animal rhodopsins, the equilibrium dark adapted state (DA) shows a 11-*cis* (i.e., C11–C12 double bond) *r*PSB chromophore, which photoisomerizes to its all-*trans* configuration (Fig. 3a). In both microbial and heliorhodopsin families the DA state is dominated by an all – *trans* (i.e., C13–C14 double bond) *r*PSB chromophore, which is usually transformed into the 13 – *cis* configuration (see Fig. 3b) upon light absorption. Finally, in heliorhodopsins, the N-terminus amino acid is exposed to the cytoplasm or intracellular (IC) part, and the C-terminus residue faces the extracellular (EC) part of the cell membrane, whereas the opposite happens for the other two families (Fig. 2) [15].

2.2 Biological Functions

Rhodopsins exhibit an extensive pool of biological functions [5, 6, 8, 9, 16, 17]. For instance, animals use the photosensory functions (i.e., visual responses) of type II rhodopsins, lower organisms utilize type I for light energy conversion and intracellular signaling, while the function of heliorhodopsins is still not well elucidated [18–20].

In particular, microbial rhodopsins (type I) are represented in many diverse microorganisms, spanning the three domains of cellular life, as well as in giant viruses [12]. As such, modifications of a single protein scaffold through evolution produced many novel, chemical, light-dependent biological functions [5–7, 9, 10, 15, 17, 21]. Such functions can be roughly divided into two categories: (1) photoenergy transducers able to convert light into electrochemical potential to energize cells (e.g., light-driven ion pumps catalyzing outward active transport of protons [H^+], inward chloride transport [Cl^-], and outward sodium transport [Na^+]); and (2) photosensory receptors that make use of light to gain information about the environment to regulate inter- or intracellular processes (e.g., photosensors with membrane-embedded or soluble transducers, ion channels [anion and cation channel rhodopsins (ChRs)], and enzyme rhodopsins) [5–7, 9, 17, 22].

2.3 Photoreactivity

Rhodopsin functions are triggered by highly stereo-selective photoisomerization events, i.e., generally *cis* → *trans* of the C11=C12 bond for type II and *trans* → *cis* of the C13=C14 for type I and heliorhodopsins (Fig. 3). The specific wavelength (hereinafter referred to as Maximum absorption wavelength (λ^a_{max})) is due to a phenomenon called "opsin-shift", which consists of spectral shifts, in a wide range of the UV/visible spectra, modulated by the interaction between the retinal chromophore and the opsin residues.

Figure 4 depicts the inter-state photoisomerization pathways of rhodopsins, representing the potential energy change, i.e., the progression along the potential energy surface (PES), along a reaction coordinate possibly involving a combination of electronic transitions between the Ground-state (S_0) and both first (S_1) and second (S_2) excited electronic states. Such a coordinate is usually considered as the

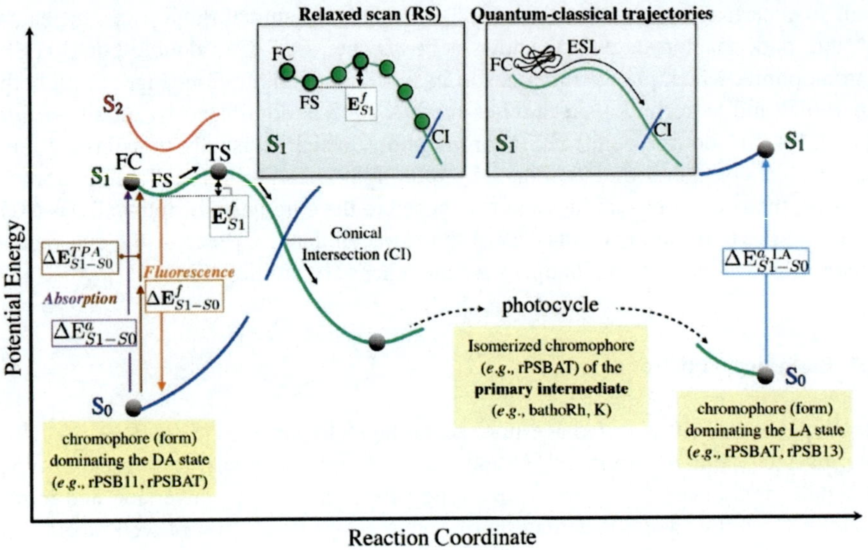

Fig. 4 Light-induced and light-emission properties of rhodopsin proteins, investigated using computational modeling. Schematic diagram displaying the photoiomerization path, including the relevant S_0, S_1 and S_2 energy profiles of a generic rhodopsin. QM/MM models are used to compute the vertical excitation energy for absorption ($\Delta E^a_{S1-S0} \equiv \lambda^a_{max}$), fluorescence ($\Delta E^f_{S1-S0} \equiv \lambda^f_{max}$) and two-photon absorption ($\Delta E^{TPA}_{S1-S0} \equiv \lambda^{TPA}_{max}$) as well as the excited state energy isomerization barrier (E^f_{S1}) associated with emission, computed as the energy difference between the fluorescent excited-state (FS) structure and the transition state (TS). *Inset* (top-center) Schematic illustration of the calculation of an excited state isomerization path providing the E^f_{S1} value via a Relaxed scan (RS) and of a Franck–Condon (FC) quantum-classical trajectory (this provides, in case of an ultrafast reaction, an estimate of the ESL associated to the double bond energy isomerization barrier). QM/MM models can also been used to investigate the structure and spectroscopy of primary photocycle intermediates (batho and K intermediates) and of photocycle intermediates corresponding to light-adapted states

skeletal dihedral angle of the double-bond highlighted in Fig. 3. The involved electronic states commonly change their electronic character and can cross [23, 24]. Different rhodopsin "functions" may take advantage of different features of the rhodopsin PES, illustrated in the following.

The primary event of light absorption of one photon of energy $h\nu$ prompts the vertical electronic π-π^* transition of the retinal chromophore from the ground to the first excited Franck–Condon[2] (FC) state. The length of the π-conjugated polyene chain in the retinal chromophore, and the protonation state of the *r*SB linkage, determine the energy gap, hereafter called Vertical Excitation energy (ΔE^a_{S1-S0}), of this process [5]. The computational evaluation of ΔE^a_{S1-S0} allows the prediction of λ^a_{max} for a given rhodopsin, and, in general, for the modeling of light absorption. Considering that the wavelength dependence of the absorption efficiency defines the colors of the rhodopsin proteins, the modulation of the ΔE^a_{S1-S0} gap makes possible the engineering

[2] Classically, the Franck–Condon principle is the approximation that an electronic transition is most likely to occur without changes in the positions of the nuclei in the molecular entity and its environment. The resulting state is called a Franck–Condon state, and the transition, a vertical transition [4].

of either blue- or red-shifted rhodopsin variants, absorbing in specific regions of the UV-Vis spectrum (i.e., color tuning, see Sect. 6.4) [19–21, 25–35]. Photoexcitation to the S_1 state can also be achieved by simultaneous absorption of two infrared photons (i.e., Two-Photon Absorption (TPA) process) of the same energy $\Delta E^{TPA}_{S1\text{-}S0}$, corresponding to half of the energy necessary for One-Photon Absorption (OPA) [23, 36–40].

After photoexcitation, the retinal chromophore leaves the FC region by relaxing along stretching and torsional modes, and starts the exploration of the S_1 PES. Depending on the surface topography, it may (1) quickly encounter a minimum, hereinafter called fluorescent excited-state (FS), or (2) continue visiting other regions of the S_1 PES. In (1), the chromophore remains in a long-lived excited state that eventually decays back to the ground state via fluorescence, i.e., spontaneous emission of radiation (luminescence) [4]. In this case, a photon of a different wavelength can be emitted after a short while (10^{-9} to 10^{-5} s). The energy gap associated to this process is denominated Vertical Emission energy (ΔE^{f}_{S1-S0}). Therefore, the computational evaluation of ΔE^{f}_{S1-S0} allows the prediction of the Maximum emission wavelength (λ^{f}_{max}) for a given rhodopsin, and in general the modeling of emission properties, such as fluorescence. In (2), that is, if there is a shallow or no FS, the retinal chromophore twists around the reactive C=C bond and reaches the Conical intersection (CI) region, where it decays to S_0. As illustrated in Fig. 4, "reacting" rhodopsins then trigger a series of sequential protein moiety conformational changes (required for their biological functions) [8, 10, 23] and returns to the initial state (e.g., process known as photocycle), whereas the "non-reacting" molecules relax back to the original S_0 state without entering such a photocycle. The described cycle allows microbial rhodopsins to repeat their functions every light stimulation since the chromophore is ultimately regenerated through the photocycle. This type of photocycle (in chemistry one would talk about type 1 photochromics) is remarkably different from that of vertebrate visual opsins (i.e., vertebrate rhodopsin and cone visual pigments), in which the retinal chromophore dissociates after the photoreaction and, therefore, additional retinal is required to regenerate the pigments [9].

2.4 Applications of Natural and Engineered Rhodopsins: Optogenetics

The modeling of primary photoproducts, photoisomerization reaction paths and bistable states in animal and microbial rhodopsins is of great interest for engineering photoswitchable fluorescent probes [41–43]. Bistable rhodopsins are rhodopsins featuring two stable isomeric forms (i.e., characterized by two chromophore isomers such all – *trans* and 13-*cis*) and thus require the sequential and independent absorption of two photons (often of different wavelengths) to complete the photocycle [43]. Certain bistable rhodopsins can be interconverted using light of different wavelengths (type-2 or P-type photochromism[3])

[3] Reversible transformation of a molecular entity between two forms, A and B, having different absorption spectra, induced in one or both directions by absorption of electromagnetic radiation. The spectral change produced is typically, but not necessarily, of visible color and is accompanied by differences in other physical properties [4].

[44]. However, more frequently the light adapted state (LA) state reverts back to the DA state state thermally (type-1 or T-type photochromism), in which case (i.e., for monostable microbial rhodopsins) the photocycle is completed after the absorption of only one photon [44]. Applications of bistable rhodopsins are related to different properties/functions of the DA and LA states and the use of light irradiation to change the rhodopsin isomeric composition passing from a DA-dominated to a LA-dominated equilibrium. Since the efficiency of such conversion is proportional to the difference between the two λ^{a}_{max} values, it is apparent that achieving bistable rhodopsins featuring one form with a λ^{a}_{max} value significantly shifted to the red, may facilitate applications where it is important to switch on-and-off (i.e., control) the rhodopsin biological function using light irradiation [43].

The photoisomerization event that initiates each rhodopsin function (see Sect. 2.2), has been studied widely in scientific areas such as physics, chemistry, and biology [5–7], with the aim of allowing for novel and modern applications in fields as diverse as medicine [45–47], bioenergetics [9, 48], biotechnology and neurosciences [7–10, 16], among others [9]. Particularly, specific microbial rhodopsins that function as either ion transporters or channels are at the heart of a new biotechnology called optogenetics.

Optogenetics (i.e., a combination of "optics" and "genetics") uses mainly genetically encoded, and often specifically engineered, microbial rhodopsins for the optical control of physiological processes [7–10, 16, 41]. The development of so-called optogenetic tools leads to the investigation of the nervous system at the cellular and tissue level, without noticeable tissue damage as well as side effects. Currently, rhodopsins with potential application as optogenetic tools are used as light-driven actuators (i.e., action potential triggers), light-driven silencers (i.e., action potential quenchers) and as fluorescent reporters (i.e., action potential probes) [7].

In order to either improve the current or develop new optogenetic tools, it is imperative to gain further insight into the different factors controlling color tuning in rhodopsins, to then be able to increase the variety of λ^{a}_{max}, thus enabling simultaneous optical control by different colors of light [19–21, 25–35]. As such, various rhodopsin genes have been screened in order to find additional colors [49, 50]. In particular, while many blue-absorbing rhodopsin at λ^{a}_{max} < 500 nm have been reported [51] and even applied to optogenetics [49], the longer absorption maxima are limited in λ^{a}_{max} < 600 nm. In this regard, there is presently an interest in screening rhodopsin variants exhibiting a longer λ^{a}_{max} and/or enhanced fluorescence (i.e., high fluorescence intensity) [52], achieved through the effects of single or multiple amino acid mutations of a template Wild-type (WT) structure.

The rational design of artificially mutated variants is necessary to identify the amino acid replacements that are effective for color tuning and for influencing fluorescent properties. A systematic experimental screening of thousands of possible candidates is not feasible, thus requiring the development of computational approaches for a fast, congruous, and rational design of in silico point mutations, to narrow down the number of tested candidates.

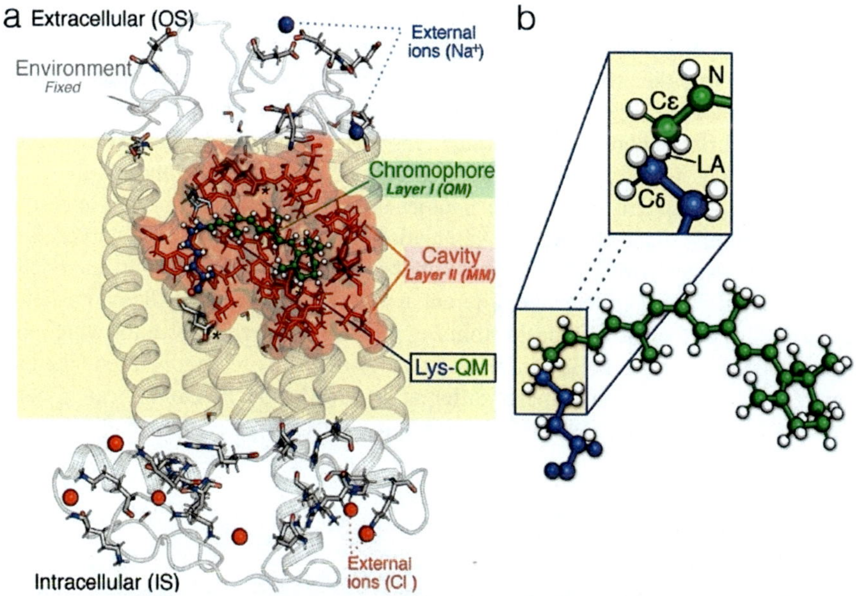

Fig. 5 General structure of a monomeric, gas-phase and globally uncharged Automatic Rhodopsin Modeling (ARM) quantum-mechanics molecular mechanics (QM/MM) model. **a** Relationship between the ARM model's three subsystems and two multiscale layers. Detailed description of the components of the three subsystems (for an animal rhodopsin from the DA state of Bovine rhodopsin). Gray cartoon Environment subsystem, green ball-and-sticks *r*PSB chromophore, blue ball-and-sticks lysine side-chain covalently linked to the *r*PSB chromophore, cyan tubes main chromophore counterion, tubes marked with * residues with non-standard protonation states, red spheres external Cl , blue spheres Na^+ counterions, red/white tubes crystallographic water molecules, red frames surface amino acid residues forming the chromophore cavity subsystem, gray tubes external OS and IS charged residues. **b** The *r*PSB chromophore (green) and the linked Lys side-chain fragment (blue) form the Lys-QM subsystem, which includes the H-link atom located along Cε–Cδ connecting blue and green atoms (inset: *LA* H-linked atom), which belong to the MM and QM parts of the model. Adapted with permission from [40]. Copyright 2019 American Chemical Society

3 The *Original* Version of ARM: a Pioneer Technology for Rhodopsin QM/MM Modeling

3.1 State-of-the-Art for QM/MM Modeling of Rhodopsins

The past few decades have witnessed a growing interest in developing hybrid QM/MM approaches to tackle problems in computational photochemistry and photobiology [27]. A remarkable advantage of using hybrid QM/MM methodologies is that in silico models featuring a high-level of complexity can be properly constructed, through the definition of subsystems, each treated at a different level of theory (or layers) according to the required level of accuracy.

As shown in Fig. 5 and further described in detail in Ref. [43], a basic QM/MM model for a biological photoreceptor (e.g., rhodopsin) should feature at least three subsystems: (1) the reactive part of the system, or prosthetic group, that carries the photochemical process (i.e., chromophore), treated with a suitable, and usually computationally expensive, QM method (see Section 2.1.1 in Ref. [43]); (2) the residues that, due to either steric or electrostatic interactions with the prosthetic group, directly influence the role of the reactive part (i.e., amino acids and water molecules forming the chromophore cavity), treated classically with a less expensive (optionally polarizable), MM force field (see Section 2.1.2 in Ref. [43]); and (3) the residues without an evident role on the photochemical process (i.e., protein environment) structurally fixed during the calculation and treated as point charges (see Fig. 5). Usually, QM/MM models are complex and, unfortunately, not univocally defined, thus Ref. [43] collates the different approaches to modeling the photochemical properties of Bovine rhodopsin from *Bos taurus* (Rh), a rhodopsin for which the X-ray structure is available [2, 53, 54] and that, also for this reason, is often taken as a reference for the benchmarking of different QM/MM models. Differences in construction protocols of the QM/MM setup lead to variations in computed $\lambda_{max}^{a,calc}$ up to 41 nm [29–31, 55–58].

3.2 ARM Scope

It is important to define a standardized protocol for the fast and automated production of congruous QM/MM models, which can subsequently be replicated in any laboratory. Such a protocol should not strictly aim at the prediction of the absolute values of observable properties, but to the description of their changes along sets of different rhodopsin variants. The ARM protocol described in this work represents our attempt to provide such a tool. The protocol is based on two well-defined phases called “generators”: the (1) input file generator and (2) QM/MM model generator. Sections 3.3 and 3.4 describe the development of the original version of the ARM protocol [59], which provided the QM/MM model generator. Furthermore, Ref. [59] provides a series of instructions for the manual preparation of the input file, which served for the subsequent development of the input file generator presented in Sect. 4.

The original ARM protocol is designed for a semi-automatic, fast and parallel building of congruous sets of QM/MM models of wild-type and mutant rhodopsin-like photoreceptors [59]. Accordingly, as illustrated in Fig. 5, this version provides specialized QM/MM models that, in general, would not be applicable to other (e.g., cytoplasmic) photoresponsive proteins, or even to rhodopsins that contain artificial retinal chromophores. In the general framework of the QM/MM model generator, one needs to consider (1) the wise division of the complex molecular system into different, simpler subsystems, and (2) the definition of particular layers that represent the approaches (i.e., levels of theory) used for the proper description of each of the subsystems.

To assess point (1), one can refer to Sect. 2.1, where it is specified that rhodopsins belonging to the three known families (i.e., animal, microbial and heliorhodopsins) share a common architecture constituted by a protein environment featuring seven transmembrane helices, which form a cavity hosting the retinal chromophore (see Fig. 5a). As illustrated in Fig. 5b, an important feature to be considered is that the chromophore is linked covalently via a specific lysine residue (e.g., located in the middle of helix VII and helix G for animal and microbial rhodopsins, respectively), via an imine (–C=N–) linkage, forming the *r*PSB.

To evaluate point (2), instead, it is crucial to identify the chemical/physical phenomena to be modeled and the target properties to be computed. Section 2.3 shows that the process driving the diverse biological functions of rhodopsins is the photoisomerization of the *r*PSB chromophore, occurring immediately after the absorption of a photon of the appropriate wavelength. As illustrated in Fig. 4, the most relevant properties to be reproduced/predicted are the rhodopsin color (or absorption wavelength), fluorescence, and photochemical reactivity.

The scope of ARM, since inception, has been that of being capable of reproducing experimental trends in rhodopsin series. In other words, it was not designed to be a predictive, but rather an investigative, tool; given a set of rhodopsins, the generated corresponding models should reflect trends in spectroscopic and/or photochemical properties. In turn, the computed models could be used to investigate, at the molecular level, the origin of such property changes.

3.3 Definition of an ARM QM/MM Model

As previously mentioned, it is possible to subdivide the rhodopsin structure into three subsystems. Figure 5a exemplifies how these subsystems are defined, using the case of Rh rhodopsin: the (protein) environment (gray cartoon), the chromophore cavity (red frames/surface), and the Lys-chromophore (blue/green ball-and-sticks). The protein environment sub-system features residues (backbone and side-chain atoms) fixed at the crystallographic or comparative (homology) structure, and incorporates external Cl^- and/or Na^+ counterions (see discussion below) also fixed at pre-optimized positions. The chromophore cavity sub-system, instead, contains residues with fixed backbone and relaxed side-chains. The Lys-QM system contains the atoms of the covalently linked lysine side-chain in contact (through Cδ) with the QM/MM frontier and the entire QM subsystem, which corresponds to a *N*-methylated retinal chromophore. All the Lys-QM atoms are free to relax during the QM/MM calculation.

Accordingly, the ARM QM/MM models illustrated in Fig. 5 can be defined as basic, monomeric, gas-phase and globally uncharged, with electrostatic embedding. The term “gas-phase” refers to the fact that, although rhodopsins are proteins exposed to trans-membrane electrostatic fields (i.e., few tenths of meV) occasioned by an asymmetric distribution of the surface ions, in the ARM basic approach, the interactions between the protein and the membrane, as well as solvation effects, are not modeled explicitly. Instead, the protocol mimics the situation of a globally uncharged protein by adding external Cl^- and/or Na^+ counterions, properly placed near the most positively and/or negatively charged surface amino acids, in both the intracellular (IS) and extracellular (OS) protein surfaces.

To this aim, ARM uses the No (net) Surface Charge (NSC) scheme in which the user must execute the following operations manually: (1) identify the number of positively and negatively charged surface residues, (2) calculate

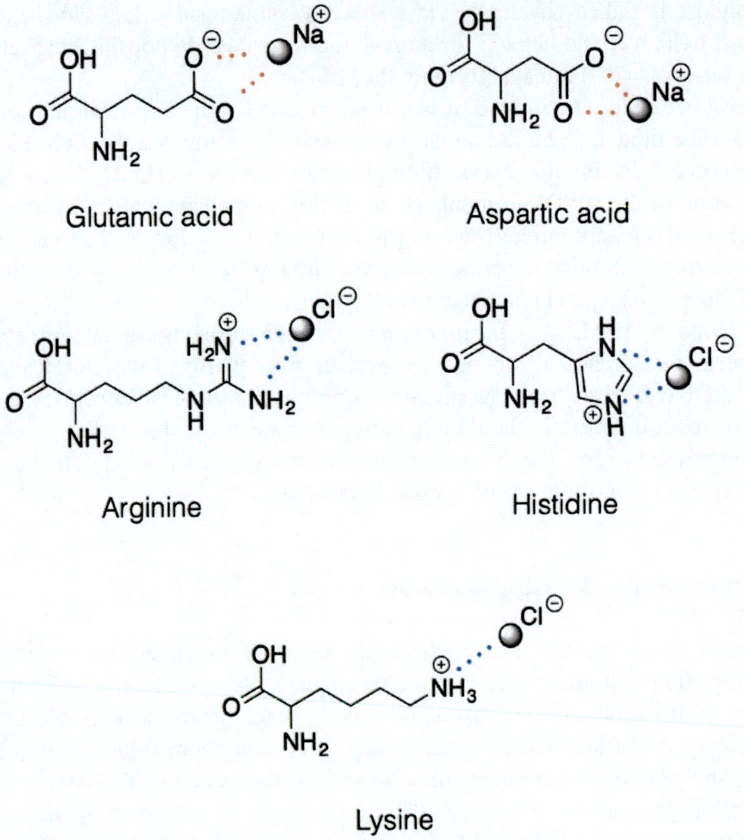

Fig. 6 External counterion positions relative to ionizable residues, according to the No Surface Charge (NSC) scheme used for the ARM QM/MM models Schematic representation of the positions of the external counterions (i.e., Na^+ and Cl) used to neutralize the IS and OS surfaces in the gas-phase ARM QM/MM models. Reproduced with permission from [59] Copyright 2016 American Chemical Society

independently the charge in the IS and OS , (3) determine the number of Cl^- and/or Na^+ to be added, and (4) position the external counterions as illustrated in Fig. 6. As described in Sect. 4, in the most updated version of the protocol, such a procedure is performed automatically by a specific algorithm. Moreover, to account for the water-mediated hydrogen-bond network (HBN) in the protein cavity, the internal crystallographic water molecules are retained in the model, while the water molecules that are not detected experimentally are assumed to be extremely mobile or just absent in the chromophore hydrophobic protein cavity.

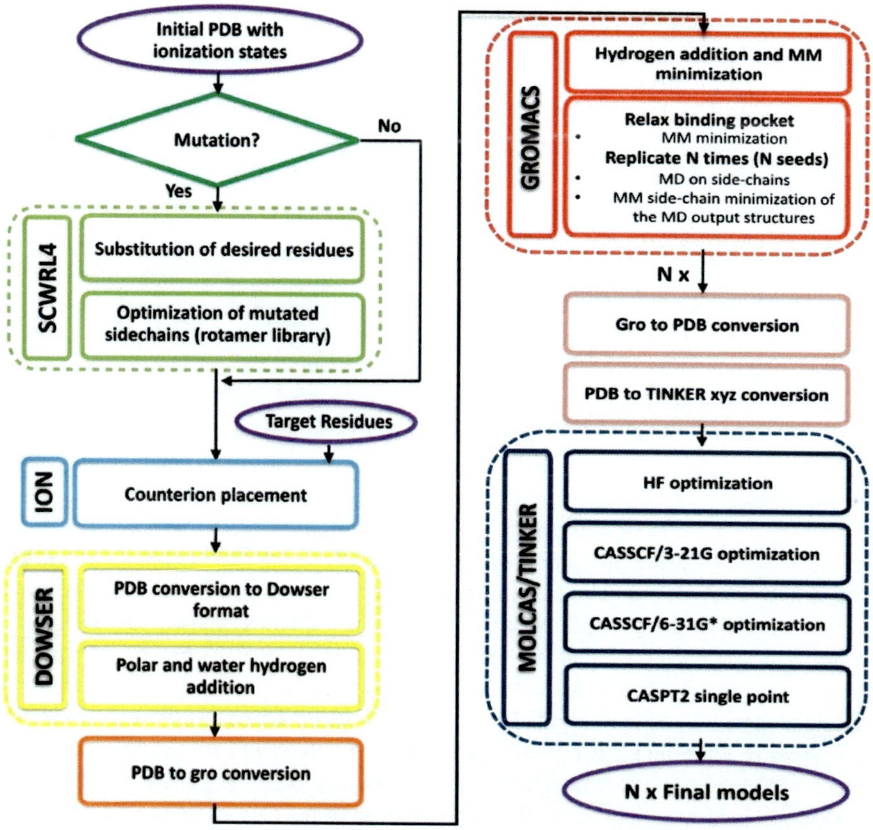

Fig. 7 General workflow of the QM/MM model generator developed in the original version of ARM. The procedure starts with the previously prepared input structure and finishes with the S_0 ARM QM/MM model, that consists on 10 replicas of the optimized structure along with the computed Maximum absorption wavelength (λ^{a}_{max}). The different steps of the protocol, as well as the level of theory and the used software are provided Adapted with permission from [59]. Copyright 2016 American Chemical Society

3.4 QM/MM Model Generator

Figure 7 illustrates the general workflow of the QM/MM model generator proposed in Ref. [59] for the semi-automatic building of ground-state ARM QM/MM models of rhodopsins and subsequent computation of the Maximum absorption wavelength, via vertical excitation energy calculations.

3.4.1 Classical Molecular Dynamics Simulations

A preliminary preparation of the input structure is required, which consists of the following steps:

(1) Selection and optimization (i.e., energy criteria) of all the water molecules, using the crystallographic/comparative positions as initial guess.
(2) Addition of hydrogen atoms to polar residues and waters.
(3) Addition of hydrogen atoms to the other residues of the protein environment, chromophore cavity and chromophore subsystems.
(4) MM energy minimization on the added hydrogen atoms.
(5) MM geometry optimization on all the side-chains of the residues of the chromophore cavity subsystem.
(6) Generation of N=10 independent models (replicas) to simulate and explore the possible relative conformational phase space of the cavity residue side-chains and retinal chromophore.

Steps 1 and 2 are performed using the program DOWSER [60]. These steps ensure proper treatment of water and Hydrogen-bond networks (HBN), which affect side-chain conformations and long-range electrostatics, thus ultimately modifying spectral and photochemical properties.

Steps 3–6 are performed using the program GROMACS [61]. Step 6 uses classical molecular dynamics simulations (MD) to perform a simulated annealing relaxation at 298 K on all side-chains of the Lys-QM and cavity subsystem, keeping the backbone fixed at the crystallographic/comparative structure; during the MD computation the retinal chromophore subsystem is also allowed to move. To warrant independent initial conditions, each of the N=10 independent MDs starts with a different, randomly chosen seed. Note that, during the MD run, the chromophore subsystem is represented using an MM parameterization and partial charges computed as AMBER-like Restrained Electrostatic Potential (RESP) charges, which are specific for each used isomer of the chromophore (e.g., 11-*cis*, all-*trans* and 13-*cis* *r*PSB). The corresponding parameterized RESP point charges, currently used in ARM, are reported in the Supplementary Information of Ref. [59]). The default heating, equilibration, and production times for the MD (selected via benchmark calculations in Ref. [59]) are 50, 150, and 800 ps, respectively, for a total length of 1 ns. In each run, the *frame closest to the average* of the 1 ns simulation is then selected as the starting geometry (i.e., guess structure) for constructing the corresponding QM/MM model. Melaccio et al. [59] have shown that, for a set of three phylogenetically diverse rhodopsins (Rh, SqRh and ASR $_{13C}$), N=10 replicas are enough to provide sufficient variability.

3.4.2 QM/MM Calculations

As shown in Fig. 7, each of the 10 replicas generated as guess structures undergoes a series of QM/MM calculations, defining a well-established protocol. In this regard, Fig. 5b illustrates the atoms that are considered as the Lys-QM layer during these calculations, corresponding to the full N-methyl retinal chromophore (*i.e*, QM subsystem; 53 atoms) and its covalently linked lysine side-chain of the MM subsystem (i.e., 9 atoms). The QM/MM frontier is treated within a link atom approach (see Fig. 5b), whose position is restrained according to the Morokuma scheme, and is placed across the lysine Cδ–Cε bond (where Cε is a

QM atom). The lysine charges are modified by setting the Cδ charge to zero to avoid hyperpolarization and to redistribute the residual fractional charge on the most electronegative atoms of the lysine, thus ensuring a +1 integer charge of the Lys-QM layer.

The following QM/MM calculations are performed sequentially, using the program [Open]Molcas/Tinker [62–64]:

(1) Geometry optimization at the HF/AMBER/3-21G level.
(2) Geometry optimization at the 2-roots single-state CASSCF(12,12)/AMBER/3-21G level.
(3) Geometry optimization at the 2-roots single-state CASSCF(12,12)/AMBER/6-31G(d) level.
(4) Inclusion of the electron correlation via a single point energy calculation at the 3-roots state-average CASPT2(12,12)/6-31G(d) level.

The sequential optimizations steps 1–3 aim at a more rapid convergence of both the molecular orbitals and geometry, and use "microiterations" that provide quicker convergence, lower energies, and a more realistic description of chromophore-environment interactions [65]. Besides, suitable level shifting values are used during the CASSCF and CASPT2 calculations, to minimize the possibility of convergence failure due to state mixing and intruder state problems. The CASPT2//CASSCF/6-31G(d)/MM treatment [55] has been investigated extensively for photobiological studies and its limitations are well understood. As previously documented [59], the rather small (ca. 3–4 kcal mol^{-1}) error in excitation energy reported in several studies for this level of theory, is somewhat due to error cancellations associated with the limited quality of single-state CASSCF/AMBER/6-31G(d) equilibrium geometries. Therefore, the different properties computed by ARM are expected to be affected by systematic error cancellations. Nevertheless, the main focus of ARM is the ability to reproduce observed trends in vertical excitation energies (i.e., the sign and magnitude of the individual differences concerning experimental data).

As observed in Fig. 7, the final output consists of 10 replicas of equilibrated QM/MM models of the type described in Sect. 3.3 and, for each replica, the vertical excitation energy values between S_0 and the first two singlet excited states S_1 and S_2 is provided.

3.5 Automation Issues

The *original* ARM protocol provides basic, gas-phase and computationally fast QM/MM models for comparative studies, to predict photochemical property trends (e.g., for large arrays of rhodopsin variants) that fit selected sets of experimental data (mainly $\lambda^{a}_{\max}$ values), within a well-established error bar. However, the general target in the formulation of an automated modeling of rhodopsins is to achieve a protocol featuring the following features for both the input file generator and QM/MM model generator phases:

(1) **transferability**, so as to properly describe rhodopsins with differences in protein sequence (i.e., organism belonging to different life domains and kingdoms; see Sect. 2.1), and different configurations of the *r*PSB;
(2) **documented accuracy**, so as to be able to translate results obtained in silico into hypothesis that can be proved experimentally;
(3) **reproducibility**, so as to be reproduced in any laboratory starting the model building from scratch;
(4) **speed** and **parallelization**, so as to achieve the fast generation of large arrays of rhodopsins variants (i.e., wild-type and mutants) simultaneously;
(5) **automation**, so as to reduce building errors (i.e., human factors) and avoid biased QM/MM modeling.

In order to assess whether or not the original ARM satisfies each of the points (1)–(5) described above, the protocol was benchmarked on the prediction of trends in λ^a_{max} for a limited set of 10 wild-type and 17 mutant rhodopsins [59]. Accordingly, points (1) and (2) were successfully achieved, while points (3), (4) and (5) were accomplished only partially. More specifically, point (1) was achieved since the predicted trends in λ^a_{max} presented a good agreement with experimental data (i.e., error bar of ca. 4.0 kcal mol^{-1}). Point (2) has been recently demonstrated via collaborative experimental and computational studies attempting enhancement of either color tuning [42] and fluorescence [66] properties for microbial rhodopsins, achieving novel applications in optogenetics. Further applications can be found in [43]. The main drawback of the original ARM is that it does not include a computational tool for the automatic, or even semi-automatic, generation of the input files (i.e., no input file generator). Input file generation is, instead, achieved through a manual manipulation of the template structure [59]. Such an input file is based on a X-ray crystallographic structure or comparative model of the protein in PDB (Protein Data Bank) format [67, 68], which contains the information specified in the caption of Fig. 5 and summarized as follows:

(i) the selected monomeric chain structure, including the *r*PSB chromophore, crystallographic/comparative water molecules, and excluding membrane lipids and non functional ions;
(ii) a list of residues forming the chromophore cavity;
(iii) the protonation states for all the internal and surface ionizable amino acid residues;
(iv) suitable external counterions (Cl^-/Na^+) needed to neutralize both IS and OS protein surfaces.

Due to possible different user choices (e.g., during the placement of IS and OS counter-ions; see Fig. 5a), reproducibility of the results described (point (3)) cannot be guaranteed. In addition, the input preparation required a few hour user's manipulation of the template protein structure (i.e., a skilled user completes the preparation of an ARM input for a new rhodopsin in not less than 3 h and after taking a series of decisions based on their chemical/physical knowledge and intuition). Such limitations, added to the human error factor, represent a serious issue when the target is the generation of hundreds rhodopsin models. Therefore, due to manual interventions of the user in the input generation, speed and parallelization (point (4)) are not guaranteed. Of course, the lacking of an automatic input file generator and the current methodological issues of the QM/MM model generator described below, made the protocol semi-automatic rather than automatic (i.e., also point (5) was not accomplished).

Additionally, as specified by Melaccio et al. [59], the code was written as a series of independent bash-shell scripts that are not interconnected by a general driver. Therefore, as explained in the Supporting Information in [59], for each step of the protocol, the user should execute each script manually and make a series of choices via a command-line assisted tool. This features the QM/MM model generator as a semi-automatic rather than an automatic tool. The following sections will show how each of these problems have been overcome.

4 *a*-ARM: the First Major Update Towards Automation

(Most of the content of the following four sections is reproduced/adapted with permission from [69], copyright 2019 American Chemical Society, while content of Sect. 4.5 is reproduced/adapted with permission from [35], open access under a CC BY license (Creative Commons Attribution 4.0 International License)).

4.1 Methodological Aspects

This section presents the *advanced* *a*-ARM [69], an updated version of the *original* semi-automatic ARM protocol [59] (Sect. 3) that, as main novelty, features an input file generator phase for either the fully automatic or semi-automatic computer-aided building of the ARM input. This updated version overcomes most of the automation issues of the original ARM, highlighted in Sect. 3.5, by including methodological improvements that lead to more reliable and reproducible QM/MM models.

The first aim of the *a*-ARM model building is to be consistent (in terms of output) with the original protocol, as described in Sect. 3.3 and shown in Fig. 5. Figure 8 shows the general workflow of *a*-ARM, which encapsulates two well-defined and automated phases (panels a and b), hereafter referred to as Phase I and Phase II, respectively. Whereas the latter is substantially the same QM/MM model generator phase reviewed in Sect. 3.3 (i.e., in terms of methodology, although not of implementation), the former is the new input file generator phase.

The combined use of the two phases achieves the automatic building of Ground-state ARM QM/MM models, starting from the structure of a rhodopsin in PDB format, either as PDB code or as a comparative homology model. As observed in Fig. 8, the initial structure is processed by Phase I to obtain the ARM input, which is subsequently processed by Phase II to obtain the ARM QM/MM model (i.e., gas-phase equilibrated optimized S_0 structure) and the predicted average λ^a_{max}. As further explained in Ref. [69], the input file generator is implemented as a user-friendly command-line interface, where the researcher

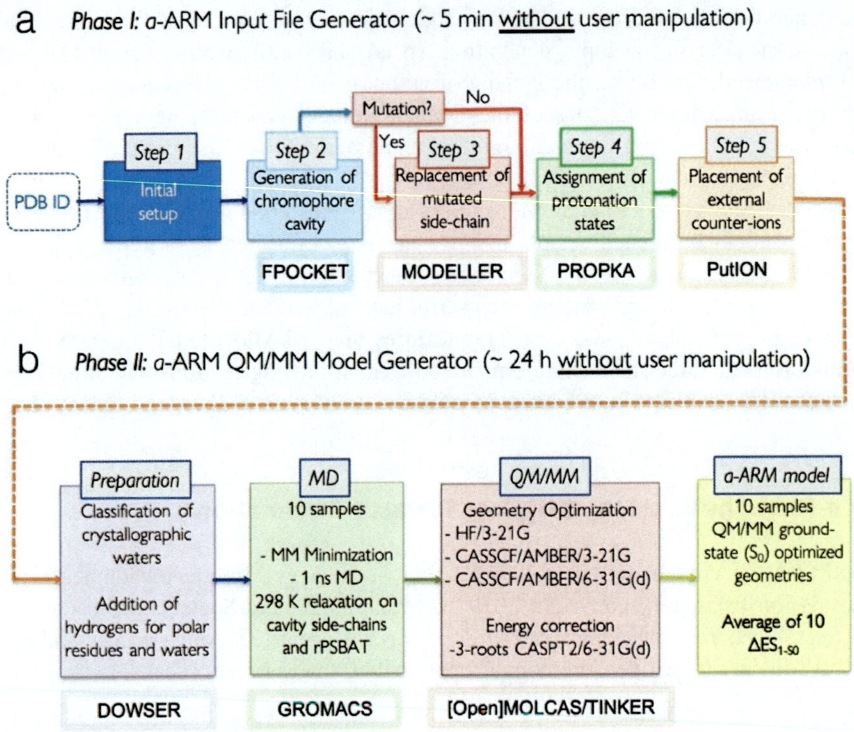

Fig. 8 General workflow of the two phases of the *a*-ARM rhodopsin model building protocol. **a** Input file generator phase. **b** QM/MM model generator phase

interacts with the program by typing information directly in the computer terminal, without the need to manipulate text files or visualize chemical structures, as previously necessary in the original "manual" strategy. Refs. [69] (see Section S1 in the SI) and [43] (see Section 3.1 in this reference) report a detailed description of both manual (i.e., *original* ARM) and automatic procedures used to pursue steps 1–5 of Fig. 8a, with particular emphasis on the improvements achieved with *a*-ARM, as well as in its higher level of automation. Figure 9 presents an overview of such improvements.

One of the most remarkable features of the new protocol is that, given the options (i.e., parameters) selected in steps 1–5 of phase I (Fig. 8a), *a*-ARM allows either the automatic or semi-automatic computer-aided production of the ARM input. Accordingly, *a*-ARM is sub-divided in *a*-ARM$_{default}$ (see Section 3.1 of Ref. [69]) and *a*-ARM$_{customized}$ (see Section 3.2 of Ref. [69]) approaches. The former refers to a fully automatic input generation, which uses default parameters as suggested by the code (i.e., chain, rotamers or side-chain conformations, pH, protonation states, residues forming the chromophore cavity), whereas the latter allows the computer-aided customization of some of such parameters when the default choices are not suitable.

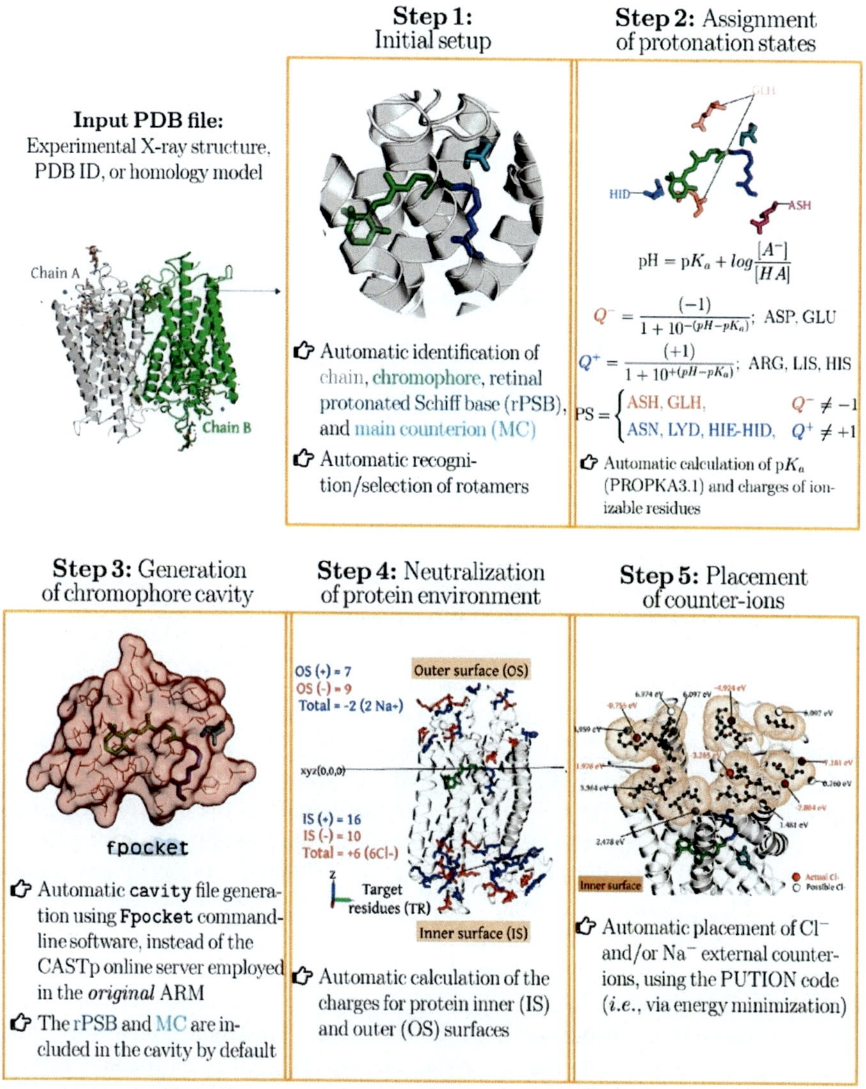

Fig. 9 Overview of the most relevant features of the input file generator, introduced in the *a*-ARM version of the protocol. Methodological and automation improvements achieved with the input file generator, in terms of: initial setup; automatic strategy adopted for the assignment of protonation states for ionizable residues; replacement of the software (i.e., fpocket instead of CASTp) for the automatic generation of the chromophore cavity; automatic approach to define the charge of the IS and OS surfaces and automatic counterion placement based on energy minimization

The customized approach is used in cases where the default choices produce outlying models that fail to reproduce trends in absorption properties. Several examples are presented in Refs. [35, 43, 69]. For instance, Fig. 10 illustrates how the customization, in terms of either selection of side-chain conformations and protonation states, is achieved for the case of the microbial rhodopsin

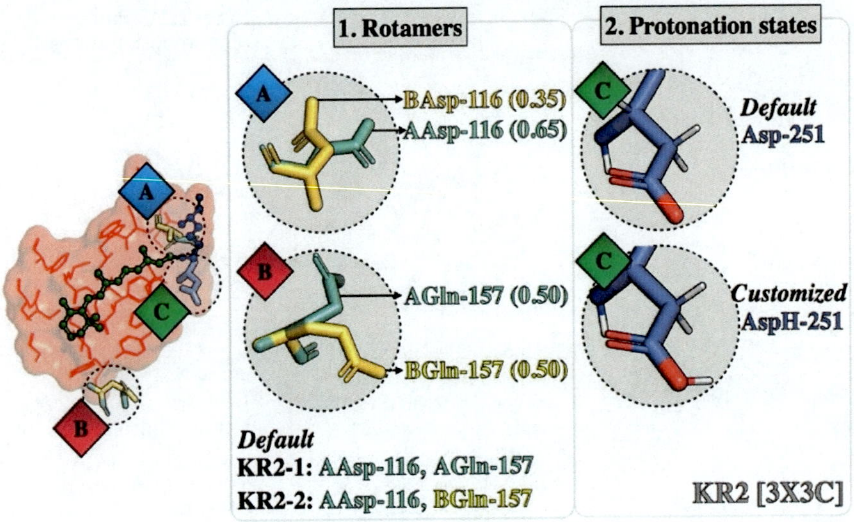

Fig. 10 Default and customized *a*-ARM models for Krokinobacter rhodopsin 2 from *Krokinobacter eikastus* (KR2) [PDB ID 3X3C [70]]. Left: conformational (the occupancy factor of the rotamers Asp-116 and Gln-157 are presented in parentheses). Right: ionization state variability

Krokinobacter rhodopsin 2 from *Krokinobacter eikastus* (KR2) [35, 69, 70]. As observed, in the 3X3C X-ray structure [70], the residue Asp-116, considered as the main counterion (MC) of the *r*PSB, exhibits two side-chain conformations, namely, AAsp and BAsp, labeled with occupancy numbers 0.65 and 0.35, respectively. Moreover, the residue Gln-157 that is part of the environment subsystem (i.e., fixed during the QM/MM calculations) presents two conformations (AGln and BGln) both with 0.5 occupancy. According to the occupancy numbers, *a*-ARM$_{default}$ selects the rotamer AAsp-116 and generates two models relative to Gln-157: KR2-1, which includes AAsp-116 and AGln-157, and KR2-2, which includes AAsp-116 and BGln-157. The computed ΔE^{a}_{S1-S0} for both default models, presented below in Fig. 11, features an error of about 15.0 kcal mol^{-1} with respect to experimental data. Since the default models are unable to provide values inside the experimental trend, the *a*-ARM$_{customized}$ approach is necessary. As shown in the right panel of Fig. 10, such customization is performed through a more rational assignment of the protonation states of the two aspartic acid residues forming the counterion complex of the *r*PSB, namely Asp-116 and Asp-251 [35]. The default model predicts that both aspartic acids are negatively charged. However, as further discussed in Refs. [35, 43, 69], the presence of these two negative charges would outbalance the single positive charge of the *r*PSB, generating the large blue-shifted effect mentioned above. Accordingly, in the customized model the secondary counterion (SC) Asp-251 is, instead, protonated (i.e., neutral) to counterbalance the charge in the vicinity of the *r*PSB. As can be seen in Fig. 11, such customization provides a model with a small error bar of about 1.5 kcal mol^{-1}. An updated ARM model of KR2 was recently reported [35]. Although such a model was constructed starting from a different

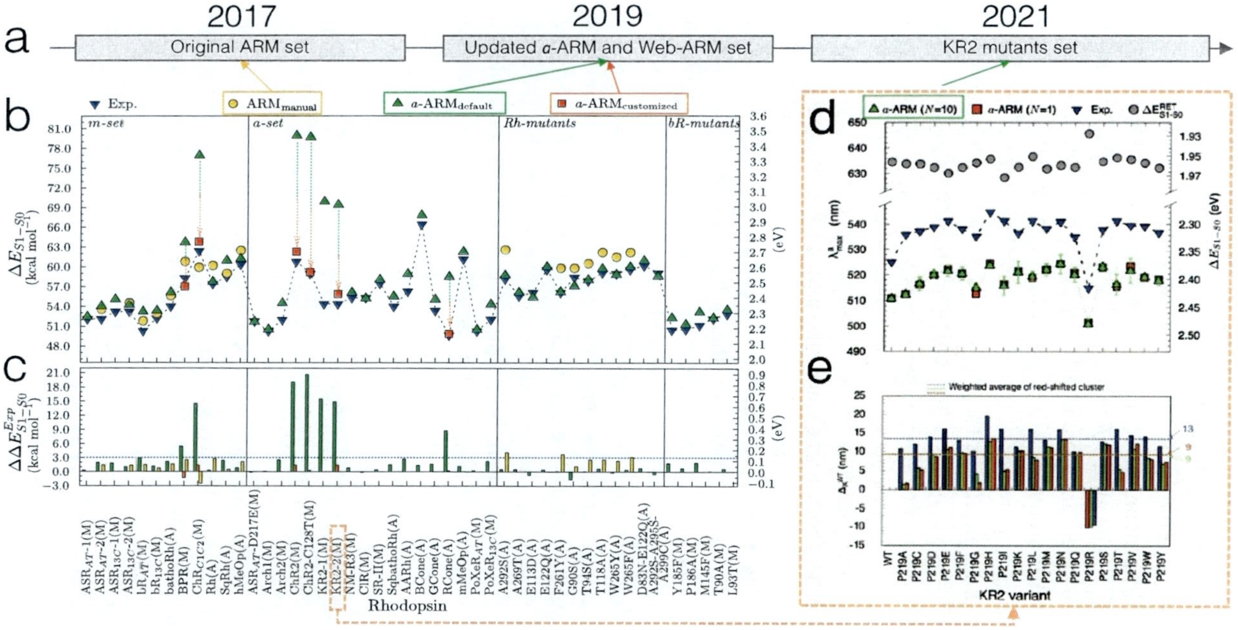

Fig. 11 Evolution and benchmarking of the ARM protocol over time, in terms of reproduction of experimental trends in λ^a_{max}. **a** Timeline for the benchmark sets used for testing each version of ARM over the years. **b** Comparison between vertical excitation energies (ΔE_{S1-S0}) computed with either *a*-ARM$_{default}$ (up green triangles) or *a*-ARM$_{customized}$ (red squares) [69] and ARM$_{original}$ [59] (yellow circles) and experimental data (down blue triangles), along with the **c** differences between computed and experimental $\Delta E_{S1\text{-}S0}$ ($\Delta\Delta E^{Exp}_{S1\text{-}S0}$). The *m*-set corresponds to wild-type (WT) rhodopsins forming the original benchmark set for the ARM protocol; *a*-set introduces new rhodopsins to the benchmark set of the *a*-ARM protocol; *Rh – mutants* set contains mutants of bovine Rhodopsin (Rh) belonging either to the benchmark set of the ARM or *a*-ARM protocols (see Refs. [59, 69]); and *bR – mutants* set contains mutants of bR evaluated with the Web-ARM interface (see Sect. 5 and Ref. [74]). **d** Comparison between ΔE_{S1-S0} computed with *a*-ARM, average value (up green triangles) and representative seed (red squares), and experimental data (down blue triangles) for WT and P219X mutants of KR2 [6REW], along with **e** the differences between these values for each mutant with respect to the WT (see Ref. [35]). Adapted with permission from [69]. Copyright 2019 American Chemical Society and [35], open access under a CC BY license (Creative Commons Attribution 4.0 International License)

template X-ray structure [PDB ID 6REW [1]], the same customized setup for the protonation states of the counterion complex of the *r*PSB (i.e., deprotonated Asp-116, protonated Asp-251) was found. Remarkably, such *a*-ARM $_{\text{customized}}$ model was used as a starting structure for modeling a set of 19 mutants and for reproducing not only experimental trends in ΔE^a_{S1-S0}, but also giving further insights into the mechanism of color tuning in the position Pro-219 of KR2 [35].

Notice that in the *a*-ARM benchmark set (Fig. 11) a similar large blue-shifting has been observed in ΔE^a_{S1-S0} for all *a*-ARM$_{\text{default}}$ models that predict two negative charges near the *r*PSB (see Tables S2 and S3 in Ref. [69]). The customization procedure (see below) to obtain a ΔE^a_{S1-S0} within the error bar of the protocol, always implies the neutralization of one of these two charges. This is, in fact, one particularity of the simplified structure/scheme of the *a*-ARM models.

In general, protonation state assignment for ionizable residues remains a basic issue for current QM/MM protein modeling. No robust method is available that guarantees a correct choice of a pKa value, due to the complexity of the protein environments and its interconnected local effects. Furthermore, a residue may be present as an equilibrium between ionized and not-ionized forms, hence carrying only a partial charge. These issues are a current research matter [35, 71–73].

Despite the difficulties mentioned above, *a*-ARM adopts the following guiding customization procedure. As shown in the latter example, customized ARM QM/MM models can be constructed according to well-defined operations that can be easily replicated. Ref. [69] proposes to focus on the selection of the ionization states and side-chain conformations only. Accordingly, the customization of the protonation states involves three phases: (1) at pH > 6 the ionization states are modified by setting the pH to 5.2 in step 4 (see Fig. 8b); (2) the protonation state of the main and secondary counterions of the rPSB are checked, and if the analysis shows them both ionized the secondary counterion is neutralized; (3) in case the model generated in step (2) does not reproduce the experimental absorption maxima, then the secondary and main counterion ionization states are exchanged (see also the Supporting Information of Ref. [35] for further details on, e.g., the pH value choice). Note that in QM/MM modeling, it is a common practice to evaluate the protonation states of the *r*PSB counterion complex by looking, as a guide, at the reproducibility of the experimental λ^a_{max} (see, for instance, Refs. [35, 71–73])

Indeed, the novelty of the default and customized approaches is that, regardless of the user or computational facility, reproducible inputs, and consequently reproducible ARM QM/MM models, are guaranteed when the same parameters are used. This represents an advancement with respect to the original version since it allows for the models to be reproduced in any laboratory and by any user, even when starting building an ARM QM/MM model from scratch (see point (3) of Sect. 3.5).

4.2 Software Implementation Aspects

The computational implementation of both the input file generator and the QM/MM model generator phases as Python-based, modular codes boosted the building of the `PyARM` software package that will be introduced in Sect. 6. Moreover, their ease of transferability, allowed their use behind the Web-ARM web page, which will be described in Sect. 5.

Furthermore, although *a*-ARM can presently build only rhodopsin models (i.e., with natural retinal), it provides a template for the development and generation of an automatic QM/MM building strategy for other, more general, systems such as rhodopsins incorporating artificial (i.e., unnatural) chromophores. This is straightforwardly achieved given the modular architecture of the `PyARM` package and the fact that any chromophore can be treated when using the appropriate force field.

Finally, it is worth stressing that the new protocol achieves all the features (1)–(5) described in Sect. 3.5, overcoming the automation limits of the original version.

4.3 Benchmark, Validation and Application Aspects

Figure 11 shows the current validation of the *a*-ARM protocol through the prediction of trends in λ^a_{max}, performed using a benchmark set of 44 animal and microbial rhodopsin variants (i.e., 25 wild type and 19 mutants) that come from different organism and are phylogenetically diverse [69, 74] (see Fig. 11b). The full benchmark set features values ranging from 458 nm (62.4 kcal mol^{-1}, 2.71 eV) to 575 nm (49.7 kcal mol^{-1}, 2.15 eV). Such a relatively wide range provides information on the method accuracy, while the rhodopsin diversity provides information on the transferability and general applicability of the generated models. Figure 11b, c are divided in four different regions: *m*-set, *a*-set, *Rh – mutants* set, and *bR – mutants* set. The *m*-set and *Rh – mutants* set are used to compare the performance of original ARM and *a*-ARM versions, while the remaining sets focus exclusively on the performance of *a*-ARM.

The *a*-ARM$_{default}$ approach proved to be capable of reproducing the ΔE^a_{S1-S0} values for 86% of cases (38/44), with an error lower than 4.0 kcal mol^{-1} (0.13 eV), whereas the other 14% cases were successfully obtained with the *a*-ARM$_{customized}$ approach (i.e., changing the side-chain conformation and/or protonation states pattern). A detailed description of the customization procedure used for reproducing the experimental λ^a_{max} values of KR2, BPR, ChR2-C128T, ChR2 and ChR$_{C1C2}$ is provided in Section 3.2 of Ref. [69] and Section 3.3 of Ref. [43].

Recently, *a*-ARM was applied to (1) reproduce the λ^a_{max} of WT KR2 and 19 mutants [35] (see Fig. 11a) and (2) to gain further insights into the origin of red- or blue-shifting. As observed in Fig. 11d, e, the performance of *a*-ARM reported for the benchmark set (see above), is maintained when modeling a set of mutants that feature λ^a_{max} spanning a red-to-blue range going from 545 nm (54.5 kcal mol^{-1}, 2.36 eV) to 515 nm (57.0 kcal mol^{-1}, 2.47 eV). Furthermore, *a*-ARM demonstrated to be useful to generate models that reproduce blue- or red-shifting effects observed experimentally (see Ref. [35]). The final S_0 optimized equilibrium structures can be then used for further excited-state optimizations (see Sect. 6.2). Indeed, some of the ARM QM/MM models produced have been used as input for sophisticated constant-pH dynamics [75], the simulation of one-/two-photon absorption spectra [40], and in combined computational/experimental studies on color tuning possibilities in KR2 [35].

Accordingly, and as stated in Sect. 3.2, we claim that the *a*-ARM protocol, in its current version, does not represent a predictive tool, but rather is designed to produce models for rhodopsins, which structure was obtained from either X-ray crystallography or comparative modeling, useful for reproducing and explaining the origin of trends in spectroscopic/photochemical properties (e.g., between sequence variability and function) appearing from sets of experimental data.

4.4 Limitations and Pitfalls of *a*-ARM

Despite the encouraging outcome of the photochemical studies based on *a*-ARM (as previously mentioned), additional work is necessary to generate a tool that can be systematically applied to larger arrays of rhodopsins. The following main issues, in part anticipated above, have to be tackled to improve the input file generator phase:

- Assignment of the protonation states: there are two main aspects that limit the confidence in the automation of the ionizable state assignment described in Refs. [43, 69]. The first is that, due to the fact that the information provided by PROPKA [76] is approximated, the computed pK_a^{Calc} value may, in certain cases, be not sufficiently realistic. The second aspect regards the assignment of the correct tautomer of histidine. *a*-ARM uses as default the histidine dipeptide (HID) tautomer (deprotonated δ-nitrogen) for the automatic assignment, or allows the user to choose between the three possible tautomers for a "not-automated selection. Therefore, when possible, the user should collect the available experimental data and/or inspect the chemical environment of the ionizable residues including the histidines, and propose the appropriate tautomer [69]. Alternatively, one has to systematically examine all sensible choices, which may not always be feasible.
- Automatic construction of comparative models: since rhodopsin structural data are rarely available, it would be important to investigate the possibility of building, automatically, the corresponding comparative models. With such an additional tool, one could achieve a protocol capable of producing QM/MM models starting directly from the constantly growing repositories of rhodopsin amino acid sequences. This target is currently pursued in our laboratory.
- Automatic prediction of side-chain conformation for mutants: recent efforts have been directed to achieve a successful technology for systematically predicting mutant structures, which provides a superior level of accuracy of the *a*-ARM models than that proposed in Ref. [69].

  More specifically, the mutations routine that used a backbone-dependent rotamer library (i.e., SCWRL4 [77]) was replaced by a software based on comparative modeling (i.e., MODELLER [78]) (see Fig. 8b). The description of the new approach as well as an example that illustrates its effectiveness for mutating a specific position with each of the 20 essential amino acids, are provided in Ref. [35] and summarized in Sect. 4.5.

- Insufficient description of possible cavity rearrangements after mutation: the updated procedure described in Ref. [35] for modeling the side-chain conformation (see point above) comprises a short MD, where the introduced side-chain is allowed to relax, whereas the rest of the cavity residues, water molecules, chromophore and protein environment remain fixed at the crystallographic/comparative structure (see Supplementary Note 13 in Ref. [35]).

  Notwithstanding the following, more sophisticated MD step related to the cavity residues (see Sect. 3.4.1), a proper description of the impact of the new side-chain on the protein environment is lacking, due to a not sufficient description of possible local steric/electronic rearrangements of those residues of the chromophore cavity surrounding the mutated one.
- Mutations only allowed in the chromophore cavity: currently, *a*-ARM only allows mutations of residues that belong to the chromophore cavity sub-system, as well as backbone relaxation is not allowed. The latter is to ensure that, during the QM/MM model generator phase, the geometry of the new modeled side-chain as well as the sidechain of its neighbors (belonging to the chromophore cavity) can be readjusted during the 1 ns GROMACS MD step, while assuming that the general structure of the protein is conserved.
- Lack of a predictive tool for mutants generation: the fact that the mutants generator relies on the use of experimental data to select the correct rotamer limits the usability of the protocol, which cannot be considered as a predictor tool.

4.5 Recent Updates and Improvements

In silico modeling of point mutations in proteins relies on the selection of a robust methodology for the prediction of the side-chain conformation of the replaced amino acid [77, 79–90]. Both *original* [59] and *advanced* [43, 64, 69, 91] versions of the ARM protocol use the software SCWRL4 [77] to predict the side-chain conformation of the mutated residues. This approach is based on backbone-dependent rotamer libraries (from public databases of experimentally resolved protein structures), and was found adequate for the production of single, double and triple point rhodopsin mutants. This was demonstrated by studies carried out by some of the authors on mutants of bovine rhodopsin (Rh) [59, 69], Anabaena Sensory rhodopsin (ASR) [42, 59], bacteriorhodopsin (bR) [74] and KR2 rhodopsins [92].

Recently, some of the authors reported the first attempt to use *a*-ARM model building to systematically and exhaustively mutate a single residue [35]. More specifically, they attempted, unsuccessfully, to perform single point mutations of KR2 at the P219 location near the β-ionone ring of the *r*PSB via SCWRL4 modelling. Despite the encouraging results reported in Ref. [42] for the cases P219A, P219G and P219T, the authors found that, for larger side-chains SCWRL4 generated conformers sterically clashing with either the *r*PSB or neighboring amino acids.

After examining the tools available for side-chain predictions (see, for instance, Ref. [79]) and evaluating them in terms of performance and accessibility as command-line tools, the authors of Ref. [35] modified the mutations routine (see Section 2.2.5. in Ref. [69]) by substituting SCWRL4 with MODELLER [78]. This alternative approach allows the production of mutants suitable for the prediction of absorption wavelengths in either an automatic or a computer-aided semi-automatic fashion.

Figure 12a illustrates the general workflow of the proposed subroutine, which replaces Step 3 of the input file generator phase of *a*-ARM (see Fig. 8a). At input level, each point mutation is generated via a customized version of the `mutate_model.py` routine implemented in Modeller, where the conformation of the modeled side-chain is optimized using a conjugate gradient method, and refined using a short MD (^{mod}).

Briefly, as reported in Ref. [78], the `mutate_model.py` script has been designed to model point mutations via side-chain replacement in a fixed environment, assuming that single mutations do not generally determine deep conformational changes of the protein backbone. Accordingly, and consistently with the structurally "conservative" approach of the *a*-ARM protocol (see Sect. 3.3), our methodology replaces only the side-chains of the mutated residues keeping the backbone atoms at fixed positions. In order to sample the conformational space of a mutated residue more extensively. and evaluate its effect on the vertical excitation energy (ΔE^{a}_{S1-S0}), the new customized setup produces 30 rotamers of the same mutated side-chain by providing the script with different initial seeds (i.e., initial velocities) for the MD^{mod} run. The obtained rotamer structures are compared with each other, in terms of root mean square deviation, and discarded if found less than 0.025 Å from another. Although not particularly efficient, this procedure allows for the quick selection of a set of non-redundant rotamers for a single mutant, which are evaluated using the scoring function discrete optimized protein energy (DOPE) [93], implemented by MODELLER, and ranked from lowest to highest. The ARM input for the three highest DOPE scored mutated side-chain rotamers is completed by phase I of the *a*-ARM protocol, and their ARM QM/MM models are produced using phase II (see Fig. 8). The corresponding computed ΔE^{a}_{S1-S0} is then used to evaluate the performance of different rotamers of the mutated side-chain in reproducing the experimental trend in line with the WT, leading to the selection to the conformer (rotamer) that better agrees with experimental data. Figure 12, panels b and c illustrate an example of the procedure for selecting a rotamer from three evaluated models. Further deatils can be found in Supplementary Note 13 in the supporting information of Ref. [35].

Although this approach relies on experimental information and does not represent a predictive tool, it automates the side-chain conformation selection during the construction of mutant QM/MM models.

5 Web-ARM, a Web-Based Interface to ARM

(Most of the content of this section is reproduced/adapted with permission from [74]. Copyright 2020 American Chemical Society).

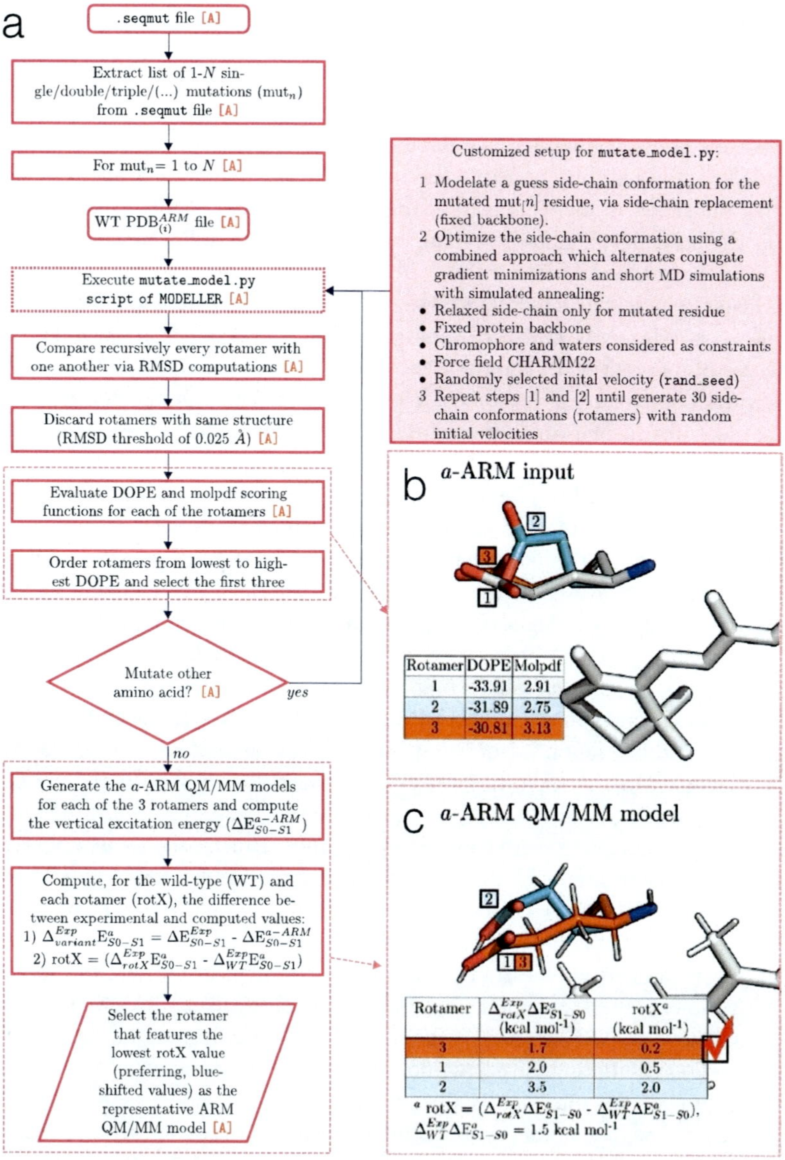

Rotamer	DOPE	Molpdf
1	-33.91	2.91
2	-31.89	2.75
3	-30.81	3.13

Rotamer	$\Delta^{Exp}_{rotX}\Delta E^a_{S1-S0}$ (kcal mol-1)	rotX[a] (kcal mol-1)
3	1.7	0.2
1	2.0	0.5
2	3.5	2.0

[a] rotX = $(\Delta^{Exp}_{rotX}\Delta E^a_{S1-S0} - \Delta^{Exp}_{WT}\Delta E^a_{S1-S0})$, $\Delta^{Exp}_{WT}\Delta E^a_{S1-S0}$ = 1.5 kcal mol-1

Fig. 12 General workflow of the novel side-chain generator. **a** Modified routine for the mutants generator of *a*-ARM, based on Modeller. This procedure is used to model, e.g., the side-chain of the E219 residue of the KR2 rhodopsin, as shown in panels **b**, **c**. **b** First, the discrete optimized protein energy (DOPE) and molpdf scoring functions for all the possible rotamers are evaluated and the three best values are ranked. **c** Then, the a-ARM QM/MM model for each rotamer is generated and the rotamer model featuring the lowest difference in vertical excitation energy with respect to experimental data (rotamer 3) is selected Adapted with permission from [35], open access under a CC BY license (Creative Commons Attribution 4.0 International License)

5.1 Interface Features

The *a*-ARM protocol (described in Sect. 4) and its most updated version `PyARM`, a Python-based software package (that will be presented in Sect. 6) represent an easy-to-use command-line interface directed to users (i.e., researchers, undergraduate students) familiar with the Linux environment. The latter is not due to a fact of usability, but rather to the technically complex initial setup of the package, since it requires the prior installation of several software and python dependencies. Moreover, it is recommended to install the `PyARM` package in a high-performance computer cluster, which is usually required to run the underlying computationally intensive tasks (i.e., MM, MD, QM/MM), rather than on a local personal machine.

The use of a user-friendly interface accessible through the web and together with the provider computational resources, avoids dealing with complicate installations and the need for local computer facilities. This would mean accessing a simple, computationally fast and automated construction and analysis of rhodopsin QM/MM models, fit for an interdisciplinary community that is interested mostly in actual applications, rather than methodological development.

Accordingly, Ref. [74] reports the web version of the *a*-ARM protocol, that is the Web-ARM interface. Web-ARM is a user-friendly interface written using Python 3, available at the following address: web-arm.org. Therefore, the potential user only needs to use an up-to-date browser on any operating system (i.e., Linux, macOS, Windows, Android, iOS) and platform to take advantage of the *a*-ARM protocol.

In order to generate an ARM QM/MM model, Web-ARM goes through the four phases explained in Section 2.1 of Ref. [74], and illustrated in Fig. 13. As described for the command-line version, the procedure starts with the initial structure of the rhodopsin variant and finishes with the generation of the S_0 equilibrium geometry along with the calculations on absorption properties. During such a procedure, the interface gives the user enough flexibility to generate either *a*-ARM $_{\text{default}}$ or *a*-ARM $_{\text{customized}}$ inputs (see Sect. 4.1), the former automatically and the latter by modifying some of the default choices. This is made on top of the implementation of the input file generator inside the framework of the web interface. Then, the so-generated ARM input is used to compute a QM/MM model, by using the QM/MM model generator. The Web-ARM internal driver takes care of performing all the necessary steps, as well as submitting the calculations to the dedicated computational facilities.

One feature of the interface is that, once a QM/MM model is generated, the user is provided with a summary of all the relevant data (i.e., energetics, oscillator strengths), along with a downloadable file (in compressed format) containing the major output files. Further information, and a complete walk-through, are provided in a Tutorial that can be accessed/downloaded from the Web-ARM main web page.

Web-ARM is intended as both a research, as well as a teaching tool. Ref. [74] shows that the interface can systematically screen rhodopsin variants, and thus obtain a qualitative check prior to, e.g., an experimental study. The interface can also be used successfully in teaching and learning activities, e.g., to introduce students to the idea of QM/MM models and corresponding computed data. Therefore, Web-ARM is envisioned as a tool used in teaching and training, as well as by non-experienced users, as previously noted, mostly for bulk production. However,

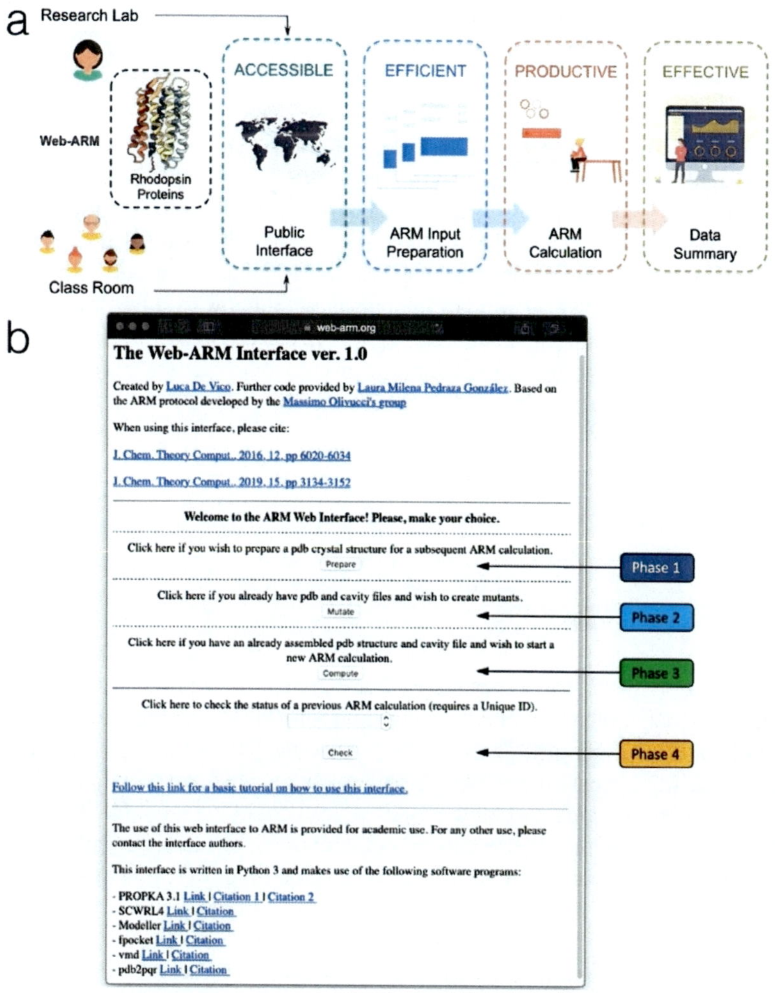

Fig. 13 General overview of the Web-ARM interface. **a** Main features and **b** home page of the Web-ARM interface Adapted with permission from [74]. Copyright 2019 American Chemical Society

also an experienced computational chemist can take advantage of the web interface, to produce rhodopsin QM/MM models in a standardized manner, being aware of the documented accuracy and rate of success. Of course, one, possibly very useful, application of such a model is to provide high quality guesses for more sophisticated subsequent calculations, e.g., as a starting substrate to which apply further, high-level refinement methods.

In conclusion, by using Web-ARM both junior researchers and trainees will be able to perform meaningful QM/MM calculations focusing on the underling research targets, methodological concepts, and data analysis, while remaining confident that the calculations are internally consistent.

5.2 Limitations and Future Development of Web-ARM

- Limited computational resources: before using the Web-ARM interface, the user is asked to provide an email address to be registered into our database. Registered users are allowed to build as many concurrent ARM QM/MM models as wished (default 10). However, given to the present threshold in the host available computational resources, guest users are allowed to build only one ARM QM/MM model model at a time on the developer's dedicated resources.
- Technical issues: the Tutorial of the Web-ARM interface reports on possible errors or issues in the execution of the interface, and how to solve or avoid them.
- Current and future implementation: presently, the capability of the Web-ARM interface is limited to the construction of ground-state models. Future work will implement inside the interface all of the features of the `PyARM` software package, illustrated in Fig. 14.

6 PyARM

6.1 Package Description

`PyARM` is a user-friendly, open-source Python-based software tool designed to facilitate the systematic, reproducible and congruous QM/MM modeling/analysis of photoexcited states of rhodopsin proteins. More specifically, `PyARM` is a development platform that implements high-level algorithms associated to specialized QM/MM protocols, under the framework of the *a*-ARM protocol [43, 59, 64, 69, 74, 91].

The package structure represents a collection of hierarchical instances here defined as: scripts, basic low-level functions, general high-level functions, modules, drivers and templates. Table 1 provides a technical definition of each of these terms.

Figure 14 depicts the general overview of `PyARM`, showing how all components of the package are modular, i.e., independent of the context in which they are used. The characteristics of the drivers and package create an application-programming tool, capable of making complex workflows/protocols available to users with minimal programming knowledge. Indeed, the inclusion of new protocols (i.e., modules and drivers) can be achieved easily with only a minimal modification of the code. In this regard, the package includes a programming interface for parsing user input, and retrieving and storing specific data. Therefore, a user can simply use a module to perform a given type of application (e.g., perform a geometry optimization) or use a driver to launch a protocol connecting different modules and, thus, executing more complex applications (e.g., generate a S_0 or S_1 QM/MM model, locate a fluorescent rhodopsin, compute an absorption band, etc.). Nevertheless, an experienced user can also create a new driver, capable of performing a new type of application/analysis based on *a*-ARM models and, therefore, within the limits of their accuracy [59, 69, 74].

Furthermore, the modular framework of the package, which was envisioned initially for the study of natural rhodopsins, presents a flexible architecture

Table 1 Different instances of the PyARM framework

Instance	Definition
Script	Python file that is intended to be executed directly. This means that scripts will often contain code written outside the scope of any classes or functions
Basic low-level function	Basic, importable function that performs a single, simple and generic task. Example: Copy/move files, create folders, calculate energy gaps, generate input file
General high-level function	General function that makes use of basic low-level functions to perform a more complex task. Can be used either as importable or as stand-alone by parsing arguments in the terminal. Example: quantum chemistry calculation
Module	Script that makes use of both basic low-level and general high-level functions to perform a complex task (i.e., semi-protocol) and that requires different stages. Can be used either as importable for the drivers or as stand-alone by parsing arguments in the terminal. Example: Input file generator << `a_arm_input_generator` >>
Driver	Script that makes use of the modules to implement a complex protocol. Works as stand-alone by parsing arguments in the terminal. Example: *a* -ARM model building << `a_arm_protocol` >>
Package	Collection of related modules that work together to provide certain functionality. These modules are contained within a folder (`/src/`) and can be imported just like any other modules. This folder contains a special `__init__`.py file that tells Python it's a package, potentially containing more modules nested within subfolders. Example: << PyARM >>

suitable to be generalized to other photoreceptors including synthetic light-responsive proteins. As a first step in such a direction, the current implementation is suitable for the study of rhodopsins featuring artificial retinal-like chromophores, which is currently pursued in our group.

6.2 Current ARM-Based QM/MM Protocols

Presently, PyARM contains four different ARM-based fully automatic protocols, each implemented as a general driver (i.e., user-friendly one-click command-line interface), for the QM/MM modeling of rhodopsin electronically excited states. A brief description of each of these protocols, illustrated in Fig. 14, is provided below.

(1) Protocol: *a*-ARM model building protocol: ARM with chromophore cavity generation, ionization state selection, and external counterion placement.

- Driver: `a_arm_qmmm_protocol`
    - Phase I: Input file generator. Module: `a_arm_input_generator` (see Fig. 8a)

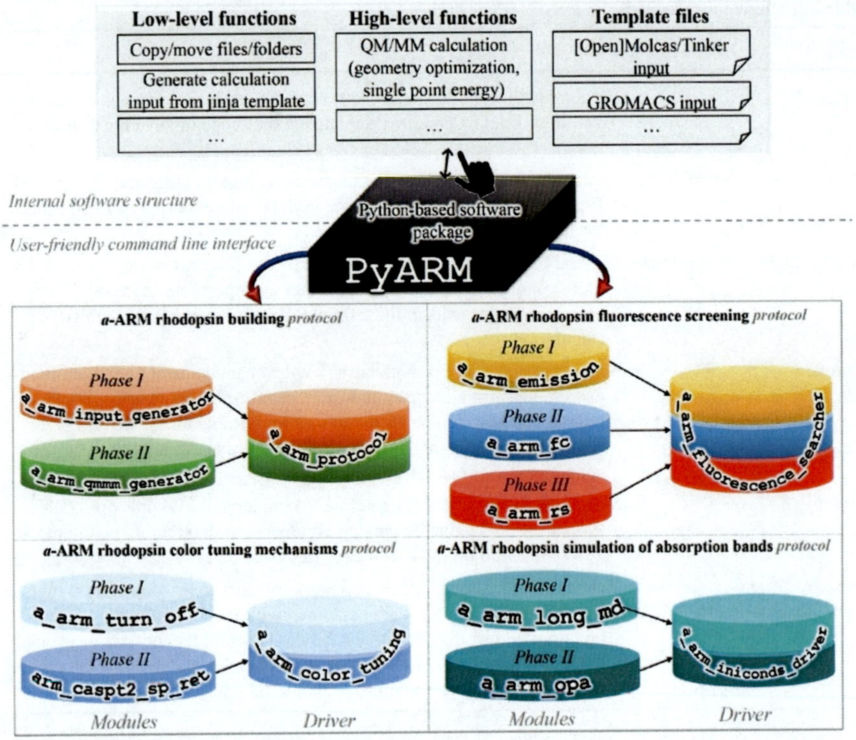

Fig. 14 Representation of the contents of the PyARM software package. The package contains various modules, which can be steered using a driver. Modules make use of high- and low-level functions, as well as template files that are also an integral part of the package

  – Phase II: QM/MM model generator. Module: a_arm_qmmm_generator (see Fig. 8b)

- Description: The first step towards the standardization and full automation of the *original* ARM was the development of *a*-ARM. This updated version not only overcame the automation drawbacks of the original version, but also included significant methodological/computational improvements. The achieved level of automation is accompanied by other features, such as speed in preparing the model building input, and standardization and reproducibility of the final model when operated by different users. In fact, the time required for preparing the input for the QM/MM model construction is reduced from around 3 h to less than 5 min (user time), with respect to the original ARM protocol. This is a consequence of the automation of the different preparatory steps, thus avoiding the user manipulation of text files and/or visualization of chemical structures. More specifically, a Python-based automatic input file generator subroutine was written to automate the assignment of the residues defining the chromophore cavity, including the chromophore linker and counter-ions, the protonation state of ionizable residues and, finally, the unambiguous placement of cytoplasmic and extracellular counter-ions. The benchmark calculations demonstrated that the resulting models for several rhodopsin sets are accurate enough for reproducing experimental trends in vertical excitation energy (i.e., the computed vertical excitation energies have an error bar of less than 4.0 kcal mol^{-1} blue-shifted), as well as transferability to rhodopsins of very different sequence. This protocol has been further described in Sect. 4.

(2) Protocol: Automated Analysis of Color Tuning Mechanisms in Rhodopsins.

- Driver: `a_arm_color_tuning`
  - Phase I: Electrostatic effects. Module: `a_arm_turn_off` (see Sects. 6.4.3 and 6.4.4)
  - Phase II: Steric effects. Module: `arm_caspt2_sp_ret` (see Sect. 6.4.2)
- Description: This protocol provides a tool to study how the protein environment (i.e., protein sequence) modulates the absorption wavelength of the retinal chromophore and, in turn, the color of the protein. In other words, it performs rhodopsin color tuning analysis so as to reveal steric and electrostatic effects between the retinal chromophore and the surrounding cavity amino acid residues. This tools aids the mechanistic description of the ways amino acid residues influence the photophysical properties of the retinal chromophore. This, in turn, builds up a reference book for helping the tuning of rhodopsin cavities (i.e., via point mutations) towards a desired effect (e.g., blue or red shifted wavelength). This protocol will be further described in Sect. 6.4.

(3) Protocol: Automated Simulation of Absorption Bands and light-induced Dynamics of Rhodopsins.

- Driver: `a_arm_iniconds_driver`
  - Phase I: Long MD from a-ARM QM/MM model. Module: `a_arm_long_md`
  - Phase II: Simulation of one photon absorption spectrum and light-induced dynamics. Module: `a_arm_opa`
- Description: This protocol allows the automatic simulation of one photon absorption (OPA) spectra, and the subsequent calculation of the initial conditions necessary for the study of the excited state dynamics, also allowing estimation of the photoisomerization quantum yield (QY). In other words, this tools provides another link between experimental quantities and simulated results. The photoisomerization QY, in particular, is of importance when interested in devising rhodopsins that are either particularly reactive or, contrarily, extremely fluorescent. This protocol will be the subject of a future publication.

(4) Protocol: Automated QM/MM Model Screening of Rhodopsin Variants Displaying Enhanced Fluorescence.

- Driver: `a_arm_fluorescence_searcher`

  - Phase I: Location of the S_1 fluorescent excited-state (FS). Module: `a_arm_emission`
  - Phase II: Computation of Franck–Condon (FC) trajectories. Module: `a_arm_fc`
  - Phase III: Calculation of the S_1 photoisomerization path (RS). Module: `a_arm_rs`

- Description: This protocol is designed for gaining insights into the possible molecular-level mechanisms behind fluorescence enhancement, when the fluorescent species correspond to the initial dark state of the rhodopsin photocycle. It is composed of three different phases able to categorize a rhodopsin as dim- or enhanced-fluorescent, with respect to a reference. This allows the fast and systematic in silico screening of potentially hundreds of rhodopsin mutants. Furthermore, each phase provides information (i.e., properties such as emission wavelength, excited state lifetime (ESL), energy isomerization barrier (E^f_{S1}), fluorescence quantum yield (ϕ^f) and structural parameters) that allows to elucidate and understand the factors determining rhodopsin fluorescence and, possibly, learn how to modulate it, with the ultimate goal of designing fluorescent candidates in silico for applications in, e.g., optogenetics [94]. This protocol will be the subject of a future publication.

6.3 PyARM Technical Details

6.3.1 PyARM Default Parameters

The setup for all the MD (see Sect. 3.4.1) and QM/MM (see Sect. 3.4.2) calculations performed by the different modules and drivers of PyARM (Sect. 6.2), is consistent with that reported for the *a*-ARM protocol in Refs. [43, 59, 64, 69] (see also Supplementary Notes 2 and 3 of Ref. [35]).

As described in Sect. 3.4.2, the default parameters for QM/MM calculations correspond to 2-roots single-state CASSCF(12,12)/AMBER optimization and state-average 3-roots CASPT2(12,12)/6-31G(d) energy correction. Such values for the active space and number of roots were selected based on benchmark calculations [59]. While the latter is not a modifiable parameter, the former could be modified via command-line arguments; however, it is strongly recommended to keep the default values.

As shown in Sect. 6.2, each instance of PyARM (see Table 1) is executed via a command-line interface, which contains a help menu facility that provides descriptions and syntax for the default parameters and how to modify them. Some of these parameters are listed below. Note that it is possible (but not recommended) to customize all the parameters for the MD run.

```
MD setup
-n    Number of independent replicas of Molecular Dynamics
      [default 10]
-ns   Specific seed for MD
      [default random number]
-mdt  Production temperature of the MD simulations (Kelvin)
      [default 298 K]
-mdh  Simulation time for the MD heating phase (ps)
      [default 50 ps]
-mde  Simulation time for the MD equilibration phase (ps)
      [default 150 ps]
-mdp  Simulation time for the MD production phase (ps)
      [default 800 ps]
QM/MM setup
-nr   Number of roots SA-CASSCF
      [default 2]
-lr   Number of roots CASPT2
      [default 3]
```

6.3.2 PyARM Installation

The PyARM code can be obtained by contacting the authors. PyARM relies on the following programs being already present and properly installed: Molcas ver. 8.4 or OpenMolcas [62, 64]; TINKER ver. 6.3 (with specific QM/MM patches) [63]; PROPKA ver. 3.1 [76]; PutIon [59]; GROMACS ver. 4.5.5 [61]; DOWSER [60]; PDB2PQR [95, 96]; Fpocket [97]; Modeller ver. 9.18 [78]. Furthermore, PyARM necessitates of a Python 3.x installation including the following modules: OpenBabel [98]; MDAnalysis [99, 100]; Matplotlib [101]; pandas [102]; NumPy [103]; SciPy [104]; Jinja2; Python-crontab; PyYAML; TextTable; cclib [105], GromacsWrapper.

PyARM contains a configuration file, the contents of which point to the installation path of the aforementioned programs. Submission template files are also provided for each type of calculation. These templates have to be tailored to the specifics of the computer cluster queuing system. Once these files are in place, and the user has write permissions for the Python installation, PyARM is installed as a normal Python 3.x package.

6.3.3 PyARM Tailoring

The modular nature of PyARM allows the relatively easy implementation of changes in the various drivers, as well as introducing a new driver. As a hypothetical example, *a*-ARM models have been used to compare excitation energy data obtained using CASPT2, MS-CASPT2 and MC-PDFT levels of theory [106]. The introduction of a different level of theory to evaluate excitation energies would imply using a new high-level function (Table 1) that performs such calculation, which in turn is based on a new template input file.

The new high-level function would be used by one or more modules, which in turn would belong to any number of drivers. If the user would like to simply swap level of theory, they would need to change the call to the original high-level function in each module with the newly made one. Otherwise, the user can simply create copies of the desired modules and drivers, which makes use of the new high-level function.

Similarly, a new driver could be implemented, taking advantage of the already present modules and functions as much as possible. For example, one could introduce a new, hypothetical, driver to locate the possible photoproduct structure of a given rhodopsin. Such a driver would combine the machinery of the previously mentioned `a_arm_fluorescence_searcher`, but extending the `a_arm_rs` module (i.e., making a new module) to perform a ca. 180 degrees isomerization, followed by a S_0 geometry optimization and energetic evaluation. In this case, all necessary high-level functions and templates are already present, they just have to be combined in a different way.

6.4 Color Tuning Analysis in Terms of Steric and Electrostatic Effects

(Part of the content of the next three sections is reproduced/adapted with permission from [35], open access under a CC BY license (Creative Commons Attribution 4.0 International License)).

As an example of a possible usage of a driver and a module as found inside `PyARM`, this section presents the driver (`a_arm_color_tuning`) behind a common kind of analysis [35, 42], which aims at discerning the contribution of nearby cavity residues onto the computed excitation energy ΔE^{a}_{S1-S0} of the *r*PSB, and assign them as blue- or red-shifting. The shifting contributions are assessed in terms of both electrostatic effects onto the charged *r*PSB (Sect. 6.4.3), and steric effects distorting the geometry of the retinal chromophore (Sect. 6.4.2). This assessment is performed based on quantities defined in Sect. 6.4.1. Figure 15 presents the results of using this analysis, as found while performing all possible mutations (P219X) for the KR2 rhodopsin [35]. Finally, Sect. 6.4.4 presents a simpler possible use of the module `a_arm_turn_off` to analyze electrostatic contributions from cavity residues onto the *r*PSB.

6.4.1 Computed Quantities

In Ref. [35], the authors define three fundamental quantities ($\Delta\Delta E^{TOT}_{S1\text{-}S0}$, $\Delta\Delta E^{STR}_{S1\text{-}S0}$, and $\Delta\Delta E^{OFF}_{S1\text{-}S0}$) whose values are a function of either the structural (both at the chromophore and protein cavity levels) or electrostatic changes of each mutant with respect to WT, using as an illustrative example the mutants P219X of the KR2 rhodopsin (Fig. 11c).

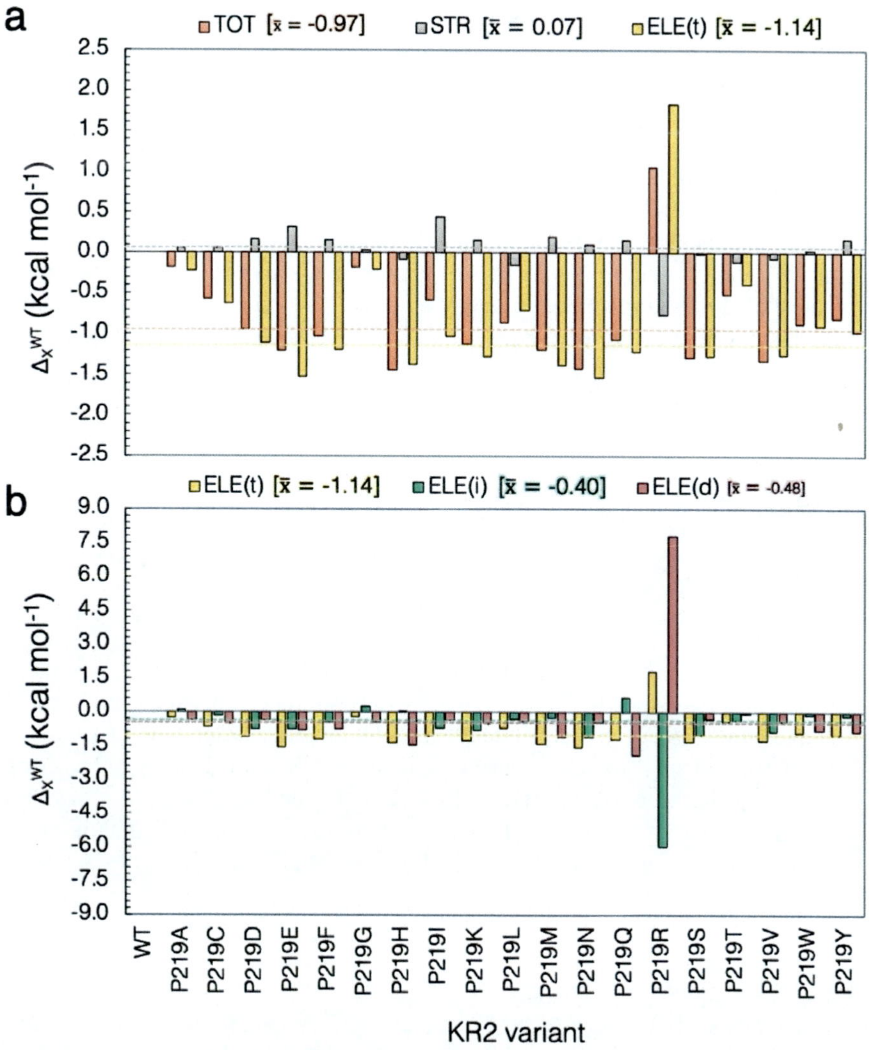

Fig. 15 Steric and electrostatic contributions to the vertical excitation energy. **a** Total ($\Delta\Delta E^{TOT}_{S1\text{-}S0}$), steric ($\Delta\Delta E^{STR}_{S1\text{-}S0}$) and electrostatic ($\Delta\Delta E^{ELE(t)}_{S1\text{-}S0}$) contributions to the interaction of the retinal with the protein environment for the possible P219X mutations. **b** Decomposition of the total electrostatic effects on its indirect ($\Delta\Delta E^{ELE(i)}_{S1\text{-}S0}$) and direct ($\Delta\Delta E^{ELE(d)}_{S1\text{-}S0}$) components. The dashed lines and corresponding numerical values refer to the weighted average values ($\overline{x}$) of the 18 residues of the red-shifted cluster exclusively (that is, excluding P219R), presented in square parenthesis. Reproduced with permission from [35], open access under a CC BY license (Creative Commons Attribution 4.0 International License)

- $\Delta\Delta E^{TOT}_{S1\text{-}S0}$ is the "Total" excitation energy change. It is computed directly as the difference between the QM/MM computed vertical excitation energies of mutant and WT models.
- $\Delta\Delta E^{STR}_{S1\text{-}S0}$ is the "Steric component" of $\Delta\Delta E^{TOT}_{S1\text{-}S0}$. It is computed directly as the difference between the QM/MM vertical excitation energies of the iso-

lated retinal chromophores, however retaining the geometries as if inside the protein environment, between mutant and WT.
- $\Delta\Delta E^{OFF}_{S1\text{-}S0}$ is used to quantify the "indirect" electrostatic component (see below) of $\Delta\Delta E^{TOT}_{S1\text{-}S0}$. It is computed as the difference between the vertical excitation energy of the mutant and WT obtained after having switched off (turned to zero) the charges of residue 219. $\Delta\Delta E^{OFF}_{S1\text{-}S0}$ can also be used directly to analyze, qualitatively, the red- or blue-shifting role of each of the residues of the chromophore cavity for a determined rhodopsin (Turn-Off Module). In such a case, it is computed as the differences between the vertical excitation energy of the rhodopsin before and after having switched off (turned to zero) the charges of specific residues.

As we will see in the following, using these quantities, we can compute three additional components.

- $\Delta\Delta E^{ELE(t)}_{S1\text{-}S0}$ is the "Total electrostatic component" that is indirectly computed as the difference between the Total and the Steric components above ($\Delta\Delta E^{ELE(t)}_{S1\text{-}S0} = \Delta\Delta E^{TOT}_{S1\text{-}S0} - \Delta\Delta E^{STR}_{S1\text{-}S0}$) for each mutant. As specified in Sect. 6.4.2, $\Delta\Delta E^{ELE(t)}_{S1\text{-}S0}$ can be decomposed into two parts:

  - $\Delta\Delta E^{ELE(i)}_{S1\text{-}S0}$ is the "Indirect electrostatic component" that is indirectly computed in two steps by first computing the differences between the vertical excitation energy of the mutant and WT obtained after having switched off (turned to zero) the charges of residue 219, and then by subtracting from such difference the steric effect $\Delta\Delta E^{STR}_{S1\text{-}S0}$ defined above.
  - Finally, $\Delta\Delta E^{ELE(d)}_{S1\text{-}S0}$ is the "Direct electrostatic component" that is computed indirectly as $\Delta\Delta E^{ELE(d)}_{S1\text{-}S0} = \Delta\Delta E^{ELE(t)}_{S1\text{-}S0} - \Delta\Delta E^{ELE(i)}_{S1\text{-}S0}$.

6.4.2 Steric Effects

In the context of the protocol, by "steric effects" we mean "indirect" or "geometrical" effects, i.e., the change in excitation energy of a chromophore due to a change in the minimum geometry. In turn, this change in geometry of the *r*PSB could be induced by both steric and electrostatic factors. Such effects are investigated by analyzing how the retinal chromophore is structurally modified by mutations near the *β*-ionone or near the Schiff base linkage (see Fig. 5).

As discussed in Ref. [35], such structural rearrangements of the *r*PSB may be due to different factors, such as a simple effect induced by the side-chain replacement, a different charge distribution due to changes in protonation states for ionizable residues, as well as water molecules addition/removal. Notice that steric effects were evaluated through an "atomistic" approach, focused on the changes in the *r*PSB geometrical and electronic structure, and therefore not directly related to steric effects evaluated on the basis of the changes in residue volume [35, 92].

6.4.3 Electrostatic Effects

As mentioned above, the total electrostatic effect ($\Delta\Delta E^{ELE(t)}_{S1\text{-}S0}$) can be decomposed in two parts: (1) the first can be considered as a direct component ($\Delta\Delta E^{ELE(d)}_{S1\text{-}S0}$) due to the variation in number, magnitude, and position of the point charges of mutated residue caused by the P to X replacement, and (2) a more indirect component ($\Delta\Delta E^{ELE(i)}_{S1\text{-}S0}$) produced from the reorganization of the local environment and hydrogen bond network induced by the same replacement and due to the fact that conserved residues and water molecules change in position or orientation. Moreover, as discussed in Ref. [35], possible changes in protonation states of conserved residues, induced by P to X replacement, have a major contribution to the indirect component. Figure 15 shows the contributions due to different effects (steric and electrostatic) and components.

6.4.4 Turn-Off Module

As a final example, we present a possible usage of the `a_arm_turn_off` module, which is part of the `a_arm_color_tuning` driver (Phase I). As previously stated (Table 1), modules can be used also independently from the parent driver. When used as a stand-alone script, the command-line menu of the module provides four different options to turn-off residues in the chromophore cavity, with different scopes, as follows:

1. < `cav` >: Turn off the charges of EACH of the residues in the cavity, independently. For instance, if the cavity is formed by *N* residues, the output will be *N* single calculations.
2. < `cav_all` >: Turn off the charges for ALL the residues in the cavity, simultaneously. Therefore, the output will be a single calculation.
3. < `single` >: Turn off the charges for ONE specific residue given by the user.
4. < `multiple` >: Turn off the charges for a list consisting of MULTIPLE residues, given by the user, simultaneously. Therefore, the output will be a single calculation.

As an example of a possible analysis, we envision the use of the module with the first option (< `cav` >), to map out the expected electrostatic influence of cavity residues onto the ΔE^{a}_{S1-S0} of the *r*PSB. Such analysis could drive subsequent mutation tests, such as all possible mutations performed onto the residue responsible for, e.g., the largest or smallest shift.

6.5 PyARM Current Accuracy and Drawbacks

A number of limitations of *a*-ARM have already been described in Sects. 3.5, 4.4 and 5.2. Due to their simplified definition, ARM models are more exposed to potential pitfalls than more complex QM/MM models. Such possible pitfalls, that concern the QM/MM model generator phase, can be summarized as:

(1) lack of a proper description of the protein environment (membrane + explicit solvent),
(2) rigid protein backbone and non-cavity side-chains,
(3) approximated protonation states for ionizable residues, and
(4) missing description of any mutual polarization effects between the QM and MM sub-systems, that can be accounted for by polarizable embedding using a polarizable force field. Since polarizable force fields are technologies still under development in the QM/MM area (see, for instance Ref. [107]), we have not adopted/benchmarked them in this version of our specialized QM/MM models.

When considering points 1–4, the different properties computed by ARM are expected to be affected by a systematic error. Our current research is aimed at dealing with those points, while maintaining reasonable computational costs, or estimating the errors due to them. Nevertheless, according to the philosophy of the ARM protocol (Sect. 3.2), the main focus of ARM is the ability to reproduce property (especially λ^a_{max}) and explain trends, rather than predicting their absolute values. The default model generation and subsequent customization revised above clearly shows that the models can fit the experimental results. In other words, the customization protocol searches for possible sources of systematic errors (i.e., different protonation state, different rotamer), due to points 1–4.

A trend deviation factor was computed [69], to evaluate the accuracy of the *a*-ARM data in a given set of rhodopsins with available experimental data. Thus, a mean absolute error of 2.5 kcal mol^{-1} was found (Table S4 in the supporting information of Ref. [69]), when considering *a*-ARM$_{default}$ results for the benchmark set of Fig. 11b. A smaller value of 0.4 kcal mol^{-1} (Supplementary Table 5 of Ref. [35]) was found for the coherent set of KR2 mutants presented in Fig. 11d. These numbers show the expected accuracy of `PyARM` in evaluating trends of photophysical properties of rhodopsin sets. Logically, a coherent set of mutants of the same rhodopsin shows a smaller deviation, with respect to a set of rhodopsins spanning different organisms.

Although outside the *a*-ARM scope, it is possible to evaluate the accuracy of `PyARM` with respect to computing absolute absorbed wavelengths.[4] For example, with respect to the data behind Fig. 11b (Table S3 in the supporting information of Ref. [69]), we found an average accuracy for the *a*-ARM$_{default}$ protocol of 90%.

[4] We here define accuracy as the absolute difference of computed and experimental absorbed wavelength, divided by the experimental number, in percentage. This is done for each of the N=10 replicas of a rhodopsin (Sect. 3.4.1) and then averaged. The final number is the average of the accuracy of all the considered rhodopsins.

7 Outlook and Concluding Remarks

In this review, we presented the ARM protocol, along with its past and current development and achievements. After a brief introduction to rhodopsins, their importance and current and potential technological uses, we presented the development of ARM. We saw how the protocol was thought from inception to be accurate enough to reproduce trends in photochemical properties (mainly ΔE^{a}_{S1-S0}), that is points (1) and (2) of Sect. 3.5.

Most of the remaining points (3)–(5) were addressed through the subsequent development of the Updated version of the Automatic Rhodopsin Modeling protocol *a*-ARM, presented in Sect. 4. The major achievement of *a*-ARM has been the complete automation of the input file generator and QM/MM model generator, and the corresponding coding as python, user-friendly, command-line interfaces. Such tools were also included in Sect. 5. The method was benchmarked (Fig. 11), and found reliable in obtaining photochemical properties of different rhodopsins, obtained from various life domains. Furthermore, the usage of the input file generator ensured reproducibility of the results, even when considering the two approaches *a*-ARM$_{\text{default}}$ and *a*-ARM$_{\text{customized}}$.

Finally, Sect. 6 presented the latest development of the ARM protocol, namely `PyARM`. As shown in Fig. 14, `PyARM` consists of different drivers capable of performing various actions and automatic analyses, combining several python modules, which in turn are composed of functions and templates (Table 1). The automation level achieved by `PyARM` allows the efficient generation of, e.g., all 19 possible mutants of a given residue (Fig. 11c). The concurrent preparation of many QM/MM models allows, in principle, the study of rhodopsin mutants arrays. The completely modular architecture of `PyARM` will permit scaling up the code (i.e., introducing novel kinds of analyses), through the implementation of new drivers, alongside those described in Sect. 6.2.

Despite the encouraging outcome of the applications of the protocol, additional work is required, for providing the scientific community with a robust tool that can be applied systematically to the study and design of sizable rhodopsin arrays. Much of its future success will depend on further improvements in the construction of the *a*-ARM model and, in particular, of rhodopsin mutants models. Some of the envisioned improvements are as follows:

- Since rhodopsin structural data are rarely available and still difficult to obtain experimentally, it would be important to integrate, in an automatic fashion, comparative modeling technologies in the protocol for *a*-ARM model building. It is possible that with such tool one could achieve a protocol capable of producing more accurate QM/MM models starting directly from the constantly growing repositories of rhodopsin sequences. This target is currently pursued in our laboratory.
- The above item is also related to the improvement of mutant models generation and screening. As a perspective of this work, we suggest to introduce a mutant

generator routine using proper comparative (homology) modeling, instead of just modifying the mutated side-chain conformation locally, as seen in Sect. 4.5.
- Both previous items could also benefit from the introduction of a machine-learning or artificial intelligence (AI)-assisted protocol for more accurate structures or overall structure predictions.
- Methods for improving the prediction of the residue protonation states during the construction of ARM QM/MM models appear of capital importance for increasing the accuracy (e.g., the percentage of success in reproducing trends in spectroscopic properties) of the final models. These are being presently investigated in our lab (see for instance a preliminary study in Ref. [75]).
- Currently, the *a* ARM model building protocol is the only ARM-based tool implemented in the Web-ARM interface and, therefore, accessible through the web. Future efforts will be devoted to the implementation of the four protocols described in Sect. 6.2 as utilities of the Web-ARM interface.

Finally, long-term development goals include introducing new drivers for `PyARM` (possibly re-using some of the already existing modules and scripts) for novel kinds of analyses. It is our intention to pursue also the possibility of simulating the use of different retinal chromophores, either natural or synthetic. We foresee the possibility of having a generalized method capable of handling any kind of chromophore. Last but not least, we would like to extend the applicability of the protocol beyond rhodopsins, and be able to apply it to many (any) photoactive protein complex.

Acknowledgements Thank to all current and past group members that have contributed to the development of the ARM protocol: Federico Melaccio, Alessio Valentini, Yoelvis Orozco-Gonzalez, Nicolas Ferré, Samira Gholami, Fabio Montisci, Silvia Rinaldi, Marco Cherubini, Xuchun Yang, Michael Stenrup, Hoi-Ling Luk, María Del Carmen Marín, Simone Bonfrate and Riccardo Palombo.

Author Contributions LPG and LDV prepared text and figures, all authors participated to their revision

Funding We acknowledge funding from the following entities: NSF (grant CHE-CLP-1710191), NIH (grant 1R15GM126627 01), Banca D'Italia, MIUR (Dipartimento di Eccellenza, 2018–2022), MUR (Rita Levi Montalcini grant, 2021-2024)

Code availabiity The code can be obtained by contacting the authors.

Declarations

Conflict of interest All authors declare that they have no competing interests.

Ethics approval Not applicable

Consent to participate Not applicable

Consent for publication Not applicable

Open Access This article is licensed under a Creative Commons Attribution 4.0 International License, which permits use, sharing, adaptation, distribution and reproduction in any medium or format, as long as you give appropriate credit to the original author(s) and the source, provide a link to the Creative

Commons licence, and indicate if changes were made. The images or other third party material in this article are included in the article's Creative Commons licence, unless indicated otherwise in a credit line to the material. If material is not included in the article's Creative Commons licence and your intended use is not permitted by statutory regulation or exceeds the permitted use, you will need to obtain permission directly from the copyright holder. To view a copy of this licence, visit http://creativecommons.org/licenses/by/4.0/.

References

1. Kovalev K, Polovinkin V, Gushchin I, Alekseev A, Shevchenko V, Borshchevskiy V, Astashkin R, Balandin T, Bratanov D, Vaganova S et al (2019) Structure and mechanisms of sodium-pumping KR2 rhodopsin. Sci Adv 5(4):2671
2. Okada T, Sugihara M, Bondar AN, Elstner M, Entel P, Buss V (2004) The retinal conformation and its environment in rhodopsin in light of a new 2.2 Å crystal structure. J Mol Biol 342(2):571–583
3. Shihoya W, Inoue K, Singh M, Konno M, Hososhima S, Yamashita K, Ikeda K, Higuchi A, Izume T, Okazaki S et al (2019) Crystal structure of heliorhodopsin. Nature 574(7776):132–136
4. Braslavsky SE (2007) Glossary of terms used in photochemistry, (IUPAC Recommendations 2006). Pure Appl Chem 79(3):293–465
5. Ernst OP, Lodowski DT, Elstner M, Hegemann P, Brown LS, Kandori H (2014) Microbial and animal rhodopsins: structures, functions, and molecular mechanisms. Chem Rev 114(1):126–163
6. Govorunova EG, Sineshchekov OA, Li H, Spudich JL (2017) Microbial rhodopsins: diversity, mechanisms, and optogenetic applications. Annu Rev Biochem 86:845–872
7. Kandori H (2020) Retinal proteins: photochemistry and optogenetics. Bull Chem Soc Jpn 93(1):76–85
8. Kurihara M, Sudo Y (2015) Microbial rhodopsins: wide distribution, rich diversity and great potential. Biophys Psychobiol 12:121–129
9. Kojima K, Shibukawa A, Sudo Y (2020) The unlimited potential of microbial rhodopsins as optical tools. Biochemistry 59(3):218–229
10. Kojima K, Kurihara R, Sakamoto M, Takanashi T, Kuramochi H, Zhang XM, Bito H, Tahara T, Sudo Y (2020) Comparative studies of the fluorescence properties of microbial rhodopsins: spontaneous emission versus photointermediate fluorescence. J Phys Chem B 124(34):7361–7367
11. Needham DM, Yoshizawa S, Hosaka T, Poirier C, Choi CJ, Hehenberger E, Irwin NA, Wilken S, Yung C-M, Bachy C et al (2019) A distinct lineage of giant viruses brings a rhodopsin photosystem to unicellular marine predators. Proc Natl Acad Sci USA 116(41):20574–20583
12. Bratanov D, Kovalev K, Machtens J-P, Astashkin R, Chizhov I, Soloviov D, Volkov D, Polovinkin V, Zabelskii D, Mager T et al (2019) Unique structure and function of viral rhodopsins. Nat Commun 10(1):4939
13. Pushkarev A, Béjà O (2016) Functional metagenomic screen reveals new and diverse microbial rhodopsins. ISME J 10(9):2331–2335
14. Luk HL, Melaccio F, Rinaldi S, Gozem S, Olivucci M (2015) Molecular bases for the selection of the chromophore of animal rhodopsins. Proc Natl Acad Sci USA 112(50):15297–15302
15. Pushkarev A, Inoue K, Larom S, Flores-Uribe J, Singh M, Konno M, Tomida S, Ito S, Nakamura R, Tsunoda SP et al (2018) A distinct abundant group of microbial rhodopsins discovered using functional metagenomics. Nature 558(7711):595–599
16. Lenahan C, Sanghavi R, Huang L, Zhang JH (2020) Rhodopsin: a potential biomarker for neurodegenerative diseases. Front Neurosci 14:14
17. Tsujimura M, Ishikita H (2020) Insights into the protein functions and absorption wavelengths of microbial rhodopsins. J Phys Chem B 124(52):11819–11826
18. Tahara S, Singh M, Kuramochi H, Shihoya W, Inoue K, Nureki O, Béjà O, Mizutani Y, Kandori H, Tahara T (2019) Ultrafast dynamics of heliorhodopsins. J Phys Chem B. 123(11):2507–2512
19. Tanaka T, Singh M, Shihoya W, Yamashita K, Kandori H, Nureki O (2020) Structural basis for unique color tuning mechanism in heliorhodopsin. Biochem Biophys Res Commun 533(3):262–267

20. Kim S-H, Chuon K, Cho S-G, Choi A, Meas S, Cho H-S, Jung K-H (2021) Color-tuning of natural variants of heliorhodopsin. Sci Rep 11(1):1–9
21. Karasuyama M, Inoue K, Nakamura R, Kandori H, Takeuchi I (2018) Understanding colour tuning rules and predicting absorption wavelengths of microbial rhodopsins by data-driven machine-learning approach. Sci Rep 8(1):15580
22. Harris A, Lazaratos M, Siemers M, Watt E, Hoang A, Tomida S, Schubert L, Saita M, Heberle J, Furutani Y, Kandori H, Bondar AN, Brown LS (2020) Mechanism of inward proton transport in an Antarctic microbial rhodopsin. J Phys Chem B 124(24):4851–4872
23. Kandori H, Shichida Y, Yoshizawa T (2001) Photoisomerization in rhodopsin. Biochemistry (Moscow) 66(11):1197–1209
24. Mai S, González L (2020) Molecular photochemistry: recent developments in theory. Angew Chem Int Ed 59(39):16832–16846
25. Luecke H, Schobert B, Lanyi JK, Spudich EN, Spudich JL (2001) Crystal structure of sensory rhodopsin: insights into color tuning and transducer interaction II at 2.4 Angstroms. Science 293(5534):1499–1503
26. Hoffmann M, Wanko M, Strodel P, König PH, Frauenheim T, Schulten K, Thiel W, Tajkhorshid E, Elstner M (2006) Color tuning in rhodopsins: the mechanism for the spectral shift between bacteriorhodopsin and sensory rhodopsin II. J Am Chem Soc 128(33):10808–10818
27. Wanko M, Hoffmann M, Frauenheim T, Elstner M (2006) Computational photochemistry of retinal proteins. J Comput Aided Mol Des 20(7–8):511–518
28. Fujimoto K, Hasegawa J-Y, Hayashi S, Kato S, Nakatsuji H (2005) Mechanism of color tuning in retinal protein: SAC-CI and QM/MM study. Chem Phys Lett 414(1–3):239–242
29. Fujimoto K, Hayashi S, Hasegawa JY, Nakatsuji H (2007) Theoretical studies on the color-tuning mechanism in retinal proteins. J Chem Theory Comput 3(2):605–618
30. Altun A, Yokoyama S, Morokuma K (2008) Mechanism of spectral tuning going from retinal in vacuo to bovine rhodopsin and its mutants: multireference ab initio quantum mechanics/molecular mechanics studies. J Phys Chem B 112(51):16883–16890
31. Altun A, Yokoyama S, Morokuma K (2008) Spectral tuning in visual pigments: an ONIOM(QM:MM) study on bovine rhodopsin and its mutants. J Phys Chem B 112(22):6814–6827
32. Kim SY, Waschuk SA, Brown LS, Jung KH (2008) Screening and characterization of proteorhodopsin color-tuning mutations in *Escherichia coli* with endogenous retinal synthesis. Biochim Biophys Acta Bioenerg 1777(6):504–513
33. Palczewska G, Vinberg F, Stremplewski P, Bircher MP, Salom D, Komar K, Zhang J, Cascella M, Wojtkowski M, Kefalov VJ et al (2014) Human infrared vision is triggered by two-photon chromophore isomerization. Proc Natl Acad Sci USA 111(50):5445–5454
34. Engqvist MKM, McIsaac RS, Dollinger P, Flytzanis NC, Abrams M, Schor S, Arnold FH (2015) Directed evolution of *Gloeobacter violaceus* rhodopsin spectral properties. J Mol Biol 427(1):205–220
35. Nakajima Y, Pedraza-González L, Barneschi L, Inoue K, Olivucci M, Kandori H (2021) Pro219 is an electrostatic color determinant in the light-driven sodium pump KR2. Commun Biol 4(1185):1–15
36. Birge RR, Murray LP, Pierce BM, Akita H, Balogh-Nair V, Findsen LA, Nakanishi K (1985) Two-photon spectroscopy of locked-11-cis-rhodopsin: evidence for a protonated schiff base in a neutral protein binding site. Proc Natl Acad Sci USA 82(12):4117–4121
37. Birge RR (1986) Two-photon spectroscopy of protein-bound chromophores. Acc Chem Res 19(5):138–146
38. Swartz TE, Szundi I, Spudich JL, Bogomolni RA (2000) New photointermediates in the two photon signaling pathway of sensory rhodopsin-i. Biochemistry 39(49):15101–15109
39. Ehrenberg D, Varma N, Deupi X, Koyanagi M, Terakita A, Schertler GF, Heberle J, Lesca E (2019) The two-photon reversible reaction of the bistable jumping spider rhodopsin-1. Biophys J 116(7):1248–1258
40. Gholami S, Pedraza-González L, Yang X, Granovsky AA, Ioffe IN, Olivucci M (2019) Multi-state multiconfiguration quantum chemical computation of the two-photon absorption spectra of bovine rhodopsin. J Phys Chem Lett 10(20):6293–6300
41. Deisseroth K (2011) Optogenetics. Nat Methods 8(1):26–29

42. Marín MdC, Agathangelou D, Orozco-Gonzalez Y, Valentini A, Kato Y, Abe-Yoshizumi R, Kandori H, Choi A, Jung KH, Haacke S, Olivucci M (2019) Fluorescence enhancement of a microbial rhodopsin via electronic reprogramming. J Am Chem Soc 141(1):262–271
43. Pedraza-González L, Marín MdC, De Vico L, Yang X, Olivucci M (2020) On the automatic construction of QM/mm models for biological photoreceptors: rhodopsins as model systems. QM/MM studies of light-responsive biological systems. Springer, Berlin, pp 1–75
44. Bouas-Laurent H, Dürr H (2001) Organic photochromism (iupac technical report). Pure Appl Chem 73(4):639–665
45. Mendes HF, Van Der Spuy J, Chapple JP, Cheetham ME (2005) Mechanisms of cell death in rhodopsin retinitis pigmentosa: implications for therapy. Trends Mol Med 11(4):177–185
46. Mendes HF, Cheetham ME (2008) Pharmacological manipulation of gain-of-function and dominant-negative mechanisms in rhodopsin retinitis pigmentosa. Hum Mol Genet 17(19):3043–3054
47. Athanasiou D, Aguila M, Bellingham J, Li W, McCulley C, Reeves PJ, Cheetham ME (2018) The molecular and cellular basis of rhodopsin retinitis pigmentosa reveals potential strategies for therapy. Prog Retin Eye Res 62:1–23
48. Skulachev VP, Bogachev A (1988) Membrane bioenergetics. Springer, Berlin
49. Klapoetke NC, Murata Y, Kim SS, Pulver SR, Birdsey-Benson A, Cho YK, Morimoto TK, Chuong AS, Carpenter EJ, Tian Z, Wang J, Xie Y, Yan Z, Zhang Y, Chow BY, Surek B, Melkonian M, Jayaraman V, Constantine-Paton M, Wong GKS, Boyden ES (2014) Independent optical excitation of distinct neural populations. Nat Methods 11(3):338–346
50. Bogomolni RA, Spudich JL (1987) The photochemical reactions of bacterial sensory rhodopsin-I. Flash photolysis study in the one microsecond to eight second time window. Biophys J 52(6):1071–1075
51. Béja O, Spudich EN, Spudich JL, Leclerc M, DeLong EF (2001) Proteorhodopsin phototrophy in the ocean. Nature 411(6839):786–789
52. Romei MG, Lin CY, Mathews II, Boxer SG (2020) Electrostatic control of photoisomerization pathways in proteins. Science 367(6473):76–79
53. Okada T, Fujiyoshi Y, Silow M, Navarro J, Landau EM, Shichida Y (2002) Functional role of internal water molecules in rhodopsin revealed by x-ray crystallography. Proc Natl Acad Sci USA 99(9):5982–5987
54. Teller DC, Okada T, Behnke CA, Palczewski K, Stenkamp RE (2001) Advances in determination of a high-resolution three-dimensional structure of rhodopsin, a model of G-protein-coupled receptors (GPCRs). Biochemistry 40(26):7761–7772
55. Andruniów T, Ferré N, Olivucci M (2004) Structure, initial excited-state relaxation, and energy storage of rhodopsin resolved at the multiconfigurational perturbation theory level. Proc Natl Acad Sci USA 101(52):17908–17913
56. Tomasello G, Gloria OG, Altoè P, Stenta M, Luis SA, Merchán M, Orlandi G, Bottoni A, Garavelli M (2009) Electrostatic control of the photoisomerization efficiency and optical properties in visual pigments: on the role of counterion quenching. J Am Chem Soc 131(14):5172–5186
57. Bravaya K, Bochenkova A, Granovsky A, Nemukhin A (2007) An opsin shift in rhodopsin: Retinal S0–S1 excitation in protein, in solution, and in the gas phase. J Am Chem Soc 129(43):13035–13042
58. Valsson O, Campomanes P, Tavernelli I, Rothlisberger U, Filippi C (2013) Rhodopsin absorption from first principles: bypassing common pitfalls. J Chem Theory Comput 9(5):2441–2454
59. Melaccio F, Marín MdC, Valentini A, Montisci F, Rinaldi S, Cherubini M, Yang X, Kato Y, Stenrup M, Orozco-Gonzalez Y, Ferré N, Luk HL, Kandori H, Olivucci M (2016) Toward automatic rhodopsin modeling as a tool for high-throughput computational photobiology. J Chem Theory Comput 12(12):6020–6034
60. Zhang L, Hermans J (1996) Hydrophilicity of cavities in proteins. Proteins J Chem Theory Comput Bioinf 24(4):433–438
61. Pronk S, Páll S, Schulz R, Larsson P, Bjelkmar P, Apostolov R, Shirts MR, Smith JC, Kasson PM, Van Der Spoel D, Hess B, Lindahl E (2013) GROMACS 4.5: a high-throughput and highly parallel open source molecular simulation toolkit. Bioinformatics 29(7):845–854
62. ...Aquilante F, Autschbach J, Carlson RK, Chibotaru LF, Delcey MG, De Vico L, Fdez Galván I, Ferré N, Frutos LM, Gagliardi L, Garavelli M, Giussani A, Hoyer CE, Li Manni G, Lischka H, Ma D, Malmqvist PÅ, Müller T, Nenov A, Olivucci M, Bondo Pedersen T, Peng D, Plasser F, Pritchard B, Reiher M, Rivalta I, Schapiro I, Segarra-Martí J, Stenrup M, Truhlar DG, Ungur L, Valentini A, Vancoillie S, Veryazov V, Vysotskiy VP, Weingart O, Zapata F, Lindh R (2016)

Molcas8: new capabilities for multiconfigurational quantum chemical calculations across the periodic table. J Comput Chem 37(5):506–541
63. Rackers JA, Wang Z, Lu C, Laury ML, Lagardére L, Schnieders MJ, Piquemal J-P, Ren P, Ponder JW (2018) Tinker 8: software tools for molecular design. J Chem Theory Comput 14(10):5273–5289
64. ...Aquilante F, Autschbach J, Baiardi A, Battaglia S, Borin VA, Chibotaru LF, Conti I, De Vico L, Delcey M, Fdez Galván I, Ferré N, Freitag L, Garavelli M, Gong X, Knecht S, Larsson E, Lindh R, Lundberg M, Malmqvist P-A, Nenov A, Norell J, Odelius M, Olivucci M, Pedersen T, Pedraza-González L, Phung Q, Pierloot K, Reiher M, Schapiro I, Segarra-Martí J, Segatta F, Seijo L, Sen S, Sergentu D-C, Stein C, Ungur L, Vacher M, Valentini A, Veryazov V (2020) Modern quantum chemistry with [Open] Molcas. J Chem Phys 152(21):214117
65. Melaccio F, Olivucci M, Lindh R, Ferré N (2011) Unique QM/MM potential energy surface exploration using microiterations. Int J Quantum Chem 111(13):3339–3346
66. Inoue K, Ito S, Kato Y, Nomura Y, Shibata M, Uchihashi T, Tsunoda SP, Kandori H (2016) A natural light-driven inward proton pump. Nat Commun 7(1):1–10
67. Bernstein FC, Koetzle TF, Williams GJB, Meyer EF Jr, Brice MD, Rodgers JR, Kennard O, Shimanouchi T, Tasumi M (1977) The protein data bank: a computer-based archival file for macromolecular structures. Eur J Biochem 80(2):319–324
68. Berman HM, Westbrook J, Feng Z, Gilliland G, Bhat TN, Weissig H, Shindyalov IN, Bourne PE (2000) The protein data bank. Nucleic Acids Res 28(1):235–242
69. Pedraza-González L, De Vico L, Marín MdC, Fanelli F, Olivucci M (2019) a-ARM: automatic rhodopsin modeling with chromophore cavity generation, ionization state selection, and external counterion placement. J Chem Theory Comput 15(5):3134–3152
70. Kato HE, Inoue K, Abe-Yoshizumi R, Kato Y, Ono H, Konno M, Hososhima S, Ishizuka T, Hoque MR, Kunitomo H, Ito J, Yoshizawa S, Yamashita K, Takemoto M, Nishizawa T, Taniguchi R, Kogure K, Maturana AD, Iino Y, Yawo H, Ishitani R, Hideki K, Nureki O (2015) Structural basis for Na(+) transport mechanism by a light-driven Na(+) pump. Nature 521(7550):48–53
71. Broser M, Spreen A, Konold PE, Peter E, Adam S, Borin V, Schapiro I, Seifert R, Kennis JT, Sierra YAB et al (2020) Neor, a near-infrared absorbing rhodopsin. Nat Commun 11(1):5682
72. Adam S, Wiebeler C, Schapiro I (2021) Structural factors determining the absorption spectrum of channelrhodopsins: a case study of the chimera c1c2. J Chem Theory Comput 17(10):6302–6313
73. Kaufmann JC, Krause BS, Adam S, Ritter E, Schapiro I, Hegemann P, Bartl FJ (2020) Modulation of light energy transfer from chromophore to protein in the channelrhodopsin reachr. Biophys J 119(3):705–716
74. Pedraza-González L, Marín MdC, Jorge AN, Ruck TD, Yang X, Valentini A, Olivucci M, De Vico L (2020) Web-ARM: a web-based interface for the automatic construction of QM/MM models of rhodopsins. J Chem Inf Model 60(3):1481–1493
75. Pieri E, Ledentu V, Sahlin M, Dehez F, Olivucci M, Ferré N (2019) CpHMD-then-QM/MM identification of the amino acids responsible for the anabaena sensory rhodopsin pH-dependent electronic absorption spectrum. J Chem Theory Comput 15(8):4535–4546
76. Olsson MHM, Søndergaard CR, Rostkowski M, Jensen JH (2011) PROPKA3: consistent treatment of internal and surface residues in empirical pKa predictions. J Chem Theory Comput 7(2):525–537
77. Krivov GG, Shapovalov MV, Dunbrack RL (2009) Improved prediction of protein side-chain conformations with SCWRL4. Proteins Struct Funct Bioinf 77(4):778–795
78. Webb B, Sali A (2016) Comparative protein structure modeling using MODELLER. Curr Protoc Bioinform 54(1):5–6
79. Ochoa R, Soler MA, Laio A, Cossio P (2018) Assessing the capability of in silico mutation protocols for predicting the finite temperature conformation of amino acids. Phys Chem Chem Phys 20(40):25901–25909
80. Ignatov A (2021) Statistical analysis of protein side-chain conformations. J Phys Conf Ser 1740:012013
81. Xiang Z, Honig B (2001) Extending the accuracy limits of prediction for side-chain conformations. J Mol Biol 311(2):421–430
82. Wilson C, Gregoret LM, Agard DA (1993) Modeling side-chain conformation for homologous proteins using an energy-based rotamer search. J Mol Biol 229(4):996–1006
83. Dunbrack RL Jr, Karplus M (1993) Backbone-dependent rotamer library for proteins application to side-chain prediction. J Mol Biol 230(2):543–574
84. Vasquez M (1996) Modeling side-chain conformation. Curr Opin Struct Biol 6(2):217–221
85. Kono H, Doi J (1996) A new method for side-chain conformation prediction using a hopfield network and reproduced rotamers. J Comput Chem 17(14):1667–1683

86. Canutescu AA, Shelenkov AA, Dunbrack RL Jr (2003) A graph-theory algorithm for rapid protein side-chain prediction. Protein Sci 12(9):2001–2014
87. Peterson LX, Kang X, Kihara D (2014) Assessment of protein side-chain conformation prediction methods in different residue environments. Proteins Struct Funct Bioinf 82(9):1971–1984
88. Nagata K, Randall A, Baldi P (2012) Sidepro: a novel machine learning approach for the fast and accurate prediction of side-chain conformations. Proteins Struct Funct Bioinf 80(1):142–153
89. Liang S, Zheng D, Zhang C, Standley DM (2011) Fast and accurate prediction of protein side-chain conformations. Bioinformatics 27(20):2913–2914
90. Dunbrack RL Jr (2002) Rotamer libraries in the 21st century. Curr Opin Struct Biol 12(4):431–440
91. Mroginski M-A, Adam S, Amoyal GS, Barnoy A, Bondar A-N, Borin VA, Church JR, Domratcheva T, Ensing B, Fanelli F et al (2021) Frontiers in multiscale modeling of photoreceptor proteins. Photochem Photobiol 97(2):243–269
92. Inoue K, Marín MdC, Tomida S, Nakamura R, Nakajima Y, Olivucci M, Kandori H (2019) Red-shifting mutation of light-driven sodium-pump rhodopsin. Nat Commun 10(1):1993
93. Shen M-Y, Sali A (2006) Statistical potential for assessment and prediction of protein structures. Protein Sci 15(11):2507–2524
94. Kandori H (2020) Retinal proteins: photochemistry and optogenetics. Bull Chem Soc Jpn 93(1):76–85
95. Dolinsky TJ, Nielsen JE, McCammon JA, Baker NA (2004) PDB2PQR: an automated pipeline for the setup of Poisson–Boltzmann electrostatics calculations. Nucleic Acids Res 32:66–667
96. Dolinsky TJ, Czodrowski P, Li H, Nielsen JE, Jensen JH, Klebe G, Baker NA (2007) PDB2PQR: expanding and upgrading automated preparation of biomolecular structures for molecular simulations. Nucleic Acids Res 35(suppl–1):522–525
97. Le Guilloux V, Schmidtke P, Tuffery P (2009) Fpocket: an open source platform for ligand pocket detection. Bioinformatics 10(1):168–179
98. O'Boyle NM, Banck M, James CA, Morley C, Vandermeersch T, Hutchison GR (2011) Open babel: an open chemical toolbox. J Cheminform 3:33
99. Michaud-Agrawal N, Denning EJ, Woolf TB, Beckstein O (2011) MDAnalysis: a toolkit for the analysis of molecular dynamics simulations. J Comput Chem 32(10):2319–2327
100. Gowers RJ, Linke M, Barnoud J, Reddy TJ, Melo MN, Seyler SL, Domanski J, Dotson DL, Buchoux S, Kenney IM, Beckstein O (2016) MDAnalysis: a Python package for the rapid analysis of molecular dynamics simulations. In: Sebastian B, Scott R (eds) Proceedings of the 15th Python in science conference, pp 98–105
101. Hunter JD (2007) Matplotlib: a 2d graphics environment. Comput Sci Eng 9(3):90–95
102. McKinney W (2010) Data structures for statistical computing in Python. In: van der Walt S, Millman J (eds) Proceedings of the 9th Python in science conference, pp 56–61
103. Harris CR, Millman KJ, van der Walt SJ, Gommers R, Virtanen P, Cournapeau D, Wieser E, Taylor J, Berg S, Smith NJ, Kern R, Picus M, Hoyer S, van Kerkwijk MH, Brett M, Haldane A, del Río JF, Wiebe M, Peterson P, Gérard-Marchant P, Sheppard K, Reddy T, Weckesser W, Abbasi H, Gohlke C, Oliphant TE (2020) Array programming with NumPy. Nature 585(7825):357–362
104. ...Virtanen P, Gommers R, Oliphant TE, Haberland M, Reddy T, Cournapeau D, Burovski E, Peterson P, Weckesser W, Bright J, van der Walt SJ, Brett M, Wilson J, Millman KJ, Mayorov N, Nelson ARJ, Jones E, Kern R, Larson E, Carey CJ, Polat İ, Feng Y, Moore EW, VanderPlas J, Laxalde D, Perktold J, Cimrman R, Henriksen I, Quintero EA, Harris CR, Archibald AM, Ribeiro AH, Pedregosa F, van Mulbregt P (2020) SciPy 1.0 contributors: SciPy 1.0: fundamental algorithms for scientific computing in Python. Nat Methods 17:261–272
105. O'boyle NM, Tenderholt AL, Langner KM (2008) cclib: a library for package-independent computational chemistry algorithms. J Comput Chem 29(5):839–845
106. Marín MdC, De Vico L, Dong SS, Gagliardi L, Truhlar DG, Olivucci M (2019) Assessment of MC-PDFT excitation energies for a set of QM/MM models of rhodopsins. J Chem Theory Comput 15(3):1915–1923
107. Loco D, Lagardère L, Caprasecca S, Lipparini F, Mennucci B, Piquemal J-P (2017) Hybrid qm/mm molecular dynamics with amoeba polarizable embedding. J Chem Theory Comput 13(9):4025–4033

Publisher's Note Springer Nature remains neutral with regard to jurisdictional claims in published maps and institutional affiliations.

Authors and Affiliations

Laura Pedraza-González[1,3] · Leonardo Barneschi[1] · Daniele Padula[1] · Luca De Vico[1] · Massimo Olivucci[1,2]

Leonardo Barneschi
leonardo.barneschi@student.unisi.it

Daniele Padula
daniele.padula@unisi.it

1 Dipartimento di Biotecnologie, Chimica e Farmacia, Università degli Studi di Siena, Via Aldo Moro 2, 53100 Siena, Italy

2 Department of Chemistry, Bowling Green State University, Bowling Green, OH 43403, USA

3 Present Address: Department of Chemistry and Industrial Chemistry, University of Pisa, Via Moruzzi 13, 56124 Pisa, Italy

Topics in Current Chemistry (2022) 380:8
https://doi.org/10.1007/s41061-021-00361-7

REVIEW

Coupled- and Independent-Trajectory Approaches Based on the Exact Factorization Using the PyUNIxMD Package

Tae In Kim[1] · Jong-Kwon Ha[1] · Seung Kyu Min[1]

Received: 27 August 2021 / Accepted: 11 November 2021 / Published online: 27 January 2022
© The Author(s), under exclusive licence to Springer Nature Switzerland AG 2022

Abstract
We present mixed quantum-classical approaches based on the exact factorization framework. The electron–nuclear correlation term in the exact factorization enables us to deal with quantum coherences by accounting for electronic and nuclear nonadiabatic couplings effectively within classical nuclei approximation. We compare coupled- and independent-trajectory approximations with each other to understand algorithms in description of the bifurcation of nuclear wave packets and the correct spatial distribution of electronic wave functions along with nuclear trajectories. Finally, we show numerical results for comparisons of coupled- and independent-trajectory approaches for the photoisomerization of a protonated Schiff base from excited state molecular dynamics (ESMD) simulations with the recently developed Python-based ESMD code, namely, the PyUNIxMD program.

Keywords Excited state molecular dynamics · Exact factorization · Chemistry software

Mathematics Subject Classification 81-08 · 81-10 · 81-04

1 Introduction

The excited state molecular dynamics (ESMD) simulation is a great tool to help understand photochemical reactions [1–4], polaritonic chemistry [5–7], solar cells [8–11], light-induced molecular machines [12–14], and photosynthesis [15–18]. As experimental techniques develop, there are many collaborative works between

Chapter 6 was originally published as Kim, T. I., Ha, J-K. & Min, S. K. Topics in Current Chemistry (2022) 380: 8. https://doi.org/10.1007/s41061-021-00361-7.

✉ Seung Kyu Min
skmin@unist.ac.kr

[1] Department of Chemistry, Ulsan National Institute of Science and Technology (UNIST), 50 UNIST-gil, Ulju-gun, Ulsan 44919, South Korea

computational approaches based on ESMD and femtosecond photochemical experiments [19–23]. The synergy of theoretical and experimental approaches sheds light on the fundamental understanding of light-matter interactions. From the computational chemistry point of view, many theoretical approaches have been developed so far to deal with ESMD [24–29]. When a molecule is promoted to an electronic excited state by light absorption, the molecule feels a different potential from the electronic ground state, which leads to different reaction pathways. Nuclear propagation after light absorption cannot usually be explained by a single adiabatic potential energy surface showing nuclear wave packet bifurcation or splitting. Moreover, corresponding electronic propagations show population transfers induced by so-called nonadiabatic transitions. Therefore, the correlation between electrons and nuclei becomes crucial in ESMD.

The most accurate way to describe correlated electron–nuclear dynamics is to treat the entire system quantum mechanically. Quantum dynamics simulation is possible only at smaller numbers of nuclear degrees of freedom. Also, we have the multiconfigurational time-dependent Hartree (MCTDH) method to simulate quantum mechanical nuclear propagation [24]. The MCTDH approach is extremely useful when we consider a model Hamiltonian based on normal mode coordinates such as linear vibronic coupling Hamiltonians. In this case, we consider a limited number of normal modes that are expected to be essential for chemical processes of interest. Many interesting phenomena such as exciton transfer, charge transfer, and vibronic couplings have been investigated with MCTDH [30–33]. In addition, the full multiple spawning approach provides exact quantum dynamics by solving time-dependent Schrödinger equations in terms of Born–Huang expansions [25, 34, 35].

For more realistic situations with larger degrees of freedom, mixed quantum-classical (MQC) and semiclassical approaches where we handle nuclear/electronic degrees of freedom with classical/quantum mechanics, respectively, are considered as more appropriate solutions. Among many MQC methods, Ehrenfest dynamics [26] and trajectory surface hopping (TSH) dynamics [27] are by far the most popular approaches due to their simplicity and efficiency. Many applications from molecular systems to extended systems have been investigated [36–40], and many improved versions of TSH have been developed [41–48]. One fundamental assumption of Ehrenfest dynamics and TSH is that classical trajectories are independent of each other. As a result, we encounter the so-called "overcoherence" problem. In Ehrenfest dynamics, the electronic state follows the time-dependent electronic Schrödinger equation, whereas classical nuclei propagate according to the Newtonian equation of motion. In a given nuclear trajectory, the electronic state becomes coherent as a result of nonadiabatic transitions with finite off-diagonal elements in the electronic density matrix. The incorrect description of electron–nuclear correlation provides an overestimation of electronic coherences as well as a wrong description of nuclear propagation. On the other hand, TSH describes nuclear propagations in a stochastic way. A nuclear trajectory chooses a single electronic state (usually an adiabatic state), which reproduces the bifurcation of nuclear trajectories after nonadiabatic transitions. However, the off-diagonal electronic density matrix elements are still finite after separation of the nuclear wave packet since the electronic equation of motion is equivalent to Ehrenfest dynamics. Thus, TSH cannot capture the correct electronic density matrix as a function of nuclear degrees of freedom, which results in the overestimation of nuclear coherence [49–53]. Within the framework of the Born–Huang expansion, many improved versions of TSH algorithms have been suggested to describe quantum coherence [43, 45]. However, we may encounter the fundamental difficulty of choosing a potential energy as we introduce a classical nuclear trajectory based on Born–Huang expansion. As a direct example, Ehrenfest dynamics chooses a mean-field potential from the coherent electronic state while TSH uses one of multiple adiabatic electronic states.

Recently, various MQC approaches based on exact factorization [54–58] have been proposed [28, 29]. Exact factorization can take into account the correct electron–nuclear correlation based on the formally exact coupled equations of motions for electronic and nuclear wave functions. One can describe nuclear propagation with a "single" time-dependent potential energy surface that can be uniquely determined by a suitable gauge choice [59]. Moreover, multiple classical trajectories along a single potential energy surface can generate the bifurcation of nuclear wave packets [60]. Based on systematic approximations for classical nuclei, one can develop a coupled-trajectory approach, the so-called coupled-trajectory mixed quantum-classical (CT-MQC) approach, to handle quantum mechanical distribution of nuclear probability [28, 61], and an independent-trajectory approach, the so-called decoherence-induced surface hopping based on exact factorization (DISH-XF) approach, with further approximations toward numerical stability and computational efficiency [29]. Many photochemical reactions, such as the ring opening reaction of oxirane [62], photoisomerization of molecules [63–65], and molecular rotary motors [66, 67], have been studied so far. Recently, the performance of DISH-XF was compared to ab initio multiple spawning [35] and other representative decoherence-corrected TSH methods on the so-called "molecular Tully model" [68]. However, due to the complexity, numerical instability, and high computational cost of CT-MQC approaches, there is no paper comparing CT-MQC and DISH-XF for realistic molecular systems. Recently, with the combination of machine learning (ML) approaches and quantum chemical calculations, one can develop ML models that can provide quantum mechanical energy levels with quantum chemical accuracy and low computational cost [69]. Therefore, ESMD based on CT-MQC is now feasible with ML approaches. In this paper, we present numerical comparisons between CT-MQC and DISH-XF for the first time.

A number of modern program packages for ESMD with various MQC methods, such as Newton-X [70], SHARC [71], Libra [72], PYXAID [73], and NEXMD [74], have been developed for various MQC approaches. The recently developed PyUNIxMD [75] is a user-friendly Python-based ESMD package that employs various MQC approaches, including DISH-XF and CT-MQC. In a previous work [75], we introduced the code structure and flow of the PyUNIxMD package with various MQC implementations. However, the CT-MQC and DISH-XF approaches have not been compared so far. In this paper, we focus on comparing the CT-MQC and DISH-XF approaches, with detailed descriptions based on working equations as implemented in PyUNIxMD. Then, we perform ESMD simulations with DISH-XF and CT-MQC to describe photoisomerization of the protonated Schiff base using the PyUNIxMD package. Numerical results from the two different approaches are discussed.

2 Exact Factorization

Since the exact factorization scheme has been explained and reviewed in many articles [55–58], in this paper we introduce the fundamental concepts and working equations only briefly. The starting point of exact factorization is to write the time-dependent molecular wave function $\Psi(\underline{\underline{\mathbf{r}}}, \underline{\underline{\mathbf{R}}}, t)$ as a single product of a nuclear wave function $\chi(\underline{\underline{\mathbf{R}}}, t)$ and an electronic wave function $\Phi_{\underline{\underline{\mathbf{R}}}}(\underline{\underline{\mathbf{r}}}, t)$ as $\Psi(\underline{\underline{\mathbf{r}}}, \underline{\underline{\mathbf{R}}}, t) = \chi(\underline{\underline{\mathbf{R}}}, t)\Phi_{\underline{\underline{\mathbf{R}}}}(\underline{\underline{\mathbf{r}}}, t)$ where $\underline{\underline{\mathbf{R}}}$ and $\underline{\underline{\mathbf{r}}}$ represent nuclear and electronic degrees of freedom, respectively. Here, we enforce that the electronic wave function $\Phi_{\underline{\underline{\mathbf{R}}}}(\underline{\underline{\mathbf{r}}}, t)$ satisfies a partial normalization condition such that $\int d\underline{\underline{\mathbf{r}}}|\Phi_{\underline{\underline{\mathbf{R}}}}(\underline{\underline{\mathbf{r}}}, t)|^2 = 1$ for any nuclear configuration $\underline{\underline{\mathbf{R}}}$. Therefore, the modulus square of a nuclear wave function $|\chi(\underline{\underline{\mathbf{R}}}, t)|^2$ has a physical meaning of the exact many-body nuclear density. Moreover, this many-body nuclear wave function provides the exact many-body current density. The exact factorization provides a unique pair of a nuclear wave function and a corresponding electronic wave function up to a gauge freedom $\Phi_{\underline{\underline{\mathbf{R}}}}(\underline{\underline{\mathbf{r}}}, t) \rightarrow \Phi_{\underline{\underline{\mathbf{R}}}}(\underline{\underline{\mathbf{r}}}, t)e^{iS(\underline{\underline{\mathbf{R}}},t)}$ where $S(\underline{\underline{\mathbf{R}}}, t)$ is a real-valued function. Rigorous mathematical derivations can be found elsewhere [55, 56]. From the Schrödinger equation for the molecular wave function, we can derive the coupled equations of motions for nuclear and electronic wave functions as, in atomic units,

$$i\frac{\partial}{\partial t}\chi\left(\underline{\underline{\mathbf{R}}}, t\right) = \left[\sum_{\nu} \frac{\left(-i\nabla_{\nu} + \mathbf{A}_{\nu}\left(\underline{\underline{\mathbf{R}}}, t\right)\right)^2}{2M_{\nu}} + \epsilon\left(\underline{\underline{\mathbf{R}}}, t\right)\right]\chi\left(\underline{\underline{\mathbf{R}}}, t\right), \tag{1}$$

and

$$i\frac{\partial}{\partial t}\Phi_{\underline{\underline{\mathbf{R}}}}\left(\underline{\underline{\mathbf{r}}}, t\right) = \left[\hat{H}_{\mathrm{BO}}\left(\underline{\underline{\mathbf{r}}}, t; \underline{\underline{\mathbf{R}}}\right) + \hat{U}_{en} - \epsilon\left(\underline{\underline{\mathbf{R}}}, t\right)\right]\Phi_{\underline{\underline{\mathbf{R}}}}\left(\underline{\underline{\mathbf{r}}}, t\right), \tag{2}$$

respectively, where $\mathbf{A}_{\nu} = \int d\underline{\underline{\mathbf{r}}}\Phi^*_{\underline{\underline{\mathbf{R}}}}(\underline{\underline{\mathbf{r}}}, t)\left\{-i\nabla_{\nu}\Phi_{\underline{\underline{\mathbf{R}}}}(\underline{\underline{\mathbf{r}}}, t)\right\}$ is the time-dependent vector potential from the electronic wave function, $\hat{H}_{\mathrm{BO}}$ is the Born–Oppenheimer (BO) Hamiltonian for electrons, and M_{ν} is the mass of the νth nucleus. Equation (1) tells us that the nuclear motion is governed by the scalar potential $\epsilon(\underline{\underline{\mathbf{R}}}, t)$ in the presence of the vector potential $\mathbf{A}_{\nu}(\underline{\underline{\mathbf{R}}}, t)$ generated by the fast electronic motion. In the BO limit, the vector potential corresponds to the Berry connection [76]. Here, we have two important terms in the coupled Eqs. (1) and (2); the electron–nuclear correlation operator $\hat{U}_{\mathrm{en}}$ and the exact potential energy surface $\epsilon(\underline{\underline{\mathbf{r}}}, t)$. From the full time-dependent Schrödinger equation, we can obtain the electron–nuclear correlation operator as

$$\hat{U}_{\text{en}} = \sum_{\nu} \frac{1}{M_{\nu}} \left[\frac{\left(-i\nabla_{\nu} - \mathbf{A}_{\nu}\left(\underline{\underline{\mathbf{R}}}, t\right)\right)^2}{2} + \left(\frac{-i\nabla_{\nu}\chi\left(\underline{\underline{\mathbf{R}}}, t\right)}{\chi\left(\underline{\underline{\mathbf{R}}}, t\right)} + \mathbf{A}_{\nu}\left(\underline{\underline{\mathbf{R}}}, t\right) \right) \cdot \left(-i\nabla_{\nu} - \mathbf{A}_{\nu}\left(\underline{\underline{\mathbf{R}}}, t\right)\right) \right], \tag{3}$$

and the exact (scalar) potential energy surface as

$$\epsilon\left(\underline{\underline{\mathbf{R}}}, t\right) = \left\langle \Phi_{\underline{\underline{\mathbf{R}}}} \middle| \hat{H}_{\text{BO}} + \hat{U}_{en} \middle| \Phi_{\underline{\underline{\mathbf{R}}}} \right\rangle_{\underline{\underline{\mathbf{r}}}} + \left\langle \Phi_{\underline{\underline{\mathbf{R}}}} \middle| -i\frac{\partial}{\partial t}\Phi_{\underline{\underline{\mathbf{R}}}} \right\rangle_{\underline{\underline{\mathbf{r}}}}, \tag{4}$$

where $\langle \cdots \rangle_{\underline{\underline{\mathbf{r}}}}$ represents the integration over the electronic degrees of freedom only. We note that the first term in Eq. (4) is gauge-invariant while the second term is gauge-dependent. One of the main features of the exact factorization is that we describe the nuclear dynamics based on the single time-dependent scalar potential energy surface and the single time-dependent vector potential while the conventional approaches describe the nuclear wave packet dynamics with multiple time-independent BO energies with multiple nonadiabatic time-independent coupling vectors and Berry connections. We can approximate the single potentials for MQC approaches straightforwardly. The electron–nuclear correlation term contains dependency on both the electronic wave function $\Phi_{\underline{\underline{\mathbf{R}}}}(\underline{\underline{\mathbf{r}}}, t)$ and the nuclear wave function $\chi(\underline{\underline{\mathbf{R}}}, t)$. Moreover, the second term in Eq. (3) has the nuclear wave function in the denominator, which may diverge as the nuclear wave function vanishes numerically. The direct time propagation with the coupled Eqs. (1) and (2) is very unstable [77], thus we need complicated algorithms to increase numerical stability.

According to the conventional Born–Huang expansion $\Psi(\underline{\underline{\mathbf{r}}}, \underline{\underline{\mathbf{R}}}, t) = \sum_n \chi(\underline{\underline{\mathbf{R}}}, t)\Phi_n(\underline{\underline{\mathbf{r}}}; \underline{\underline{\mathbf{R}}})$, we understand the nuclear dynamics in electronic excited states as the nuclear wave packet on each electronic state, namely the BO electronic state $\Phi_n(\underline{\underline{\mathbf{r}}}; \underline{\underline{\mathbf{R}}})$ by solving the time-independent Schrödinger equation for electrons $\hat{H}_{\text{BO}}(\underline{\underline{\mathbf{r}}}; \underline{\underline{\mathbf{R}}})\Phi_n(\underline{\underline{\mathbf{r}}}; \underline{\underline{\mathbf{R}}}) = \epsilon_n^{BO}(\underline{\underline{\mathbf{R}}})\Phi_n(\underline{\underline{\mathbf{r}}}; \underline{\underline{\mathbf{R}}})$ with the state index n follows the BO potential energies $\epsilon_n^{\text{BO}}(\underline{\underline{\mathbf{R}}})$ with nonadiabatic transitions that generate population transfers of nuclear wave packets on different electronic states. The major sources of nonadiabatic transitions are nonzero values of electronic nonadiabatic coupling matrix elements $\langle \Phi_i | \frac{\partial}{\partial t} \Phi_j \rangle_{\underline{\underline{\mathbf{r}}}}$ and nuclear nonadiabatic coupling $\chi_i^* \nabla_R \chi_j$.

The approximated mixed quantum-classical approaches such as surface hopping and Ehrenfest dynamics usually omit the nuclear nonadiabatic couplings. Thus, we cannot describe the correct nuclear wave packet splitting and coherences in electronic wave functions. In fact, in Ehrenfest dynamics, the classical nuclear trajectory propagates along an averaged potential energy surface with electronic propagation as well as electronic nonadiabatic coupling matrix elements, which shows overestimation of electronic coherences. Instead, TSH exploits stochastic hopping algorithms to account for nuclear nonadiabatic coupling. However, the incorrect account generates overestimation of nuclear coherences [51]. In the framework of exact factorization, the electron–nuclear correlation contains both electronic and nuclear nonadiabatic interactions due to terms containing spatial derivatives of electronic and nuclear wave functions. Moreover, the nuclear wave function $\chi(\underline{\underline{\mathbf{R}}}, t)$ is not a projected nuclear wave function in the Born–Huang expansion but the true nuclear wave function, which provides the correct many-body nuclear density and the many-body nuclear current density. Therefore, exact factorization claims that a suitable and systematic approximation for classical nuclei for the nuclear wave function can provide the correct equation of motions for mixed quantum-classical approaches. Indeed, it is shown that the classical trajectories governed by Newtonian equations from the exact potential energy surfaces show the correct nuclear wave packet splitting after a nonadiabatic coupling region [60].

3 Mixed Quantum-Classical Approaches Based on the Exact Factorization

The coupled equation of motions for electrons and nuclei can be derived by introducing the Lagrangian frame [61] while we solve the electronic equation along with the moving classical trajectory $\underline{\underline{\mathbf{R}}}(t)$. From the electron–nuclear correlation term, writing the nuclear wave function $\chi(\underline{\underline{\mathbf{R}}}, t)$ as $\chi(\underline{\underline{\mathbf{R}}}, t) = |\chi(\underline{\underline{\mathbf{R}}}, t)|e^{iS(\underline{\underline{\mathbf{R}}}, t)}$ provides the term with classical canonical momentum $\mathbf{P}_\nu = \mathbf{A}_\nu + \nabla_\nu S$ which naturally provides the total time derivative of the electronic wave function as

$$\frac{\mathrm{d}}{\mathrm{d}t}\Phi_{\underline{\underline{\mathbf{R}}}}\left(\underline{\underline{\mathbf{r}}}, t\right) = \frac{\partial}{\partial t}\Phi_{\underline{\underline{\mathbf{R}}}}\left(\underline{\underline{\mathbf{r}}}, t\right) + \sum_\nu \frac{\mathbf{P}_\nu}{M_\nu} \cdot \nabla_\nu \Phi_{\underline{\underline{\mathbf{R}}}}\left(\underline{\underline{\mathbf{r}}}, t\right). \tag{5}$$

Moreover, from the quantum mechanical Hamilton–Jacobi equation, we can obtain the classical equation of motion for nuclei as

$$\frac{\partial}{\partial t} S\left(\underline{\underline{\mathbf{R}}}, t\right) = -\sum_\nu \frac{\left[\nabla_\nu S\left(\underline{\underline{\mathbf{R}}}, t\right) + \mathbf{A}_\nu\left(\underline{\underline{\mathbf{R}}}, t\right)\right]^2}{2M_\nu} - \epsilon\left(\underline{\underline{\mathbf{R}}}, t\right) + \sum_\nu \frac{1}{2M_\nu} \frac{\nabla_\nu^2 \left|\chi\left(\underline{\underline{\mathbf{R}}}, t\right)\right|}{\left|\chi\left(\underline{\underline{\mathbf{R}}}, t\right)\right|} \tag{6}$$

which can be approximated as the classical Newtonian equation by omitting the quantum potential $\nabla_\nu |\chi|^2 / |\chi|^2$ as

$$\mathbf{F}_\nu = -\nabla_\nu \left(\epsilon\left(\underline{\underline{\mathbf{R}}}, t\right) + \sum_{\nu'} \mathbf{A}_{\nu'}\left(\underline{\underline{\mathbf{R}}}, t\right) \cdot \frac{\mathbf{P}_{\nu'}}{M_\nu} \right)\Bigg|_{\underline{\underline{\mathbf{R}}}(t)} + \dot{\mathbf{A}}_\nu(\underline{\underline{\mathbf{R}}}, t)|_{\underline{\underline{\mathbf{R}}}(t)}. \tag{7}$$

Therefore, within this approximation, the quantum tunneling effect cannot be described. Including the quantum potential during nuclear propagation to deal with nuclear quantum effects is another future research topic in the mixed quantum-classical dynamics based on exact factorization. Semiclassical analysis [60] of the time-dependent scalar and vector potentials in Eq. (7) suggest that the time-dependent scalar potential energy surface induces nonadiabatic transition and spatial separation of nuclear wave packets on different BO states, and that the time-dependent vector potential adjusts the nuclear momenta of nuclear densities on different BO states. While such effects of time-dependent potentials are encoded in the nuclear force in CT-MQC, a stochastic hop with a momentum adjustment from the potential energy difference replaces the time-dependent potentials approximately in DISH-XF. We note that the above coupled Eqs. (1) and (2) have gauge dependency whereas the observables such as the classical nuclear momentum and the nuclear force are gauge-invariant. Thus, any gauge choice generates the identical nuclear propagation as it should be. Once we choose the gauge as $\epsilon(\underline{\underline{\mathbf{R}}}, t) + \sum_{\nu'} \mathbf{A}(\underline{\underline{\mathbf{R}}}, t) \cdot \mathbf{P}_{\nu'}/M_{\nu} = 0$, we can derive the final coupled equations of motions for the electronic wave function $\Phi_{\underline{\underline{\mathbf{R}}}}(\underline{\underline{\mathbf{r}}}, t)$ and the nuclear trajectory $\underline{\underline{\mathbf{R}}}(t)$ as

$$\begin{aligned} i\frac{\mathrm{d}}{\mathrm{d}t}\Phi_{\underline{\underline{\mathbf{R}}}}\left(\underline{\underline{\mathbf{r}}}, t\right) =& \hat{H}_{\mathrm{BO}}\Phi_{\underline{\underline{\mathbf{R}}}}\left(\underline{\underline{\mathbf{r}}}, t\right) \\ &+ \sum_{\nu} \frac{1}{M_{\nu}} \frac{i\nabla_{\nu}\left|\chi\left(\underline{\underline{\mathbf{R}}}, t\right)\right|}{\left|\chi\left(\underline{\underline{\mathbf{R}}}, t\right)\right|} \cdot \left(\mathbf{A}_{\nu}\left(\underline{\underline{\mathbf{R}}}, t\right) + i\nabla_{\nu}\right)\Phi_{\underline{\underline{\mathbf{R}}}}\left(\underline{\underline{\mathbf{r}}}, t\right), \end{aligned} \tag{8}$$

and

$$\begin{aligned} \mathbf{F}_{\nu} =& \left\langle \Phi_{\underline{\underline{\mathbf{R}}}} \middle| -\nabla_{\nu}\hat{H}_{\mathrm{BO}} \middle| \Phi_{\underline{\underline{\mathbf{R}}}} \right\rangle_{\underline{\underline{\mathbf{r}}}} \\ &+ \sum_{\nu'} \frac{2}{M_{\nu'}} \frac{\nabla_{\nu'}\left|\chi\left(\underline{\underline{\mathbf{R}}}, t\right)\right|}{\left|\chi\left(\underline{\underline{\mathbf{R}}}, t\right)\right|} \left(\mathbf{A}_{\nu'}\left(\underline{\underline{\mathbf{R}}}, t\right)\mathbf{A}_{\nu}\left(\underline{\underline{\mathbf{R}}}, t\right) - \Re\left\langle \nabla_{\nu'}\Phi_{\underline{\underline{\mathbf{R}}}} | \nabla_{\nu}\Phi_{\underline{\underline{\mathbf{R}}}} \right\rangle_{\underline{\underline{\mathbf{r}}}}\right), \end{aligned} \tag{9}$$

respectively. Here, we should emphasize that we introduce another important approximation. Since the second derivatives in the first term of the electron–nuclear correlation operator $\hat{U}_{en}$ and the corresponding term in the exact potential energy surface $\epsilon(\underline{\underline{\mathbf{R}}}, t)$ require complicated numerical calculations, we use only the second term in the $\hat{U}_{\mathrm{en}}$. Since we have $\langle \Phi_{\underline{\underline{\mathbf{R}}}} | \hat{U}_{\mathrm{en}} | \Phi_{\underline{\underline{\mathbf{R}}}} \rangle_{\underline{\underline{\mathbf{r}}}} = 0$, there is no contribution from $\hat{U}_{\mathrm{en}}$ for the scalar potential. One can notice that the first terms in Eqs. (8) and (9) correspond to the equations of motions in Ehrenfest dynamics. The remaining terms come from the electron–nuclear correlation term $\hat{U}_{\mathrm{en}}$ which contains the nuclear quantum momentum $-\nabla_{\nu}|\chi(\underline{\underline{\mathbf{R}}}, t)|/|\chi(\underline{\underline{\mathbf{R}}}, t)|$. If we consider purely classical nuclei by neglecting quantum momentum, the coupled equations of motions collapse to the Eherenfest dynamics. In addition, one can note that the electronic Eq. (8) satisfies the norm conservation of the electronic wave function naturally.

3.1 Coupled-Trajectory Approach

We first introduce the CT-MQC approach to deal with nonadiabatic transitions as well as nuclear wave packet splitting. The major approximations of CT-MQC are as follows: (1) the nuclear density in Eq. (6) is constructed by multi-dimensional Gaussian functions centered on each classical trajectory, and (2) the spatial derivative of electronic wave function is approximated by the spatial derivatives of electronic phases based on BO approximations. During the propagation, the width of Gaussian varies in time by calculating the standard deviation of nuclear configurations. Therefore, the nuclear quantum momentum $\mathcal{P}^I_\nu$ for the Ith trajectory $\underline{\underline{\mathbf{R}}}^I(t)$, $\mathcal{P}^I_\nu = -\nabla_\nu|\chi(t)|/|\chi(t)|$ at $\underline{\underline{\mathbf{R}}}^I(t)$, can be written as

$$\mathcal{P}^I_\nu = \sum_J^{N_{\mathrm{traj}}} W^{IJ}_\nu(t)\left(\mathbf{R}^I_\nu(t) - \mathbf{R}^J_\nu(t)\right) \tag{10}$$

where

$$W^{IJ}_\nu = \frac{\prod_{\nu'} g_{\sigma^J_{\nu'}(t)}\left(\mathbf{R}^I_{\nu'}(t) - \mathbf{R}^J_{\nu'}(t)\right)}{2\sigma^{J^2}_\nu(t)\sum_K \prod_{\nu'} g_{\sigma^K_{\nu'}(t)}\left(\mathbf{R}^I_{\nu'}(t) - \mathbf{R}^K_{\nu'}(t)\right)}, \tag{11}$$

N_{traj} is the total number of nuclear trajectories. Here I, J, and K represent trajectory indices while ν and ν' represent nuclear indices. $g_{\sigma^J_\nu(t)}(\mathbf{R}_\nu - \mathbf{R}^J_\nu(t))$ is a normalized Gaussian function centered at $\mathbf{R}^J_\nu(t)$ with the variance $\sigma^I_\nu(t)$ where we calculate the variance from the nearby nuclear trajectories. Further, in the construction of nuclear density, we enforce that overall BO population exchanges are allowed only in the presence of electronic nonadiabatic couplings. We calculate the spatial derivative of the electronic wave function $\nabla_\nu\Phi_{\underline{\underline{\mathbf{R}}}}(\underline{\underline{\mathbf{r}}}, t)$ as $\nabla_\nu\Phi_{\underline{\underline{\mathbf{R}}}}(\underline{\underline{\mathbf{r}}}, t) = \sum_i \mathbf{f}_{i\nu} C_i(t)\Phi_i(\underline{\underline{\mathbf{r}}};\underline{\underline{\mathbf{R}}}(t))$ where we introduce Born–Huang expansion for the electronic wave function as $\Phi_{\underline{\underline{\mathbf{R}}}}(\underline{\underline{\mathbf{r}}}, t) = \sum_i C_i(t)\Phi_i(\underline{\underline{\mathbf{r}}};\underline{\underline{\mathbf{R}}}(t))$. Here, the term $\mathbf{f}_{i,\nu}$ is the spatial gradient of the phase of $C_i(t)$, which we further approximate as the time integration of the energy gradient of the ith BO potential energy E_i as $\int^t \mathrm{d}t - \nabla_\nu E_i(\underline{\underline{\mathbf{R}}}(t))$. Since the overall nuclear wave packet is constructed by all trajectories, the entire trajectories are coupled with each other.

As we focus on the ESMD for molecules, we exploit quantum chemical electronic structure calculations to obtain BO many-body electronic states $\Phi_i(\underline{\underline{\mathbf{r}}};\underline{\underline{\mathbf{R}}})$ and nonadiabatic couplings among them. Thus, the working equation of motions for electronic and nuclear propagation within the CT-MQC approach can be written as

$$\begin{aligned}\frac{\mathrm{d}}{\mathrm{d}t}C_i(t) = &- i\epsilon^{\mathrm{BO}}_i C_i(t) - \sum_j\sum_\nu \mathbf{d}_{ij,\nu}\cdot\dot{\mathbf{R}}_\nu(t)c_j(t)\\ &- \sum_\nu \frac{\mathcal{P}_\nu(t)}{M_\nu}\cdot\left[\sum_j \rho_{jj}\mathbf{f}_{j,\nu} - \mathbf{f}_{i,\nu}\right]c_i,\end{aligned} \tag{12}$$

and

$$\mathbf{F}_\nu = -\sum_i \rho_{ii}(t)\nabla_\nu \epsilon_i^{\mathrm{BO}} - \sum_{j,i} \rho_{ji}(t)\left(\epsilon_i^{\mathrm{BO}} - \epsilon_j^{\mathrm{BO}}\right)\mathbf{d}_{ji,\nu}$$
$$- \sum_j \rho_{jj}(t)\left(\sum_{\nu'} \frac{2}{M_{\nu'}}\mathcal{P}_{\nu'}(t)\cdot \mathbf{f}_{j,\nu'}(t)\right)\left[\sum_i \rho_{ii}(t)\mathbf{f}_{i,\nu}(t) - \mathbf{f}_{j,\nu}(t)\right], \tag{13}$$

respectively, where $\rho_{ij}(t) = C_i^*(t)C_j(t)$ and $\mathbf{d}_{ij,\nu}$ are nonadiabatic coupling vectors (NACVs) as $\mathbf{d}_{ij,\nu} = \langle \Phi_i|\nabla_\nu \Phi_j\rangle_{\underline{\underline{\mathbf{r}}}}$. Compared with conventional surface hopping dynamics or Ehrenfest dynamics, to evaluate electron–nuclear correlation term with nuclear quantum momentum, $\mathcal{P}_\nu$, we require additional steps only to calculate nuclear density and the phase term $\mathbf{f}_{i,\nu}$, which can be obtained from the quantum chemical calculation directly. Since the quantum chemical calculations are usually the most time-consuming step in excited state calculations, the computational efficiency of CT-MQC is fairly efficient compared with the surface hopping approach and Ehrenfest dynamics. The major benefit of CT-MQC or any algorithm based on the exact factorization framework is that the norm conservation of electronic wave function always holds. Other mixed quantum-classical approaches to deal with quantum coherence exploit ad hoc modification of electronic wave function to introduce decoherence of electronic reduced density matrix, thus inevitably we should introduce renormalization of electronic density matrix [47, 48].

3.2 Independent-Trajectory Approach

Since the previous CT-MQC approach propagates nuclear trajectories at the same time, CT-MQC may encounter several practical problems. For example, the excited state calculations for BO electronic states are not stable at certain nuclear configurations. In this situation, the overall coupled nuclear trajectories fail to provide suitable propagation. In addition, the self-consistent field loop for converged electronic density sometimes fails, which gives numerical instability. Therefore, the CT-MQC approach may be suitable when we have an analytic formula for potential energy surfaces such as model Hamiltonian or potential energy surfaces from ML [78–80].

In this sense, another approximated approach, the so-called decoherence-induced surface hopping based on the exact factorization (DISH-XF), has been proposed. While the CT-MQC approach exploits all nuclear trajectories to construct the nuclear density, the DISH-XF approach exploits auxiliary nuclear trajectories to construct the nuclear density for each nuclear propagation. When the nuclear trajectory starts to feel a finite electronic nonadiabatic coupling, we introduce an auxiliary trajectory on the electronic state where the nonzero BO population appears. The initial momentum of the auxiliary trajectory is computed by rescaling of the original momentum to conserve the total energy. Then, the auxiliary trajectory propagates with constant momentum within the nuclear time step. At each time, we calculate a new nuclear momentum from the total energy conservation. The auxiliary trajectory collapses as the BO population becomes numerically negligible. DISH-XF calculates the nuclear quantum momentum $\mathcal{P}_\nu$ as

$$\mathcal{P}_\nu = \frac{1}{2\sigma^2}\left(\mathbf{R}_\nu(t) - \sum_n^{\text{states}} \rho_{nn}(t)\mathbf{R}_{n,\nu}(t)\right), \tag{14}$$

where σ is the variance of a Gaussian centered on the auxiliary trajectory which is an input parameter. Each nuclear trajectory has its own auxiliary trajectory depending on different electronic and nuclear propagations. Therefore, DISH-XF is an independent-trajectory approach. The nuclear trajectory follows the BO potential energy surfaces as

$$\mathbf{F}_\nu = -\nabla_\nu E_r\left(\underline{\underline{\mathbf{R}}}\right), \tag{15}$$

where $E_r(\underline{\underline{\mathbf{R}}})$ is the BO potential energy surface of the running state r whereas the electronic wave function follows Eq. (12). Compared with the CT-MQC approach, DISH-XF may not provide the correct nuclear behavior near nonadiabatic coupling regions. In the vicinity of nonadiabatic coupling, the exact potential energy surface shows an averaged potential energy surface, i.e., Ehrenfest region, while the exact potential energy surface shows clear step-wise function connecting different BO potential energy surfaces [59]. However, the surface hopping-based algorithm cannot deal with the mean-field potential. In this sense, the Ehrenfest approach coupled with the exact factorization framework may be another interesting topic to develop a new mixed quantum-classical algorithm.

Both CT-MQC and DISH-XF approaches exploit an ensemble of nuclear trajectory to mimic quantum mechanical nuclear wave packets along multiple BO surfaces. Usually we use Wigner samples or Boltzmann samples to construct initial distributions in phase space. Table 1 provides comparisons between the CT-MQC and DISH-XF approaches. Compared with CT-MQC, which uses time-dependent variance from nuclear trajectory, which runs simultaneously, DISH-XF usually uses a constant variance for constructing nuclear density. Therefore, the quality of simulations may depend on the shape of the potential energy surfaces. However, the overall behavior of BO populations and quantum coherence are similar to quantum mechanical results [29]. A semi-classical approximation with time-dependent variance may be applied in future study. In addition, CT-MQC requires the full information for the electronic structure calculations while

Table 1 Comparisons between the coupled-trajectory mixed quantum-classical (CT-MQC) and decoherence-induced surface hopping based on exact factorization (DISH-XF) approaches. i represents the index for electronic states while r represents the index for the running state

	CT-MQC	DISH-XF
Electronic equation	Equation (8)	Equation (8)
Nuclear equation	Equation (9)	Equation (15)
Quantum momentum	Equation (10)	Equation (14)
Gaussian width	Time-dependent	Time-independent
Requirements	E_i's, $-\nabla_\nu E_i$'s, $\langle\Phi_i\vert - \nabla_\nu \Phi_j\rangle_{\underline{\underline{\mathbf{r}}}}$'s	E_i's, $-\nabla_\nu E_r$, $\langle\Phi_i\vert - \nabla_\nu \Phi_r\rangle$'s

DISH-XF requires the gradients and nonadiabatic coupling matrix elements only for the running state.

3.3 Available Programs

Currently, available codes for mixed quantum-classical approaches are based on exact factorization are CPMD [81], G-CTMQC [82], and PyUNIxMD [75]. The CPMD program package provides various molecular dynamics algorithms based on first-principle electronic structure calculations with plane-wave basis functions. CPMD is the first program to provide a CT-MQC approach for realistic molecular systems [62]. We calculate BO electronic states and nonadiabatic couplings based on linear-response time-dependent density functional theory (LR-TDDFT). Since LR-TDDFT has limitations on calculating electronic excited states in several situations, caution is needed for the simulation. Within the CPMD program, the molecular dynamics module and the electronic structure calculations are connected internally. Thus, CT-MQC simulations can be done efficiently if the electronic structure calculations are stable for running trajectories. In addition, G-CTMQC contains many variations of coupled-trajectory approaches, such as CT-MQC, CT-eMQC, and generalized CT-MQC, including spin-orbit interactions [83–85], and we have many benchmarks with various model Hamiltonians. Since both CPMD and G-CTMQC programs are written in Fortran language, we have difficulty in implementation in other programs. PyUNIxMD program [75] is a Python-based ESMD package designed to simulate various mixed quantum-classical dynamics coupled to conventional ab initio and semiempirical quantum chemistry programs. PyUNIxMD contains the conventional Ehrenfest and surface hopping dynamics with various decoherence schemes and velocity-rescaling algorithms in addition to CT-MQC and DISH-XF. In addition, PyUNIxMD is actively under development by the authors. Fundamentally, PyUNIxMD is written in Python. Numerous scientific libraries and environments are optimized for Python, which is the most popular programming language due to its simplicity and user-friendly data structure. Since Python is suitable for object-oriented programming, modularization can be done directly. Furthermore, automation of multiple ESMD simulations can be achieved easily by simple scripts with PyUNIxMD. Such automation is important in terms of congruency, reproducibility, and user-independency as addressed in molecular model construction studies [86, 87].

4 The PyUNIxMD Package

The main theme of PyUNIxMD is to provide a universal tool for ESMD simulations with multiple algorithms that can be coupled easily to modern quantum chemistry programs. Figure 1 shows the general flow and functionalities of the PyUNIxMD program via object-oriented programming.

The PyUNIxMD package requires three essential functions to perform molecular dynamics simulations. First, *Molecule* object has the fundamental molecular

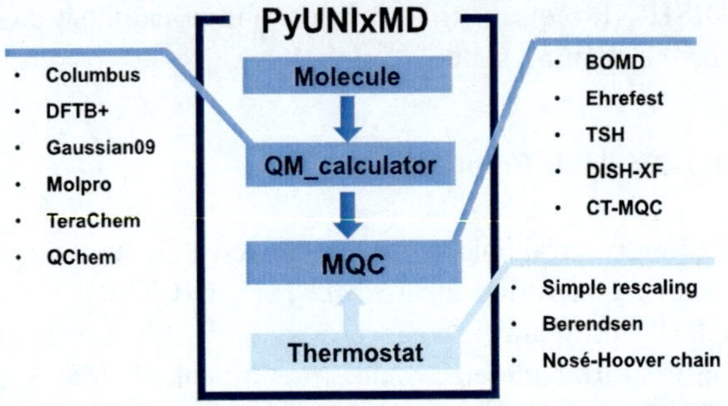

Fig. 1 General flow and functionality of PyUNIxMD. The *Molecule*, *QM_calculator*, and *MQC* functions are essential to run molecular dynamic (MD) simulations while the *Thermostat* object is optional for the constant-temperature, constant-volume ensemble (NVT) ensemble

information such as the number of atoms, the number of electronic states, atomic positions, and atomic velocities. During the molecular dynamics simulation, the *Molecule* object is updated. Then, we construct a *QM_calculator* object to define the quantum chemical BO calculation method with available quantum chemistry program packages. The current version of PyUNIxMD is interfaced with Columbus [88], DFTB+ [89], Gaussian09 [90], Molpro [91], TeraChem [92], and Q-Chem [93]. Of course, one can easily make an interface with any other quantum chemistry/physics programs by writing a simple Python code, which will be explained later. Finally, we propagate a trajectory via an *MQC* object. The *MQC* object defines the mixed quantum classical approach for the electron–nuclear propagation. The available mixed quantum classical approaches are BOMD, Ehrenfest, TSH with various decoherence schemes, DISH-XF, and CT-MQC. Compared with the procedure of independent-trajectory approaches, the CT-MQC requires a list of *Molecule* objects at the initial stage to perform the dynamics simultaneously. The remaining steps are identical to independent-trajectory approaches. In addition, the PyUNIxMD supports the numerical evaluation of nonadiabatic coupling matrix elements (NACMEs) [94] in case the direct calculation of NACVs is unavailable. Moreover, the PyUNIxMD provides thermostats with simple velocity rescaling, Berendsen thermostat [95], and Nosé–Hoover chain thermostat [96].

Figure 2 shows an example script to run PyUNIxMD. The steps for independent-trajectory simulations is summarized as in Fig. 2a: (1) Define *mol* as a *Molecule* object to include molecular information for a target trajectory, (2) define *bo* as a *QM_calculator* object to determine a quantum chemical method and a quantum chemical program for the on-the-fly calculation, (3) assign *md* to an *MQC* object which involves the information for the mixed quantum-classical dynamics simulations and the pre-defined *mol*, and (4) perform MD simulations via *run* method with the pre-defined *bo*. During the ESMD simulations, the nuclear trajectories are propagated with updating atomic positions, electronic energies, and BO electronic density matrix elements. Instead, the CT-MQC requires to create a *Molecule* object with multiple nuclear trajectories. Figure 2b represents a simple code to define a list of *Molecule* objects. We initially define an empty list *mols* and the total number of coupled trajectories *ntrajs*. Then, we construct *mols* by appending a *Molecule* object for each trajectory.

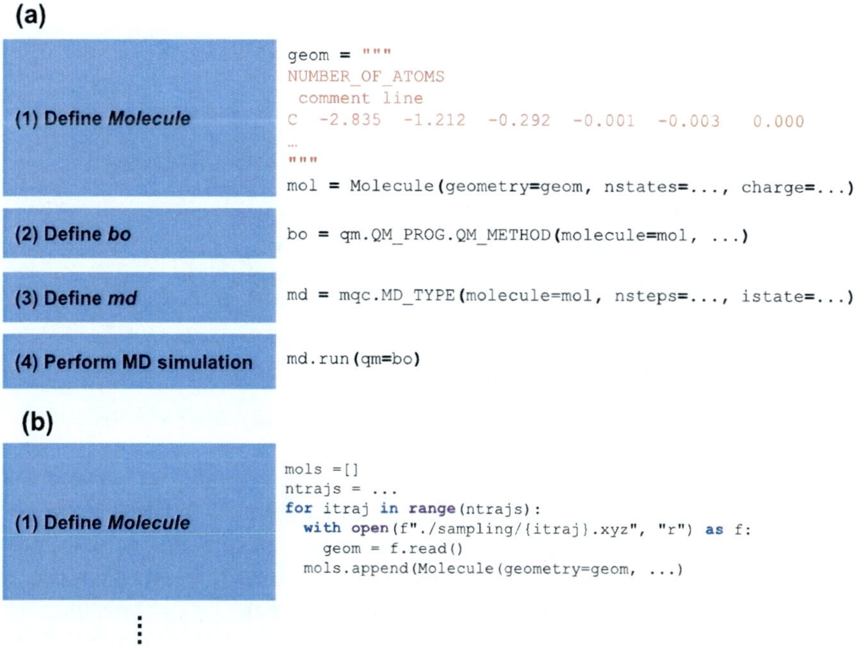

Fig. 2 **a** An example script to run PyUNIxMD and **b** a part of script for the coupled-trajectory mixed quantum-classical (CT-MQC) approach

4.1 Interfacing with Quantum Chemistry Programs

Since the PyUNIxMD package is designed to easily interface with any quantum chemistry electronic structure calculation programs as well as in-house model Hamiltonian solvers and ML models, we present steps to construct the interface between the molecular dynamics simulations and the quantum chemistry calculations. During the molecular dynamics simulation, we pass electronic energies, nuclear forces, and NACVs from the quantum chemistry calculation object to the dynamics object. In practice, PyUNIxMD uses the *get_data* method from *QM_calculator* to extract the information above in the *run* method. Hence, one can build an interface with any additional quantum chemistry software by adding a class with customized *get_data* method. Figure 3 shows a structure of class for the interface with an additional quantum chemistry program outputs, namely the *Additional_QC* class. Thus, the *Additional_QC* can replace the *QM_PROG* in Fig. 2a. First, the *Additional_QC* should inherit from *QM_calculator* class since the *QM_calculator* should know all information of the quantum chemistry calculation. In the *Additional_QC* class, we just need the *get_data* method, which calculates or obtains the electronic structure at a given nuclear configuration via *molecule*. In the example, we further define *get_input*, *run_QM*, and *extract_QM* methods to specify steps for quantum chemistry calculations. Of course, one can skip this part to get electronic structure directly with *molecule*. The *get_input* is to build an input file for quantum chemical calculations. At this stage, one can define global parameters such as basis functions,

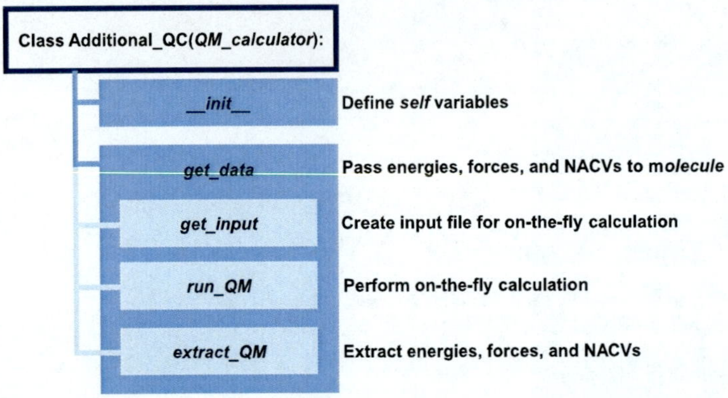

Fig. 3 A structure of additional class to the interface code with quantum chemistry program packages

active spaces, and the level of theory in the *self* variable. The *run_QM* is to execute a quantum chemistry program, which may be a running script for external programs. The *extract_QM* method extracts the necessary information such as electronic energies, gradients, and nonadiabatic couplings from the output of the quantum chemistry program. One can customize the code depending on the output structure. The extracted information is stored in the *molecule* object. In case of using the Python-based quantum chemistry program such as PySCF [97], we can use the provided modules directly in the *Additional_QC* class by including the suitable libraries or modules.

4.2 Procedures for ESMD Simulations with PyUNIxMD

A typical ESMD study involves three major stages (Fig. 4). First, one prepares initial conditions for nuclear configurations and velocities on a certain electronic state. Here, we determine the initial electronic state depending on the excitation source. As an example, we usually construct the initial nuclear configurations in thermal equilibrium from the ground state, and we choose a specific excited state to simulate the Franck-Condon excitation. Second, we perform the ESMD simulations from

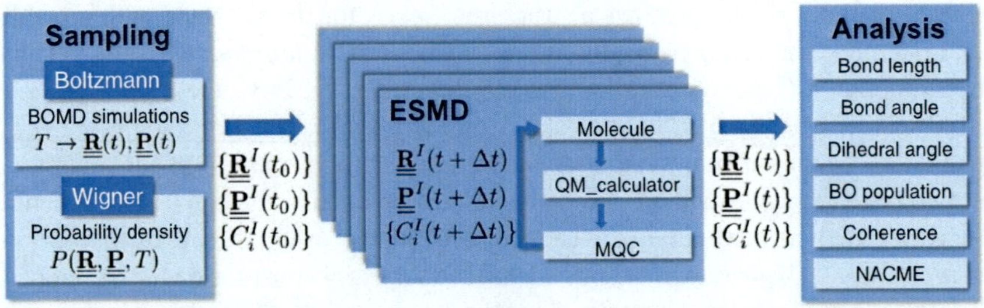

Fig. 4 A schematic flow for a complete excited state molecular dynamics (ESMD) simulation procedure using PyUNIxMD

the prepared initial conditions. Finally, after the simulations, we analyze observables from the trajectories via post-processing. The overall process can be easily done with PyUNIxMD since the program is well modularized by Python, the most popular modern programming language.

We sample the initial conditions from the Boltzmann or Wigner sampling [98]. The former suppose that the target system is in classical thermal equilibrium with the surrounding environment while the latter consider that the molecule populates the vibrational states usually obtained from the harmonic approximation. In the Boltzmann sampling, single or multiple nuclear trajectories are equilibrated to a target temperature/pressure with BO molecular dynamics using a thermostat/barostat, then the trajectories are propagated further for a long time so that the trajectories can traverse across the most of the accessible phase space at the equilibrium state [98]. We sample the nuclear configurations and velocities randomly or periodically from the trajectories after equilibrium. We can define the initial electronic state in two different ways. The simplest electronic state initialization method is to populate the electronic states for all trajectories to a single excited state. Alternatively, the electronic state can be determined by transition probabilities calculated from energy gap and oscillator strength for a given initial condition [99]. In Wigner sampling, the initial conditions are sampled from the Wigner distribution which maps the quantum mechanical probability densities for nuclear positions and momenta on the classical phase space [98, 100, 101]. In practice, we obtain the normal modes for nuclear degrees of freedom by the frequency calculations at a ground state minimum. Then, we calculate the probability density from the vibrational states of the harmonic oscillator, for example, zero-point vibrational state, i.e., Gaussian function. Finally, we can sample the initial conditions according to the probability density in the phase space. Boltzmann sampling with NVT ensemble is available via BO molecular dynamics with a thermostat in the current version of PyUNIxMD. A script for Wigner sampling is currently under development.

The ESMD simulations for sampled initial conditions can be performed via a simple Python script with PyUNIxMD as described in Fig. 1. In PyUNIxMD, the nuclei and electronic state are propagated by velocity-Verlet and 4th order Runge-Kutta algorithms, respectively, by default. Since electrons are significantly faster than nuclei, the electronic state is more sensitive to the time interval than the nuclear motion. Thus, we usually use a smaller time step for electronic state evolution. During electronic propagation, we require electronic structure calculations that are accessible at the nuclear time step. Thus, we exploit the linear interpolation scheme to approximate electronic energies and nonadiabatic couplings between the two nuclear time steps. The nuclear and electronic variables as a function of a time are printed as a standard output while the important variables are written in various output files for future analysis. Dynamics with multiple trajectories can be executed readily by a simple loop in Python without any external script or compilation of codes. For example, users can write a script that reads initial conditions from the movie file generated during the BOMD simulations for Boltzmann sampling and run dynamics for multiple initial conditions by adding a loop block for the trajectories.

After the simulations, the evolution of molecular properties are analyzed on the individual trajectory level or in an averaged way. Assuming that the total nuclear density can be represented as a sum of delta functions centered at nuclear configuration in time, the expectation values are expressed as a averaged value over the ensemble of trajectories [61]. The PyUNIxMD package provides scripts for statistical analysis of the electronic states and nuclear motion including the scripts for the statistical analysis of BO populations, coherences, the absolute values of NACMEs, and nuclear trajectories. The analysis of nuclear trajectory extracts the behaviors of specified bond lengths, bond angles, and dihedral angles from output trajectories. In addition, along with the default outputs from PyUNIxMD, one can store any arbitrary quantity on-the-fly to calculate additional physical properties further such as transient absorption spectra [102] and time-resolved photoelectron spectra [103].

5 Numerical Results

We perform ESMD for photoisomerization of *trans*-penta-2,4-dieniminium cation (PSB3) with the CT-MQC and DISH-XF, as implemented in PyUNIxMD, based on ML models [69].

The molecular configurations of PSB3 for stationary points are optimized with our ML models (Fig. 5). The structures and energies show good agreement with structures and energies, which are identified in [65] based on the state-interaction state-averaged spin-restricted ensemble-referenced Kohn–Sham method with an active space including two electrons in two fractionally occupied orbitals [SSR(2,2)] using ωPBEh functional and 6-31G* basis set (Table 2). $S_{0\text{min,trans}}$ and $S_{0\text{min,cis}}$ indicate equilibrium structure of trans-PSB3 and cis-PSB3 in the ground state (S_0). In the $S_{0\text{min,trans}}$, the bond lengths of the terminal C=C bond, the central C=C bond, and the terminal C=N bond are 1.346 Å, 1.369 Å, and 1.305 Å, respectively, and in

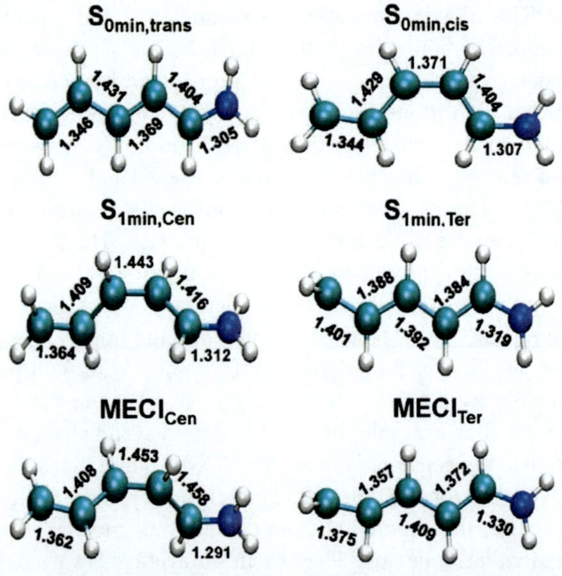

Fig. 5 Molecular configurations of PSB3 for the ground state minima ($S_{0\text{min,trans}}$, $S_{0\text{min,cis}}$), the excited state minima ($S_{1\text{min,Cen}}$, $S_{1\text{min,Ter}}$), and conical intersections (MECI$_{\text{Cen}}$, MECI$_{\text{Ter}}$). Bond lengths are in black text

Table 2 Relative energies (kcal mol^{-1}) for important stationary points of PSB3 optimized by machine learning (ML) models

ΔE	$S_{0\text{min,trans}}$	$S_{0\text{min,cis}}$	$S_{1\text{min,Cen}}$	$S_{1\text{min,Ter}}$	MECI_{Cen}	MECI_{Ter}
S_0	0	2.767	57.114	64.788	66.259	77.529
S_1	95.356	96.467	63.989	74.153		

the $S_{0\text{min,cis}}$, the bond lengths of the terminal C=C bond, the central C=C bond, and the terminal C=N bond correspond to 1.344 Å, 1.371 Å, and 1.307 Å, respectively. Bond length alternation (BLA) of the $S_{0\text{min,trans}}$ and the $S_{0\text{min,cis}}$ is 0.078 and 0.076 Å, respectively. The energy of the $S_{0\text{min,trans}}$ is lower than that of the $S_{0\text{min,cis}}$ as 2.77 kcal mol^{-1}. $S_{1\text{min,Cen}}$ and $S_{1\text{min,Ter}}$ are the structure of the local minima in the lowest excited state (S_1) with torsion and elongation of C=C double bonds, respectively. For the $S_{1\text{min,Cen}}$, the length of the central C=C bond elongates to 1.443 Å with torsion as 86.42° and BLA is 0.039 Å. The $S_{1\text{min,Ter}}$ shows the length of the terminal C=C bond is 1.401 Å with torsion as 69.22° and BLA is 0.015 Å. The $S_{1\text{min,Cen}}$ is more stable than the $S_{1\text{min,Ter}}$ as 10.16 kcal mol^{-1}. The two S_0/S_1 minimum energy conical intersections (MECIs) are labeled MECI_{Cen}, MECI_{Ter}. These correspond to configurations that are twisted around the central C=C bond as 80.06°, and the terminal C=C bond as 90.79 °, respectively. BLAs of the MECI_{Cen}, and the MECI_{Ter} are 0.064 Å, and −0.007 Å, respectively. The MECI_{Cen} is the lower energy structure than the MECI_{Ter} by 11.27 kcal mol^{-1}. The energies of two MECIs are lower than the energy of S_1 in the Franck–Condon region, leading to the fact that these two channels can be accessible during photoisomerization. From Wigner distribution at 300 K around $\text{S}_{0\text{min,trans}}$, we sample total 100 distinct initial molecular configurations and momenta for DISH-XF and CT-MQC simulations. After the Franck-Condon excitation to S_1, we propagate classical nuclear trajectories up to 300 fs with a time interval of 0.24 fs within the NVE ensemble.

Figure 6 shows population dynamics and averaged coherence from DISH-XF and CT-MQC. The averaged populations are computed from BO populations (P_i) or running states (p_i) from $P_i(t) = \sum_I^{N_{\text{traj}}} \rho_{ii}^{(I)}(t)/N_{\text{traj}}$ and $p_i(t) = N_i(t)/N_{\text{traj}}$, respectively, where $\rho_{ii}^{(I)}(t)$ is the BO population at the ith electronic state for the Ith nuclear trajectory, $N_i(t)$ is the number of trajectories running on the ith electronic state in time t for DISH-XF, and N_{traj} is the total number of trajectories. The coherence indicator, $\langle|\rho_{ij}|^2\rangle$, is calculated as $\langle|\rho_{ij}|^2\rangle = \sum_I \rho_{ii}^{(I)}(t)\rho_{jj}^{(I)}(t)/N_{\text{traj}}$. The overall behaviors of the averaged population from DISH-XF and CT-MQC are identical to each other. The population transfer from S_1 to S_0 is not initiated immediately. The trajectories take time to approach coupling region, with torsion and elongation of C=C bonds. This latency time is around 20 fs in both DISH-XF and CT-MQC, implying that the change of population occurs after 20 fs. After 20 fs, the population of S_1 decays exponentially with a time constant of around 90 fs, which is in a good agreement with the S_1 lifetimes of around 110 fs obtained from previous ESMD simulations [104, 105]. In the same manner, coherence patterns reveal around 20 fs. Additionally, the coherence patterns of DISH-XF and CT-MQC are similar while CT-MQC

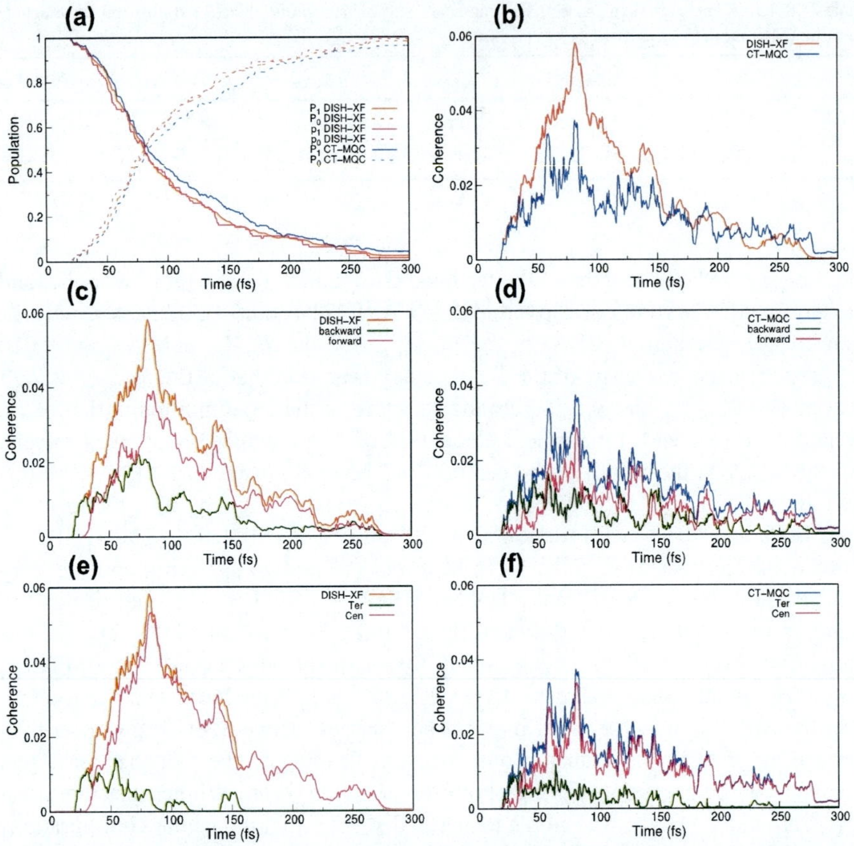

Fig. 6 **a** Born–Oppenheimer (BO) populations and **b** coherence indicators as a function of time with respect to different approaches. **c**, **d** Coherence indicators for trans to cis (forward) and trans to trans (backward). **e**, **f** Coherences with respect to different internal conversion channels via $MECI_{Cen}$ and $MECI_{Ter}$

shows smaller coherences indicating faster decoherences for an individual trajectory. The coherence becomes large up to around 90 fs. Then, after leaving coupling region, the coherence value goes to nearly zero. This fact means that both DISH-XF and CT-MQC describe the electron–nuclear correlation correctly. Additionally, the internal consistency for DISH-XF supports the fact that DISH-XF can explain the correlation between nuclei and electrons.

We further analyze the coherence indicators with different reaction pathways. With respect to direction of isomerization, the fast (< 25 fs) component corresponds to backward-going trajectories, which return trans-PSB3, i.e., trans-PSB3 to trans-PSB3 and the slow (> 25 fs) component indicates forward-going trajectories, which form cis-PSB3, i.e., trans-PSB3 to cis-PSB3. The coherence shows different patterns along groups. From DISH-XF simulations, the coherence begins at 20.4 fs for the backward-going trajectories, and reaches the maximum at 74.88 fs. While coherence begins at 29.52 fs and reaches maximum coherence of 87.12 fs for the forward-going trajectories. The amount of coherence is dominant for the forward-going trajectories. Also, the fast and slow components correspond to trajectories passing $MECI_{Ter}$ and $MECI_{Cen}$, respectively. The trajectories with internal conversions from S_1 to S_0 through $MECI_{Ter}$ show the early coherence from 20.4 fs. With maximum at 52.24 fs, whereas the trajectories via $MECI_{Cen}$ show the slower and broader coherences after 29.28 fs. For CT-MQC, we notice that the coherence strength of CT-MQC is smaller than that of DISH-XF. One of the major differences between CT-MQC and DISH-XF is the treatment of variances in quantum momenta. The CT-MQC exploits the time-dependent variance from multiple coupled nuclear trajectories at a given time *t* while the DISH-XF uses a constant value, which is usually determined from the initial distribution. In the PSB3 simulation, we observe that the quantum momentum in CT-MQC is larger than in DISH-XF, resulting in weak coherence in CT-MQC compared with DISH-XF. The overall behavior of coherence is similar to DISH-XF, where the second coherence appears slightly earlier than that of DISH-XF.

We also perform structural analysis by calculating the averaged torsional angle of the central C=C bond, the BLA, and the length of the terminal C=N bond for DISH-XF and CT-MQC (Fig. 7). The BLA decreases quickly and becomes the minimum at 13.68 fs, indicating that the trajectories approach nonadiabatic coupling region with elongation and torsion of the central C=C bond. After 13.68 fs, the averaged BLA increases with oscillation by forming trans-PSB3 or cis-PSB3. The small amplitude of change of the terminal C=N bond means that the channel via MECI by the torsion around the terminal C=N bond ($MECI_{CN}$) is not activated because $MECI_{CN}$

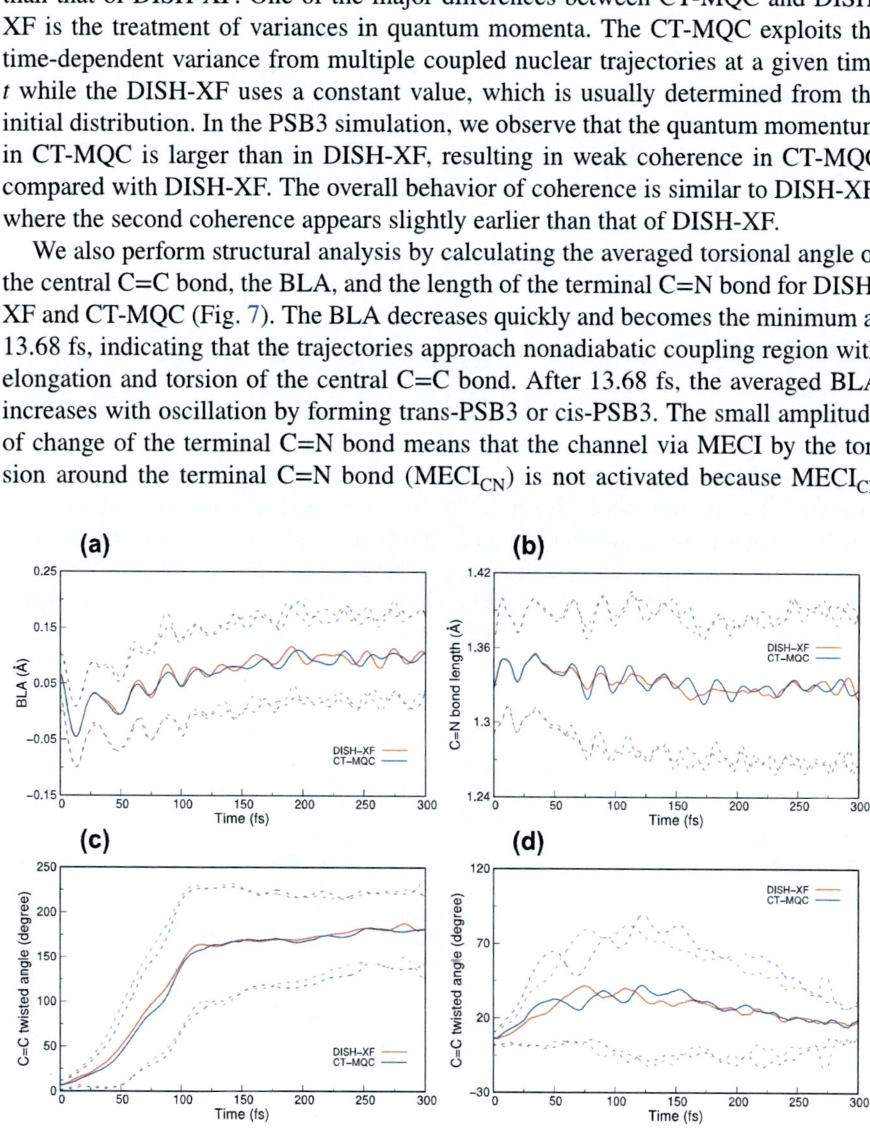

Fig. 7 Change in **a** averaged bond length alternation (BLA) and **b** terminal C=N bond lengths as a function of time. **c, d** Change in averaged twist angle of the central C=C bond for forward-going and backward-going trajectories, respectively. *Dotted lines* Standard deviations

has higher energy than S_1 state after Franck–Condon excitation. Compared with the result based on SSR-ωPBEh, the dynamics based on multi-state second-order perturbation theory (MSPT2) shows the activation of $\mathrm{MECI_{CN}}$ channel [65, 106]. The change in torsional motion along the central C=C bond for forward-going and backward-going trajectories implies the trans-to-cis isomerization and the trans-to-trans internal conversion via $\mathrm{MECI_{Cen}}$. The overall behaviors of the amplitude of motion with their standard deviation represent that both DISH-XF and CT-MQC show a similar swarm of trajectories, meaning that the nuclear wave packet are described similarly.

Figure 8 shows quantum yields and photoisomerization pathways. Both DISH-XF and CT-MQC show the majority of trajectories (~ 80%) chooses the internal conversion via $\mathrm{MECI_{Cen}}$ for trans-to-trans (~ 20%) and trans-to-cis (~ 60%) pathways while the minority (~ 20%) passes $\mathrm{MECI_{Ter}}$, which activates only the trans-to-trans pathway. As a result, CT-MQC and DISH-XF provide ~ 60% of the isomerization quantum yield. Overall, the similar behaviors from CT-MQC and DISH-XF simulations suggest that DISH-XF may be a good choice for efficient simulations.

6 Conclusion

We present mixed quantum-classical dynamics based on the exact factorization framework. Since exact factorization is able to deal with the correct electron–nuclear correlation beyond the electronic nonadiabatic couplings, the correct quantum coherence can be described properly. The electron–nuclear correlation term contains the nuclear quantum momentum, which depends on the shape of nuclear density. The electron–nuclear correlation term generates BO population transfers in the absence of electronic nonadiabatic coupling. Moreover, the addition of electron–nuclear correlation guarantees the norm conservation naturally without any ad hoc correction of electronic density matrix. Thus, any artificial renormalization is not necessary. Based on exact factorization, we explain the CT-MQC and DISH-XF approaches, which approximate nuclear density based on nuclear trajectories.

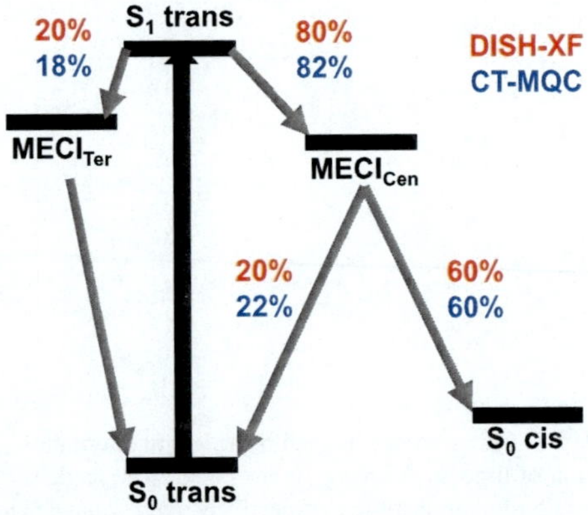

Fig. 8 Nonadiabatic transition pathways and population ratios for the photoisomerization dynamics

The CT-MQC is the first approximate mixed quantum-classical approach from the exact factorization that can run an ensemble of nuclear trajectories simultaneously with the approximated nuclear density being constructed from the coupled nuclear trajectories. Instead, the DISH-XF approach exploits a more efficient surface hopping algorithm for nuclear propagation while electrons propagate along the equation of motions from the exact factorization. Here, the nuclear quantum momentum is calculated from auxiliary trajectories for each independent nuclear trajectory. Both the CT-MQC and DISH-XF methods are highly efficient; however, the DISH-XF is more stable numerically due to the benefit from independent simulations. As a result, we perform the ESMD with ML-assisted electronic state calculations for PSB3 by using PyUNIxMD program, the Python-based ESMD simulations package. Based on the analysis of nuclear trajectories, BO populations, and coherence indicators, both CT-MQC and DISH-XF give results that are in good agreement with each other.

Acknowledgements This work was supported by the research fund (NRF-2019R1A2C1007744 and NRF-2019R1A4A1027934) of the National Research Foundation of Korea (NRF) by the Ministry of Science and ICT.

Data availability PyUNIxMD is distributed via GitHub repository (https://github.com/skmin-lab/unixmd) under the MIT license.

Declarations

Conflict of interest The authors declare that they have no conflict of interest.

References

1. Li X, Hu D, Xie Y, Lan Z (2018) Analysis of trajectory similarity and configuration similarity in on-the-fly surface-hopping simulation on multi-channel nonadiabatic photoisomerization dynamics. J Chem Phys 149:244104. https://doi.org/10.1063/1.5048049
2. Fukushima T, Gupta S, Rad B, Cornejo JA, Petzold CJ, Chan LJG, Mizrahi RA, Ralston CY, Ajo-Franklin CM (2017) The molecular basis for binding of an electron transfer protein to a metal oxide surface. J Am Chem Soc 139:12647–12654. https://doi.org/10.1021/jacs.7b06560
3. Cháber P et al (2017) A low-spin Fe (III) complex with 100-ps ligand-to-metal charge transfer photoluminescence. Nature 543:695–699. https://doi.org/10.1038/nature21430
4. Zhang M, Wang L, Shu S, Sancar A, Zhong D (2016) Bifurcating electron-transfer pathways in DNA photolyases determine the repair quantum yield. Science 354:209–213. https://doi.org/10.1126/science.aah6071
5. Fabini DH, Siaw TA, Stoumpos CC, Laurita G, Olds D, Page K, Hu JG, Kanatzidis MG, Han S, Seshadri R (2017) Universal dynamics of molecular reorientation in hybrid lead iodide perovskites. J Am Chem Soc 139:16875–16884. https://doi.org/10.1021/jacs.7b09536
6. Christiansen D, Selig M, Berghäuser G, Schmidt R, Niehues I, Schneider R, Arora A, de Vasconcellos SM, Bratschitsch R, Malic E, Knorr A (2017) Phonon sidebands in monolayer transition metal dichalcogenides. Phys Rev Lett 119:187402. https://doi.org/10.1103/PhysRevLett.119.187402
7. Park M, Neukirch AJ, Reyes-Lillo SE, Lai M, Ellis SR, Dietze D, Neaton JB, Yang P, Tretiak S, Mathies RA (2017) Excited-state vibrational dynamics toward the polaron in methylammonium lead iodide perovskite. Nat Commun 9:2525. https://doi.org/10.1038/s41467-018-04946-7
8. Blancon JC, Tsai H, Nie W, Stoumpos CC, Pedesseau L, Katan C, Kepenekian M, Soe CMM, Appavoo K, Sfeir MY, Tretiak S, Ajayan PM, Kanatzidis MG, Even J, Crochet JJ, Mohite AD

(2017) Extremely efficient internal exciton dissociation through edge states in layered 2D perovskites. Science 355:1288–1292. https://doi.org/10.1126/science.aal4211

9. Nah S, Spokoyny B, Stoumpos C, Soe CMM, Kanatzidis M, Harel E (2017) Spatially segregated free-carrier and exciton populations in individual lead halide perovskite grains. Nat Photon 11:285–288. https://doi.org/10.1038/nphoton.2017.36
10. Huang J, Yuan Y, Shao Y, Yan Y (2017) Understanding the physical properties of hybrid perovskites for photovoltaic applications. Nat Rev Mater 2:17042. https://doi.org/10.1038/natrevmats.2017.42
11. Qiao L, Fang W-H, Long R, Prezhdo OV (2017) Photoinduced dynamics of charge carriers in metal halide perovskites from an atomistic perspective. J Phys Chem Lett 11:7066–7082. https://doi.org/10.1021/acs.jpclett.0c01687
12. Koumura N, Zijlstra RW, van Delden RA, Harada N, Feringa BL (1999) Light-driven monodirectional molecular rotor. Nature 401:152–155. https://doi.org/10.1038/43646
13. Suda M, Thathong Y, Promarak V, Kojima H, Nakamura M, Shiraogawa T, Ehara M, Yamamoto HM (2019) Light-driven molecular switch for reconfigurable spin filters. Nat Commun 10:2455. https://doi.org/10.1038/s41467-019-10423-6
14. Chen J, Leung FK-C, Stuart MCA, Kajitani T, Fukushima T, van der Giessen E, Feringa BL (2018) Artificial muscle-like function from hierarchical supramolecular assembly of photo-responsive molecular motors. Nat Chem 10:132–138. https://doi.org/10.1038/nchem.2887
15. Scholes GD, Fleming GR, Chen LX, Aspuru-Guzik A, Buchleitner A, Coker DF, Engel GS, van Grondelle R, Ishizaki A, Jonas DM, Lundeen JS, McCusker JK, Mukamel S, Ogilvie JP, Olaya-Castro A, Ratner MA, Spano FC, Whaley KB, Zhu X (2017) Using coherence to enhance function in chemical and biophysical systems. Nature 543:647–656. https://doi.org/10.1038/nature21425
16. Romero E, Novoderezhkin VI, van Grondelle R (2017) Quantum design of photosynthesis for bio-inspired solar-energy conversion. Nature 543:355–365. https://doi.org/10.1038/nature22012
17. Kaucikas M, Maghlaoui K, Barber J, Renger T, van Thor JJ (2017) Ultrafast infrared observation of exciton equilibration from oriented single crystals of photosystem II. Nat Commun 7:13977. https://doi.org/10.1038/ncomms13977
18. Nürnberg DJ, Morton J, Santabarbara S, Telfer A, Joliot P, Antonaru LA, Ruban AV, Cardona T, Krausz E, Boussac A, Fantuzzi A, Rutherford AW (2018) Photochemistry beyond the red limit in chlorophyll f-containing photosystems. Science 360:1210–1213. https://doi.org/10.1126/science.aar8313
19. ...Wolf TJA, Sanchez DM, Yang J, Parrish JRM, Nunes JPF, Centurion M, Coffee R, Cryan JP, Gühr M, Hegazy K, Kirrander A, Li RK, Ruddock J, Shen X, Vecchione T, Weathersby SP, Weber PM, Wilkin K, Yong H, Zheng Q, Wang XJ, Martínez Minitti MPTJ (2019) The photochemical ring-opening of 1,3-cyclohexadiene imaged by ultrafast electron diffraction. Nat Chem 11:504–509. https://doi.org/10.1038/s41557-019-0252-7
20. McFarland BK, Farrell JP, Miyabe S, Tarantelli F, Aguilar A, Berrah N, Bostedt C, Bozek JD, Bucksbaum PH, Castagna JC, Coffee RN, Cryan JP, Fang L, Feifel R, Gaffney KJ, Glownia JM, Martínez TJ, Mucke M, Murphy B, Natan A, Osipov T, Petrović VS, Schorb S, Schultz Th, Spector LS, Swiggers M, Tenney I, Wang S, White JL, White W, Gühr M (2014) Ultrafast X-ray Auger probing of photoexcited molecular dynamics. Nat Commun 5:4235. https://doi.org/10.1038/ncomms5235
21. Coquelle N, Sliwa M, Woodhouse J, Schirò G, Adam V, Aquila A, Barends TRM, Boutet S, Byrdin M, Carbajo S, De la Mora E, Doak RB, Feliks M, Fieschi F, Foucar L, Guillon V, Hilpert M, Hunter MS, Jakobs S, Koglin JE, Kovacsova G, Lane TJ, Lévy B, Liang M, Nass K, Ridard J, Robinson JS, Roome CM, Ruckebusch C, Seaberg M, Thepaut M, Cammarata M, Demachy I, Field M, Shoeman RL, Bourgeois D, Colletier J-P, Schlichting I, Martin Weik M (2018) Chromophore twisting in the excited state of a photoswitchable fluorescent protein captured by time-resolved serial femtosecond crystallography. Nat Chem 10:31–37. https://doi.org/10.1038/nchem.2853
22. Gozem S, Johnson PJM, Halpin A, Luk HL, Morizumi T, Prokhorenko VI, Ernst OP, Olivucci M, Miller RJD (2020) Excited-state vibronic dynamics of bacteriorhodopsin from two dimensional electronic photon echo spectroscopy and multiconfigurational quantum chemistry. J Phys Chem Lett 11:3889–3896. https://doi.org/10.1021/acs.jpclett.0c01063
23. Pooler DRS, Pierron R, Crespi S, Costil R, Pfeifer L, Léonard J, Olivucci M, Feringa BL (2021) Effect of charge-transfer enhancement on the efficiency and rotary mechanism of an oxindolebased molecular motor. Chem Sci 12:7486–7497. https://doi.org/10.1039/D1SC01105G

24. Meyer H-D, Manthe U, Cederbaum LS (1990) The multi-configurational time-dependent Hartree approach. Chem Phys Lett 165:73–78. https://doi.org/10.1016/0009-2614(90)87014-I
25. Martínez TJ, Ben-Nun M, Levine RD (1996) Multi-electronic-state molecular dynamics: a wave function approach with applications. J Phys Chem 100:7884–7895. https://doi.org/10.1021/jp953105a
26. Prezhdo OV (1999) Mean field approximation for the Stochastic Schrödinger equation. J Chem Phys 111:8366–8377. https://doi.org/10.1063/1.480178
27. Tully JC (1990) Molecular dynamics with electronic transitions. J Chem Phys 93:1061–1071. https://doi.org/10.1063/1.459170
28. Min SK, Agostini F, Gross EKU (2015) Coupled-trajectory quantum-classical approach to electronic decoherence in non-adiabatic processes. Phys Rev Lett 115:073001. https://doi.org/10.1103/PhysRevLett.115.073001
29. Ha J-K, Lee IS, Min SK (2018) Surface hopping dynamics beyond nonadiabatic couplings for quantum coherence. J Phys Chem Lett 9:1097–1104. https://doi.org/10.1021/acs.jpclett.8b00060
30. Wjatscheslaw P, Matthias P, Keith HH, Rocco M, Irene B (2019) Vibronic coupling models for donor-acceptor aggregates using an effective-mode scheme: application to mixed Frenkel and charge-transfer excitons in oligothiophene aggregates. J Chem Phys 150:244114. https://doi.org/10.1063/1.5100529
31. Konstantin F, Hiroyuki T, Irene B, Miquel H-R (2018) Ultrafast carbon monoxide photolysis and heme spin-crossover in myoglobin via nonadiabatic quantum dynamics. Nat Commun 9:4502. https://doi.org/10.1038/s41467-018-06615-1
32. Martha YJ, Yanli L, Roberto I, Fabrizio S (2020) Ultrafast dynamics of the two lowest bright excited states of cytosine and 1-methylcytosine: a quantum dynamical study. J Chem Theory Comput 16:5792–5808. https://doi.org/10.1021/acs.jctc.0c00455
33. Maria F, Clémence C (2020) Direct, mediated, and delayed intramolecular singlet fission mechanism in donor-acceptor copolymers. J Phys Chem Lett 11:9788–9794. https://doi.org/10.1021/acs.jpclett.0c03076
34. Martínez TJ, Ben-Nun M, Levine RD (1997) Molecular collision dynamics on several electronic states. J Phys Chem A 101:6389–6402. https://doi.org/10.1021/jp970842t
35. Martínez TJ, Levine RD (1997) Non-adiabatic molecular dynamics: split-operator multiple spawning with applications to photodissociation. J Chem Soc Faraday Trans 93:941–947. https://doi.org/10.1039/A605958I
36. Uwe M, Gerhard S (1997) Surface-hopping modeling of photoinduced relaxation dynamics on coupled potential-energy surfaces. J Chem Phys 107:6230–6245. https://doi.org/10.1063/1.474288
37. Wenfang H, György L, Biswajit M, George CS (2007) Trajectory surface hopping study of the O(3P) + ethylene reaction dynamics. J Phys Chem A 112:2093–2103. https://doi.org/10.1021/jp076716z
38. Farnaz AS, Pengfei H (2017) Ring polymer surface hopping: incorporating nuclear quantum effects into nonadiabatic molecular dynamics simulations. J Phys Chem Lett 8:3072–3080. https://doi.org/10.1021/acs.jpclett.7b01343
39. Wang L, Beljonne D (2013) Charge transport in organic semiconductors: assessment of the mean field theory in the hopping regime. J Chem Phys 139:064316. https://doi.org/10.1063/1.4817856
40. Levi L, Patrick T, Deyan M, Svetlana K (2020) Phonon-driven energy relaxation in PbS/CdS and PbSe/CdSe Core/ shell quantum dots. J Phys Chem Lett 11:4269–4278. https://doi.org/10.1021/acs.jpclett.0c00845
41. Wang L, Akimov A, Prezhdo OV (2016) Recent progress in surface hopping: 2011–2015. J Phys Chem Lett 7:2100–2112. https://doi.org/10.1021/acs.jpclett.6b00710
42. Jaeger HM, Fischer S, Prezhdo OV (2012) Decoherence-induced surface hopping. J Chem Phys 137:22A545. https://doi.org/10.1063/1.4757100
43. Subotnik JE, Shenvi N (2011) A new approach to decoherence and momentum rescaling in the surface hopping algorithm. J Chem Phys 134:024105. https://doi.org/10.1063/1.3506779
44. Subotnik JE, Ouyang W, Landry BR (2013) Can we derive Tully's surface-hopping algorithm from the semiclassical quantum Liouville equation? Almost, but only with decoherence. J Chem Phys 139:214107. https://doi.org/10.1063/1.4829856
45. Gao X, Thiel W (2017) Non-Hermitian surface hopping. Phys Rev E 95:013308. https://doi.org/10.1103/PhysRevE.95.013308
46. Granucci G, Persico M, Zoccante A (2010) Including quantum decoherence in surface hopping. J Chem Phys 133:134111. https://doi.org/10.1063/1.3489004

47. Nelson T, Fernandez-Alberti S, Roitberg AE, Tretiak S (2013) Nonadiabatic excited-state molecular dynamics: treatment of electronic decoherence. J Chem Phys 138:224111. https://doi.org/10.1063/1.4809568
48. Granucci G, Persico M (2007) Critical appraisal of the fewest switches algorithm for surface hopping. J Chem Phys 126:134114. https://doi.org/10.1063/1.2715585
49. Hammes-Schiffer S, Tully JC (1994) Proton-transfer in solution-molecular-dynamics with quantum transitions. J Chem Phys 101:4657–4667. https://doi.org/10.1063/1.467455
50. Bittner ER, Rossky PJ (1995) Quantum decoherence in mixed quantum-classical systems: nonadiabatic processes. J Chem Phys 103:8130–8143. https://doi.org/10.1063/1.470177
51. Prezhdo OV, Rossky PJ (1997) Evaluation of quantum transition rates from quantum-classical molecular dynamics simulations. J Chem Phys 107:5863–5878. https://doi.org/10.1063/1.474312
52. Schwartz BJ, Bittner ER, Prezhdo OV, Rossky PJ (1996) Quantum decoherence and the isotope effect in condensed phase nonadiabatic molecular dynamics simulations. J Chem Phys 104:5942–5955. https://doi.org/10.1063/1.471326
53. Jasper AW, Truhlar DG (2007) Non-Born–Oppenheimer molecular dynamics of Na...FH photodissociation. J Chem Phys 127:194306. https://doi.org/10.1063/1.2798763
54. Cederbaum LS (2013) The exact molecular wavefunction as a product of an electronic and a nuclear wavefunction. J Chem Phys 138:224110. https://doi.org/10.1063/1.4807115
55. Abedi A, Maitra NT, Gross EKU (2010) Exact factorization of the time-dependent electron-nuclear wave function. Phys Rev Lett 105:123002. https://doi.org/10.1103/PhysRevLett.105.123002
56. Abedi A, Maitra NT, Gross EKU (2012) Correlated electron-nuclear dynamics: exact factorization of the molecular wave-function. J Chem Phys 137:22A530. https://doi.org/10.1063/1.4745836
57. Gidopoulos NI, Gross EKU (2014) Electronic non-adiabatic states: towards a density functional theory beyond the Born–Oppenheimer approximation. Phil Trans R Soc A 372:20130059. https://doi.org/10.1098/rsta.2013.0059
58. Abedi A, Agostini F, Gross EKU (2014) Mixed quantum-classical dynamics from the exact decomposition of electron-nuclear motion. EPL 106:33001. https://doi.org/10.1209/0295-5075/106/33001
59. Abedi A, Agostini F, Suzuki Y, Gross EKU (2013) Dynamical steps that bridge piecewise adiabatic shapes in the exact time-dependent potential energy surface. Phys Rev Lett 110:263001. https://doi.org/10.1103/PhysRevLett.110.263001
60. Agostini F, Abedi A, Yasumitsu S, Min SK, Marita NT, Gross EKU (2015) The exact forces on classical nuclei in non-adiabatic charge transfer. J Chem Phys 142:084303. https://doi.org/10.1063/1.4908133
61. Agostini F, Min SK, Abedi A, Gross EKU (2016) Quantum-classical nonadiabatic dynamics: coupled- vs independent-trajectory methods. J Chem Theory Comput 12:2127–2143. https://doi.org/10.1021/acs.jctc.5b01180
62. Min SK, Agostini F, Tavernelli I, Gross EKU (2017) Ab initio nonadiabatic dynamics with coupled trajectories: a rigorous approach to quantum (De)coherence. J Phys Chem Lett 8:3048–3055. https://doi.org/10.1021/acs.jpclett.7b01249
63. Filatov M, Min SK, Choi CH (2019) Theoretical modelling of the dynamics of primary photoprocess of cyclopropanone. Phys Chem Chem Phys 21:2489–2498. https://doi.org/10.1039/C8CP07104G
64. Filatov M, Min SK, Kim KS (2019) Non-adiabatic dynamics of ring opening in cyclohexa-1,3-diene described by anensemble density-functional theory method. Mol Phys 117:1128–1141. https://doi.org/10.1080/00268976.2018.1519200
65. Filatov M, Min SK, Kim KS (2018) Direct nonadiabatic dynamics by mixed quantum-classical formalism connected with ensemble density functional theory method: application to trans-penta-2,4-dieniminium cation. J Chem Theory Comput 14:4499–4512. https://doi.org/10.1021/acs.jctc.8b00217
66. Filatov M, Paolino M, Min SK, Choi CH (2019) Design and photoisomerization dynamics of a new family of synthetic 2-stroke light driven molecular rotary motors. Chem Commun 55:5247–5250. https://doi.org/10.1039/C9CC01955C
67. Filatov M, Paolino M, Min SK, Kim KS (2018) Fulgides as light-driven molecular rotary motors: computational design of a prototype compound. J Phys Chem Lett 9:4995–5001. https://doi.org/10.1021/acs.jpclett.8b02268

68. Vindel-Zandbergen PV, Ibele LM, Ha JK, Min SK, Curchod BFE, Maitra NT (2021) Study of the decoherence correction derived from the exact factorization approach for nonadiabatic dynamics. J Chem Theory Comput 17:3852–3862. https://doi.org/10.1021/acs.jctc.1c00346
69. Ha J-K, Kim K, Min SK (2021) Machine learning-assisted excited state molecular dynamics with the state-interaction state-averaged spin-restricted ensemble-referenced Kohn-Sham approach. J Chem Theory Comput 17:694–702. https://doi.org/10.1021/acs.jctc.0c01261
70. Barbatti M, Ruckenbauer M, Plasser F, Pittner J, Granucci G, Persico M, Lischka H (2014) NEWTON-X: a surface-hopping program for nonadiabatic molecular dynamics. WIREs Comput Mol Sci 4:26–33. https://doi.org/10.1002/wcms.1158
71. Mai S, Marquetand P, González L (2018) Nonadiabatic dynamics: the SHARC approach. WIREs Comput Mol Sci 8:e1370. https://doi.org/10.1002/wcms.1370
72. Akimov AV (2016) Libra: an open-source "Methodology Discovery'' library for quantum and classical dynamics simulations. J Comput Chem 37:1626–1649. https://doi.org/10.1002/jcc.24367
73. Akimov AV, Prezhdo OV (2013) The PYXAID program for non-adiabatic molecular dynamics in condensed matter systems. J Chem Theory Comput 9:4959–4972. https://doi.org/10.1021/ct400641n
74. Sifain AE, Bjorgaard JA, Nelson TR, Nebgen BT, White AJ, Gifford BJ, Gao DW, Prezhdo OV, Fernandez-Alberti S, Roitberg AE, Tretiak S (2018) Photoexcited nonadiabatic dynamics of solvated push-pull pi-conjugated oligomers with the NEXMD software. J Chem Theory Comput 14:3955–3966. https://doi.org/10.1021/acs.jctc.8b00103
75. Lee IS, Ha J-K, Han D, Kim TI, Moon SW, Min SK (2021) PyUNIxMD: a Python-based excited state molecular dynamics package. J Comput Chem 42(24):1755–1766. https://doi.org/10.1002/jcc.26711
76. Min SK, Abedi A, Kim KS, Gross EKU (2014) Is the molecular berry phase an artifact of the Born–Oppenheimer approximation? Phys Rev Lett 113:263004. https://doi.org/10.1103/PhysRevLett.113.263004
77. Gossel GH, Lacombe L, Maitra NT (2019) On the numerical solution of the exact factorization equations. J Chem Phys 150:154112. https://doi.org/10.1063/1.5090802
78. Dral PO, Barbatti M, Thiel W (2018) Nonadiabatic excited-state dynamics with machine learning. J Phys Chem Lett 9:5660–5663. https://doi.org/10.1021/acs.jpclett.8b02469
79. Hu D, Xie Y, Li X, Li L, Lan Z (2018) Inclusion of machine learning Kernel ridge regression potential energy surfaces in on-the-fly nonadiabatic molecular dynamics simulation. J Phys Chem Lett 9:2725–2732. https://doi.org/10.1021/acs.jpclett.8b00684
80. Chen W-K, Liu X-Y, Fang WH, Dral PO, Cui G (2018) Deep learning for nonadiabatic excited-state dynamics. J Phys Chem Lett 9:6702–6708. https://doi.org/10.1021/acs.jpclett.8b03026
81. CPMD. http://www.cpmd.org/, Copyright IBM Corp 1990-2015, Copyright MPI für Festkörperforschung Stuttgart (1997–2001)
82. GCTMQC. https://gitlab.com/agostini.work/g-ctmqc.git
83. Marsili E, Olivucci M, Lauvergnat D, Agostini F (2020) Quantum and quantum-classical studies of the photoisomerization of a retinal chromophore model. J Chem Theory Comput 16:6032–6048. https://doi.org/10.1021/acs.jctc.0c00679
84. Talotta F, Morisset S, Rougeau N, Lauvergnat D, Agostini F (2020) Internal conversion and intersystem crossing with the exact factorization. J Chem Theory Comput 16:4833–4848. https://doi.org/10.1021/acs.jctc.0c00493
85. Pieroni C, Marsili E, Lauvergnat D, Agostini F (2021) Relaxation dynamics through a conical intersection: quantum and quantum-classical studies. J Chem Phys 154:034104. https://doi.org/10.1063/5.0036726
86. Pedraza-Gonzalez L, De Vico L, Marin MDC, Fanelli F, Olivucci M (2019) a-ARM: automatic rhodopsin modeling with chromophore cavity generation, ionization state selection, and external counterion placement. J Chem Theory Comput 15:3134–3152. https://doi.org/10.1021/acs.jctc.9b00061
87. Brunken C, Reiher M (2021) Automated construction of quantum-classical hybrid models. J Chem Theory Comput 17:3797–3813. https://doi.org/10.1021/acs.jctc.1c00178
88. Lischka H, Müller T, Szalay PG, Shavitt I, Pitzer RM, Shepard E (2011) Columbus-a program system for advanced multireference theory calculations. WIREs Comput Mol Sci 1:191–199. https://doi.org/10.1002/wcms.25
89. Hourahine B, Aradi B, Blum V, Bonafé F, Buccheri A, Camacho C, Cevallos C, Deshaye MY, Dumitrică T, Dominguez A, Ehlert S, Elstner M, van der Heide T, Hermann J, Irle S, Kranz JJ,

Köhler C, Kowalczyk T, Kubař T, Lee IS, Lutsker V, Maurer RJ, Min SK, Mitchell I, Negre C, Niehaus TA, Niklasson AMN, Page AJ, Pecchia A, Penazzi G, Persson MP, Řezáč J, Sánchez CG, Sternberg M, Stöhr M, Stuckenberg F, Tkatchenko A, Yu VW-Z, Frauenheim T (2020) DFTB+, a software package for efficient approximate density functional theory based atomistic simulations. J Chem Phys 152:124101. https://doi.org/10.1063/1.5143190

90. Frisch MJ, Trucks GW, Schlegel HB, Scuseria GE, Robb MA, Cheese-Man JR, Scalmani G, Barone V, Mennucci B, Petersson GA, Nakatsuji H, Caricato M, Li X, Hratchian HP, Izmaylov AF, Bloino J, Zheng G, Sonnenberg JL, Hada M, Ehara M, Toyota K, Fukuda R, Hasegawa J, Ishida M, Nakajima T, Honda Y, Kitao O, Nakai H, Vreven T, Montgomery JA, Peralta JE, Ogliaro F, Bearpark M, Heyd J, Brothers E, Kudin KN, Staroverov VN, Kobayashi R, Normand J, Raghavachari K, Rendell A, Burant JC, Iyengar SS, Tomasi J, Cossi M, Rega N, Millam M, Klene M, Knox JE, Cross JB, Bakken V, Adamo C, Jaramillo J, Gomperts R, Stratmann RE, Yazyev O, Austin AJ, Cammi R, Pomelli C, Ochterski JW, Martin RL, Morokuma K, Zakrzewski VG, Voth GA, Salvador P, Dannenberg JJ, Dapprich S, Daniels AD, Farkas O, Foresman JB, Ortiz JV, Cioslowski J, Fox DJ (2009) Gaussian 09 Revision A.02. Gaussian Inc, Wallingford
91. Werner H-J, Knowles PJ, Knizia G, Manby FR, Schütz M (2012) Molpro: a general-purpose quantum chemistry program package. WIREs Comput Mol Sci 2:242–253. https://doi.org/10.1002/wcms.82
92. Seritan S, Bannwarth C, Fales BS, Hohenstein EG, Kokkila-Schumacher SIL, Luehr N, Snyder JW, Song C, Titov AV, Ufimtsev IS, Martínez TJ (2020) TeraChem: accelerating electronic structure and ab initio molecular dynamics with graphical processing units. J Chem Phys 152:224110. https://doi.org/10.1063/5.0007615
93. Shao Y, Gan Z, Epifanovsky E, Gilbert AT, Wormit M, Kussmann J, Lange AW, Behn A, Deng J, Feng X, Ghosh D, Goldey M, Horn PR, Jacobson LD, Kaliman I, Khaliullin RZ, Kuś T, Landau A, Liu J, Proynov EI, Rhee YM, Richard RM, Rohrdanz MA, Steele RP, Sundstrom EJ, Zimmerman PM, Zuev D, Albrecht B, Alguire E, Austin B, Beran GJO, Bernard YA, Berquist E, Brandhorst K, Bravaya KB, Brown ST, Casanova D, Chang C-M, Chen Y, Chien SH, Closser KD, Crittenden DL, Diedenhofen M, Do H, Dutoi AD, Edgar RG, Fatehi S, Fusti-Molnar L, Ghysels A, Golubeva-Zadorozhnaya A, Gomes J, Hanson-Heine MW, Harbach PH, Hauser AW, Hohenstein EG, Holden ZC, Jagau T-C, Ji H, Kaduk B, Khistyaev K, Kim J, Kim J, King RA, Klunzinger P, Kosenkov D, Kowalczyk T, Krauter CM, Lao KU, Laurent AD, Lawler KV, Levchenko SV, Lin CY, Liu F, Livshits E, Lochan RC, Luenser A, Manohar A, Manzer SF, Mao S-P, Mardirossian N, Marenich AV, Maurer SA, Mayhall NJ, Neuscamman E, Oana CM, Olivares-Amaya R, O'Neill DP, Parkhill JA, Perrine TM, Peverati R, Prociuk A, Rehn DR, Rosta E, Russ NJ, Sharada SM, Sharma S, Small DW, Sodt A, Stein T, Stück D, Su Y-C, Thom AJ, Tsuchimochi T, Vanovschi V, Vogt L, Vydrov O, Wang T, Watson MA, Wenzel J, White A, Williams CF, Yang J, Yeganeh S, Yost SR, You Z-Q, Zhang IY, Zhang X, Zhao Y, Brooks BR, Chan GK, Chipman DM, Cramer CJ, Gordon MS, Hehre WJ, Klamt A, Schaefer HF III, Schmidt MW, Sherrill CD, Truhlar DG, Warshel A, Xu X, Aspuru-Guzik A, Baer R, Bell AT, Besley NA, Chai J-D, Dreuw A, Dunietz BD, Furlani TR, Gwaltney SR, Hsu c-P, Jung Y, Kong J, Lambrecht DS, Liang W, Ochsenfeld C, Rassolov VA, Slipchenko LV, Subotnik JE, Voorhis TV, Herbert JM, Krylov AI, Gill PM, Head-Gordon M (2015) Advances in molecular quantum chemistry contained in the Q-Chem 4 program package. Mol Phys 113:184–215. https://doi.org/10.1080/00268976.2014.952696
94. Ryabinkin IG, Nagesh J, Izmaylov AF (2015) Fast numerical evaluation of time-derivative nonadiabatic couplings for mixed quantum-classical methods. J Phys Chem Lett 6:4200–4203. https://doi.org/10.1021/acs.jpclett.5b02062
95. Berendsen HJC, Postma JPM, van Gunsteren WF, DiNola A, Haak JR (1984) Molecular dynamics with coupling to an external bath. J Chem Phys 81:3684. https://doi.org/10.1063/1.448118
96. Martyna GJ, Tuckerman ME, Tobias DJ, Klein ML (1996) Explicit reversible integrators for extended systems dynamics. Mol Phys 87:1117–1157. https://doi.org/10.1080/00268979600100761
97. Sun Q, Zhang X, Banerjee S, Bao P, Barbry M, Blunt NS, Bogdanov NA, Booth GH, Chen J, Cui Z-H, Eriksen JJ, Gao Y, Guo S, Hermann J, Hermes MR, Koh K, Koval P, Lehtola S, Li Z, Liu J, Mardirossian N, McClain JD, Motta M, Mussard B, Pham HQ, Pulkin A, Purwanto W, Robinson PJ, Ronca E, Sayfutyarova ER, Scheurer M, Schurkus HF, Smith JET, Sun C, Sun S-N, Upadhyay S, Wagner LK, Wang X, White A, Whitfield JD, Williamson MJ, Wouters S, Yang J, Yu JM, Zhu T, Berkelbach TC, Sharma S, Sokolov AY, Chan GK-L (2020) Recent developments in the PySCF program package. J Chem Phys 153:024109. https://doi.org/10.1063/5.0006074

98. Barbatti M, Sen K (2015) Effects of different initial condition samplings on photodynamics and spectrum of pyrrole. Int J Quantum Chem 116:762–771. https://doi.org/10.1002/qua.25049
99. Barbatti M, Granucci G, Persico M, Ruckenbauer M, Vazdar M, Eckert-Maksić Lischka H (2007) The on-the-fly surface-hopping program system Newton-X: Application to ab initio simulation of the nonadiabatic photodynamics of benchmark systems. J Photochem Photobiol A 190:228–240. https://doi.org/10.1016/j.jphotochem.2006.12.008
100. Wigner E (1932) On the quantum correction for thermodynamic equilibrium. Phys Rev 40:749–759. https://doi.org/10.1103/PhysRev.40.749
101. Zobel JP, Nogueira JJ, González L (2018) Finite-temperature Wigner phase-space sampling and temperature effects on the excited-state dynamics of 2-nitronaphthalene. Phys Chem Chem Phys 21:13906–13915. https://doi.org/10.1039/C8CP03273D
102. Borrego-Varillas R, Teles-Ferreira DC, Nenov A, Conti I, Ganzer L, Manzoni C, Garavelli M, Maria de Paula A, Cerullo G (2018) Observation of the Sub-100 femtosecond population of a dark state in a thiobase mediating intersystem crossing. J Am Chem Soc 140:16087–16093. https://doi.org/10.1021/jacs.8b07057
103. Williams M, Forbes R, Weir H, Veyrinas K, MacDonell RJ, Boguslavskiy AE, Schuurman MS, Stolow A, Martinez TJ (2021) Unmasking the cis-stilbene phantom state via vacuum ultraviolet time-resolved photoelectron spectroscopy and Ab initio multiple spawning. J Phys Chem Lett 12:6363–6369. https://doi.org/10.1021/acs.jpclett.1c01227
104. Barbatti M, Ruckenbauer M, Szymczak JJ, Aquino AJA, Lischka H (2008) Nonadiabatic excited-state dynamics of polar π-systems and related model compounds of biological relevance. Phys Chem Chem Phys 10:482–494. https://doi.org/10.1039/B709315M
105. Keal TW, Wanko M, Thiel W (2009) Assessment of semiempirical methods for the photoisomerisation of a protonated Schiff base. Theoret Chem Acc 123:145–156. https://doi.org/10.1007/s00214-009-0546-8
106. Lihong L, Jian L, Martínez TJ (2016) Dynamical correlation effects on photoisomerization: Ab initio multiple spawning dynamics with MS-CASPT2 for a model trans-protonated Schiff base. J Phys Chem 120:1940–1949. https://doi.org/10.1021/acs.jpcb.5b09838

Publisher's Note Springer Nature remains neutral with regard to jurisdictional claims in published maps and institutional affiliations.

Topics in Current Chemistry (2021) 379:43
https://doi.org/10.1007/s41061-021-00351-9

REVIEW

The Static–Dynamic–Static Family of Methods for Strongly Correlated Electrons: Methodology and Benchmarking

Yangyang Song[1] · Yang Guo[1] · Yibo Lei[2] · Ning Zhang[3] · Wenjian Liu[1]

Received: 15 July 2021 / Accepted: 15 September 2021 / Published online: 1 November 2021
© The Author(s), under exclusive licence to Springer Nature Switzerland AG 2021

Abstract

A series of methods (SDSCI, SDSPT2, iCI, iCIPT2, iCISCF(2), iVI, and iCAS) is introduced to accurately describe strongly correlated systems of electrons. Born from the (restricted) static–dynamic–static (SDS) framework for designing many-electron wave functions, SDSCI is a minimal multireference (MR) configuration interaction (CI) approach that constructs and diagonalizes a $3N_P \times 3N_P$ matrix for N_P states, regardless of the numbers of orbitals and electrons to be correlated. If the full molecular Hamiltonian H in the QHQ block (which describes couplings between functions of the first-order interaction space Q) of the SDSCI CI matrix is replaced with a zeroth-order Hamiltonian H_0 before the diagonalization is taken, we obtain SDSPT2, a CI-like second-order perturbation theory (PT2). Unlike most variants of MRPT2, SDSPT2 treats single and multiple states in the same way and is particularly advantageous in the presence of near degeneracy. On the other hand, if the SDSCI procedure is repeated until convergence, we will have iterative CI (iCI), which can converge quickly from the above to the exact solutions (full CI) even when starting with a poor guess. When further combined with the selection of important configurations followed by a PT2 treatment of dynamic correlation, iCI becomes iCIPT2, which is a near-exact theory for medium-sized systems. The microiterations of iCI for relaxing the coefficients of contracted many-electron functions can be generalized to an iterative vector interaction (iVI) approach for finding exterior or interior roots of a given matrix, in which the dimension of the search subspace is fixed by either the number of target roots or the user-specified energy window. Naturally, iCIPT2 can be employed as the active space solver of the complete active space (CAS) self-consistent field, leading to iCISCF(2), which can further be combined with iCAS for automated selection of active orbitals and assurance of the same CAS for all states and all geometries. The methods are calibrated by taking the Thiel set of benchmark systems as examples. Results for the corresponding cations, a new set of benchmark systems, are also reported.

Chapter 7 was originally published as Song, Y., Guo, Y., Lei, Y., Zhang, N. & Liu, W. Topics in Current Chemistry (2021) 379: 43. https://doi.org/10.1007/s41061-021-00351-9.

Extended author information available on the last page of the article

Keywords Excited states · Strong correlation · Static–dynamic–static · SDSCI · SDSPT2 · iCI · iCIPT2 · iCISCF(2) · iCAS · iVI

1 Introduction

In principle, relativistic, correlation, and even quantum electrodynamics (QED) effects ought to be accounted for simultaneously in electronic structure calculations of atoms and molecules to predict experimental observations precisely, i.e., "relativity + correlation + QED = experiment." With the advent of effective QED [1, 2], quasi-four-component (Q4C) [3, 4], exact two-component (X2C) [5–8], and spin-separated X2C [9, 10] relativistic Hamiltonians [11–20], relativity and QED problems in quantum chemistry can be regarded as solved because the "Hamiltonian ladder" has been made complete [2, 10, 13], such that one can just pick the right Hamiltonian according to the target physics/chemistry and accuracy. As such, the remaining challenge is still the correlation problem, especially for so-called strongly correlated systems of electrons. Literally, a strongly correlated system features multiple open-shell orbitals or nearly degenerate electronic states due to the existence of a dense set of energetically adjacent frontier orbitals, such that there does not exist a leading component in the wave functions. Because of this, single-reference methods are bound to fail, especially for low-spin states. Instead, multireference methods are needed for even a qualitative description of such states. A unanimous assumption underlying standard multireference methods [21–26] is that the full many-electron Hilbert space can be decomposed into a model space P and the external complement $Q = 1 - P$, which are responsible for the static and dynamic components of the overall correlation, respectively. Then, according to when the static and dynamic correlations are handled, such methods can be classified into three families [27, 28], viz. static-then-dynamic, dynamic-then-static, and static–dynamic–static (SDS), where the first family of methods refers to those that start with diagonalization of the bare Hamiltonian projected onto the pre-chosen model space P to yield the zeroth-order (reference) functions $\Psi_k^{(0)} = \sum_{\mu\in P} \Phi_\mu \tilde{C}_{\mu k}^{(0)}$ (which describe static correction) and then build up corrections to $\Psi_k^{(0)}$ for dynamic correlation. A characteristic feature of such methods lies in the fact that the zeroth-order coefficients $\tilde{C}_{\mu k}^{(0)}$ are not revised by dynamic correlation; That is, dynamic correlation sees static correlation, but static correlation does not see dynamic correlation. Classic examples of such methods are multistate complete active space (CAS) second-order perturbation theory (PT2) (MS-CASPT2) [29–31] and multistate n-electron valence state PT2 (MS-NEVPT2) [32]; even though these are often called diagonalize–perturb–diagonalize, they are still static-then-dynamic because the second diagonalization does not give rise to sufficient revisions of the coefficients $\tilde{C}_{\mu k}^{(0)}$. By contrast, the dynamic-then-static family of methods refers to those that build up dynamic corrections to each model function Φ_μ within P (without knowledge of the coefficients $\tilde{C}_{\mu k}^{(0)}$) and then diagonalize the effective Hamiltonian for static correlation. When the orbitals are optimized for each (non-Aufbau) model function Φ_μ, the method has been termed "nonorthogonal state interaction" (NOSI) [11], which differs from "nonorthogonal configuration interaction" (NOCI) in that the latter has no dynamic correlation to each model function. Different realizations of this Ansatz have been presented [33–37], among which the multistate density functional theory [35–37], which builds dynamic correlation into individually optimized model functions through (transition) density functionals, is particularly effective. It is clear that static correlation sees dynamic correlation but dynamic correlation does not see static correlation in such dynamic-then-static approaches, just opposite to static-then-dynamic ones. Naturally, an SDS type of approach is required for greater robustness, where the zeroth-order coefficients $\tilde{C}^{(0)}_{\mu k}$ are sufficiently relaxed; that is, static and dynamic correlations do see each other. There exists a number of methods belonging to this family [27, 28], among which partially contracted MRCI [24–26] is arguably the simplest.

Despite the great success of the above standard multireference methods in describing low-lying electronic states, they still have strong limitations. For instance, it is by no means trivial to maintain the same model space P so as to produce smooth potential energy surfaces. Even if this can indeed be achieved, there is no guarantee that the chosen model space will be equally good for all geometries and all target states. Even more seriously, those states that have only small projections in the chosen model space cannot be described well. Taking all these facts into account, one can classify chemical systems into weakly correlated (which can be described well by single-reference methods), moderately correlated [which can well be described by standard multireference methods with a moderate model space, typically much smaller than CAS(18e,18o)], and strongly correlated (which require a large model space and are therefore beyond the capability of standard multireference methods). In fact, the concept of model space loses its original meaning for so-defined strongly correlated systems because the states in a large model space P and those formally in the orthogonal complement Q are strongly interspersed in both energy and composition. In other words, the static and dynamic correlations in such cases are strongly entangled and interchangeable. Moreover, solving the large-model-space problem itself is already a demanding task [38–55], thus clearly more so the subsequent treatment of dynamic correlation. So, the question is how to design some adaptive method that can adapt to the variable static correlation automatically while also handling the residual dynamic correlation efficiently. It appears that the most promising approach in this regard is to introduce some selection procedure, so as to build up progressively a compact yet accurate variational space P, in full accordance with the nature of the system under consideration. In this spirit, a number of near-exact approaches have been designed in the past, including the density matrix renormalization group technique [56–66], full configuration interaction (FCI) quantum Monte Carlo (FCIQMC) [67–77], cluster-analysis-driven FCIQMC [78], full coupled-cluster (CC) reduction [79, 80], deterministically projected FCI [81], many-body expanded FCI [82–85], incremental FCI [86, 87], rank-reduction FCI [88], fast randomized interaction-based FCI [89], intrinsic scaling-based correlation expansion [90, 91], as well as selected CI (sCI) [92–114]. Among these, the sCI approaches (which are largely a revival of old ideas [115–119] in one way or another) are the simplest in structure and rival more sophisticated approaches, as shown in a recent blind test [120]. In the present account, we briefly introduce our own (restricted) SDS family of methods, including SDSCI [106], SDSPT2 [106, 107], iCI [27, 28], iCIPT2 [108, 121], iVI [122–125], iCISCF(2) [55], and iCAS [126], with the acronyms explained in Table 1. The (restricted) SDS Ansatz [106] for designing many-electron wave functions is first recapitulated in Sect. 2.1, from which the SDSCI, SDSPT2, iCI, iCIPT2, and iVI approaches follow naturally. The iCAS and iCISCF(2) approaches are discussed in Sect. 2.2. Unless otherwise stated, the following convention is used: $\{i, j, k, l, \ldots\}$, $\{t, u, v, w, \ldots\}$, $\{a, b, c, d, \ldots\}$, and $\{p, q, r, s, \ldots\}$ denote doubly occupied (core), active, virtual, and unspecified orbitals, respectively; repeated indices are always summed.

2 The SDS Family of Methods

2.1 SDSCI, SDSPT2, iCI, iCIPT2, and iVI

For brevity, we only discuss solutions of a spin-free quantum-mechanical equation, but the ideas can readily be extended to the spin-dependent case. For both cases, FCI provides the exact solutions for a given one-particle basis. Without loss of generality, such solutions can be classified into three portions: N_P low-lying, N_S intermediate, and N_Q high-lying states. The specific values of N_P, N_S, and N_Q are of course system dependent. If N_P is just 1 and the state is well separated from the rest, the system is only weakly correlated and can be well described by a single-reference prescription. On the other hand, if N_P is exceedingly large (i.e., $N_\mathrm{P} \simeq N_\mathrm{P} + N_\mathrm{S} + N_\mathrm{Q}$), the system would be ultrastrongly correlated, which exists only in the thermodynamic limit. What is particularly interesting is the intermediate case, i.e., what are commonly called strongly correlated systems, with N_P being a limited but not necessarily small number. Given such a general feature of the spectrum, we write a FCI solution $|\Psi_I\rangle$ $(I \leq N_\mathrm{P})$ as

Table 1 The SDS family of methods

Acronym	Full description	References
SDSCI	Static–dynamic–static configuration interaction	[106]
SDSPT2	Static–dynamic–static second-order perturbation theory	[106, 107]
iCI	Iterative configuration interaction	[27, 28]
SiCI	iCI with selection	[108, 121]
iCIPT2	iCI with selection and PT2	[108, 121]
iVI	Iterative vector interaction	[122–125]
iCAS	Imposed automatic selection and localization of complete active space	[126]
iCISCF(2)	iCI-based multiconfigurational self-consistent field theory with PT2 in the variational space only	[55]
iOI	Iterative orbital interaction as a bottom-up solver of the SCF problem	[127]

$$|\Psi_I\rangle = \sum_{\mu=1}^{M} |\Phi_\mu\rangle C_{\mu I}, \quad C_{\mu I} = \sum_{k=1}^{N} \tilde{C}_{\mu k}\bar{C}_{kI}, \quad N = \mathrm{rank}(\mathbf{C}) \tag{1}$$

$$= \sum_{k=1}^{N} |\tilde{\Phi}_k\rangle \bar{C}_{kI}, \quad |\tilde{\Phi}_k\rangle = \sum_{\mu=1}^{M} |\Phi_\mu\rangle \tilde{C}_{\mu k}. \tag{2}$$

$$\approx \sum_{k=1}^{\tilde{N}} |\tilde{\Phi}_k\rangle \bar{C}_{kI}. \tag{3}$$

Here, $\{\Phi_\mu;\mu = 1,\dots,M\}$ represent all possible n-electron Slater determinants (SDs) or equivalently configuration state functions (CSFs), while $\{\tilde{\Phi}_k;k = 1,\dots,N \le M\}$ are called contracted n-electron functions or simply "states." The second equation in Eq. (1) is nothing but a singular value decomposition of the CI vector $\mathbf{C}$, i.e., $\mathbf{C} = \tilde{\mathbf{C}}\boldsymbol{\sigma}\mathbf{R}^\dagger = \tilde{\mathbf{C}}\bar{\mathbf{C}}$, such that nothing has really happened when going from Eqs. (1) to (2). However, the latter allows one to choose just a small number of contracted functions $\{\tilde{\Phi}_k;k = 1,\dots,\tilde{N}\}$ to represent accurately the desired state Ψ_I, thereby leading to Eq. (3). (Indeed, if $\tilde{\mathbf{C}} = \mathbf{C}$ and hence $\bar{\mathbf{C}} = \mathbf{I}$, $\tilde{N} = 1$ would be exact.) Stated differently, when the coefficients $\bar{C}_{kI}$ are arranged in descending order of magnitude, the first $\tilde{N}$ contracted functions $\tilde{\Phi}_k$ in Eq. (2) can be regarded as most important. In contrast, when working with individual SDs/CSFs, the selection of a small number of them is usually impossible as they are not well structured in energy. In fact, as an approximation to Eq. (2), Eq. (3) has three degrees of freedom to bring into play: (a) what SDs/CSFs Φ_μ are to be used, (b) how to determine their contraction coefficients $\tilde{C}_{\mu k}$, and (c) how many states $\tilde{\Phi}_k$ are to be used. Regarding (c), the *restricted* SDS Ansatz [106] takes the most democratic option, $\tilde{N} = 3N_\mathrm{P}$; that is, the same number (N_P) of primary, secondary, and external states is to be employed for $\tilde{\Phi}_k$ in Eq. (3). The procedure goes as follows:

As usually done, a model space $P = \{\Phi_\mu;\mu = 1,\dots,d_R\}$ is first constructed (following, e.g., the selection procedure [118]), which need not be CAS nor that used to determine the one-particle orbitals. The lowest N_P solutions of the Hamiltonian in this space, i.e., $\Psi_k^{(0)} = \sum_{\mu=1}^{d_R} \Phi_\mu \tilde{C}_{\mu k}^{(0)}$, which provide either semiquantitatively or qualitatively correct descriptions of the N_P exact states, are to be taken as the primary states. For convenience, we introduce the following projectors:

$$P_\mathrm{m} = \sum_{k=1}^{N_\mathrm{P}} |\Psi_k^{(0)}\rangle\langle\Psi_k^{(0)}|, \tag{4}$$

$$P_\mathrm{s} = \sum_{\mu=1}^{d_R} |\Phi_\mu\rangle\langle\Phi_\mu| - P_\mathrm{m}, \tag{5}$$

where P_{m} and P_{s} characterize the primary and secondary parts of the model space P, respectively. As known from many-body perturbation theory (MBPT), the first-order corrections to $\Psi_k^{(0)}$, viz.

$$|\Xi_k^{(1)}\rangle = Q\frac{1}{E_k^{(0)} - H_0}QH|\Psi_k^{(0)}\rangle = \sum_{q\in Q}|\Phi_q\rangle\tilde{C}_{qk}^{(1)}, \tag{6}$$

$$Q = 1 - P, \tag{7}$$

are very effective for describing dynamic correlation and hence good candidates for the second (external) N_{P} functions $\tilde{\Phi}_k$ in Eq. (3). To account for changes in the static correlation (described by $\Psi_k^{(0)}$) due to the inclusion of dynamic correlation (described by $\Xi_k^{(1)}$), the following *not-energy-biased* Lanczos-type functions

$$|\Theta_k^{(2)}\rangle = P_{\mathrm{s}}H|\Xi_k^{(1)}\rangle \tag{8}$$

$$\approx P'_{\mathrm{s}}H|\Xi_k^{(1)}\rangle = \sum_{\mu=1}^{d_R}|\Phi_\mu\rangle\tilde{C}_{\mu k}^{(2)}, \quad P'_{\mathrm{s}} = \sum_{l=N_{\mathrm{P}}+1}^{N_{\mathrm{P}}+M_{\mathrm{P}}}|\Psi_l^{(0)}\rangle\langle\Psi_l^{(0)}| \tag{9}$$

can be introduced [106] to mimic the third (secondary) N_{P} functions $\tilde{\Phi}_k$ in Eq. (3). Therefore, as a first attempt, Eq. (3) reads

$$|\Psi_I\rangle = \sum_{k=1}^{N_{\mathrm{P}}}|\Psi_k^{(0)}\rangle\bar{C}_{kI} + \sum_{k=1}^{N_{\mathrm{P}}}|\Xi_k^{(1)}\rangle\bar{C}_{(k+N_{\mathrm{P}})I} + \sum_{k=1}^{N_{\mathrm{P}}}|\Theta_k^{(2)}\rangle\bar{C}_{(k+2N_{\mathrm{P}})I}. \tag{10}$$

The yet-unknown coefficients $\bar{\mathbf{C}}$ are to be determined by the generalized secular equation

$$\bar{\mathbf{H}}\bar{\mathbf{C}} = \bar{\mathbf{S}}\bar{\mathbf{C}}\mathbf{E}, \tag{11}$$

where the reduced Hamiltonian $\bar{\mathbf{H}}$ and metric $\bar{\mathbf{S}}$ have the following structure:

$$\bar{\mathbf{H}} = \begin{pmatrix} P_{\mathrm{m}}HP_{\mathrm{m}} & P_{\mathrm{m}}HQ & P_{\mathrm{m}}HP_{\mathrm{s}} \\ QHP_{\mathrm{m}} & QHQ & QHP_{\mathrm{s}} \\ P_{\mathrm{s}}HP_{\mathrm{m}} & P_{\mathrm{s}}HQ & P_{\mathrm{s}}HP_{\mathrm{s}} \end{pmatrix}_{3N_{\mathrm{p}}\times 3N_{\mathrm{p}}} \tag{12}$$

$$= \begin{pmatrix} E_k^{(0)}\delta_{kl} & \langle\Psi_k^{(0)}|H|\Xi_l^{(1)}\rangle & 0 \\ \langle\Xi_l^{(1)}|H|\Psi_k^{(0)}\rangle & \langle\Xi_k^{(1)}|H|\Xi_l^{(1)}\rangle & \langle\Xi_k^{(1)}|H|\Theta_l^{(2)}\rangle \\ 0 & \langle\Theta_l^{(2)}|H|\Xi_k^{(1)}\rangle & \langle\Theta_k^{(2)}|H|\Theta_l^{(2)}\rangle \end{pmatrix}, \quad k,l\in[1,N_{\mathrm{P}}], \tag{13}$$

$$\bar{\mathbf{S}} = \begin{pmatrix} \delta_{kl} & 0 & 0 \\ 0 & \langle\Xi_k^{(1)}|\Xi_l^{(1)}\rangle & 0 \\ 0 & 0 & \langle\Theta_k^{(2)}|\Theta_l^{(2)}\rangle \end{pmatrix}, \quad k,l\in[1,N_{\mathrm{P}}]. \tag{14}$$

It has been shown [106] that the weights of $\Psi_k^{(0)}$, $\Xi_k^{(1)}$, and $\Theta_k^{(2)}$ in $|\Psi_I\rangle$ (10) do decrease, thereby justifying the static–dynamic–static characterization. Note in passing that P'_s (9) with $N_P \leq M_P \ll d_R - N_P$ is a very good approximation to the full P_s (5) but renders the evaluation of the matrix elements much easier.

Several remarks can be made here:

(1) The three sets of functions, $\{|\Psi_k^{(0)}\rangle\}_{k=1}^{N_P}$, $\{|\Xi_k^{(1)}\rangle\}_{k=1}^{N_P}$, and $\{|\Theta_k^{(2)}\rangle\}_{k=1}^{N_P}$, are treated independently, so that the dimension of Eq. (11) is $3N_P$, irrespective of the numbers of correlated electrons and orbitals. This Ansatz is best characterized as "internally and externally contracted minimal MRCI with singles and doubles and augmented with secondary states" (ixc-MRCISD+s), denoted as SDSCI for short. Note in passing that the three-subspace philosophy has long been used in configuration selection–extrapolation approaches [118, 128] but in a different way. The aim therein is to construct as small as possible a primary subspace P_m as well as a secondary (intermediate) subspace P_s, by adding in progressively individual configurations based on the magnitude of their first-order coefficients or second-order energies. Once this is done, the corrections of P_s to P_m are accounted for either variationally or at a high order of perturbation theory, whereas the corrections of the external singles and doubles to P_m are accounted for at second order. Such approaches are intrinsically state-specific MRPT2 in the spirit of intermediate Hamiltonians [129] but belong to the static-then-dynamic family of methods. They miss the interaction P_sHQ between the secondary and external subspaces as compared with the *variational* SDSCI.

(2) In principle, any state-specific external functions for dynamic correlation can be used in lieu of $\{|\Xi_k^{(1)}\rangle\}_{k=1}^{N_P}$. Therefore, the restricted SDS framework is only restricted in the structure (10) of the wave function and has sufficient degrees of freedom for improvement [106]. The "buffering" role of $\{|\Theta_k^{(2)}\rangle\}_{k=1}^{N_P}$ is obvious from the structure of $\bar{\mathbf{H}}$ (Eq. (13)): they interact with $\{|\Psi_k^{(0)}\rangle\}_{k=1}^{N_P}$ only indirectly, through interactions with $\{|\Xi_k^{(1)}\rangle\}_{k=1}^{N_P}$. Except for the particular form (8), other types of secondary states are also possible [122, 123]. For instance, the M_P ($\geq N_P$) solutions $\{\Psi_{k+N_P}^{(0)}; k = 1, \ldots, M_P\}$ of the projected Hamiltonian PHP are also good candidates in view of intermediate Hamiltonian theory [129] and are computationally very efficient.

(3) SDSCI can be reduced to MS-MRPT2 [29–32] by setting $\bar{C}_{kI} = \bar{C}_{(k+N_P)I}$ and $\bar{C}_{(k+2N_P)I} = 0$ in Eq. (10), leading to

$$\Psi_I^{(1)} = \sum_{k=1}^{N_P} (\Psi_k^{(0)} + \Xi_k^{(1)}) \bar{C}_{kI}. \tag{15}$$

The Hamiltonian matrix (11) correct to second order is then of dimension N_P, viz.

$$
\begin{aligned}
(\bar{H}^{[2]})_{kl} =& \frac{1}{2}\{\langle\Psi_k^{(0)} + \Xi_k^{(1)}|H|\Psi_l^{(0)}\rangle + \langle\Psi_k^{(0)}|H|\Psi_l^{(0)} + \Xi_l^{(1)}\rangle\} \\
=& \langle\Psi_k^{(0)}|H|\Psi_l^{(0)}\rangle + \frac{1}{2}\{\langle\Xi_k^{(1)}|H|\Psi_l^{(0)}\rangle + \langle\Psi_k^{(0)}|H|\Xi_l^{(1)}\rangle\}.
\end{aligned}
\tag{16}
$$

Yet another approximate variant of SDSCI can be obtained by simply replacing the QHQ block in Eq. (12) with QH_0Q before the diagonalization step is taken. This variant, termed SDSPT2 [106, 107], is a CI-like MRPT2, which treats single and multiple states in the same way, at a cost of $\mathcal{O}(n^5)$. From the computational point of view, all MS-MRPT2 based on Eq. (16) can be obtained from SDSPT2 for free, as long as the same zeroth-order Hamiltonian H_0 and perturbers (zeroth-order states allowed to interact directly with a reference state) are used. This is because the former amounts to just picking the $P_\mathrm{m}HP_\mathrm{m}$ and $P_\mathrm{m}HQ$ blocks of $\bar{\mathbf{H}}$ (Eq. (12)). It should be clear from Eq. (16) that the fixed, one-to-one ratio between $|\Psi_k^{(0)}\rangle$ and $|\Xi_k^{(1)}\rangle$ as well as the limited dimensionality of $\bar{\mathbf{H}}^{[2]}$ cannot give rise to sufficient revisions of the coefficients of the model functions $|\Phi_\mu\rangle$ (i.e., $C_{\mu I} = \sum_{k=1}^{N_\mathrm{P}} \tilde{C}_{\mu k}\bar{C}_{kI}$ are not much different from $\tilde{C}_{\mu k}$) on the one hand, and leads to reduced dynamic correlation on the other (i.e., the coefficient $\bar{C}_{kI}^{(1)}$ of $\Xi_k^{(1)}$ is less than 1 in magnitude, as compared with the intermediate normalization). SDSPT2 avoids both side effects by introducing the secondary states $\{\Theta_k^{(2)}\}$ and meanwhile treating the external states $\{\Xi_k^{(1)}\}$ as independent basis functions. To support this argument, we just mention that the avoided crossing point (closest approach distance) between the $1^1\Sigma^+$ and $2^1\Sigma^+$ states of LiF (resulting from the swap of the ionic and covalent characters in the wave functions) determined by SDSPT2 (7.0 Å) is much closer to those obtained by internally contracted (ic) MRCISD (7.0 Å) and even uncontracted (uc) MRCISD (6.9 Å) than those obtained by MS-CASPT2 (7.3 Å) or MS-NEVPT2 (7.3 Å). As a result, SDSPT2 has much smaller variations than MS-CASPT2 and MS-NEVPT2 in the gap error (relative to uc-MRCISD) as a function of geometry (for more details, see Ref. [107]).

(4) Both SDSCI and SDSPT2 are not size consistent. However, the errors can readily be removed by using the Pople [130] or renormalized Davidson [131] correction. Note in passing that MS-NEVPT2 [132] is also not size consistent, albeit with small size-consistency errors.

(5) At variance with the SDSPT2 approximation to SDSCI, the eigenvectors of the SDSCI equation (11) can be taken as new primary states $\{|\Psi_k^{(0)}\rangle\}_{k=1}^{N_\mathrm{P}}$, such that the SDS procedure can be iterated until convergence. Such iterative configuration interaction (iCI) [27, 28] accesses in each iteration (defined as macroiteration) a space that is higher by two ranks than that of the preceding iteration. Up to $2i$-tuple excitations (relative to the initial primary space) can be accessed if i macroiterations are carried out. A few microiterations j can be invoked in each macroiteration so as to relax the contraction coefficients. As such, every iCI(i, j) corresponds to a physically meaningful model. It has been shown both theoretically and numerically that iCI can converge quickly from above to FCI even when starting with a very poor initial guess. More generally, iCI can be viewed as a particular sequential, exact partial diagonalization of a huge matrix, by

getting first the roots of one portion of the matrix and then those of an enlarged portion, until the full matrix has been sampled, whereas its microiterations can be generalized to an iterative vector interaction (iVI) approach [122, 123] for arbitrary roots of a given matrix treated as a whole. In particular, by combining with the energy-vector following technique, iVI can directly access interior roots belonging to a predefined window, without knowing the number and characters of the roots [124, 125].

(6) As an exact FCI solver, the computational cost of iCI [27, 28] stems solely from the construction of the Hamiltonian matrix $\mathbf{H}$ in the basis of SDs/CSFs before contracting to $\tilde{\mathbf{H}}$ (Eq. (13)) for diagonalization. Given the sparsity of CI vectors, it is natural to combine iCI with configuration selection (i.e., SiCI) so as to construct a variational space P_{m} that is as compact as possible. The dynamic correlation from the complementary space $Q = 1 - P_{\mathrm{m}}$ can then be accounted for by PT2. The resulting iCIPT2 [108, 121] has the following features:

(a) The configuration selection is carried out with a combined integral- and coefficient-driven algorithm over the entire Hilbert space or a large active space P (which can in this context be viewed as a coarse-grained selection step). Specifically, starting with a guess space P_0, if there exists an orbital configuration $|J\rangle \in P_0$ for which $f(|I\mu\rangle, |J\rangle, C_{\mathrm{min}})$ is true, then the CSF $|I\mu\rangle \in Q = 1 - P_0$ is selected. The Boolean function $f(|I\mu\rangle, |J\rangle, C_{\mathrm{min}})$ is defined as

(A) If $|I\rangle$ is identical to or singly excited from $|J\rangle$, then

$$f(|I\mu\rangle, |J\rangle, C_{\mathrm{min}}) = \left(\max_{\nu}(|H^{IJ}_{\mu\nu}C_{J\nu}|) \geq C_{\mathrm{min}}\right) \\ \text{and} \left(\max_{\nu}\left(\left|\frac{H^{IJ}_{\mu\nu}C_{J\nu}}{E_0 - H^{II}_{\mu\mu}}\right|\right) \geq C_{\mathrm{min}}\right); \tag{17}$$

(B) If $|I\rangle$ is doubly excited from $|J\rangle$, then

$$f(|I\mu\rangle, |J\rangle, C_{\mathrm{min}}) = \left(\max_{\nu}(|\tilde{H}^{IJ}C_{J\nu}|) \geq C_{\mathrm{min}}\right) \\ \text{and} \left(\max_{\nu}(|H^{IJ}_{\mu\nu}C_{J\nu}|) \geq C_{\mathrm{min}}\right) \\ \text{and} \left(\max_{\nu}\left(\left|\frac{H^{IJ}_{\mu\nu}C_{J\nu}}{E_0 - H^{II}_{\mu\mu}}\right|\right) \geq C_{\mathrm{min}}\right). \tag{18}$$

Specifically, for case (A), loop over all CSFs $|I\mu\rangle$ in Q and evaluate $H^{IJ}_{\mu\nu}$ $(= \langle I\mu|H|J\nu\rangle)$ for all CSFs $|J\nu\rangle \in P_0$. If $\max_{\nu}(|H^{IJ}_{\mu\nu}C_{J\nu}|)$ is larger than C_{min}, then evaluate $H^{II}_{\mu\mu}$; otherwise discard $|I\mu\rangle$. If $\max_{\nu}(|\frac{H^{IJ}_{\mu\nu}C_{J\nu}}{E_0 - H^{II}_{\mu\mu}}|)$ is larger than C_{min}, then $|I\mu\rangle$ is selected. For case (B), only those doubly excited orbital configurations $|I\rangle$ with upper bounds $\tilde{H}^{IJ}$ [108] larger than $C_{\mathrm{min}}/\max_{\nu}|C_{J\nu}|$

need to be generated (i.e., the unimportant ones are never touched, as in heat-bath CI [92] but working with SDs). For such $\{|I\rangle\}$, the remaining step is the same as case (A). The so-selected $|I\mu\rangle$ and those in P_0 form an expanded space P_1, which is further diagonalized by iVI [122–125] to prune away those CSFs of coefficients smaller in magnitude than C_{min}, thereby reducing P_1 to P_{m} upon convergence (for more details, see Ref. [121]). The cost of such selection is merely ca. 20% of the total central processing unit (CPU) time of iCIPT2.

(b) A particle-hole representation [121] of orbital configurations is invoked to establish connections between the selected CSFs in P_{m}, which greatly facilitates the construction and update of **H** in CSR format.

(c) A tabulated unitary group approach (TUGA) [108] is invoked to compute and reuse the basic coupling coefficients (matrix elements of $U(n)$ generators E_{p}^{q} and E_{pq}^{rs} over CSF pairs), which is more flexible and efficient than the symmetric group-based table CI technique [133–136].

(d) In each iteration for the growth of the wave function, the first-order interacting space (FOIS) is decomposed into disjoint subspaces via the residue-based algorithm [104, 105, 137], so as to reduce memory requirements on the one hand and facilitate parallelization on the other.

(e) Thanks to the use of CSFs as the many-electron basis, full spin symmetry is always maintained, which is of vital importance for describing low-spin states of general open-shell systems as well as subsequent treatment of spin–orbit couplings [138].

(f) Upon termination of the selection (which is controlled by a single parameter C_{min}), the residual dynamic correlation is estimated by using the state-specific, constraint-based [105] Epstein–Nesbet [139, 140] type of PT2 (ENPT2).

(g) The linear relationship between the total versus PT2 energies allows for an accurate extrapolation, thereby rendering iCIPT2 a near-exact theory [120].

2.2 iCAS and iCISCF(2)

While sCI can automatically build up a compact variational space through an iterative procedure, there are still demands on calculations with fixed active spaces. The complete active space self-consistent field (CASSCF) method [25, 141–146] is most popular in this context due to its operational simplicity (n_e electrons are distributed in n_o orbitals in all possible ways, subject to spin and spatial symmetry constraints). Yet, CASSCF has been plagued by several factors: (1) it is by no means trivial to select which and how many orbitals as active orbitals, given some rules of thumb [147]; (2) It is often the case that some desired active orbitals run out of the CAS during SCF iterations, thereby resulting in undesired solutions or no convergence; (3) It is particularly difficult to maintain the same CAS when scanning potential energy surfaces; (4) The size of CAS grows combinatorially with respect to

n_o and n_e, to name just the major ones. Herewith, we introduce our iCAS [126] and iCISCF(2) [55] approaches to handle problems (1–3) and (4), respectively.

Like the atomic valence active space (AVAS) approach [148], iCAS assumes that the target low-lying electronic states have large projections on the space spanned by a small set of valence atomic orbitals (VAOs; e.g., *d* orbitals of transition metals, *f* orbitals of lanthanides and actinides, p_z orbitals forming aromatic rings, etc.) that can be chosen according to (minimal) chemical/physical intuitions. Such VAOs are then converted to an equivalent set of pre-canonical molecular orbitals (CMOs), by diagonalizing the pre-Fock matrix obtained by projecting the converged molecular Fock operator onto the VAOs. The occupation numbers (2, 1, or 0) of the pre-CMOs can readily be determined by examining their overlaps with the doubly, singly, and zero occupied restricted open-shell Hartree–Fock molecular orbitals (ROHF-MOs) separately. Then, the pre-CMOs are localized to yield pre-localized molecular orbitals (LMOs). The doubly/singly/zero occupied pre-LMOs are then taken as probes to select precisely the same numbers of doubly/singly/zero occupied ROHF-LMOs as guess orbitals. A clear difference between iCAS and AVAS lies in that the number of active orbitals is just that of the VAOs in iCAS but at least twice that of the same VAOs in AVAS. The form has been achieved by making use of the "energetic locality" (i.e., transferability of the occupations of fragment orbitals [149], with pre-CMOs/LMOs as special cases). A second feature of iCAS lies in that the localized CASSCF orbitals of two adjacent iterations are enforced to match most, so as to guarantee the same core, active, and virtual subspaces throughout the iterations. Herein, the localization of HF/ROHF/CASSCF orbitals can readily be achieved by means of the noniterative "top-down, least-change" algorithm [149–151]. Since the VAOs and hence the pre-LMOs do not change with geometric displacements, iCAS can always select the same guess orbitals for the whole geometry. On the other hand, the orbital matching procedure helps to reduce the possibility for the orbital optimization to converge to different CASSCF solutions at different geometries. Numerous examples have shown [126] that iCAS can indeed produce smooth potential energy surfaces, even in very difficult situations.

Regarding the CAS size problem, the above-described selected iCI can be employed as the active space solver, leading to iCISCF. Just like other approximate CASSCF approaches [38–54], the active–active orbital rotations must be accounted for in iCISCF, in addition to the core–active, core–virtual, and active–virtual orbital rotations in exact CASSCF. It turns out that the Jacobi rotation (JR) algorithm is particularly robust for the active–active orbital rotations, whereas the simple Broyden–Fletcher–Goldfarb–Shanno (BFGS)-based quasi-Newton (QN) method can be adopted for the other types of orbital rotations. Upon convergence, an ENPT2 correction can be performed within the P_s space (left over from the CAS P after configuration selection; $P = P_m + P_s$), leading to iCISCF(2). In contrast, if the ENPT2 correction is to be carried out in the whole complementary space of P_m, $Q = 1 - P_m = 1 - P + P_s$, the method would be the original iCIPT2. It has been shown [55] that LMO-iCISCF(2) can handle active spaces up to CAS(50e,50o). As such, the combination of iCAS and LMO-iCISCF(2) is a valuable approach for investigating complicated systems with multiple unpaired electrons.

For clarity, the internal relations between the SDS family of methods are further depicted in Fig. 1.

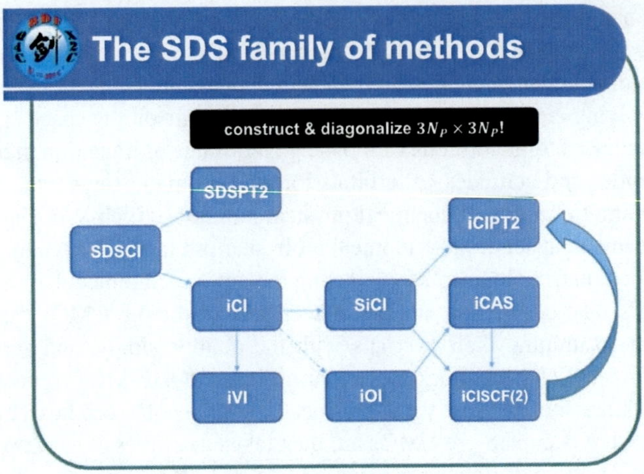

Fig. 1 The SDS family of methods (cf. Table 1)

3 Benchmarking SDSCI and SDSPT2

3.1 Choice of H_0 in SDSCI and SDSPT2

Since iCI is exact and iCIPT2 is nearly exact, there is no need to benchmark them. Instead, both SDSCI and SDSPT2 remain to be benchmarked systematically). To this end, we consider the Thiel set of systems [152]. Yet, both the zeroth-order Hamiltonian H_0 and perturbers to define $|\Xi_k^{(1)}\rangle$ (6) remain to be specified. Like NEVPT2 [32, 153, 154], we adopt here the Dyall CAS/A Hamiltonian H^{D} [155] as H_0, viz.

$$H^{\mathrm{D}} = H_{\mathrm{I}}^{D} + H_{\mathrm{A}}^{D}, \tag{19}$$

$$H_{\mathrm{I}}^{D} = \sum_i \epsilon_i E_i^i + \sum_a \epsilon_a E_a^a + C_{\mathrm{I}}^{D}, \tag{20}$$

$$H_{\mathrm{A}}^{D} = \sum_{tu} f_{tu}^{c} E_u^t + \frac{1}{2}\sum_{tuvw} (tu|vw) E_{uw}^{tv}, \tag{21}$$

$$E_q^p = \sum_\sigma a^{p\sigma} a_{q\sigma}, \quad E_{qs}^{pr} = E_q^p E_{\mathrm{s}}^r - \delta_{qr} E_{\mathrm{s}}^p, \tag{22}$$

$$C_{\mathrm{I}}^{D} = E^c - 2\sum_i \epsilon_i, \quad E^c = \sum_i (h_{ii} + f_{ii}^c), \tag{23}$$

$$f_{pq}^{c} = h_{pq} + \sum_i [2(pq|ii) - (pi|iq)], \tag{24}$$

$$\epsilon_i = \sum_k W_k \left[\langle \Psi_k^{(0)} | H | \Psi_k^{(0)} \rangle - \frac{1}{2} \sum_\sigma \langle a_{i\sigma} \Psi_k^{(0)} | H | a_{i\sigma} \Psi_k^{(0)} \rangle \right] \tag{25}$$

$$= \frac{1}{2} \sum_k W_k \sum_\sigma \langle \Psi_k^{(0)} | a^{i\sigma} [a_{i\sigma}, H] | \Psi_k^{(0)} \rangle = F_{ii}, \tag{26}$$

$$\epsilon_a = \sum_k W_k \left[\frac{1}{2} \sum_\sigma \langle a^{a\sigma} \Psi_k^{(0)} | H | a^{a\sigma} \Psi_k^{(0)} \rangle - \langle \Psi_k^{(0)} | H | \Psi_k^{(0)} \rangle \right] \tag{27}$$

$$= \frac{1}{2} \sum_k W_k \sum_\sigma \langle \Psi_k^{(0)} | [a_{a\sigma}, H] a^{a\sigma} | \Psi_k^{(0)} \rangle = F_{aa}, \tag{28}$$

$$\hat{F}_{pq} = \frac{1}{2} \sum_\sigma [a^{q\sigma}, [a_{p\sigma}, H]]_+ = h_{pq} + \sum_{rs} [(pq|rs) - \frac{1}{2}(ps|rq)] E_s^r, \tag{29}$$

$$F_{pq} = \sum_k W_k \langle \Psi_k^{(0)} | \hat{F}_{pq} | \Psi_k^{(0)} \rangle \tag{30}$$

$$= h_{pq} + \sum_{rs} [(pq|rs) - \frac{1}{2}(ps|rq)] D_{rs} \tag{31}$$

$$= f_{pq}^c + \sum_{tu} [(pq|tu) - \frac{1}{2}(pu|tq)] D_{tu}, \tag{32}$$

$$D_{rs} = \sum_k W_k \langle \Psi_k^{(0)} | E_s^r | \Psi_k^{(0)} \rangle. \tag{33}$$

The particular choice of constant C_{I}^{D} (23) is to make $\langle \Phi_\mu | H^D | \Phi_\nu \rangle$ equal to $\langle \Phi_\mu | H | \Phi_\nu \rangle$ for all CSFs in P, such that

$$\langle \Psi_k^{(0)} | H^D | \Psi_l^{(0)} \rangle = \langle \Psi_k^{(0)} | H | \Psi_l^{(0)} \rangle = E_k^{(0)} \delta_{kl}, \tag{34}$$

where

$$E_k^{(0)} = E^c + \sum_{tu} f_{tu}^c \langle \Psi_k^{(0)} | E_u^t | \Psi_k^{(0)} \rangle + \frac{1}{2} \sum_{tuvw} (tu|vw) \langle \Psi_k^{(0)} | E_{uw}^{tv} | \Psi_k^{(0)} \rangle. \tag{35}$$

Equation (34) implies that there is no first-order correction with H^{D} as H_0. Yet, there exists one caveat here: the CAS must be maintained the same for all states at all geometries to avoid discontinuities. This can be achieved by iCAS [126] outlined above. As for the perturbers, the union of linearly independent, spin-adapted, and fully internally contracted single and double excitations from all the reference functions is used, viz. $\{E^p_q|\Psi^{(0)}_k\rangle, E^p_q E^r_s|\Psi^{(0)}_k\rangle\}_{k=1}^{N_P} \backslash \{E^t_u|\Psi^{(0)}_k\rangle, E^t_u E^v_w|\Psi^{(0)}_k\rangle\}_{k=1}^{N_P}$. This is different from MS-NEVPT2 [32], where the perturbers are state specific. For more technical details, see Ref. [107].

Since the first-order functions $\{\Xi^{(1)}_k\}$ are treated in SDSCI/SDSPT2 as independent functions, some double excitations to high-lying virtual orbitals can be excluded so as to speed up the calculation. This can be done in two different ways: (1) Choose two level-thresholds ϵ_1 and ϵ_2 ($> \epsilon_1$): those virtual orbitals below ϵ_1, between ϵ_1 and ϵ_2, and above ϵ_2 can accept 2, 1, and 0 electrons, respectively; (2) More preferentially, use a small basis $\{\chi^S_\mu\}_{\mu=1}^{N_S}$ [e.g., double zeta plus polarization (DZP)] to map out a reduced set of virtual orbitals $\{\tilde{\psi}^L_a\}$ from a large basis $\{\chi^L_\mu\}_{\mu=1}^{N_L}$ calculation:

$$|\tilde{\chi}^S_\mu\rangle = |\chi^S_\mu\rangle - \sum_{i=1}^{N_{\rm occ}} |\psi^L_i\rangle\langle\psi^L_i|\chi^S_\mu\rangle, \tag{36}$$

$$\tilde{S}^{SL}_{\mu a} = \langle\tilde{\chi}^S_\mu|\psi^L_a\rangle = (\mathbf{L}\sigma\mathbf{R}^\dagger)_{\mu a}, \tag{37}$$

where $\{|\psi^L_i\rangle\}$ and $\{|\psi^L_a\rangle\}$ are the occupied and unoccupied ROHF orbitals in the basis $\{\chi^L_\mu\}$, respectively, such that $\{|\tilde{\chi}^S_\mu\rangle\}$ span the virtual space of the basis $\{\chi^S_\mu\}_{\mu=1}^{N_S}$. The singular value decomposition of the cross overlap matrix $\tilde{\mathbf{S}}^{\rm SL}$ can have at most $N_S - N_{\rm occ}$ singular values σ_a that are close to 1. The right vectors $\mathbf{R}$ can then be used to rotate the virtual orbitals,

$$|\tilde{\psi}^L_a\rangle = \sum_{b=1}^{N_L - N_{\rm occ}} |\psi^L_b\rangle R_{ba}, \quad a \in [1, N_L - N_{\rm occ}], \tag{38}$$

$N_S - N_{\rm occ}$ of which correspond to large σ_a, whereas the rest have (nearly) zero σ_a. The former can each accept two electrons, whereas the latter can each accept only one electron. Moreover, those singles and doubles with first-order coefficients smaller in magnitude than the threshold $C_{\rm min}$ can further be neglected in both cases. As such, the computational cost will not grow too much as the basis is enlarged. Suppose that SDSPT2 can be performed without any cutoff, then the SDSCI and ic-MRCISD energies can be estimated as

$$E^{\rm X}_I \approx E^{\rm X'}_I + (E^{\rm SDSPT2}_I - E^{\rm SDSPT2'}_I), \quad \mathrm{X} = \mathrm{SDSCI}, \text{ ic-MRCISD}, \tag{39}$$

where the primed values refer to those with the cutoff.

It should be clear from the formalisms that the computational cost of SDSCI is virtually the same as that of a single iteration of ic-MRCISD contracted in the same way [156, 157], whereas SDSPT2 and MS-NEVPT2 can be obtained from SDSCI and SDSPT2, respectively, at no extra cost. To make close comparisons of SDSPT2, MS-NEVPT2, and SDSCI against ic-MRCISD, all calculations were performed using the BDF program package [158–161] (see Ref. [162] for a recent review). Moreover, to enable a direct comparison with previous MS-CASPT2 and CC3 results [152], the same geometries, symmetries, active spaces, numbers of averaged states, numbers of correlated electrons, and one-particle basis set (TZVP [163]) were used here. In particular, the vertical excitation energies (VEEs) were calculated as follows [152]: the ground state was calculated with orbitals from state-specific CASSCF, whereas the energetic separations of states within an irreducible representation (irrep) were calculated with orbitals from CASSCF averaged over all the states.

3.2 Results and Discussion

3.2.1 Size Consistency

Like MRCISD, both SDSCI and SDSPT2 are not size consistent. It can be seen from Table 2 that the ground-state size-consistency errors (i.e., $\Delta E = E(\mathrm{X}\ldots\mathrm{Y}) - E(\mathrm{X}) - E(\mathrm{Y})$; $\mathrm{X} = \mathrm{Y} = \mathrm{Li2}, \mathrm{Be}, \mathrm{N}$ with the def2-TZVP basis set [164]) decrease from ic-MRCISD to SDSCI then SDSPT2. Such errors are largely removed by the (renormalized) Davidson correction [131] and especially the Pople correction [130]. In fact, it can be deduced from the underlying derivations that the Pople correction [130] is more suited for internally contracted wave functions, whereas both the original [165] and renormalized [131] Davidson corrections are more suited for uncontracted wave functions. As can be seen from Table 3, the same findings hold also for valence π–π^* excitation energies of benzene … Rg (Rg = He, Ne, Ar, Kr) relative to benzene. Here, the rage gas atom is set to 100 Å from the center of benzene, the geometry of which is taken from Ref. [152]. Hereinafter, we use X+P and X+Q to denote the Pople and (renormalized) Davidson corrections to method X, respectively. For comparison, the results by the second-order dynamic correlation dressed CAS (DCD-CAS(2)) and its bias-corrected (bc) version [166] are also given in Tables 2 and 3. It can be seen that the size-consistency errors of SDSPT2 (SDSPT2+P) are larger for the ground states but become smaller for the excited states than those by DCD-CAS(2) (bc-DCD-CAS(2)). In short, the size-consistency errors of X +P (X = SDSPT2, SDSCI, ic-MRCISD) are tolerable for both ground and excited states.

3.2.2 Excited States of Closed-Shell Systems

Thiel and coworkers reported VEEs for 149 singlet and 72 triplet states of 28 organic closed-shell molecules [152]. In fact, there are several other states that are within the reported energy ranges: the $3\,^1B_{1g}$ and $4\,^1B_{2u}$ states of naphthalene, the $4\,^1A_1$ state of pyridazine, the $3\,^1B_{2u}$ state of *s*-tetrazine, the $2\,^1B_{1g}$, $3\,^1B_{1u}$, $2\,^1A_{1u}$, $2\,^1B_{3u}$, $2\,^3B_{1g}$, $2\,^3B_{3g}$, $2\,^3A_u$, and $2\,^3B_{1u}$ states of *p*-benzoquinone, the $2\,^3A'$ state of formamide, the $2\,^3A'$ state of acetamide, and the $2\,^3A'$ state of propanamide, totaling eight singlet and seven triplet states. These states are also counted in our statistics. It should be emphasized that, to compare excited states by different methods, it is the alignment of the wave functions rather than simply the energies that should be adopted, to avoid comparing apples with oranges. For this reason, we closely inspected the results reported in Ref. [152] and found that (1) the CC3 excitation energy for the 3^1A_1 state of cyclopentadiene should be 8.69 eV instead of 6.69 eV according to the

Table 2 Ground-state size-consistency errors (in meV) of SDSPT2, SDSCI, and ic-MRCISD using the def2-TZVP basis set

System	DCD-CAS(2)[a]		SDSPT2[b]			SDSCI[b]			ic-MRCISD[b]			uc-MRCISD[b]		
	Original	bc	Original	Pople	Davidson	Original	Pople	Davidson	Original	Pople	Davidson	Original	Pople	Davidson
$Li_2 \ldots Li_2^c$	0.00	0.00	41.55	0.94	2.62	68.77	1.49	8.15	215.35	126.19	181.24	216.72	123.42	180.52
$Be \ldots Be^d$	0.11	0.01	3.16	0.01	0.04	4.65	0.02	0.10	6.85	0.49	0.68	6.62	0.23	0.41
$N \ldots N^e$	0.00	0.00	122.20	0.01	9.18	199.88	0.00	20.97	208.57	6.03	29.53	166.19	-61.68	-37.52

The noninteracting systems were placed 100 Å away in supermolecular calculations

[a]Ref. [166]

[b]This work [c]CAS(2,2) for Li_2 and CAS(4,4) for $Li_2 \ldots Li_2$, Li–Li = 1.0 Å

[d]CAS(2,4) for Be and CAS(4,8) for Be … Be

[e]CAS(5,4) for N and CAS(10,8) for N … N

Table 3 Size-consistency errors (in meV) of SDSPT2, SDSCI, and ic-MRCISD for valence π–π^* excitation energies of benzene … Rg (Rg = He, Ne, Ar, Kr; 100 Å away from the center of benzene) with respect to benzene[a].

Excited state	DCD-CAS(2)[b]				SDSPT2[c]				SDSCI[c]				ic-MRCISD[c]			
	He	Ne	Ar	Kr	He	Ne	Ar	Kr	He	Ne	Ar	Kr	He	Ne	Ar	Kr
	Original				Original											
$^1B_{2u}$	22.1	158.7	278.4	306.5	−1.3	−11.8	−10.0	−13.1	−1.5	−4.0	−1.4	−0.4	−0.5	−3.5	−4.2	−4.7
$^1B_{1u}$	35.0	252.8	436.6	480.6	12.9	124.0	84.1	119.1	17.7	107.7	78.9	94.6	8.9	59.7	59.5	70.3
$^1E_{1u}$	40.9	296.2	507.5	558.5	21.0	203.5	131.2	189.3	28.9	181.7	127.8	156.3	12.8	86.5	83.2	99.3
$^1E_{2g}$	36.1	260.6	448.9	494.1	−1.2	−10.8	−11.1	−13.8	−1.3	−1.4	−0.3	1.5	−0.7	−4.4	−5.9	−6.5
$^3B_{1u}$	−0.1	−1.1	−1.9	−2.0	−3.3	−30.5	−23.5	−31.7	−4.1	−20.0	−13.5	−14.9	−1.7	−11.3	−12.0	−13.9
$^3E_{1u}$	5.2	37.5	67.0	73.8	2.0	19.7	11.7	17.4	3.0	20.8	15.8	19.7	2.1	14.7	13.1	16.0
$^3B_{2u}$	15.1	108.6	192.7	212.2	15.7	151.1	100.2	143.0	21.5	127.4	88.2	105.8	8.7	58.3	58.4	68.8
$^3E_{2g}$	15.5	111.7	197.6	217.6	−2.4	−21.6	−18.6	−24.2	−2.7	−10.8	−7.2	−6.8	−1.3	−8.7	−9.8	−11.2
ME[d]	183.2				34.9				34.6				19.4			
MAE[d]	183.5				49.2				40.2				25.6			
	Bias correction				Pople correction											
$^1B_{2u}$	0.8	5.3	16.9	18.9	0.2	0.5	1.3	2.5	−2.2	4.8	−4.5	3.0	0.4	2.9	2.7	3.5
$^1B_{1u}$	1.8	10.4	37.3	41.3	−2.4	−9.0	−16.9	−30.2	4.7	−15.9	2.8	−27.4	2.6	40.3	9.9	18.4
$^1E_{1u}$	2.6	15.8	52.1	57.5	−3.2	−6.8	−22.9	−44.1	7.1	−12.8	7.6	−36.6	6.5	78.6	27.8	45.2
$^1E_{2g}$	2.4	15.0	45.2	50.0	0.6	3.8	3.6	5.7	−2.3	9.4	−3.2	6.7	1.0	7.1	5.9	7.7
$^3B_{1u}$	−0.1	−0.5	−0.8	−0.9	0.0	−2.7	0.4	2.4	−3.0	2.6	−6.9	3.2	0.0	−1.7	0.8	0.6
$^3E_{1u}$	0.2	1.3	2.8	3.1	−0.1	1.0	−0.7	−2.1	−0.3	1.3	−1.5	−1.6	1.6	17.1	7.4	11.3
$^3B_{2u}$	0.2	0.7	5.8	6.6	−2.8	−9.1	−19.4	−35.9	8.0	−15.3	12.6	−27.4	2.4	38.2	9.0	17.2
$^3E_{2g}$	0.6	4.0	11.1	12.4	0.2	−0.2	1.5	3.5	−2.9	5.5	−6.0	4.3	0.6	3.4	4.0	5.0
ME[d]	13.1				−5.7				−2.7				11.8			

Table 3 (continued)

Excited state	DCD-CAS(2)[b]				SDSPT2[c]				SDSCI[c]				ic-MRCISD[c]			
	He	Ne	Ar	Kr	He	Ne	Ar	Kr	He	Ne	Ar	Kr	He	Ne	Ar	Kr
MAE[d]	13.3				7.4				7.9				11.9			
					Davidson correction											
$^1B_{2u}$					−0.6	−5.5	−5.8	−7.3	−1.8	2.6	−1.7	3.7	0.0	0.5	−0.7	−0.6
$^1B_{1u}$					6.3	58.6	54.4	71.2	12.9	43.4	60.2	53.2	7.5	56.6	52.5	65.4
$^1E_{1u}$					11.3	105.5	94.2	123.6	21.5	86.1	105.9	99.8	12.8	95.6	84.5	106.3
$^1E_{2g}$					−0.3	−2.5	−3.9	−4.6	−1.6	7.6	1.6	9.0	0.4	3.0	0.6	1.1
$^3B_{1u}$					−2.0	−17.3	−16.8	−21.4	−3.9	−7.1	−12.9	−7.7	−1.0	−6.2	−7.1	−8.6
$^3E_{1u}$					1.2	10.8	9.4	12.4	1.8	12.2	12.9	15.7	2.5	18.9	15.8	20.0
$^3B_{2u}$					7.9	73.7	67.4	88.1	16.7	52.6	72.5	60.5	7.2	54.5	50.9	63.3
$^3E_{2g}$					−1.2	−11.0	−11.5	−14.4	−2.9	−0.1	−6.2	0.3	−0.3	−1.5	−3.3	−3.7
ME[d]					20.9				22.1				21.5			
MAE[d]					28.8				24.9				23.5			

The computations were performed at the CAS(6,6)/def2-TZVP level with valence π electrons and π orbitals in the active space

[a]The geometry of benzene is taken from Ref. [152]

[b]Ref. [166]

[c]This work

[d]*ME* mean size-consistency error, *MAE* mean absolute size-consistency error

supporting information provided therein; (2) the CC3 excitation energy for the $2\,^1A'$ state of formamide should be 7.24 eV instead of 8.27 eV, while the latter should be assigned [167] to $3\,^1A'$; (3) according to the CC3 calculations carried out in the present study, the CC3 excitation energy for the $2\,^1A_1$ state of formaldehyde should be 9.53 eV instead of 10.45 eV, while the latter should be assigned to $3\,^1A_1$, which has a much reduced π–$\pi*$ component compared with $2\,^1A_1$; (4) the MS-CASPT2 $5\,^1A'$ ($6\,^1A'$) state of adenine assigned originally to the 10th (13th) root of equation-of-motion coupled cluster with singles and doubles (EOM-CCSD) must be reassigned to the 10th (8th) state of EOM-CCSD carried out in the present study. Having done such careful assignment, we collect the calculated VEEs in Tables 4 and 5 for the 157 singlet and 79 triplet states, respectively, which are to be compared with the previous CC3 results (139 singlet and 71 triplet excited states) [152, 167]. The missing states by SDSCI and ic-MRCISD are due to lack of memory, as these two modules have not yet been parallelized. Given so many values, we decided to carry out separate statistical analyses of the singlet and triplet states.

3.2.3 Statistical Analysis of Singlet States

MS-CASPT2, MS-NEVPT2, SDSPT2, SDSCI, and ic-MRCISD should first be compared because they all share the same function space and differ only in how the expansion coefficients are determined. The statistical measures of MS-CASPT2 [152], MS-NEVPT2, SDSPT2, and SDSCI against ic-MRCISD are given in Table 6 and further plotted in Fig. 2. Here, ic-MRCISD+P/Q and ic-MRCISD+P are taken as the references for SDSCI/SDSPT2+P/Q and MS-CASPT2/NEVPT2, respectively. As expected, SDSCI+P is the most accurate approximation to ic-MRCISD. SDSPT2+P is also closer to ic-MRCISD than both MS-NEVPT2 and MS-CASPT2.

The statistical measures of the above MR methods against CC3 [152, 167] are given in Table 7 and further plotted in Fig. 3. It appears that MS-CAPPT2/NEVPT2, SDSPT2/SDSCI+P, and ic-MRCISD+P all differ from CC3 by ca. 0.2–0.3 eV on average. Yet, there exist a number of states for which ic-MRCISD differs from CC3 by more than 0.5 eV (Table 8). To confirm which is better, ic-MRCISD or CC3, we further performed iCIPT2 [108, 121] calculations on these states. Here, the natural orbitals determined by iCIPT2 with $C_{\text{min}} = 1.0 \times 10^{-4}$ are utilized for subsequent iCIPT2 calculations with smaller C_{min}. The total iCIPT2 versus PT2 energies are then linearly extrapolated to obtain the best estimates of the VEEs. It is can be seen from Table 8 that CC3 is clearly better than ic-MRCISD.

Apart from internal comparisons between theoretical methods, the calculated results should eventually be compared with experimental measures. However, as pointed out repeatedly [168–172], theoretically calculated VEEs are always blue-shifted as compared with the peak positions of experimental spectra. This arises from differences between the vibrations on the excited- and ground-state surfaces. Such shifts can readily be estimated by means of, e.g., the nuclear ensemble approach [173]. Among the considered states of organic molecules [173], 25 singlet states appear in the Thiel test set. The 25 experimental results [173–191] before and after the vibrational corrections can be found from Table 9. Again, both

Table 4 Vertical excitation energies (in eV) of singlet states

Molecule	State	Type	MS-CASPT2[a]	MS-NEVPT2[b]	SDSPT2[b]		SDSCI[b]		ic-MRCISD[b]		CC3[c]
					Pople	Davidson	Pople	Davidson	Pople	Davidson	
Ethene	$1\,^1B_{1u}$	$\pi \rightarrow \pi^*$	8.62	8.64	8.70	8.70	8.83	8.83	8.45	8.45	8.37
E-butadiene	$2\,^1A_g$	$\pi \rightarrow \pi^*$	6.63	6.84	6.80	6.79	6.50	6.51	6.63	6.63	6.77
	$1\,^1B_u$	$\pi \rightarrow \pi^*$	6.47	6.16	6.32	6.44	7.14	7.23	6.76	6.85	6.58
All-*E*-hexatriene	$2\,^1A_g$	$\pi \rightarrow \pi^*$	5.42	5.57	5.56	5.57	5.42	5.45	5.48	5.50	5.72
	$1\,^1B_u$	$\pi \rightarrow \pi^*$	5.31	4.81	5.16	5.45	6.21	6.40	5.97	6.10	5.58
All-*E*-octatetraene	$2\,^1A_g$	$\pi \rightarrow \pi^*$	4.64	4.73	4.72	4.73	4.62	4.65	4.70	4.71	4.97
	$1\,^1B_u$	$\pi \rightarrow \pi^*$	4.70	4.02	4.47	4.91	5.56	5.86	5.45	5.61	4.94
Cyclopropene	$1\,^1B_1$	$\sigma \rightarrow \pi^*$	6.76	6.87	6.91	6.92	7.07	7.08	6.92	6.93	6.90
	$1\,^1B_2$	$\pi \rightarrow \pi^*$	7.06	7.10	7.22	7.26	7.55	7.59	7.28	7.33	7.10
Cyclopentadiene	$2\,^1A_1$	$\pi \rightarrow \pi^*$	6.31	6.48	6.66	6.67	6.51	6.53	6.54	6.55	6.61
	$3\,^1A_1$	$\pi \rightarrow \pi^*$	8.52	8.38	8.44	8.73	9.19	9.42	8.87	9.05	8.69
	$1\,^1B_2$	$\pi \rightarrow \pi^*$	5.51	5.16	5.34	5.54	6.25	6.38	5.85	5.98	5.73
Norbornadiene	$1\,^1A_2$	$\pi \rightarrow \pi^*$	5.34	4.98	5.21	5.58	6.23	6.47	5.95	6.15	5.64
	$2\,^1A_2$	$\pi \rightarrow \pi^*$	7.44	7.10	7.28	7.76	8.44	8.78	8.14	8.39	7.71
	$1\,^1B_2$	$\pi \rightarrow \pi^*$	6.11	5.78	6.01	6.53	7.27	7.63	6.93	7.21	6.49
	$2\,^1B_2$	$\pi \rightarrow \pi^*$	7.32	6.99	7.22	7.74	8.40	8.77	8.06	8.34	7.64
Benzene	$1\,^1E_{2g}$	$\pi \rightarrow \pi^*$	8.21	8.42	8.40	8.39	8.13	8.14	8.24	8.24	8.43
	$1\,^1B_{1u}$	$\pi \rightarrow \pi^*$	6.45	6.39	6.55	6.68	7.00	7.10	6.79	6.89	6.68
	$1\,^1B_{2u}$	$\pi \rightarrow \pi^*$	5.05	5.22	5.21	5.19	4.99	4.99	5.07	5.07	5.07
	$1\,^1E_{1u}$	$\pi \rightarrow \pi^*$	7.07	7.10	7.27	7.52	7.94	8.14	7.66	7.82	7.45
Naphthalene	$2\,^1A_g$	$\pi \rightarrow \pi^*$	5.90	6.24	6.22	6.17	5.82	5.85	5.97	5.96	5.98
	$3\,^1A_g$	$\pi \rightarrow \pi^*$	6.71	6.87	6.85	6.88	6.67	6.74	6.80	6.81	6.90

Table 4 (continued)

Molecule	State	Type	MS-CASPT2[a]	MS-NEVPT2[b]	SDSPT2[b]		SDSCI[b]		ic-MRCISD[b]		CC3[c]
					Pople	Davidson	Pople	Davidson	Pople	Davidson	
	$1\,^1B_{1g}$	$\pi \rightarrow \pi^*$	6.00	5.96	6.05	6.18	6.17	6.29	6.20	6.25	6.07
	$2\,^1B_{1g}$	$\pi \rightarrow \pi^*$	6.48	6.23	6.86	7.18	7.24	7.60	7.25	7.42	6.79
	$3\,^1B_{1g}$	$\pi \rightarrow \pi^*$		7.74	7.22	7.42	7.52	7.60	7.43	7.51	
	$1\,^1B_{2u}$	$\pi \rightarrow \pi^*$	4.77	4.33	4.52	4.96	5.47	5.74	5.21	5.41	5.03
	$2\,^1B_{2u}$	$\pi \rightarrow \pi^*$	6.33	5.97	6.17	6.62	7.01	7.32	6.78	6.99	6.57
	$3\,^1B_{2u}$	$\pi \rightarrow \pi^*$	8.18	7.83	8.20	8.79	8.91	9.39	8.94	9.16	8.44
	$4\,^1B_{2u}$	$\pi \rightarrow \pi^*$		9.06	9.07	9.06	8.74	8.79	8.84	8.85	
	$1\,^1B_{3u}$	$\pi \rightarrow \pi^*$	4.24	4.36	4.35	4.35	4.23	4.25	4.27	4.28	4.27
	$2\,^1B_{3u}$	$\pi \rightarrow \pi^*$	6.07	5.55	5.86	6.46	7.29	7.55	6.71	6.94	6.33
	$3\,^1B_{3u}$	$\pi \rightarrow \pi^*$	7.76	7.97	7.87	7.91	7.41	7.62	7.79	7.83	8.12
Furan	$2\,^1A_1$	$\pi \rightarrow \pi^*$	6.52	6.79	6.74	6.74	6.57	6.59	6.67	6.67	6.62
	$3\,^1A_1$	$\pi \rightarrow \pi^*$	8.22	8.25	8.38	8.62	8.93	9.11	8.70	8.84	8.53
	$1\,^1B_2$	$\pi \rightarrow \pi^*$	6.43	6.30	6.41	6.61	7.05	7.20	6.81	6.93	6.60
Pyrrole	$2\,^1A_1$	$\pi \rightarrow \pi^*$	6.31	6.60	6.56	6.57	6.37	6.40	6.45	6.46	6.40
	$3\,^1A_1$	$\pi \rightarrow \pi^*$	8.17	8.34	8.33	8.50	8.60	8.75	8.47	8.57	8.17
	$1\,^1B_2$	$\pi \rightarrow \pi^*$	6.33	6.58	6.64	6.79	7.06	7.18	6.93	7.02	6.71
Imidazole	$2\,^1A'$	$\pi \rightarrow \pi^*$	6.19	6.54	6.72	6.75	6.68	6.72	6.73	6.75	6.58
	$3\,^1A'$	$\pi \rightarrow \pi^*$	6.93	7.11	7.01	7.13	7.26	7.37	7.20	7.27	7.10
	$4\,^1A'$	$\pi \rightarrow \pi^*$	8.15	8.44	8.46	8.65	8.79	8.96	8.65	8.76	8.45

Table 4 (continued)

Molecule	State	Type	MS-CASPT2[a]	MS-NEVPT2[b]	SDSPT2[b]		SDSCI[b]		ic-MRCISD[b]		CC3[c]
					Pople	Davidson	Pople	Davidson	Pople	Davidson	
Pyridine[d]	$1\,^1A''$	$n \rightarrow \pi^*$	6.81	6.94	6.98	6.99	6.84	6.87	6.86	6.88	6.82
	$2\,^1A''$	$\pi \rightarrow \pi^*$	7.91	8.03	8.07	8.11	8.04	8.08	8.03	8.06	7.93
	$2\,^1A_1$	$\pi \rightarrow \pi^*$	6.39	6.39	6.58	6.74	7.16	7.26	7.00	7.08	6.85
	$3\,^1A_1$	$\pi \rightarrow \pi^*$	7.46	7.25	8.01	8.17	8.23	8.46	8.02	8.18	7.70
	$4\,^1A_1$	$\pi \rightarrow \pi^*$	8.70	8.80	8.24	8.34	8.37	8.39	8.41	8.43	8.68
	$1\,^1A_2$	$n \rightarrow \pi^*$	5.47	5.47	5.57	5.62	5.74	5.78	5.66	5.70	5.50
	$1\,^1B_1$	$n \rightarrow \pi^*$	5.14	5.26	5.29	5.29	5.17	5.19	5.20	5.21	5.05
	$1\,^1B_2$	$\pi \rightarrow \pi^*$	5.02	5.26	5.30	5.29	5.08	5.09	5.12	5.13	5.15
	$2\,^1B_2$	$\pi \rightarrow \pi^*$	7.29	6.96	7.76	7.92	8.21	8.25	8.06	8.12	7.59
	$3\,^1B_2$	$\pi \rightarrow \pi^*$	8.62	8.73	8.18	8.34	8.39	8.63	8.29	8.41	8.77
Pyrazine	$2\,^1A_g$	$\pi \rightarrow \pi^*$	8.61	8.89	8.85	8.83	8.46	8.47	8.59	8.58	8.69
	$1\,^1B_{1g}$	$n \rightarrow \pi^*$	6.41	6.50	6.68	6.77	6.85	6.93	6.88	6.94	6.75
	$1\,^1B_{2g}$	$n \rightarrow \pi^*$	5.68	5.87	5.85	5.86	5.66	5.69	5.80	5.81	5.74
	$1\,^1B_{3g}$	$\pi \rightarrow \pi^*$	8.47	8.74	8.71	8.69	8.43	8.43	8.54	8.53	8.77
	$1\,^1A_u$	$n \rightarrow \pi^*$	4.70	4.72	4.88	5.04	5.30	5.42	5.27	5.36	5.05
	$1\,^1B_{1u}$	$\pi \rightarrow \pi^*$	6.89	6.75	6.95	7.10	7.45	7.57	7.19	7.31	7.07
	$2\,^1B_{1u}$	$\pi \rightarrow \pi^*$	7.79	7.71	7.91	8.23	8.74	8.97	8.35	8.55	8.06
	$1\,^1B_{2u}$	$\pi \rightarrow \pi^*$	4.85	5.11	5.24	5.24	5.01	5.03	5.01	5.03	5.02
	$2\,^1B_{2u}$	$\pi \rightarrow \pi^*$	7.65	7.61	7.75	8.06	8.55	8.79	8.26	8.42	8.05

Table 4 (continued)

Molecule	State	Type	MS-CASPT2[a]	MS-NEVPT2[b]	SDSPT2[b]		SDSCI[b]		ic-MRCISD[b]		CC3[c]
					Pople	Davidson	Pople	Davidson	Pople	Davidson	
	$1\,^1B_{3u}$	$n \rightarrow \pi^*$	4.12	4.21	4.19	4.27	4.36	4.42	4.43	4.47	4.24
Pyrimidine	$2\,^1A_1$	$\pi \rightarrow \pi^*$	6.64	6.67	6.86	7.01	7.41	7.50	7.23	7.31	7.06
	$3\,^1A_1$	$\pi \rightarrow \pi^*$	7.21	7.40	7.93	8.08	8.34	8.40	8.16	8.23	7.74
	$1\,^1A_2$	$n \rightarrow \pi^*$	4.81	4.82	4.85	4.93	5.06	5.13	5.10	5.15	4.93
	$1\,^1B_1$	$n \rightarrow \pi^*$	4.44	4.53	4.52	4.57	4.58	4.63	4.65	4.68	4.50
	$1\,^1B_2$	$\pi \rightarrow \pi^*$	5.24	5.53	5.58	5.57	5.31	5.32	5.30	5.32	5.36
	$2\,^1B_2$	$\pi \rightarrow \pi^*$	7.64	7.53	8.19	8.40	8.65	8.89	8.30	8.48	8.01
Pyridazine	$2\,^1A_1$	$\pi \rightarrow \pi^*$	5.18	5.53	5.49	5.45	5.10	5.09	5.22	5.21	5.22
	$3\,^1A_1$	$\pi \rightarrow \pi^*$	7.62	7.44	8.44	8.59	8.57	8.81	8.36	8.51	7.82
	$4\,^1A_1$	$\pi \rightarrow \pi^*$		9.18	8.30	8.41	8.50	8.52	8.52	8.54	
	$1\,^1A_2$	$n \rightarrow \pi^*$	4.32	4.43	4.44	4.53	4.63	4.70	4.67	4.71	4.49
	$2\,^1A_2$	$n \rightarrow \pi^*$	5.77	6.04	5.99	6.04	5.89	5.97	5.98	6.03	5.74
	$1\,^1B_1$	$n \rightarrow \pi^*$	3.78	3.88	3.86	3.95	4.06	4.14	4.10	4.14	3.92
	$2\,^1B_1$	$n \rightarrow \pi^*$	6.52	6.73	6.70	6.78	6.67	6.76	6.76	6.81	6.41
	$1\,^1B_2$	$\pi \rightarrow \pi^*$	6.31	6.35	6.69	6.86	7.41	7.50	7.09	7.19	6.93
	$2\,^1B_2$	$\pi \rightarrow \pi^*$	7.29	7.10	7.74	7.91	8.20	8.27	7.97	8.06	7.55
s-Triazine	$2\,^1A_1'$	$\pi \rightarrow \pi^*$	7.25	7.14	7.25	7.40	7.79	7.89	7.58	7.66	7.41
	$1\,^1A_1''$	$n \rightarrow \pi^*$	4.60	4.62	4.75	4.87	4.97	5.07	5.02	5.09	4.78
	$1\,^1A_2'$	$\pi \rightarrow \pi^*$	5.79	5.92	5.89	5.87	5.59	5.59	5.71	5.70	5.71

Table 4 (continued)

Molecule	State	Type	MS-CASPT2[a]	MS-NEVPT2[b]	SDSPT2[b]		SDSCI[b]		ic-MRCISD[b]		CC3[c]
					Pople	Davidson	Pople	Davidson	Pople	Davidson	
	$1\,^1A_2''$	$n \rightarrow \pi^*$	4.68	4.89	4.86	4.90	4.83	4.88	4.95	4.97	4.76
	$1\,^1E'$	$\pi \rightarrow \pi^*$	7.49	7.65	8.39	8.52	8.67	8.71	8.56	8.61	8.04
	$2\,^1E'$	$\pi \rightarrow \pi^*$	8.99	9.59	9.08	9.20	9.14	9.33	9.05	9.16	9.44
	$1\,^1E''$	$n \rightarrow \pi^*$	4.71	4.80	4.58	4.77	5.04	5.19	5.07	5.17	4.81
	$2\,^1E''$	$n \rightarrow \pi^*$	7.72	8.06	8.00	8.07	7.87	7.95	7.99	8.03	7.80
s-Tetrazine	$1\,^1B_{1g}$	$n \rightarrow \pi^*$	4.73	4.62	4.89	4.99	4.98	5.06	5.11	5.15	4.97
	$2\,^1B_{1g}$	$n \rightarrow \pi^*$	6.38	6.65	6.66	6.75	6.63	6.73	6.83	6.88	6.87
	$3\,^1B_{1g}$	$n \rightarrow \pi^*$	6.74	6.93	6.93	7.03	6.89	7.01	7.04	7.11	7.08
	$1\,^1B_{2g}$	$n \rightarrow \pi^*$	5.18	5.48	5.46	5.48	5.17	5.22	5.38	5.39	5.34
	$2\,^1B_{2g}$	$n \rightarrow \pi^*$	6.07	6.35	6.22	6.32	6.21	6.32	6.42	6.47	6.23
	$1\,^1B_{3g}$	$n, n \rightarrow \pi *, \pi^*$	5.79	6.15	6.16	6.27	6.32	6.42	6.45	6.50	
	$2\,^1B_{3g}$	$\pi \rightarrow \pi^*$	8.32	8.67	8.63	8.61	8.03	8.06	8.33	8.32	8.47
	$1\,^1A_u$	$\pi \rightarrow \pi^*$	3.48	3.65	3.65	3.80	3.91	4.04	4.03	4.11	3.79
	$2\,^1A_u$	$n \rightarrow \pi^*$	5.47	5.78	5.67	5.78	5.64	5.76	5.82	5.89	5.46
	$1\,^1B_{1u}$	$\pi \rightarrow \pi^*$	6.96	6.40	6.90	7.20	7.83	8.03	7.62	7.78	7.45
	$2\,^1B_{1u}$	$\pi \rightarrow \pi^*$	7.43	7.15	7.55	7.92	8.26	8.60	8.11	8.32	7.79
	$1\,^1B_{2u}$	$\pi \rightarrow \pi^*$	4.91	5.03	5.34	5.35	4.94	4.97	5.12	5.13	5.12
	$2\,^1B_{2u}$	$\pi \rightarrow \pi^*$	8.15	7.67	8.64	8.99	9.08	9.51	9.06	9.27	8.51
	$3\,^1B_{2u}$	$\pi \rightarrow \pi^*$		8.34	9.23	9.34	8.12	8.22	8.29	8.35	

Table 4 (continued)

Molecule	State	Type	MS-CASPT2[a]	MS-NEVPT2[b]	SDSPT2[b]		SDSCI[b]		ic-MRCISD[b]		CC3[c]
					Pople	Davidson	Pople	Davidson	Pople	Davidson	
	$1\,^1B_{3u}$	$n \rightarrow \pi^*$	2.24	2.37	2.33	2.47	2.60	2.71	2.71	2.77	2.53
	$2\,^1B_{3u}$	$n \rightarrow \pi^*$	6.77	7.17	7.07	7.19	6.96	7.10	7.17	7.25	6.67
Formaldehyde	$2\,^1A_1^{e}$	$\pi \rightarrow \pi^*$	9.31	9.23	9.37	9.39	10.17	10.17	9.53	9.54	9.53[b,f]
	$1\,^1A_2$	$n \rightarrow \pi^*$	3.98	4.22	4.26	4.27	4.03	4.03	4.00	4.00	3.95
	$1\,^1B_1$	$\sigma \rightarrow \pi^*$	9.14	9.40	9.43	9.44	9.24	9.24	9.23	9.23	9.18
Acetone	$2\,^1A_1$	$\pi \rightarrow \pi^*$	9.31	9.30	9.39	9.50	9.68	9.80	9.40	9.50	9.65
	$1\,^1A_2$	$n \rightarrow \pi^*$	4.42	4.47	4.49	4.50	4.58	4.59	4.51	4.53	4.40
	$1\,^1B_1$	$\sigma \rightarrow \pi^*$	9.27	9.50	9.53	9.56	9.60	9.63	9.47	9.51	9.17
p-Benzoquinone	$1\,^1B_{1g}$	$n \rightarrow \pi^*$	2.78	2.52	2.85	2.99	3.07	3.17	3.01	3.08	2.75
	$2\,^1B_{1g}$	$n \rightarrow \pi^*$		6.35	6.34	6.43	6.31	6.41	6.38	6.42	
	$1\,^1B_{3g}$	$\pi \rightarrow \pi^*$	4.25	3.89	4.17	4.48	5.01	5.20	4.82	4.97	4.59
	$2\,^1B_{3g}$	$\pi \rightarrow \pi^*$	6.98	6.67	6.87	7.16	7.54	7.73	7.43	7.55	7.27
	$1\,^1A_u$	$n \rightarrow \pi^*$	2.80	2.51	2.84	2.98	3.08	3.18	3.02	3.08	2.85
	$2\,^1A_u$	$n \rightarrow \pi^*$		6.33	6.32	6.41	6.29	6.39	6.36	6.40	
	$1\,^1B_{1u}$	$\pi \rightarrow \pi^*$	5.29	4.83	5.03	5.52	6.19	6.52	5.92	6.16	5.62
	$2\,^1B_{1u}$	$\pi \rightarrow \pi^*$	7.91	7.59	7.71	8.22	8.74	9.12	8.54	8.78	7.82
	$3\,^1B_{1u}$	$\pi \rightarrow \pi^*$		7.15	7.60	7.74	7.86	7.95	7.81	7.88	
	$1\,^1B_{3u}$	$n \rightarrow \pi^*$	5.60	5.85	5.81	5.96	5.91	6.05	5.99	6.06	5.82
	$2\,^1B_{3u}$	$n \rightarrow \pi^*$		6.21	6.28	6.86	7.33	7.76	7.25	7.49	

Table 4 (continued)

Molecule	State	Type	MS-CASPT2[a]	MS-NEVPT2[b]	SDSPT2[b]		SDSCI[b]		ic-MRCISD[b]		CC3[c]
					Pople	Davidson	Pople	Davidson	Pople	Davidson	
Formamide	$2\,^1A'$	$\pi \rightarrow \pi^*$	7.44	7.59	7.71	7.77	8.05	8.11	7.72	7.78	7.24
	$3\,^1A'$	$\pi \rightarrow \pi^*$	10.54	10.78	10.88	10.95	11.14	11.20	10.89	10.93	10.93
	$1\,^1A''$	$n \rightarrow \pi^*$	5.63	5.93	5.97	5.96	5.57	5.57	5.64	5.64	5.65
Acetamide	$2\,^1A'$	$\pi \rightarrow \pi^*$	7.27	7.49	7.61	7.74	8.03	8.13	7.76	7.84	7.67
	$3\,^1A'$	$\pi \rightarrow \pi^*$	10.09	10.30	10.43	10.58	10.87	10.99	10.60	10.68	10.50
	$1\,^1A''$	$n \rightarrow \pi^*$	5.80	5.95	5.99	5.97	5.61	5.61	5.70	5.70	5.69
Propanamide	$2\,^1A'$	$\pi \rightarrow \pi^*$	7.20	7.43	7.55	7.74	8.02	8.17	7.77	7.87	7.62
	$3\,^1A'$	$\pi \rightarrow \pi^*$	9.94	10.16	10.29	10.52	10.77	10.96	10.53	10.65	10.06
	$1\,^1A''$	$n \rightarrow \pi^*$	5.72	5.99	6.04	6.01	5.66	5.66	5.73	5.73	5.72
Cytosine	$2\,^1A'$	$\pi \rightarrow \pi^*$	4.68	4.79	4.72	4.83	4.81	4.91	4.85	4.89	4.72
	$3\,^1A'$	$\pi \rightarrow \pi^*$	5.54	5.59	5.60	5.78	5.84	5.98	5.88	5.95	5.61
	$4\,^1A'$	$\pi \rightarrow \pi^*$	6.40	6.56	6.57	6.91	6.95	7.22	6.93	7.08	6.61
	$5\,^1A'$	$\pi \rightarrow \pi^*$	6.98	7.00	6.94	7.19	7.24	7.45	7.24	7.35	
	$6\,^1A'$	$\pi \rightarrow \pi^*$	8.23	8.36	8.24	8.42	8.35	8.52	8.44	8.51	
	$1\,^1A''$	$n \rightarrow \pi^*$	5.12	5.42	5.36	5.43	5.22	5.31	5.34	5.38	5.16
	$2\,^1A''$	$n \rightarrow \pi^*$	5.54	5.82	5.74	5.80	5.54	5.63	5.67	5.70	5.52
Thymine	$2\,^1A'$	$\pi \rightarrow \pi^*$	5.06	5.12	5.15	5.47	5.61	5.84			5.34
	$3\,^1A'$	$\pi \rightarrow \pi^*$	6.15	6.18	6.26	6.53	6.64	6.84			6.34
	$4\,^1A'$	$\pi \rightarrow \pi^*$	6.52	6.55	6.56	6.96	7.05	7.35			6.71
	$5\,^1A'$	$\pi \rightarrow \pi^*$	7.43	7.46	7.47	7.93	7.96	8.32			
	$1\,^1A''$	$n \rightarrow \pi^*$	4.94	4.90	4.93	5.08	5.02	5.16	4.96	5.13	4.98
	$2\,^1A''$	$n \rightarrow \pi^*$	6.38	6.30	6.45	6.60	6.51	6.65	6.45	6.63	6.45

Table 4 (continued)

Molecule	State	Type	MS-CASPT2[a]	MS-NEVPT2[b]	SDSPT2[b]		SDSCI[b]		ic-MRCISD[b]		CC3[c]
					Pople	Davidson	Pople	Davidson	Pople	Davidson	
	$3\,^1A''$	$n \rightarrow \pi^*$	6.86	6.71	6.65	7.11	7.30	7.66	7.14	7.44	
	$4\,^1A''$	$n \rightarrow \pi^*$	7.43	7.50	7.44	7.73	7.70	7.96	7.67	7.90	
Uracil	$2\,^1A'$	$\pi \rightarrow \pi^*$	5.23	5.30	5.34	5.57	5.67	5.84	5.66	5.76	5.44
	$3\,^1A'$	$\pi \rightarrow \pi^*$	6.15	6.24	6.24	6.46	6.57	6.73	6.58	6.66	6.29
	$4\,^1A'$	$\pi \rightarrow \pi^*$	6.75	6.74	6.76	7.07	7.15	7.39	7.15	7.29	6.84
	$5\,^1A'$	$\pi \rightarrow \pi^*$	7.42	7.46	7.49	7.85	7.87	8.16	7.86	8.02	
	$1\,^1A''$	$n \rightarrow \pi^*$	4.90	4.85	4.88	5.01	4.97	5.09	5.00	5.06	4.90
	$2\,^1A''$	$n \rightarrow \pi^*$	6.27	6.24	6.37	6.50	6.44	6.56	6.47	6.54	6.32
	$3\,^1A''$	$n \rightarrow \pi^*$	6.97	6.74	6.74	7.12	7.36	7.65	7.27	7.44	6.87
	$4\,^1A''$	$n \rightarrow \pi^*$	7.28	7.31	7.23	7.50	7.49	7.74	7.58	7.70	7.12
Adenine	$2\,^1A'$	$\pi \rightarrow \pi^*$	5.20	5.49	5.43	5.43					5.18
	$3\,^1A'$	$\pi \rightarrow \pi^*$	5.30	5.18	5.21	5.54					5.39
	$4\,^1A'$	$\pi \rightarrow \pi^*$	6.35	6.58	6.86	7.11					6.53
	$5\,^1A'$	$\pi \rightarrow \pi^*$	6.88	7.07	6.70	7.07					
	$6\,^1A'$	$\pi \rightarrow \pi^*$	6.64	6.93	6.96	7.04					
	$7\,^1A'$	$\pi \rightarrow \pi^*$	7.56	7.94	7.85	8.03					

Table 4 (continued)

Molecule	State	Type	MS-CASPT2[a]	MS-NEVPT2[b]	SDSPT2[b]		SDSCI[b]		ic-MRCISD[b]		CC3[c]
					Pople	Davidson	Pople	Davidson	Pople	Davidson	
	$1\,^1A''$	$n \to \pi^*$	5.21	5.27	5.35	5.55	5.51	5.68			5.34
	$2\,^1A''$	$n \to \pi^*$	5.97	6.10	6.11	6.27	6.17	6.32			5.96

[a]Ref. [152]

[b]This work

[c]Ref. [152, 167]

[d]CAS(6,6) for $2\,^1B_2$ and $3\,^1B_2$

[e]CASSCF averaged over the lowest three $^1A_1$ states

[f]10.45 eV in Ref. [152] (see text)

Table 5 Vertical excitation energies (in eV) of triplet states

Molecule	State	Type	MS-CASPT2[a]	MS-NEVPT2[b]	SDSPT2[b]		SDSCI[b]		ic-MRCISD[b]		CC3[c]
					Pople	Davidson	Pople	Davidson	Pople	Davidson	
Ethene	$1\,^3B_{1u}$	$\pi \rightarrow \pi*$	4.60	4.61	4.63	4.63	4.43	4.43	4.46	4.46	4.48
E-butadiene	$1\,^3A_g$	$\pi \rightarrow \pi*$	5.16	5.27	5.29	5.28	5.13	5.13	5.17	5.17	5.17
	$1\,^3B_u$	$\pi \rightarrow \pi*$	3.34	3.38	3.38	3.38	3.33	3.33	3.34	3.35	3.32
All-*E*-hexatriene	$1\,^3A_g$	$\pi \rightarrow \pi*$	4.31	4.39	4.39	4.39	4.31	4.32	4.35	4.35	4.32
	$1\,^3B_u$	$\pi \rightarrow \pi*$	2.71	2.73	2.72	2.73	2.72	2.73	2.73	2.74	2.69
All-*E*-octatetraene	$1\,^3A_g$	$\pi \rightarrow \pi*$	3.69	3.72	3.72	3.73	3.68	3.70	3.72	3.73	3.67
	$1\,^3B_u$	$\pi \rightarrow \pi*$	2.33	2.32	2.31	2.34	2.34	2.37	2.36	2.37	2.30
Cyclopropene	$1\,^3B_1$	$\sigma \rightarrow \pi*$	6.51	6.57	6.63	6.64	6.75	6.76	6.63	6.63	6.62
	$1\,^3B_2$	$\pi \rightarrow \pi*$	4.35	4.53	4.57	4.56	4.35	4.35	4.32	4.33	4.34
Cyclopentadiene	$1\,^3A_1$	$\pi \rightarrow \pi*$	5.10	5.22	5.24	5.23	5.06	5.07	5.10	5.10	5.09
	$1\,^3B_2$	$\pi \rightarrow \pi*$	3.28	3.32	3.32	3.32	3.25	3.26	3.26	3.27	3.25
Norbornadiene	$1\,^3A_2$	$\pi \rightarrow \pi*$	3.75	3.80	3.80	3.82	3.78	3.80	3.80	3.81	3.72
	$1\,^3B_2$	$\pi \rightarrow \pi*$	4.22	4.31	4.33	4.33	4.22	4.23	4.24	4.24	4.16
Benzene	$1\,^3E_{2g}$	$\pi \rightarrow \pi*$	7.41	7.59	7.59	7.57	7.29	7.29	7.38	7.37	7.49
	$1\,^3B_{1u}$	$\pi \rightarrow \pi*$	4.17	4.32	4.32	4.29	4.04	4.03	4.11	4.09	4.12
	$1\,^3B_{2u}$	$\pi \rightarrow \pi*$	5.76	5.46	5.63	5.80	6.38	6.49	6.10	6.20	6.04
	$1\,^3E_{1u}$	$\pi \rightarrow \pi*$	4.90	4.97	4.98	5.00	4.91	4.94	4.89	4.91	4.90

Table 5 (continued)

Molecule	State	Type	MS-CASPT2[a]	MS-NEVPT2[b]	SDSPT2[b]		SDSCI[b]		ic-MRCISD[b]		CC3[c]
					Pople	Davidson	Pople	Davidson	Pople	Davidson	
Naphthalene	$1\,^3A_g$	$\pi \rightarrow \pi *$	5.76	5.57	5.58	5.62	5.53	5.58	5.56	5.58	5.52
	$2\,^3A_g$	$\pi \rightarrow \pi *$	6.83	6.10	6.28	6.71	7.16	7.45	6.94	7.13	6.47
	$3\,^3A_g$	$\pi \rightarrow \pi *$	6.94	6.52	6.51	6.53	6.36	6.42	6.43	6.44	6.79
	$1\,^3B_{1g}$	$\pi \rightarrow \pi *$	4.64	4.55	4.56	4.56	4.46	4.48	4.48	4.49	4.47
	$2\,^3B_{1g}$	$\pi \rightarrow \pi *$	6.44	5.78	5.96	6.47	6.97	7.30	6.73	6.94	6.48
	$3\,^3B_{1g}$	$\pi \rightarrow \pi *$	7.23	6.81	6.79	6.80	6.61	6.66	6.68	6.70	6.76
	$1\,^3B_{2u}$	$\pi \rightarrow \pi *$	3.27	3.19	3.21	3.20	3.10	3.11	3.13	3.13	3.11
	$2\,^3B_{2u}$	$\pi \rightarrow \pi *$	4.88	4.77	4.72	4.73	4.61	4.65	4.65	4.66	4.64
	$1\,^3B_{3u}$	$\pi \rightarrow \pi *$	4.38	4.24	4.24	4.28	4.21	4.25	4.22	4.25	4.18
	$2\,^3B_{3u}$	$\pi \rightarrow \pi *$	5.11	4.43	4.61	4.98	5.48	5.70	5.22	5.40	5.11
Furan	$1\,^3A_1$	$\pi \rightarrow \pi *$	5.49	5.65	5.62	5.62	5.45	5.46	5.52	5.52	5.48
	$1\,^3B_2$	$\pi \rightarrow \pi *$	4.17	4.35	4.34	4.32	4.07	4.06	4.14	4.13	4.17
Pyrrole	$1\,^3A_1$	$\pi \rightarrow \pi *$	5.53	5.68	5.65	5.66	5.50	5.51	5.57	5.57	5.51
	$1\,^3B_2$	$\pi \rightarrow \pi *$	4.52	4.73	4.69	4.67	4.38	4.37	4.47	4.46	4.48

Table 5 (continued)

Molecule	State	Type	MS-CASPT2[a]	MS-NEVPT2[b]	SDSPT2[b]		SDSCI[b]		ic-MRCISD[b]		CC3[c]
					Pople	Davidson	Pople	Davidson	Pople	Davidson	
Imidazole	$1\,^3A'$	$\pi \rightarrow \pi*$	4.65	4.73	4.75	4.76	4.71	4.72	4.63	4.65	4.69
	$2\,^3A'$	$\pi \rightarrow \pi*$	5.74	5.81	5.86	5.88	5.85	5.87	5.77	5.79	5.79
	$3\,^3A'$	$\pi \rightarrow \pi*$	6.44	6.29	6.39	6.49	6.79	6.85	6.56	6.62	6.55
	$4\,^3A'$	$\pi \rightarrow \pi*$	7.43	7.42	7.41	7.48	7.55	7.61	7.39	7.44	7.42
	$1\,^3A''$	$n \rightarrow \pi*$	6.36	6.44	6.48	6.49	6.39	6.40	6.36	6.38	6.37
	$2\,^3A''$	$n \rightarrow \pi*$	7.51	7.58	7.63	7.67	7.63	7.67	7.58	7.62	7.51
Pyridine	$1\,^3A_1$	$\pi \rightarrow \pi*$	4.27	4.45	4.46	4.42	4.17	4.15	4.20	4.19	4.25
	$2\,^3A_1$	$\pi \rightarrow \pi*$	5.03	5.14	5.09	5.11	5.05	5.07	5.05	5.07	5.05
	$3\,^3A_1$	$\pi \rightarrow \pi*$	7.56	7.84	7.84	7.81	7.48	7.48	7.54	7.53	7.66
	$1\,^3A_2$	$n \rightarrow \pi*$	5.48	5.46	5.55	5.60	5.71	5.75	5.62	5.66	5.46
	$1\,^3B_1$	$n \rightarrow \pi*$	4.55	4.58	4.63	4.64	4.61	4.63	4.59	4.60	4.50
	$1\,^3B_2$	$\pi \rightarrow \pi*$	4.72	4.70	4.83	4.87	4.91	4.94	4.87	4.89	4.86
	$2\,^3B_2$	$\pi \rightarrow \pi*$	6.02	5.91	6.06	6.23	6.79	6.88	6.50	6.59	6.40
	$3\,^3B_2$	$\pi \rightarrow \pi*$	7.88	7.95	7.89	7.88	7.57	7.61	7.60	7.62	7.83

Table 5 (continued)

Molecule	State	Type	MS-CASPT2[a]	MS-NEVPT2[b]	SDSPT2[b]		SDSCI[b]		ic-MRCISD[b]		CC3[c]
					Pople	Davidson	Pople	Davidson	Pople	Davidson	
s-Tetrazine	$1\,^3B_{1g}$	$n \rightarrow \pi *$	4.14	4.34	4.29	4.35	4.21	4.28	4.40	4.43	4.21
	$2\,^3B_{1g}$	$n \rightarrow \pi *$	6.31	6.55	6.48	6.57	6.42	6.52	6.63	6.68	6.60
	$1\,^3B_{2g}$	$n \rightarrow \pi *$	4.93	5.15	5.13	5.15	4.85	4.89	5.06	5.08	4.93
	$2\,^3B_{2g}$	$n \rightarrow \pi *$	5.97	6.15	6.03	6.14	6.01	6.13	6.23	6.28	6.04
	$1\,^3B_{3g}$	$n, n \rightarrow \pi *, \pi *$	5.50	7.81	7.83	7.78	7.40	7.38	7.53	7.50	
	$1\,^3A_u$	$n \rightarrow \pi *$	3.26	3.37	3.38	3.62	3.62	3.82	3.72	3.85	3.52
	$2\,^3A_u$	$n \rightarrow \pi *$	5.02	5.08	4.99	5.25	5.22	5.46	5.35	5.49	5.03
	$1\,^3B_{1u}$	$\pi \rightarrow \pi *$	4.36	4.55	4.58	4.52	4.18	4.15	4.24	4.21	4.33
	$2\,^3B_{1u}$	$\pi \rightarrow \pi *$	5.40	5.50	5.43	5.45	5.37	5.39	5.41	5.42	5.38
	$1\,^3B_{2u}$	$\pi \rightarrow \pi *$	4.56	4.13	4.45	4.52	4.50	4.56	4.59	4.62	4.54
	$2\,^3B_{2u}$	$\pi \rightarrow \pi *$	7.10	7.01	7.03	7.23	7.56	7.73	7.41	7.54	7.36
	$1\,^3B_{3u}$	$n \rightarrow \pi *$	1.56	1.59	1.59	1.73	1.97	2.08	2.05	2.12	1.89
	$2\,^3B_{3u}$	$n \rightarrow \pi *$	6.54	6.77	6.71	6.82	6.72	6.83	6.90	6.96	6.53
Formaldehyde	$1\,^3A_1$	$\pi \rightarrow \pi *$	5.84	6.07	6.14	6.15	5.69	5.69	5.71	5.78	5.83
	$1\,^3A_2$	$n \rightarrow \pi *$	3.58	3.76	3.78	3.78	3.64	3.64	3.59	3.59	3.55
Acetone	$1\,^3A_1$	$\pi \rightarrow \pi *$	6.03	6.06	6.08	6.08	6.07	6.07	6.05	6.05	6.03
	$1\,^3A_2$	$n \rightarrow \pi *$	4.08	4.10	4.12	4.13	4.22	4.23	4.13	4.15	4.05

Table 5 (continued)

Molecule	State	Type	MS-CASPT2[a]	MS-NEVPT2[b]	SDSPT2[b]		SDSCI[b]		ic-MRCISD[b]		CC3[c]
					Pople	Davidson	Pople	Davidson	Pople	Davidson	
p-Benzoquinone	$1\,^3B_{1g}$	$n \rightarrow \pi*$	2.63	2.81	2.75	2.87	2.83	2.95	2.89	2.95	2.51
	$2\,^3B_{1g}$	$n \rightarrow \pi*$		6.29	6.27	6.35	6.17	6.27	6.26	6.30	
	$1\,^3B_{3g}$	$\pi \rightarrow \pi*$	3.31	3.16	3.34	3.43	3.55	3.61	3.51	3.55	3.41
	$2\,^3B_{3g}$	$\pi \rightarrow \pi*$		6.18	6.22	6.62	7.10	7.39	6.94	7.13	
	$1\,^3A_u$	$n \rightarrow \pi*$	2.68	2.82	2.76	2.88	2.85	2.96	2.91	2.96	2.62
	$2\,^3A_u$	$n \rightarrow \pi*$		6.28	6.26	6.34	6.14	6.25	6.24	6.29	
	$1\,^3B_{1u}$	$\pi \rightarrow \pi*$	2.99	2.90	2.90	2.96	3.01	3.05	3.05	3.07	2.96
	$2\,^3B_{1u}$	$\pi \rightarrow \pi*$		5.51	5.50	5.56	5.50	5.57	5.59	5.61	
Formamide	$1\,^3A'$	$\pi \rightarrow \pi*$	5.58	5.73	5.81	5.82	5.81	5.83	5.73	5.74	5.74
	$2\,^3A'$	$\pi \rightarrow \pi*$		9.08	9.07	9.11	9.11	9.15	8.97	9.00	
	$1\,^3A''$	$n \rightarrow \pi*$	5.40	5.41	5.51	5.51	5.40	5.41	5.32	5.33	5.36
Acetamide	$1\,^3A'$	$n \rightarrow \pi*$	5.75	5.58	5.71	5.75	6.02	6.05	5.83	5.86	5.88
	$2\,^3A'$	$n, \pi \rightarrow \pi*, \pi*$		8.34	8.51	8.63	9.13	9.22	8.79	8.88	
	$1\,^3A''$	$n \rightarrow \pi*$	5.53	5.46	5.58	5.59	5.51	5.53	5.39	5.42	5.42
Propanamide	$1\,^3A'$	$\pi \rightarrow \pi*$	5.79	5.85	5.94	5.98	5.99	6.04	5.90	5.93	5.90
	$2\,^3A'$	$\pi \rightarrow \pi*$		8.83	8.84	8.96	8.98	9.10	8.86	8.96	
	$1\,^3A''$	$n \rightarrow \pi*$	5.44	5.48	5.59	5.62	5.55	5.58	5.44	5.47	5.45

[a]Ref. [152]

[b]This work

[c]Ref. [152, 167]

Table 6 Deviations (in eV) of MS-CASPT2, MS-NEVPT2, SDSPT2, and SDSCI from ic-MRCISD for 145 singlet statesa

	MS-CASPT2[b,d]	MS-NEVPT2[c,d]	SDSPT2[c]		SDSCI[c]	
			Pople[d]	Davidson[e]	Pople[d]	Davidson[e]
Count	137	145	145	145	145	145
MD	− 0.33	− 0.28	− 0.18	− 0.11	0.05	0.10
MAD	0.35	0.39	0.28	0.21	0.14	0.16
SD	0.42	0.52	0.36	0.27	0.18	0.20
Max	− 1.07	− 1.43	− 0.98	0.99	0.64	0.63

[a]*MD* mean deviation, *MAD* mean absolute deviation, *SD* standard deviation, *Max* maximum deviation

[b]Ref. [152]

[c]This work

[d]Take ic-MRCISD+P[c] as reference

[e]Take ic-MRCISD+Q[c] as reference

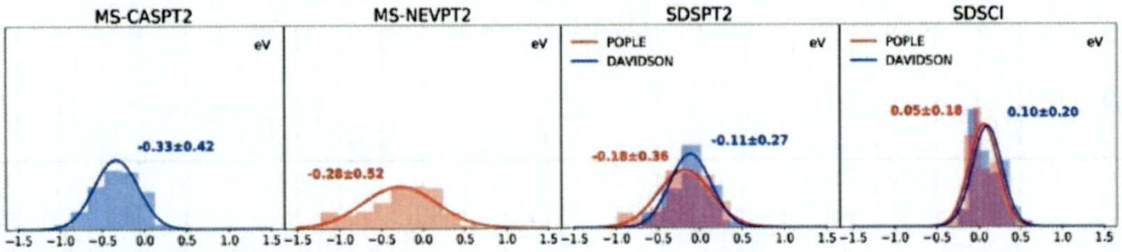

Fig. 2 Histogram and normal distribution of deviations of MS-CASPT2 [152], MS-NEVPT2, SDSPT2, and SDSCI from ic-MRCISD for the singlet states (cf. Table 6)

Table 7 Deviations (in eV) of MS-CASPT2, MS-NEVPT2, SDSPT2, SDSCI, and ic-MRCISD from CC3 [152, 167] for 139 singlet states[a]

	MS-CASPT2[b]	MS-NEVPT2[c]	SDSPT2[c]		SDSCI[c]		ic-MRCISD[c]	
			Pople	Davidson	Pople	Davidson	Pople	Davidson
Count	139	139	139	139	136	136	131	131
MD	− 0.17	− 0.13	− 0.03	0.13	0.22	0.34	0.16	0.24
MAD	0.20	0.27	0.19	0.18	0.31	0.41	0.21	0.28
SD	0.24	0.34	0.24	0.23	0.38	0.51	0.26	0.34
Max	− 0.62	− 1.05	0.62	0.77	0.96	1.30	0.72	0.96

[a]*MD* mean deviation, *MAD* mean absolute deviation, *SD* standard deviation, *Max* maximum deviation

[b]Ref. [152]

[c]This work

ic-MRCISD+P and CC3 tend to overestimate the VEEs, but CC3 is clearly better (Fig. 4). This conclusion likely holds for all the systems considered here. Even better results can be obtained by iCIPT2. Nevertheless, iCIPT2 still overestimates

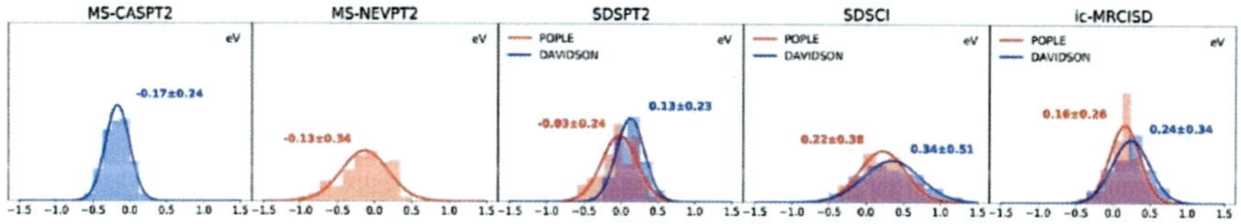

Fig. 3 Histogram and normal distribution of deviations of MS-CASPT2 [152], MS-NEVPT2, SDSPT2, SDSCI, and ic-MRCISD from CC3 [152, 167] for the singlet states (cf. Table 7)

Table 8 Vertical excitation energies (in eV) of excited states with differences between ic-MRCISD and CC3 larger than 0.50 eV

Molecule	State	ic-MRCISD[a]		CC3[b]	iCIPT2[a,c]
		Pople	Davidson		
All-*E*-octatetraene	$1\,^1B_u$	5.45	5.61	4.94	5.42
Pyridazine	$3\,^1A_1$	8.36	8.51	7.82	7.56
s-Triazine	$1\,^1E'$	8.56	8.61	8.04	8.07
s-Tetrazine	$2\,^1B_{2u}$	9.06	9.27	8.51	8.54
	$2\,^1B_{3u}$	7.17	7.25	6.67	6.60
p-Benzoquinone	$2\,^1B_{1u}$	8.54	8.78	7.82	7.82

[a]This work

[b]Refs. [152, 167]

[c]Linearly extrapolated values

significantly (by 0.5–0.7 eV) the excitation energies of the $1\,^1B_{1u}$ state of ethene as well as the $1\,^1B_u$ state of butadiene, hexatriene, and octatetraene. Yet, such overestimate arises from the deficiency of the chosen one-particle basis (TZVP [163]) in describing such valence–Rydberg mixed states, as already noted before [152, 192–195]. It is obvious that additional diffuse functions must further be added. It turns out that the corrections to the VEEs going from TZVP [163] to aug-cc-pVTZ [196] are rather insensitive to the correlation treatments (e.g., 0.1–0.2 eV [192, 193] for CC3, CCSDR(3), CC2, and MS-CASPT2). As such, the basis set corrections adopted by CC3 [193] can be added directly to iCIPT2. As seen from Table 9, the thus-corrected iCIPT2 VEEs are in excellent agreement with the experimental data.

3.2.4 Statistical Analysis of Triplet States

Regarding the triplet states, MS-CASPT2 [152], MS-NEVPT2, SDSPT2+P, and SDSCI+P all differ from ic-MRCISD+P by less than 0.2 eV on average, as seen from Table 10 and Fig. 5. Moreover, ic-MRCISD+P is also very close to CC3, as seen from Table 11 and Fig. 6. The better agreement between the methods for the triplet than for the singlet states should be ascribed to lesser correlation in triplet than singlet states.

Very recently, a subset of the Thiel systems [152] was used to benchmark MR-EOMPT (a perturbative approximation of MR-EOM-CCSD) [197]. As seen from Table 12, SDSPT2+P performs slightly better than MR-EOMPT.

Table 9 Vertical excitation energies by ic-MRCISD, CC3, iCIPT2, and vibrationally corrected experimental results ($E'_{exp} = E_{exp} + E_{shift}$)

Molecule	State	E_{exp}	E^{a}_{shift}	E'_{exp}	ic-MRCISD[b]		CC3[c]	CC3[d]	iCIPT2[b,e]	iCIPT2[b,e,f]
					Pople	Davidson				
Ethene	$1\,^1B_{1u}$	7.66[k]	0.14	7.80	8.45	8.45	8.37	7.89[g]	8.35	7.87[g]
All-*E*-butadiene	$1\,^1B_u$	5.92[l]	0.13	6.05	6.76	6.85	6.58	6.21[h]	6.57	6.20[h]
All-*E*-hexatriene	$1\,^1B_u$	4.93[m]	0.10	5.03	5.97	6.10	5.58	5.32[h]	5.54	5.28[h]
All-*E*-octa-tetraene	$1\,^1B_u$	4.41[n]	0.06	4.47	5.45	5.61	4.94	4.84[i]	5.22	4.95[j]
Cyclopropene	$1\,^1B_2$	7.19[o]	0.14	7.33	7.28	7.33	7.10	6.68[g]	7.53	7.11[g]
Cyclopentadiene	$1\,^1B_2$	5.33[p]	0.09	5.42	5.85	5.98	5.73	5.49[h]	5.74	5.50[h]
Norbornadiene	$1\,^1B_2$	5.95[q]	0.18	6.13	6.93	7.21	6.49	6.21[i]	6.17	5.89[i]
Benzene	$1\,^1E_{1u}$	6.96[r]	0.15	7.11	7.66	7.82	7.45	7.14[g]	7.50	7.19[g]
Naphthalene	$1\,^1B_{2u}$	4.45[s]	0.13	4.58	5.21	5.41	5.03	4.82[h]	4.92	4.71[h]
Furan	$1\,^1B_2$	6.06[t]	0.10	6.16	6.81	6.93	6.60	6.26[h]	6.58	6.24[h]
Pyrrole	$1\,^1B_2$	5.98[u]	0.21	6.19	6.93	7.02	6.71	6.20[h]	6.66	6.15[h]
Imidazole	$2\,^1A'$	6.00[v]	0.03	6.03	6.73	6.75	6.58	6.25[i]	6.27	5.94[i]
Pyridine	$1\,^1B_1$	4.59[w]	0.27	4.86	5.20	5.21	5.05	4.95[h]	4.88	4.78[h]
	$1\,^1B_2$	4.99[w]	0.05	5.04	5.12	5.13	5.15	5.12[h]	5.21	5.18[h]
Pyrazine	$1\,^1B_{3u}$	3.83[w]	0.14	3.97	4.43	4.47	4.24	4.13[h]	4.21	4.10[h]
	$1\,^1B_{2u}$	4.81[w]	0.05	4.86	5.01	5.03	5.02	4.97[h]	5.02	4.97[h]
Pyrimidine	$1\,^1B_1$	4.20[w]	0.07	4.27	4.65	4.68	4.50	4.43[h]	4.40	4.33[h]
	$1\,^1B_2$	5.12[w]	0.11	5.23	5.30	5.32	5.36	5.34[h]	5.35	5.33[h]
Pyridazine	$1\,^1B_1$	3.30[x]	0.08	3.38	4.10	4.14	3.92	3.85[h]	3.81	3.74[h]
	$2\,^1A_1$	4.90[x]	0.06	4.96	5.22	5.21	5.22	5.20[h]	5.20	5.18[h]
s-Triazine	$1\,^1A''_2$	4.59[w]	0.06	4.65	4.95	4.97	4.76	4.71[h]	4.60	4.55[h]
s-Tetrazine	$1\,^1B_{2u}$	5.00[y]	0.09	5.09	5.12	5.13	5.12	5.08[g]	5.08	5.04[g]
p-Benzoquinone	$1\,^1B_{1u}$	5.40[z]	0.14	5.54	5.92	6.16	5.62	5.47[h]	5.41	5.26[h]
Formamide	$2\,^1A'$	7.40[α]	0.05	7.45	7.72	7.78	7.24	7.35[g]	7.22	7.33[g]
Acetamide	$1\,^1A''$	5.44[β]	0.00	5.44	5.70	5.70	5.69	5.62[h]	5.66	5.59[h]
Count					25	25	25	25	25	25
MD					0.46	0.53	0.28	0.10	0.24	0.05
MAD					0.46	0.53	0.32	0.17	0.28	0.15
SD					0.54	0.63	0.36	0.22	0.33	0.18
Max					0.98	1.14	0.57	− 0.65	0.75	0.48

[a]Ref. [173]

[b]This work (TZVP [163] results)

[c]Ref. [152, 167] (TZVP [163] results)

[d]Estimated CC3/aug-cc-pVTZ resutls [193]

[e]Linearly extrapolated values

[f]Estimated iCIPT2/aug-cc-pVTZ results using the same basis set corrections from Ref. [193]

Table 9 (continued)

[g]Basis set correction = CC3/aug-cc-pVTZ-CC3/TZVP [193]

[h]Basis set correction = CCSDR(S)/aug-cc-pVTZ-CCSDR(S)/TZVP [193]

[i]Basis set correction = CC2/aug-cc-pVTZ-CC2/TZVP [193]

[j]iCIPT2/aug-cc-pVDZ

[k]Ref. [174]

[l]Ref. [175]

[m]Ref. [176]

[n]Ref. [177]

[o]Ref. [178]

[p]Ref. [179]

[q]Ref. [180]

[r]Ref. [181]

[s]Ref. [182]

[t]Ref. [183]

[u]Ref. [184]

[v]Ref. [185]

[w]Ref. [186]

[x]Ref. [187]

[y]Ref. [188]

[z]Ref. [189]

[α]Ref. [190]

[β]Ref. [191]

3.3 Excited States of Open-Shell Systems

The Thiel test set [152] is composed only of closed-shell molecules. It would be helpful if a similar set of open-shell systems were also available. Instead of constructing such a test set, herein we just remove one electron from each molecule in the Thiel test set to produce a new test set containing 24 organic cations (cytosine, thymine, uracil, and adenine are excluded because ic-MRCISD cannot be performed). The CAS(n, m) chosen in Ref. [152] is changed simply to CAS($n-1$, m) for most cations, except for pyridine: the original CAS(6,6) for $\pi \rightarrow \pi^*$ excitations [152] has to be revised to CAS(7,7) instead of CAS(5,6) because the singly occupied molecular orbital (SOMO) of pyridine cation is a lone-pair orbital of N, which has to be included along with six π orbitals for all doublet and quartet states.

Before investigating the excited states of the 24 cations, we must determine their ground states. For each molecule, we performed state-specific calculations with all possible spatial symmetries to determine the lowest doublet state. All the MR methods predict the same doublet ground states for the 24 cations (Table 13). It turns out that the electron-rich five-membered heterorings (furan, pyrrole, and imidazole) tend to lose one electron from the π instead of lone-pair orbital, whereas the other systems do the opposite. As seen from Table 13, the vertical ionization potentials (IPs) predicted by all the methods agree very well with each other.

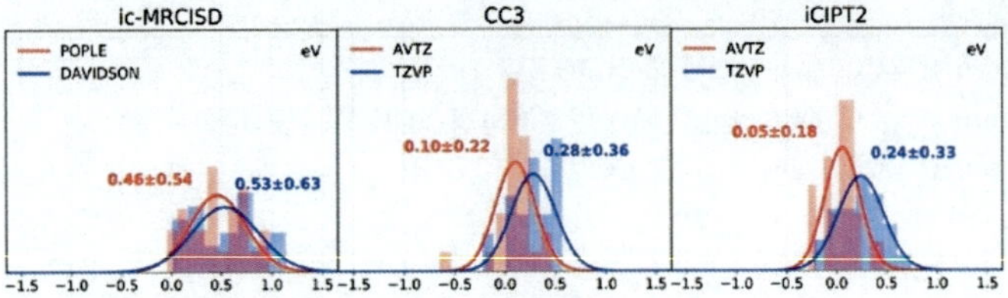

Fig. 4 Histogram and normal distribution of deviations of ic-MRCISD, CC3 [152], and iCIPT2 from vibrationally corrected experimental results [173–191] (cf. Table 9)

Table 10 Deviations (in eV) of MS-CASPT2, MS-NEVPT2, SDSPT2, and SDSCI from ic-MRCISD for 79 triplet states[a]

	MS-CASPT2[b,d]	MS-NEVPT2[c,d]	SDSPT2[c]		SDSCI[c]	
			Pople[d]	Davidson[e]	Pople[d]	Davidson[e]
Count	72	79	79	79	79	79
MD	− 0.07	− 0.05	− 0.02	0.01	0.01	0.03
MAD	0.16	0.18	0.17	0.14	0.09	0.09
SD	0.31	0.28	0.24	0.19	0.12	0.12
Max	− 2.03	− 0.95	− 0.77	− 0.51	0.34	0.36

[a] *MD* mean deviation, *MAD* mean absolute deviation, *SD* standard deviation, *Max* maximum deviation

[b] Ref. [152]

[c] This work

[d] Take ic-MRCISD+P[c] as reference

[e] Take ic-MRCISD+Q[c] as reference

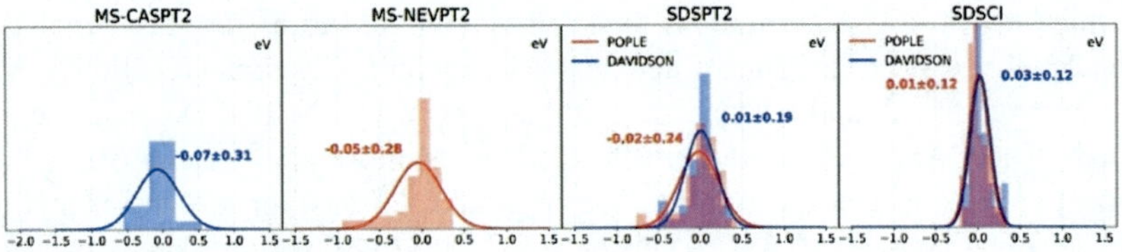

Fig. 5 Histogram and normal distribution of deviations of MS-CASPT2 [152], MS-NEVPT2, SDSPT2, and SDSCI from ic-MRCISD for the triplet states (cf. Table 10)

To determine the number of doublet/quartet states to be computed by the various MR methods, tentative ic-MRCISD calculations were performed for each irrep of every cation with enough roots. The number of roots for each irrep was then determined according to two rules: (a) If the VEE of the *n*th excited state exceeds 10.0 eV, only the first $n-1$ excited states will be considered; (b) If the leading configuration of the *n*th state is dominated by double excitations, only the first $n-1$ singly

Table 11 Deviations (in eV) of MS-CASPT2, MS-NEVPT2, SDSPT2, SDSCI, and ic-MRCISD from CC3 [152, 167] for 71 triplet states[a]

	MS-CASPT2[b]	MS-NEVPT2[c]	SDSPT2[c]		SDSCI[c]		ic-MRCISD[c]	
			Pople	Davidson	Pople	Davidson	Pople	Davidson
Count	71	71	71	71	71	71	71	71
MD	0.00	0.00	0.03	0.09	0.05	0.10	0.04	0.08
MAD	0.09	0.16	0.14	0.12	0.12	0.16	0.09	0.11
SD	0.14	0.22	0.18	0.15	0.18	0.24	0.13	0.18
Max	0.47	−0.70	−0.52	0.36	0.69	0.98	0.47	0.66

[a] *MD* mean deviation, *MAD* mean absolute deviation, *SD* standard deviation, *Max* maximum deviation

[b] Ref. [152]

[c] This work

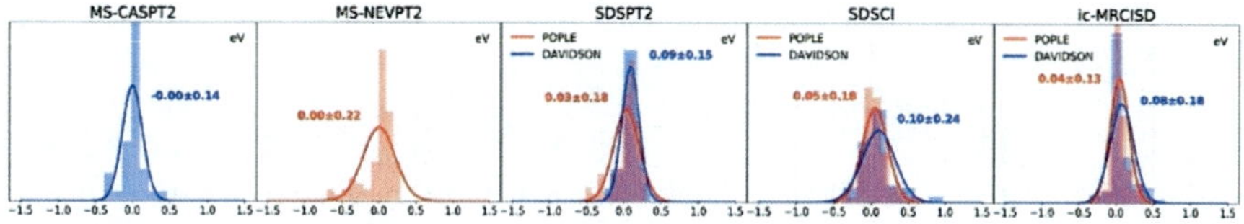

Fig. 6 Histogram and normal distribution of deviations of MS-CASPT2 [152], MS-NEVPT2, SDSPT2, SDSCI, and ic-MRCISD from CC3 [152] for the triplet states (cf. Table 11)

Table 12 Deviations (in eV) of MR-EOMPT, SDSPT2, and SDSCI from CC3 [152] for 23 states[a]

	MR-EOMPT[b]	SDSPT2[c]		SDSCI[c]	
		Pople	Davidson	Pople	Davidson
Count	23	23	23	23	23
MD	0.17	0.05	0.09	0.11	0.15
MAD	0.21	0.14	0.13	0.21	0.25
SD	0.32	0.17	0.16	0.28	0.33
Max	0.68	−0.41	0.33	0.56	0.69

[a] 12 singlet and 11 triplet excited states of ethene, butadiene, cyclopropene, benzene, and pyrrole

[b] Ref. [197]

[c] This work

excited states will be considered. Following these rules, the calculated VEEs of 168 doublet and 54 quartet states are documented in Tables 14 and 15, respectively. The statistical analysis of the doublet states is summarized in Table 16 and Fig. 7, showing that MS-NEVPT2, SDSPT2+P, and SDSCI+P all perform very well as compared with ic-MRCISD+P. The same finding holds also for the quartet states, as seen from Table 17 and Fig. 8.

Table 13 Vertical ionization potentials (in eV) and deviations (in eV) of MS-NEVPT2, SDSPT2, and SDSCI from ic-MRCISD for 24 first ionized statesa

Molecule	State	MS-NEVPT2	SDSPT2		SDSCI		ic-MRCISD	
			Pople	Davidson	Pople	Davidson	Pople	Davidson
Ethene	$1\,^2B_{3u}$	10.52	10.54	10.54	10.40	10.40	10.43	10.42
E-Butadiene	$1\,^2B_g$	8.83	8.87	8.86	8.99	8.98	8.92	8.91
All-*E*-hexatriene	$1\,^2A_u$	7.98	8.02	8.01	8.20	8.17	8.13	8.10
All-*E*-octatetraene	$1\,^2B_g$	7.42	7.47	7.48	7.69	7.67	7.61	7.59
Cyclopropene	$1\,^2B_1$	9.96	9.99	9.98	9.90	9.89	9.78	9.77
Cyclopentadiene	$1\,^2A_2$	8.36	8.41	8.39	8.52	8.49	8.44	8.42
Norbornadiene	$1\,^2B_1$	8.24	8.31	8.38	8.71	8.74	8.59	8.59
Benzene	$1\,^2B_{3g}$	9.14	9.17	9.12	9.14	9.07	9.10	9.06
Naphthalene	$1\,^2A_{1u}$	7.82	7.86	7.82	7.94	7.89	7.90	7.85
Furan	$1\,^2A_2$	8.65	8.69	8.66	8.81	8.77	8.71	8.68
Pyrrole	$1\,^2A_2$	8.12	8.08	8.03	7.96	7.92	7.99	7.95
Imidazole	$1\,^2A''$	8.82	8.79	8.74	8.64	8.60	8.67	8.63
Pyridine[a]	$1\,^2A_1$	9.55	9.56	9.52	9.50	9.46	9.48	9.45
Pyrazine	$1\,^2A_{1g}$	9.23	9.26	9.29	9.45	9.46	9.47	9.47
Pyrimidine	$1\,^2B_2$	9.37	9.39	9.40	9.49	9.49	9.51	9.50
Pyridazine	$1\,^2B_2$	8.95	8.97	8.96	9.01	9.00	9.03	9.02
s-Triazine	$1\,^2B_2$	10.16	10.18	10.17	10.21	10.20	10.22	10.21
s-Tetrazine	$1\,^2B_{3g}$	9.37	9.41	9.43	9.44	9.46	9.56	9.56
Formaldehyde	$1\,^2B_2$	10.81	10.84	10.84	10.76	10.76	10.58	10.58
Acetone	$1\,^2B_2$	9.75	9.74	9.72	9.70	9.68	9.58	9.57
p-Benzoquinone	$1\,^2B_{3g}$	9.73	9.80	9.86	9.95	10.00	9.91	9.92
Formamide	$1\,^2A'$	10.19	10.20	10.18	10.02	10.00	9.99	9.98
Acetamide	$1\,^2A'$	9.81	9.81	9.78	9.63	9.60	9.59	9.57
Propanamide	$1\,^2A'$	9.77	9.78	9.72	9.58	9.53	9.51	9.48
Count		24	24	24	24	24		
MD		−0.01	0.02	0.03	0.04	0.04		
MAD		0.15	0.13	0.12	0.06	0.06		
SD		0.17	0.15	0.14	0.08	0.07		
Max		−0.35	−0.28	0.26	0.18	0.18		

[a]*MD* mean deviation, *MAD* mean absolute deviation, *SD* standard deviation, *Max* maximum deviation

4 Conclusions and Outlook

Both noniterative (SDSCI and SDSPT2) and iterative (iCI, iCIPT2, and iCISCF(2)) realizations of the restricted yet generic SDS framework for constructing many-electron wave functions have been presented in a concise manner. The salient feature of such methods is the construction and diagonalization of a small

Table 14 Vertical excitation energies (in eV) of doublet states

Molecule	State	MS-NEVPT2	SDSPT2		SDSCI		ic-MRCISD	
			Pople	Davidson	Pople	Davidson	Pople	Davidson
Ethene	$1\,^2B_{2g}$	6.89	6.91	6.90	6.95	6.95	6.85	6.85
E-Butadiene	$2\,^2B_g$	6.76	6.79	6.80	6.65	6.68	6.68	6.70
	$3\,^2B_g$	7.99	7.98	8.05	8.17	8.24	8.05	8.11
	$1\,^2A_u$	2.47	2.48	2.49	2.45	2.46	2.46	2.46
	$2\,^2A_u$	4.86	4.87	4.90	4.86	4.90	4.83	4.86
All-*E*-hexatriene	$1\,^2B_g$	2.04	2.06	2.08	2.08	2.10	2.07	2.09
	$2\,^2B_g$	3.89	3.90	3.96	3.90	3.97	3.91	3.95
	$2\,^2A_u$	3.53	3.53	3.53	3.44	3.45	3.49	3.49
	$3\,^2A_u$	5.68	5.67	5.72	5.57	5.63	5.63	5.67
	$4\,^2A_u$	6.58	6.55	6.67	6.65	6.78	6.64	6.72
All-*E*-octatetraene	$2\,^2B_g$	3.16	3.17	3.18	3.10	3.12	3.14	3.14
	$3\,^2B_g$	4.83	4.84	4.90	4.75	4.83	4.82	4.86
	$1\,^2A_u$	1.72	1.73	1.77	1.78	1.81	1.77	1.79
	$2\,^2A_u$	3.25	3.26	3.34	3.29	3.38	3.31	3.36
	$3\,^2A_u$	4.00	4.01	4.04	3.96	4.00	3.97	4.00
Cyclopropene	$1\,^2A_1$	5.83	5.82	5.82	5.64	5.64	5.74	5.74
	$1\,^2A_2$	6.26	6.28	6.28	6.28	6.28	5.87	5.90
Cyclopentadiene	$2\,^2A_2$	6.52	6.54	6.55	6.35	6.38	6.38	6.40
	$1\,^2B_1$	2.33	2.33	2.35	2.32	2.34	2.33	2.34
Norbornadiene	$1\,^2A_1$	0.80	0.82	0.89	0.95	1.00	0.91	0.95
	$1\,^2A_2$	4.86	4.91	4.99	5.00	5.07	4.92	4.97
	$2\,^2A_2$	6.10	6.17	6.53	6.25	6.62	6.36	6.55
	$1\,^2B_2$	4.16	4.19	4.24	4.26	4.30	4.19	4.22
	$2\,^2B_2$	6.09	6.19	6.49	6.28	6.59	6.31	6.48
Benzene	$2\,^2B_{1g}$	8.43	8.42	8.45	8.26	8.31	8.34	8.36
	$2\,^2B_{3g}$	8.38	8.41	8.42	8.19	8.22	8.27	8.28
	$1\,^2A_u$	5.30	5.32	5.35	5.20	5.25	5.21	5.25
	$2\,^2A_u$	6.05	6.03	6.07	6.02	6.06	6.03	6.06
	$3\,^2A_u$	6.36	6.35	6.51	6.70	6.85	6.63	6.72
	$1\,^2B_{2u}$	2.89	2.89	2.90	2.84	2.85	2.88	2.88
Naphthalene	$1\,^2B_{2g}$	2.91	2.90	2.91	2.85	2.87	2.88	2.89
	$2\,^2B_{2g}$	3.81	3.80	3.81	3.65	3.68	3.70	3.72
	$3\,^2B_{2g}$	5.42	5.41	5.45	5.30	5.37	5.36	5.39
	$1\,^2B_{3g}$	1.95	1.95	1.97	1.93	1.96	1.94	1.96
	$2\,^2B_{3g}$	4.53	4.51	4.54	4.40	4.46	4.45	4.48
	$3\,^2B_{3g}$	5.01	4.99	5.05	4.91	4.99	4.97	5.01
	$1\,^2B_{1u}$	0.76	0.74	0.72	0.65	0.64	0.66	0.66
	$2\,^2B_{1u}$	4.40	4.38	4.37	4.24	4.25	4.29	4.28
	$3\,^2B_{1u}$	5.47	5.45	5.47	5.28	5.34	5.37	5.39

Table 14 (continued)

Molecule	State	MS-NEVPT2	SDSPT2		SDSCI		ic-MRCISD	
			Pople	Davidson	Pople	Davidson	Pople	Davidson
Furan	$2\,^2A_2$	6.72	6.82	6.84	6.72	6.75	6.73	6.76
	$3\,^2A_2$	7.06	7.03	7.11	7.26	7.33	7.15	7.20
	$1\,^2B_1$	1.47	1.44	1.45	1.35	1.36	1.38	1.38
	$2\,^2B_1$	4.96	4.98	4.99	4.85	4.87	4.86	4.87
	$3\,^2B_1$	6.75	6.65	6.72	6.67	6.74	6.66	6.71
Pyrrole	$2\,^2A_2$	6.46	6.53	6.57	6.53	6.57	6.51	6.55
	$3\,^2A_2$	7.23	7.16	7.21	7.18	7.24	7.15	7.18
	$1\,^2B_1$	0.74	0.81	0.82	0.82	0.83	0.76	0.77
	$2\,^2B_1$	4.62	4.67	4.68	4.57	4.59	4.54	4.55
	$3\,^2B_1$	6.27	6.37	6.45	6.57	6.65	6.49	6.55
	$4\,^2B_1$	8.08	8.30	8.37	8.41	8.48	8.24	8.30
	$5\,^2B_1$	8.30	8.12	8.30	8.60	8.76	8.35	8.47
Imidazole	$1\,^2A'$	1.12	1.17	1.19	1.22	1.23	1.17	1.19
	$2\,^2A'$	6.12	6.24	6.26	6.21	6.23	6.17	6.19
	$3\,^2A'$	6.70	6.76	6.78	6.74	6.77	6.69	6.72
	$2\,^2A''$	1.53	1.50	1.48	1.27	1.26	1.35	1.34
	$3\,^2A''$	5.10	5.08	5.06	4.79	4.80	4.88	4.87
	$4\,^2A''$	6.68	6.62	6.66	6.59	6.64	6.57	6.61
Pyridine	$2\,^2A_1$	4.80	4.81	4.77	4.49	4.48	4.57	4.56
	$3\,^2A_1$	5.61	5.58	5.60	5.47	5.49	5.52	5.53
	$1\,^2A_2$	0.02	0.04	0.04	0.06	0.05	0.02	0.01
	$2\,^2A_2$	5.56	5.62	5.63	5.44	5.47	5.44	5.46
	$1\,^2B_1$	0.70	0.73	0.72	0.72	0.72	0.70	0.69
	$2\,^2B_1$	3.42	3.46	3.45	3.38	3.39	3.38	3.37
	$1\,^2B_2$	4.68	4.71	4.73	4.70	4.72	4.72	4.72
	$2\,^2B_2$	5.42	5.42	5.44	5.31	5.34	5.34	5.36
Pyrazine	$2\,^2A_g$	6.25	6.31	6.28	6.07	6.06	6.13	6.11
	$1\,^2B_{1g}$	0.58	0.55	0.56	0.57	0.58	0.57	0.57
	$2\,^2B_{1g}$	6.24	6.30	6.34	6.38	6.41	6.27	6.31
	$1\,^2B_{2g}$	2.30	2.34	2.26	2.03	1.98	1.99	1.95
	$1\,^2B_{3g}$	5.83	5.84	5.82	5.64	5.64	5.68	5.67
	$2\,^2B_{3g}$	6.74	6.76	6.75	6.43	6.45	6.51	6.51
	$1\,^2A_u$	5.39	5.43	5.52	5.71	5.77	5.57	5.63
	$2\,^2A_u$	6.24	6.26	6.27	6.02	6.05	6.10	6.12
	$1\,^2B_{1u}$	1.58	1.59	1.59	1.59	1.59	1.57	1.56
	$2\,^2B_{1u}$	4.69	4.69	4.69	4.46	4.49	4.51	4.53
	$3\,^2B_{1u}$	5.93	5.97	5.97	5.79	5.81	5.81	5.82
	$1\,^2B_{2u}$	4.28	4.28	4.31	4.32	4.35	4.33	4.34

Table 14 (continued)

Molecule	State	MS-NEVPT2	SDSPT2		SDSCI		ic-MRCISD	
			Pople	Davidson	Pople	Davidson	Pople	Davidson
	$2\,^2B_{2u}$	5.10	5.10	5.13	5.00	5.03	5.03	5.05
	$1\,^2B_{3u}$	4.54	4.57	4.49	4.21	4.16	4.19	4.15
Pyrimidine	$1\,^2A_1$	1.48	1.47	1.51	1.55	1.59	1.48	1.52
	$2\,^2A_1$	4.92	4.92	4.95	4.96	5.00	4.97	4.99
	$3\,^2A_1$	5.56	5.57	5.59	5.42	5.46	5.44	5.47
	$1\,^2A_2$	1.68	1.69	1.70	1.61	1.63	1.64	1.65
	$2\,^2A_2$	4.99	5.03	5.03	4.93	4.94	4.87	4.89
	$3\,^2A_2$	6.19	6.27	6.29	6.24	6.27	6.15	6.18
	$1\,^2B_1$	0.85	0.82	0.84	0.86	0.87	0.85	0.86
	$2\,^2B_1$	4.45	4.46	4.45	4.27	4.27	4.32	4.32
	$3\,^2B_1$	5.27	5.33	5.36	5.34	5.37	5.25	5.28
	$4\,^2B_1$	5.97	6.01	6.04	5.97	6.01	5.88	5.92
	$2\,^2B_2$	4.90	4.92	4.91	4.67	4.67	4.74	4.74
	$3\,^2B_2$	6.14	6.19	6.17	5.86	5.87	5.92	5.92
Pyridazine	$1\,^2A_1$	2.06	2.07	2.10	2.10	2.13	2.04	2.07
	$2\,^2A_1$	4.83	4.85	4.87	4.78	4.80	4.79	4.80
	$3\,^2A_1$	5.57	5.57	5.58	5.49	5.51	5.48	5.50
	$1\,^2A_2$	1.44	1.45	1.46	1.46	1.47	1.45	1.46
	$2\,^2A_2$	4.67	4.72	4.75	4.75	4.77	4.67	4.70
	$1\,^2B_1$	1.99	2.01	1.96	1.86	1.83	1.79	1.77
	$2\,^2B_1$	4.91	4.94	4.89	4.69	4.66	4.65	4.63
	$2\,^2B_2$	5.12	5.17	5.17	5.01	5.01	5.05	5.05
	$3\,^2B_2$	5.65	5.67	5.66	5.42	5.44	5.48	5.49
	$4\,^2B_2$	6.26	6.29	6.33	6.35	6.38	6.35	6.37
s-Triazine	$2\,^2A_1$	2.94	2.91	2.97	2.98	3.04	2.93	2.98
	$4\,^2A_1$	5.98	5.96	5.99	5.84	5.89	5.92	5.95
	$2\,^2A_2$	4.27	4.27	4.32	4.34	4.39	4.34	4.37
	$3\,^2A_2$	4.94	4.94	4.98	4.94	4.99	4.96	4.99
	$1\,^2B_1$	1.56	1.61	1.58	1.45	1.42	1.45	1.45
	$2\,^2B_1$	4.32	4.34	4.38	4.37	4.42	4.36	4.39
	$3\,^2B_1$	4.69	4.69	4.69	4.53	4.55	4.60	4.61
	$4\,^2B_1$	4.99	4.99	5.03	5.05	5.10	5.03	5.07
	$2\,^2B_2$	5.28	5.33	5.29	4.95	4.95	5.04	5.03
	$3\,^2B_2$	6.21	6.24	6.23	5.94	5.95	6.02	6.02

Table 14 (continued)

Molecule	State	MS-NEVPT2	SDSPT2		SDSCI		ic-MRCISD	
			Pople	Davidson	Pople	Davidson	Pople	Davidson
s-Tetrazine	$1\,^2A_g$	2.95	2.94	2.98	3.00	3.03	2.98	3.01
	$2\,^2A_g$	6.93	7.00	6.97	6.67	6.68	6.75	6.75
	$1\,^2B_{1g}$	3.70	3.74	3.71	3.50	3.49	3.51	3.50
	$2\,^2B_{1g}$	5.01	5.01	5.06	5.13	5.16	5.12	5.14
	$3\,^2B_{1g}$	5.52	5.53	5.55	5.46	5.49	5.48	5.49
	$4\,^2B_{1g}$	6.03	5.97	6.05	6.12	6.20	6.10	6.15
	$1\,^2B_{2g}$	2.56	2.57	2.56	2.48	2.47	2.45	2.45
	$2\,^2B_{2g}$	4.19	4.17	4.21	4.27	4.30	4.25	4.26
	$3\,^2B_{2g}$	5.79	5.79	5.85	5.76	5.81	5.75	5.80
	$4\,^2B_{2g}$	6.53	6.52	6.56	6.58	6.61	6.59	6.60
	$1\,^2A_u$	1.98	1.94	2.01	2.32	2.36	2.27	2.30
	$2\,^2A_u$	6.83	6.84	6.91	6.86	6.93	6.89	6.94
	$1\,^2B_{1u}$	2.41	2.40	2.41	2.39	2.41	2.37	2.38
	$2\,^2B_{1u}$	4.71	4.69	4.71	4.65	4.68	4.68	4.69
	$3\,^2B_{1u}$	5.42	5.42	5.43	5.16	5.20	5.21	5.23
	$1\,^2B_{2u}$	3.08	3.08	3.09	3.04	3.06	3.05	3.06
	$2\,^2B_{2u}$	5.07	5.11	5.08	4.81	4.81	4.86	4.86
	$3\,^2B_{2u}$	5.84	5.82	5.84	5.69	5.73	5.74	5.76
	$1\,^2B_{3u}$	3.92	3.92	4.01	4.21	4.28	4.14	4.19
	$2\,^2B_{3u}$	4.86	4.90	4.99	5.20	5.26	5.12	5.17
	$3\,^2B_{3u}$	5.36	5.34	5.43	5.57	5.65	5.48	5.54
	$4\,^2B_{3u}$	6.12	6.10	6.08	5.84	5.85	5.91	5.90
Formaldehyde	$1\,^2B_1$	3.54	3.58	3.58	3.40	3.40	3.60	3.60
	$2\,^2B_1$	5.87	5.80	5.79	6.05	6.05	5.70	5.70
Acetone	$1\,^2A_2$	8.94	9.00	9.08	9.35	9.44	9.19	9.28
	$2\,^2A_2$	9.03	9.08	9.15	9.51	9.59	9.40	9.48
	$1\,^2B_1$	2.57	2.61	2.64	2.98	3.01	2.96	2.99
	$2\,^2B_1$	5.31	5.29	5.41	6.40	6.48	5.93	6.04
	$2\,^2B_2$	8.42	8.46	8.47	8.07	8.11	7.95	8.03
p-Benzoquinone	$1\,^2A_g$	3.27	3.22	3.21	3.05	3.06	3.16	3.15
	$2\,^2A_g$	4.03	3.96	3.99	3.87	3.91	3.98	3.98
	$1\,^2B_{2g}$	1.17	1.08	1.08	0.96	0.98	1.06	1.06
	$2\,^2B_{2g}$	4.64	4.65	4.62	4.49	4.48	4.53	4.52
	$1\,^2A_u$	4.62	4.55	4.51	4.25	4.24	4.39	4.36
	$2\,^2A_u$	4.72	4.72	4.70	4.55	4.54	4.60	4.58
	$3\,^2A_u$	5.84	5.74	5.72	5.43	5.45	5.59	5.58
	$1\,^2B_{2u}$	0.23	0.19	0.20	0.11	0.13	0.15	0.16
	$2\,^2B_{2u}$	3.52	3.52	3.51	3.31	3.32	3.41	3.39

Table 14 (continued)

Molecule	State	MS-NEVPT2	SDSPT2		SDSCI		ic-MRCISD	
			Pople	Davidson	Pople	Davidson	Pople	Davidson
	$3\,^2B_{2u}$	5.44	5.43	5.40	5.13	5.15	5.25	5.23
	$4\,^2B_{2u}$	5.66	5.62	5.67	5.51	5.58	5.61	5.63
	$1\,^2B_{3u}$	1.04	0.97	0.94	0.74	0.73	0.85	0.83
	$2\,^2B_{3u}$	2.28	2.29	2.29	2.23	2.24	2.24	2.24
	$3\,^2B_{3u}$	4.88	4.79	4.78	4.48	4.51	4.62	4.62
	$4\,^2B_{3u}$	5.54	5.54	5.56	5.46	5.50	5.53	5.53
Formamide	$2\,^2A'$	6.30	6.21	6.21	5.96	5.96	6.09	6.08
	$3\,^2A'$	7.60	7.56	7.60	7.63	7.67	7.31	7.37
	$1\,^2A''$	0.16	0.06	0.07	0.11	0.12	0.23	0.23
	$2\,^2A''$	3.64	3.65	3.65	3.62	3.63	3.65	3.66
	$3\,^2A''$	7.92	7.95	8.01	8.57	8.61	8.20	8.25
Acetamide	$2\,^2A'$	6.53	6.42	6.40	6.10	6.09	6.26	6.24
	$3\,^2A'$	7.56	7.48	7.55	7.46	7.55	7.36	7.43
	$1\,^2A''$	0.28	0.22	0.23	0.25	0.25	0.33	0.33
	$2\,^2A''$	3.47	3.39	3.41	3.34	3.36	3.37	3.39
Propanamide	$2\,^2A'$	6.55	6.43	6.39	6.10	6.09	6.27	6.24
	$3\,^2A'$	7.52	7.43	7.54	7.41	7.54	7.39	7.47
	$1\,^2A''$	0.23	0.17	0.19	0.22	0.24	0.33	0.32
	$2\,^2A''$	3.38	3.30	3.34	3.28	3.33	3.36	3.38

Hamiltonian matrix (in each iteration), the dimension of which is determined by the number of target states instead of the numbers of electrons and many orbitals to be correlated. While iCI is exact (relative to FCI) and iCIPT2 and iCISCF(2) are nearly exact (relative to iCI and CASSCF, respectively), SDSCI+P and even SDSPT2+P are very good approximations to ic-MRCISD+P, as demonstrated by the extensive benchmarking. Further combination with iCAS for automatic construction of active spaces renders SDSPT2+P and SDSCI+P valuable and user-friendly methods for routine applications. Future developments of the methods will include (1) the use of LMOs to reduce the overall computational scaling and enable the wave function theory (WFT)-in-WFT type of embedding, (2) the formulation of analytic energy gradients of SDSPT2/SDSCI, and (3) the treatment of relativistic effects, especially spin–orbit couplings.

Table 15 Vertical excitation energies (in eV) of quartet states

Molecule	State	MS-NEVPT2	SDSPT2		SDSCI		ic-MRCISD	
			Pople	Davidson	Pople	Davidson	Pople	Davidson
E-butadiene	$1\,^4B_g$	5.22	5.23	5.21	4.84	4.82	4.95	4.94
	$1\,^4A_u$	8.17	8.20	8.20	7.96	7.97	7.95	7.95
All-*E*-hexatriene	$1\,^4B_g$	5.58	5.58	5.55	5.34	5.33	5.42	5.41
	$1\,^4A_u$	4.19	4.20	4.19	4.08	4.08	4.10	4.10
All-*E*-octatetraene	$1\,^4B_g$	3.53	3.53	3.54	3.49	3.51	3.52	3.52
	$1\,^4A_u$	4.71	4.73	4.75	4.68	4.71	4.71	4.72
	$2\,^4A_u$	6.04	6.06	6.12	5.99	6.07	6.05	6.09
Cyclopropene	$1\,^4A_1$	5.43	5.42	5.41	5.20	5.20	5.31	5.33
Cyclopentadiene	$1\,^4A_2$	5.24	5.25	5.22	4.92	4.91	4.99	4.98
	$1\,^4B_1$	7.65	7.70	7.71	7.48	7.51	7.52	7.54
Norbornadiene	$1\,^4A_2$	4.46	4.47	4.40	4.00	3.99	4.07	4.07
	$1\,^4B_2$	5.82	5.85	5.82	5.43	5.46	5.51	5.53
Benzene	$1\,^4B_{1g}$	7.28	7.30	7.27	7.02	7.01	7.09	7.08
	$1\,^4A_u$	4.52	4.53	4.53	4.43	4.43	4.45	4.46
Naphthalene	$1\,^4B_{2g}$	4.60	4.60	4.58	4.44	4.45	4.50	4.49
	$1\,^4B_{3g}$	3.93	3.93	3.91	3.80	3.80	3.85	3.84
	$1\,^4A_u$	5.26	5.28	5.24	5.05	5.05	5.13	5.11
	$2\,^4A_u$	5.53	5.49	5.50	5.38	5.42	5.45	5.45
	$1\,^4B_{1u}$	4.63	4.61	4.54	4.34	4.31	4.41	4.38
Furan	$1\,^4A_2$	5.31	5.35	5.35	5.26	5.26	5.29	5.29
	$1\,^4B_1$	7.03	6.96	6.93	6.49	6.49	6.64	6.63
Pyrrole	$1\,^4A_2$	5.14	5.16	5.18	5.16	5.17	5.14	5.15
	$2\,^4A_2$	9.24	9.21	9.22	8.98	9.00	9.01	9.02
	$1\,^4B_1$	6.30	6.26	6.24	5.89	5.89	6.01	6.00
Imidazole	$1\,^4A'$	5.73	5.82	5.84	5.88	5.91	5.84	5.86
	$2\,^4A'$	6.44	6.50	6.50	6.35	6.37	6.35	6.37
	$1\,^4A''$	5.95	5.90	5.91	5.74	5.75	5.77	5.78
	$2\,^4A''$	6.73	6.70	6.67	6.30	6.30	6.38	6.38
	$3\,^4A''$	9.88	9.78	9.78	9.44	9.47	9.52	9.53
Pyridine	$1\,^4A_1$	4.54	4.54	4.51	4.25	4.24	4.31	4.30
	$2\,^4A_1$	4.86	4.83	4.85	4.84	4.87	4.85	4.86
	$1\,^4B_2$	4.23	4.29	4.30	4.32	4.33	4.31	4.31

Table 15 (continued)

Molecule	State	MS-NEVPT2	SDSPT2		SDSCI		ic-MRCISD	
			Pople	Davidson	Pople	Davidson	Pople	Davidson
s-Tetrazine	$1\,^4B_{1g}$	4.04	4.03	4.07	4.15	4.18	4.15	4.16
	$2\,^4B_{1g}$	5.39	5.35	5.42	5.50	5.55	5.49	5.52
	$3\,^4B_{1g}$	7.16	7.13	7.16	7.14	7.16	7.18	7.19
	$1\,^4B_{2g}$	3.15	3.11	3.17	3.39	3.42	3.38	3.40
	$2\,^4B_{2g}$	6.63	6.61	6.66	6.56	6.62	6.62	6.65
	$1\,^4A_{u}$	6.26	6.29	6.36	6.46	6.52	6.47	6.50
	$1\,^4B_{1u}$	4.01	4.01	4.01	4.02	4.03	4.03	4.03
	$1\,^4B_{2u}$	4.26	4.30	4.27	4.05	4.04	4.09	4.08
	$2\,^4B_{2u}$	5.32	5.29	5.30	5.17	5.19	5.22	5.22
	$1\,^4B_{3u}$	4.29	4.28	4.35	4.50	4.55	4.50	4.53
Formaldehyde	$1\,^4B_{2}$	6.56	6.59	6.60	6.29	6.29	6.38	6.38
Acetone	$1\,^4A_{2}$	8.64	8.66	8.72	9.08	9.13	8.77	8.84
	$1\,^4B_{2}$	6.47	6.51	6.54	6.82	6.84	6.76	6.79
p-Benzoquinone	$1\,^4A_{g}$	2.99	2.94	2.95	2.84	2.87	2.93	2.93
	$2\,^4A_{g}$	6.34	6.30	6.30	6.07	6.10	6.19	6.18
	$1\,^4B_{2g}$	5.76	5.73	5.64	5.34	5.29	5.49	5.44
	$1\,^4B_{2u}$	3.00	3.01	3.01	2.89	2.91	2.98	2.97
	$1\,^4B_{3u}$	1.83	1.83	1.83	1.78	1.78	1.79	1.78
	$2\,^4B_{3u}$	5.91	5.92	5.77	5.27	5.21	5.45	5.38
Formamide	$1\,^4A'$	5.62	5.62	5.62	5.38	5.38	5.47	5.47
Acetamide	$1\,^4A'$	5.87	5.85	5.84	5.58	5.59	5.77	5.75
Propanamide	$1\,^4A'$	5.85	5.86	5.85	5.63	5.63	5.74	5.73

Table 16 Deviations (in eV) of MS-NEVPT2, SDSPT2, and SDSCI from ic-MRCISD for 168 doublet states[a]

	MS-NEVPT2[b]	SDSPT2		SDSCI	
		Pople[b]	Davidson[c]	Pople[b]	Davidson[c]
Count	168	168	168	168	168
MD	0.04	0.04	0.05	0.00	0.01
MAD	0.11	0.11	0.10	0.07	0.06
SD	0.15	0.15	0.14	0.10	0.09
Max	− 0.62	− 0.64	− 0.63	0.47	0.45

[a]*MD* mean deviation, *MAD* mean absolute deviation, *SD* standard deviation, *Max* maximum deviation

[b]Taking ic-MRCISD+P as reference

[c]Taking ic-MRCISD+Q as reference

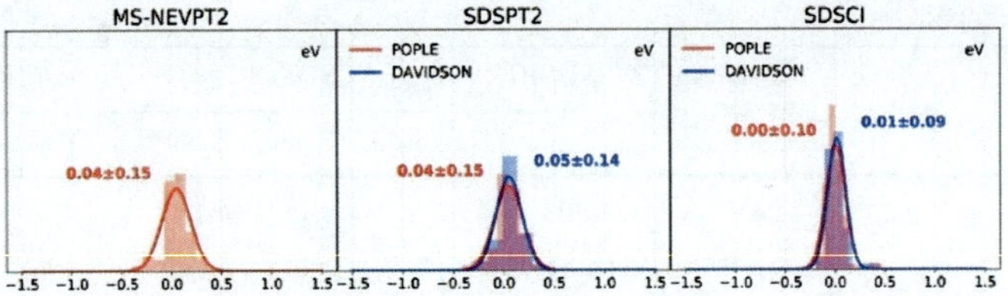

Fig. 7 Histogram and normal distribution of deviations of MS-NEVPT2, SDSPT2, and SDSCI from ic-MRCISD for the doublet states (cf. Table 16)

Table 17 Deviations (in eV) of MS-NEVPT2, SDSPT2, and SDSCI from ic-MRCISD for 54 quartet states[a]

	MS-NEVPT2[b]	SDSPT2		SDSCI	
		Pople[b]	Davidson[c]	Pople[b]	Davidson[c]
Count	54	54	54	54	54
MD	0.10	0.10	0.09	−0.05	−0.04
MAD	0.15	0.15	0.14	0.07	0.06
SD	0.19	0.18	0.17	0.09	0.08
Max	0.46	0.47	0.39	0.31	0.29

[a] *MD* mean deviation, *MAD* mean absolute deviation, *SD* standard deviation, *Max* maximum deviation

[b] Taking ic-MRCISD+P as reference

[c] Taking ic-MRCISD+Q as reference

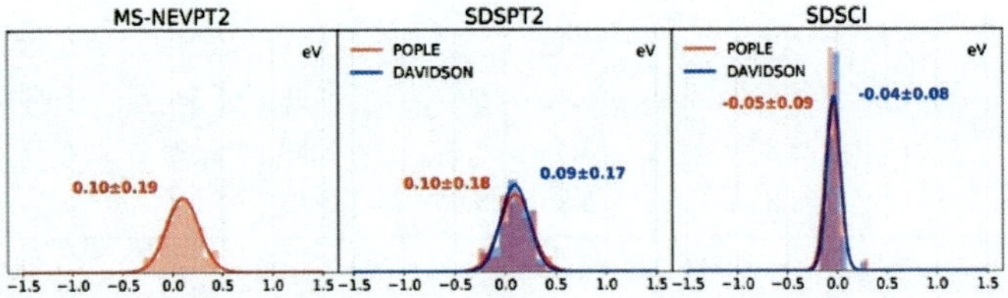

Fig. 8 Histogram and normal distribution of deviations of MS-NEVPT2, SDSPT2, and SDSCI from ic-MRCISD for the quartet states (cf. Table 17)

Acknowledgements This work was supported by the National Natural Science Foundation of China (Grant nos. 21833001 and 21973054), Mountain Tai Climb Program of Shandong Province, and Key-Area Research and Development Program of Guangdong Province (Grant no. 2020B0101350001).

Data Availability Statement All data are contained in the text.

Declarations

Conflict of interest All authors declare that they have no competing interests.

References

1. Liu W, Lindgren I (2013) Going beyond "no-pair relativistic quantum chemistry". J Chem Phys 139(1):014108
2. Liu W (2014) Advances in relativistic molecular quantum mechanics. Phys Rep 537:59–89
3. Liu W, Peng D (2006) Infinite-order quasirelativistic density functional method based on the exact matrix quasirelativistic theory. J Chem Phys 125(044):102
4. Peng D, Liu W, Xiao Y, Cheng L (2007) Making four- and two-component relativistic density functional methods fully equivalent based on the idea of from atoms to molecule. J Chem Phys 127(104):106
5. Kutzelnigg W, Liu W (2005) Quasirelativistic theory equivalent to fully relativistic theory. J Chem Phys 123(241):102
6. Kutzelnigg W, Liu W (2006) Quasirelativistic theory I. Theory in terms of a quasi-relativistic operator. Mol Phys 104(13–14):2225–2240
7. Liu W, Kutzelnigg W (2007) Quasirelativistic theory. II. Theory at matrix level. J Chem Phys 126(11):114,107
8. Liu W, Peng D (2009) Exact two-component Hamiltonians revisited. J Chem Phys 131(031):104
9. Li Z, Xiao Y, Liu W (2012) On the spin separation of algebraic two-component relativistic Hamiltonians. J Chem Phys 137(154):114
10. Li Z, Xiao Y, Liu W (2014) On the spin separation of algebraic two-component relativistic Hamiltonians: molecular properties. J Chem Phys 141(5):054,111
11. Liu W (2010) Ideas of relativistic quantum chemistry. Mol Phys 108:1679–1706
12. Liu W (2012) Perspectives of relativistic quantum chemistry: the negative energy cat smiles. Phys Chem Chem Phys 14(1):35–48
13. Liu W (2014) Perspective: relativistic hamiltonians. Int J Quantum Chem 114(15):983–986
14. Liu W (2015) Effective quantum electrodynamics Hamiltonians: a tutorial review. Int J Quantum Chem 115(10):631–640
15. Liu W (2016) Big picture of relativistic molecular quantum mechanics. Natl Sci Rev 3(2):204–221
16. Liu W (2017) With-pair relativistic Hamiltonians. Springer, Berlin, pp 345–373
17. Liu W (2017) No-pair relativistic Hamiltonians: Q4C and X2C. Springer, Berlin, pp 375–393
18. Li Z, Liu W (2017) Spin separation of relativistic Hamiltonians. Springer, Berlin, pp 411–447
19. Liu W (2020) Essentials of relativistic quantum chemistry. J Chem Phys 152(18):180,901
20. Liu W (2020) Relativistic quantum chemistry: today and tomorrow. Sci Sin Chim 50(11):1672–1696
21. Chaudhuri RK, Freed KF, Hose G, Piecuch P, Kowalski K, Włoch M, Chattopadhyay S, Mukherjee D, Rolik Z, Szabados Á, Tóth G, Surján PR (2005) Comparison of low-order multireference many-body perturbation theories. J Chem Phys 122(13):134,105
22. Hoffmann MR, Datta D, Das S, Mukherjee D, Szabados Á, Rolik Z, Surján PR (2009) Comparative study of multireference perturbative theories for ground and excited states. J Chem Phys 131(20):204,104
23. Lyakh DI, Musiał M, Lotrich VF, Bartlett RJ (2012) Multireference nature of chemistry: the coupled-cluster view. Chem Rev 112:182–243
24. Sherrill CD, Schaefer HF III (1999) The configuration interaction method: advances in highly correlated approaches. Adv Quantum Chem 34:143–269
25. Szalay PG, Muller T, Gidofalvi G, Lischka H, Shepard R (2012) Multiconfiguration self-consistent field and multireference configuration interaction methods and applications. Chem Rev 112(1):108–181
26. Lischka H, Nachtigallova D, Aquino AJ, Szalay PG, Plasser F, Machado FB, Barbatti M (2018) Multireference approaches for excited states of molecules. Chem Rev 118(15):7293–7361
27. Liu W, Hoffmann MR (2016) iCI: Iterative CI toward full CI. J Chem Theory Comput 12:1169–1178
28. Liu W, Hoffmann MR (2016) iCI: iterative CI toward full CI. J Chem Theory Comput 12:3000
29. Finley J, Malmqvist PA, Roos BO, Serrano-Andrés L (1998) The multi-state CASPT2 method. Chem Phys Lett 288:299–306
30. Granovsky AA (2011) Extended multi-configuration quasi-degenerate perturbation theory: the new approach to multi-state multi-reference perturbation theory. J Chem Phys 134(214):113

31. Shiozaki T, Győrffy W, Celani P, Werner HJ (2011) Communication: extended multi-state complete active space second-order perturbation theory: energy and nuclear gradients. J Chem Phys 135(081):106
32. Angeli C, Borini S, Cestari M, Cimiraglia R (2004) A quasidegenerate formulation of the second order n-electron valence state perturbation theory approach. J Chem Phys 121:4043–4049
33. Yost SR, Kowalczyk T, Van Voorhis T (2013) A multireference perturbation method using non-orthogonal Hartree–Fock determinants for ground and excited states. J Chem Phys 139(17):174,104
34. Yost SR, Head-Gordon M (2016) Size consistent formulations of the perturb-then-diagonalize Møller–Plesset perturbation theory correction to non-orthogonal configuration interaction. J Chem Phys 145(5):054,105
35. Gao J, Grofe A, Ren H, Bao P (2016) Beyond Kohn–Sham approximation: hybrid multistate wave function and density functional theory. J Phys Chem Lett 7(24):5143–5149
36. Grofe A, Chen X, Liu W, Gao J (2017) Spin-multiplet components and energy splittings by multistate density functional theory. J Phys Chem Lett 8(19):4838–4845
37. Yang L, Grofe A, Reimers J, Gao J (2019) Multistate density functional theory applied with 3 unpaired electrons in 3 orbitals: the singdoublet and tripdoublet states of the ethylene cation. Chem Phys Lett 736(136):803
38. Gidofalvi G, Mazziotti DA (2008) Active-space two-electron reduced-density-matrix method: complete active-space calculations without diagonalization of the N-electron Hamiltonian. J Chem Phys 129(13):134,108
39. Fosso-Tande J, Nguyen TS, Gidofalvi G, DePrince AE (2016) Large-scale variational two-electron reduced-density-matrix-driven complete active space self-consistent field methods. J Chem Theory Comput 12(5):2260–2271
40. Zgid D, Nooijen M (2008) The density matrix renormalization group self-consistent field method: orbital optimization with the density matrix renormalization group method in the active space. J Chem Phys 128(14):144,116
41. Ghosh D, Hachmann J, Yanai T, Chan GKL (2008) Orbital optimization in the density matrix renormalization group, with applications to polyenes and beta-carotene. J Chem Phys 128(14):144,117
42. Yanai T, Kurashige Y, Ghosh D, Chan GKL (2009) Accelerating convergence in iterative solution for large-scale complete active space self-consistent-field calculations. Int J Quantum Chem 109(10):2178–2190
43. Ma Y, Ma H (2013) Assessment of various natural orbitals as the basis of large active space density-matrix renormalization group calculations. J Chem Phys 138(22):224,105
44. Wouters S, Bogaerts T, Van Der Voort P, Van Speybroeck V, Van Neck D (2014) Communication: DMRG-SCF study of the singlet, triplet, and quintet states of oxo-Mn(Salen). J Chem Phys 140(24):241,103
45. Ma Y, Knecht S, Keller S, Reiher M (2017) Second-order self-consistent-field density-matrix renormalization group. J Chem Theory Comput 13(6):2533–2549
46. Sun Q, Yang J, Chan GKL (2017) A general second order complete active space self-consistent-field solver for large-scale systems. Chem Phys Lett 683:291–299
47. Thomas RE, Sun Q, Alavi A, Booth GH (2015) Stochastic multiconfigurational self-consistent field theory. J Chem Theory Comput 11(11):5316–5325
48. Li Manni G, Smart SD, Alavi A (2016) Combining the complete active space self-consistent field method and the full configuration interaction quantum Monte Carlo within a super-CI framework, with application to challenging metal-porphyrins. J Chem Theory Comput 12(3):1245–1258
49. Smith JET, Mussard B, Holmes AA, Sharma S (2017) Cheap and near exact CASSCF with large active spaces. J Chem Theory Comput 13(11):5468–5478
50. Yao Y, Umrigar C (2021) Orbital optimization in selected configuration interaction methods. arXiv:2104.02587 (arXiv preprint)
51. Zimmerman PM, Rask AE (2019) Evaluation of full valence correlation energies and gradients. J Chem Phys 150(24):244,117
52. Levine DS, Hait D, Tubman NM, Lehtola S, Whaley KB, Head-Gordon M (2020) CASSCF with extremely large active spaces using the adaptive sampling configuration interaction method. J Chem Theory Comput 16(4):2340–2354
53. Park JW (2021) Second-order orbital optimization with large active spaces using adaptive sampling configuration interaction (ASCI) and its application to molecular geometry optimization. J Chem Theory Comput 17(3):1522–1534

54. Dobrautz W, Weser O, Bogdanov N, Alavi A, Manni GL (2021) Spin-pure stochastic-CASSCF via GUGA-FCIQMC applied to iron sulfur clusters. arXiv:2106.07775 (arXiv preprint)
55. Guo Y, Zhang N, Lei Y, Liu W (2021) iCISCF: an iterative configuration interaction-based multiconfigurational self-consistent field theory for large active spaces. arXiv:2106.10223 (arXiv preprint)
56. White SR (1992) Density matrix formulation for quantum renormalization groups. Phys Rev Lett 69(19):2863
57. White SR (1993) Density-matrix algorithms for quantum renormalization groups. Phys Rev B 48:10,345-10,356
58. White SR, Martin RL (1999) Ab initio quantum chemistry using the density matrix renormalization group. J Chem Phys 110(9):4127–4130
59. Mitrushenkov AO, Fano G, Ortolani F, Linguerri R, Palmieri P (2001) Quantum chemistry using the density matrix renormalization group. J Chem Phys 115(15):6815–6821
60. Chan GKL, Head-Gordon M (2002) Highly correlated calculations with a polynomial cost algorithm: a study of the density matrix renormalization group. J Chem Phys 116(11):4462–4476
61. Chan GKL, Sharma S (2011) The density matrix renormalization group in quantum chemistry. Annu Rev Phys Chem 62:465–481
62. Sharma S, Chan GKL (2012) Spin-adapted density matrix renormalization group algorithms for quantum chemistry. J Chem Phys 136(12):124,121
63. Wouters S, Van Neck D (2014) The density matrix renormalization group for ab initio quantum chemistry. Eur Phys J D 68(9):272
64. Olivares-Amaya R, Hu W, Nakatani N, Sharma S, Yang J, Chan GKL (2015) The ab-initio density matrix renormalization group in practice. J Chem Phys 142(3):034,102
65. Li Z, Chan GKL (2017) Spin-projected matrix product states: versatile tool for strongly correlated systems. J Chem Theory Comput 13(6):2681–2695
66. Baiardi A, Reiher M (2020) The density matrix renormalization group in chemistry and molecular physics: recent developments and new challenges. J Chem Phys 152(4):040,903
67. Booth GH, Thom AJW, Alavi A (2009) Fermion Monte Carlo without fixed nodes: a game of life, death, and annihilation in Slater determinant space. J Chem Phys 131(5):054,106
68. Cleland D, Booth GH, Alavi A (2010) Communications: survival of the fittest: accelerating convergence in full configuration-interaction quantum Monte Carlo. J Chem Phys 132(4):041,103
69. Blunt N, Booth GH, Alavi A (2017) Density matrices in full configuration interaction quantum Monte Carlo: excited states, transition dipole moments, and parallel distribution. J Chem Phys 146(24):244,105
70. Petruzielo FR, Holmes AA, Changlani HJ, Nightingale MP, Umrigar CJ (2012) Semistochastic projector Monte Carlo method. Phys Rev Lett 109(23):230,201
71. Holmes AA, Changlani HJ, Umrigar CJ (2016) Efficient heat-bath sampling in Fock space. J Chem Theory Comput 12(4):1561–1571
72. Ten-no S (2013) Stochastic determination of effective Hamiltonian for the full configuration interaction solution of quasi-degenerate electronic states. J Chem Phys 138(16):164,126
73. Ghanem K, Lozovoi AY, Alavi A (2019) Unbiasing the initiator approximation in full configuration interaction quantum Monte Carlo. J Chem Phys 151(22):224,108
74. Blunt NS (2019) A hybrid approach to extending selected configuration interaction and full configuration interaction quantum Monte Carlo. J Chem Phys 151(17):174,103
75. Dobrautz W, Smart SD, Alavi A (2019) Efficient formulation of full configuration interaction quantum Monte Carlo in a spin eigenbasis via the graphical unitary group approach. J Chem Phys 151(9):094,104
76. Neufeld VA, Thom AJ (2020) Accelerating convergence in Fock space quantum Monte Carlo methods. J Chem Theory Comput 16(3):1503–1510
77. Ghanem K, Guther K, Alavi A (2020) The adaptive shift method in full configuration interaction quantum Monte Carlo: development and applications. J Chem Phys 153(22):224,115
78. Deustua JE, Magoulas I, Shen J, Piecuch P (2018) Communication: approaching exact quantum chemistry by cluster analysis of full configuration interaction quantum Monte Carlo wave functions. J Chem Phys 149(15):151,101
79. Xu E, Uejima M, Ten-no SL (2018) Full coupled-cluster reduction for accurate description of strong electron correlation. Phys Rev Lett 121(11):113,001

80. Xu E, Uejima M, Ten-No SL (2020) Towards near-exact solutions of molecular electronic structure: full coupled-cluster reduction with a second-order perturbative correction. J Phys Chem Lett 11:9775–9780
81. Zhang T, Evangelista FA (2016) A deterministic projector configuration interaction approach for the ground state of quantum many-body systems. J Chem Theory Comput 12(9):4326–4337
82. Eriksen JJ, Lipparini F, Gauss J (2017) Virtual orbital many-body expansions: a possible route towards the full configuration interaction limit. J Phys Chem Lett 8(18):4633–4639
83. Eriksen JJ, Gauss J (2018) Many-body expanded full configuration interaction. I. Weakly correlated regime. J Chem Theory Comput 14(10):5180–5191
84. Eriksen JJ, Gauss J (2019) Many-body expanded full configuration interaction. II. Strongly correlated regime. J Chem Theory Comput 15(9):4873–4884
85. Eriksen JJ, Gauss J (2019) Generalized many-body expanded full configuration interaction theory. J Phys Chem Lett 10(24):7910–7915
86. Zimmerman PM (2017) Incremental full configuration interaction. J Chem Phys 146(10):104,102
87. Zimmerman PM (2017) Strong correlation in incremental full configuration interaction. J Chem Phys 146(22):224,104
88. Fales BS, Seritan S, Settje NF, Levine BG, Koch H, Martínez TJ (2018) Large-scale electron correlation calculations: rank-reduced full configuration interaction. J Chem Theory Comput 14(8):4139–4150
89. Greene SM, Webber RJ, Weare J, Berkelbach TC (2019) Beyond walkers in stochastic quantum chemistry: reducing error using fast randomized iteration. J Chem Theory Comput 15(9):4834–4850
90. Bytautas L, Ruedenberg K (2004) Correlation energy extrapolation by intrinsic scaling. I. Method and application to the neon atom. J Chem Phys 121(22):10,905-10,918
91. Bytautas L, Ruedenberg K (2004) Correlation energy extrapolation by intrinsic scaling. II. The water and the nitrogen molecule. J Chem Phys 121(22):10919–10934
92. Holmes AA, Tubman NM, Umrigar CJ (2016) Heat-bath configuration interaction: an efficient selected configuration interaction algorithm inspired by heat-bath sampling. J Chem Theory Comput 12:3674–3680
93. Garniron Y, Scemama A, Loos PF, Caffarel M (2017) Hybrid stochastic-deterministic calculation of the second-order perturbative contribution of multireference perturbation theory. J Chem Phys 147(3):034,101
94. Holmes AA, Umrigar CJ, Sharma S (2017) Excited states using semistochastic heat-bath configuration interaction. J Chem Phys 147(16):164,111
95. Sharma S, Holmes AA, Jeanmairet G, Alavi A, Umrigar CJ (2017) Semistochastic heat-bath configuration interaction method: selected configuration interaction with semistochastic perturbation theory. J Chem Theory Comput 13(4):1595–1604
96. Chien AD, Holmes AA, Otten M, Umrigar CJ, Sharma S, Zimmerman PM (2018) Excited states of methylene, polyenes, and ozone from heat-bath configuration interaction. J Phys Chem A 122(10):2714–2722
97. Yao Y, Giner E, Li J, Toulouse J, Umrigar C (2020) Almost exact energies for the Gaussian-2 set with the semistochastic heat-bath configuration interaction method. J Chem Phys 153(12):124,117
98. Evangelista FA (2014) Adaptive multiconfigurational wave functions. J Chem Phys 140(12):124,114
99. Schriber JB, Evangelista FA (2016) Communication: an adaptive configuration interaction approach for strongly correlated electrons with tunable accuracy. J Chem Phys 144(16):161,106
100. Schriber JB, Evangelista FA (2017) Adaptive configuration interaction for computing challenging electronic excited states with tunable accuracy. J Chem Theory Comput 13(11):5354–5366
101. Schriber JB, Hannon KP, Li C, Evangelista FA (2018) A combined selected configuration interaction and many-body treatment of static and dynamical correlation in oligoacenes. J Chem Theory Comput 14(12):6295–6305
102. Tubman NM, Lee J, Takeshita TY, Head-Gordon M, Whaley KB (2016) A deterministic alternative to the full configuration interaction quantum Monte Carlo method. J Chem Phys 145(4):044,112
103. Lehtola S, Tubman NM, Whaley KB, Head-Gordon M (2017) Cluster decomposition of full configuration interaction wave functions: a tool for chemical interpretation of systems with strong correlation. J Chem Phys 147(15):154,105

104. Tubman NM, Freeman CD, Levine DS, Hait D, Head-Gordon M, Whaley KB (2020) Modern approaches to exact diagonalization and selected configuration interaction with the adaptive sampling CI method. J Chem Theory Comput 16(4):2139–2159
105. Tubman NM, Levine DS, Hait D, Head-Gordon M, Whaley KB (2018) An efficient deterministic perturbation theory for selected configuration interaction methods. arXiv:1808.02049 (arXiv preprint)
106. Liu W, Hoffmann MR (2014) SDS: the 'static–dynamic–static' framework for strongly correlated electrons. Theor Chem Acc 133:1481
107. Lei Y, Liu W, Hoffmann MR (2017) Further development of SDSPT2 for strongly correlated electrons. Mol Phys 115:2696–2707
108. Zhang N, Liu W, Hoffmann MR (2020) Iterative configuration interaction with selection. J Chem Theory Comput 16(4):2296–2316
109. Garniron Y, Scemama A, Giner E, Caffarel M, Loos PF (2018) Selected configuration interaction dressed by perturbation. J Chem Phys 149(6):064,103
110. Wang Z, Li Y, Lu J (2019) Coordinate descent full configuration interaction. J Chem Theory Comput 15(6):3558–3569
111. Giner E, Assaraf R, Toulouse J (2016) Quantum Monte Carlo with reoptimised perturbatively selected configuration-interaction wave functions. Mol Phys 114(7–8):910–920
112. Scemama A, Benali A, Jacquemin D, Caffarel M, Loos PF (2018) Excitation energies from diffusion Monte Carlo using selected configuration interaction nodes. J Chem Phys 149(3):034,108
113. Chilkuri VG, Neese F (2021) Comparison of many-particle representations for selected-CI I: a tree based approach. J Comput Chem 42(14):982–1005
114. Chilkuri VG, Neese F (2021) Comparison of many-particle representations for selected configuration interaction: II. Numerical benchmark calculations. J Chem Theory Comput 17(5):2868–2885
115. Bender CF, Davidson ER (1969) Studies in configuration interaction: the first-row diatomic hydrides. Phys Rev 183(1):23
116. Whitten JL, Hackmeyer M (1969) Configuration interaction studies of ground and excited states of polyatomic molecules. I. The CI formulation and studies of formaldehyde. J Chem Phys 51(12):5584–5596
117. Huron B, Malrieu JP, Rancurel P (1973) Iterative perturbation calculations of ground and excited state energies from multiconfigurational zeroth-order wave functions. J Chem Phys 58:5745–5759
118. Evangelisti S, Daudey JP, Malrieu JP (1983) Convergence of an improved CIPSI algorithm. Chem Phys 75:91–102
119. Buenker RJ, Peyerimhoff SD (1974) Individualized configuration selection in CI calculations with subsequent energy extrapolation. Theor Chem Acta 35:33–58
120. Eriksen JJ, Anderson TA, Deustua JE, Ghanem K, Hait D, Hoffmann MR, Lee S, Levine DS, Magoulas I, Shen J, Tubman NM, Whaley KB, Xu E, Yao Y, Zhang N, Alavi A, Chan GKL, Head-Gordon M, Liu W, Piecuch P, Sharma S, Ten-no SL, Umrigar CJ, Gauss J (2020) The ground state electronic energy of benzene. J Phys Chem Lett 11:8922–8929
121. Zhang N, Liu W, Hoffmann MR (2021) Further development of iCIPT2 for strongly correlated electrons. J Chem Theory Comput 17(2):949–964
122. Huang C, Liu W, Xiao Y, Hoffmann MR (2017) iVI: an iterative vector interaction method for large eigenvalue problems. J Comput Chem 38(29):2481–2499
123. Huang C, Liu W, Xiao Y, Hoffmann MR (2018) iVI: an iterative vector interaction method for large eigenvalue problems. J Comput Chem 39:338
124. Huang C, Liu W (2019) iVI-TD-DFT: An iterative vector interaction method for exterior/interior roots of TD-DFT. J Comput Chem 40(9):1023–1037. (E) **2018**, *39*, 338
125. Huang C, Liu W (2018) iVI-TD-DFT: an iterative vector interaction method for exterior/interior roots of TD-DFT. J Comput Chem 39:338
126. Lei Y, Suo B, Liu W (2021) iCAS: imposed automatic selection and localization of complete active spaces. J Chem Theor Comput 17(8):4846–4859
127. Wang Z, Liu W (2021) iOI: an iterative orbital interaction approach for solving the self-consistent field problem. J Chem Theory Comput 17(8):4381–4845
128. Bruna PJ, Peyerimhoff SD, Buenker RJ (1980) The ground-state of the CN+ ion—a multi-reference ci study. Chem Phys Lett 72:278–284
129. Malrieu JP, Durand PH, Daudey JP (1985) Intermediate Hamiltonians as a new class of effective-Hamiltonians. J Phys A Math Gen 18:809–826

130. Pople J, Seeger R, Krishnan R (1977) Variational configuration interaction methods and comparison with perturbation theory. Int J Quantum Chem 12(S11):149–163
131. Siegbahn PE (1978) Multiple substitution effects in configuration interaction calculations. Chem Phys Lett 55(2):386–394
132. Angeli C, Borini S, Cestari M, Cimiraglia R (2004) J Chem Phys 121:4043
133. Buenker RJ (1980) Molecular Physics and quantum chemistry: into the 80's. University of Wollongong Press, Wollongong
134. Buenker RJ (1986) Combining perturbation theory techniques with variational CI calculations to study molecular excited states. Int J Quantum Chem 29(3):435–460
135. Krebs S, Buenker RJ (1995) A new table-direct configuration interaction method for the evaluation of Hamiltonian matrix elements in a basis of linear combinations of spin-adapted functions. J Chem Phys 103(13):5613–5629
136. Khait YG, Hoffmann MR (2002) Hamiltonian matrix elements for the table-CI method using genealogical configuration state functions., vol 8, pp 176–198
137. Stampfu P, Wenzel W (2005) Improved implementation and application of the individually selecting configuration interaction method. J Chem Phys 122(2):024,110
138. Zhang N, Xiao Y, Liu W SOC-iCI and iCI-SOC: combining iCI with SOC in two ways **(unpublished)**
139. Epstein PS (1926) The stark effect from the point of view of Schroedinger's quantum theory. Phys Rev 28(4):695
140. Nesbet RK (1955) Configuration interaction in orbital theories. Proc R Soc Lond Ser A 230(1182):312–321
141. Cheung L, Sundberg K, Ruedenberg K (1978) Dimerization of carbene to ethylene. J Am Chem Soc 100(25):8024–8025
142. Roos BO, Taylor PR, Siegbahn PEM (1980) A complete active space SCF method (CASSCF) using a density-matrix formulated super-ci approach. Chem Phys 48:157–173
143. Werner HJ (1987) Matrix-formulated direct multiconfiguration self-consistent field and multiconfiguration reference configuration-interaction methods. Adv Chem Phys 69:1–62
144. Roos BO (1987) The complete active space self-consistent field method and its applications in electronic structure calculations. Adv Chem Phys 69:399–445
145. Shepard R (1987) The multiconfiguration self-consistent field method. Adv Chem Phys 69:63–200
146. Schmidt MW, Gordon MS (1998) The construction and interpretation of MCSCF wavefunctions. Annu Rev Phys Chem 49(1):233–266
147. Veryazov V, Malmqvist PÅ, Roos BO (2011) How to select active space for multiconfigurational quantum chemistry? Int J Quantum Chem 111(13):3329–3338
148. Sayfutyarova ER, Sun Q, Chan GKL, Knizia G (2017) Automated construction of molecular active spaces from atomic valence orbitals. J Chem Theory Comput 13(9):4063–4078
149. Wu F, Liu W, Zhang Y, Li Z (2011) Linear-scaling time-dependent density functional theory based on the idea of "from fragments to molecule''. J Chem Theory Comput 7(11):3643–3660
150. Li Z, Li H, Suo B, Liu W (2014) Localization of molecular orbitals: from fragments to molecule. Acc Chem Res 47:2758–2767
151. Li H, Liu W, Suo B (2017) Localization of open-shell molecular orbitals via least change from fragments to molecule. J Chem Phys 146(104):104
152. Schreiber M, Silva-Junior MR, Sauer SP, Thiel W (2008) Benchmarks for electronically excited states: CASPT2, CC2, CCSD, and CC3. J Chem Phys 128(13):134,110
153. Angeli C, Cimiraglia R, Evangelisti S, Leininger T, Malrieu JP (2001) Introduction of n-electron valence states for multireference perturbation theory. J Chem Phys 114:10,252-10,264
154. Angeli C, Cimiraglia R, Malrieu JP (2001) N-electron valence state perturbation theory: a fast implementation of the strongly contracted variant. Chem Phys Lett 350:297–305
155. Dyall KG (1995) The choice of a zeroth-order Hamiltonian for second-order perturbation theory with a complete active space self-consistent-field reference function. J Chem Theory Comput 102(12):4909–4918
156. Wang Y, Han H, Lei Y, Suo B, Zhu H, Song Q, Wen Z (2014) New schemes for internally contracted multi-reference configuration interaction. J Chem Phys 141(164):114
157. Suo B, Lei Y, Han H, Wang Y (2018) Development of Xi'an-CI package—applying the hole—particle symmetry in multi-reference electronic correlation calculations. Mol Phys 116(7–8):1051–1064

158. Liu W, Hong G, Dai D, Li L, Dolg M (1997) The Beijing 4-component density functional theory program package (BDF) and its application to EuO, EuS, YbO and YbS. Theor Chem Acc 96(2):75–83
159. Liu W, Wang F, Li L (2003) J Theor Comput Chem 2:257–272
160. Liu W, Wang F, Li L (2004) Relativistic density functional theory: the BDF program package. World Scientific, Singapore, pp 257–282
161. Liu W, Wang F, Li L (2004) Recent advances in relativistic density functional methods. Wiley, Chichester
162. Zhang Y, Suo B, Wang Z, Zhang N, Li Z, Lei Y, Zou W, Gao J, Peng D, Pu Z, Xiao Y, Sun Q, Wang F, Ma Y, Wang X, Guo Y, Liu W (2020) BDF: a relativistic electronic structure program package. J Chem Phys 152(6):064,113
163. Schäfer A, Horn H, Ahlrichs R (1992) Fully optimized contracted Gaussian basis sets for atoms Li to Kr. J Chem Phys 97(4):2571–2577
164. Weigend F, Ahlrichs R (2005) Balanced basis sets of split valence, triple zeta valence and quadruple zeta valence quality for H to Rn: design and assessment of accuracy. Phys Chem Chem Phys 7(18):3297–3305
165. Langhoff SR, Davidson ER (1974) Configuration interaction calculations on the nitrogen molecule. Int J Quantum Chem 8(1):61–72
166. Pathak S, Lang L, Neese F (2017) A dynamic correlation dressed complete active space method: theory, implementation, and preliminary applications. J Chem Phys 147(23):234,109
167. Kánnár D, Szalay PG (2014) Benchmarking coupled cluster methods on valence singlet excited states. J Chem Theory Comput 10(9):3757–3765
168. Bomble YJ, Sattelmeyer KW, Stanton JF, Gauss J (2004) On the vertical excitation energy of cyclopentadiene. J Chem Phys 121(11):5236–5240
169. Barbatti M, Aquino AJ, Lischka H (2010) The UV absorption of nucleobases: semi-classical ab initio spectra simulations. Phys Chem Chem Phys 12(19):4959–4967
170. Avila Ferrer FJ, Cerezo J, Stendardo E, Improta R, Santoro F (2013) Insights for an accurate comparison of computational data to experimental absorption and emission spectra: beyond the vertical transition approximation. J Chem Theory Comput 9(4):2072–2082
171. Zobel JP, Nogueira JJ, González L (2017) The IPEA dilemma in CASPT2. Chem Sci 8(2):1482–1499
172. Barton D, König C, Neugebauer J (2014) Vibronic-structure tracking: a shortcut for vibrationally resolved UV/Vis-spectra calculations. J Chem Phys 141(16):164,115
173. Bai S, Mansour R, Stojanović L, Toldo JM, Barbatti M (2020) On the origin of the shift between vertical excitation and band maximum in molecular photoabsorption. J Mol Model 26(5):1–9
174. Mulliken RS (1977) The excited states of ethylene. J Chem Phys 66(6):2448–2451
175. Doering J, McDiarmid R (1980) Electron impact study of the energy levels of trans-1, 3-butadiene: II. Detailed analysis of valence and Rydberg transitions. J Chem Phys 73(8):3617–3624
176. Leopold D, Pendley R, Roebber J, Hemley R, Vaida V (1984) Direct absorption spectroscopy of jet-cooled polyenes. II. The $1\,^1B_u^+ \leftarrow 1\,^1A_g^-$ transitions of butadienes and hexatrienes. J Chem Phys 81(10):4218–4229
177. Allan M, Neuhaus L, Haselbach E (1984) (all-E)-1, 3, 5, 7-Octatetraene: electron-energy-loss and electron-transmission spectra. Helv Chim Acta 67(7):1776–1782
178. Robin M, Basch H, Kuebler N, Wiberg K, Ellison G (1969) Optical spectra of small rings. II. The unsaturated three-membered rings. J Chem Phys 51(1):45–52
179. Sabljić A, McDiarmid R (1990) Analysis of the absorption spectrum of the NV 1 transition of cyclopentadiene. J Chem Phys 93(6):3850–3855
180. Frueholz RP, Flicker WM, Mosher OA, Kuppermann A (1979) Excited electronic states of cyclohexene, 1, 4-cyclohexadiene, norbornene, and norbornadiene as studied by electron-impact spectroscopy. J Chem Phys 70(4):1986–1993
181. Wilkinson P (1956) Absorption spectra of benzene and benzene-d 6 in the vacuum ultraviolet. Can J Phys 34(6):596–615
182. George G, Morris G (1968) The intensity of absorption of naphthalene from 30,000 cm^{-1} to 53,000 cm^{-1}. J Mol Spectrosc 26(1):67–71
183. Flicker WM, Mosher OA, Kuppermann A (1976) Electron impact investigation of electronic excitations in furan, thiophene, and pyrrole. J Chem Phys 64(4):1315–1321
184. Bavia M, Bertinelli F, Taliani C, Zauli C (1976) The electronic spectrum of pyrrole in the vapour and crystal. Mol Phys 31(2):479–489

185. Grebow PE, Hooker TM Jr (1975) Conformation of histidine model peptides. II. Spectroscopic properties of the imidazole chromophore. Biopolymers 14(4):871–881
186. Bolovinos A, Tsekeris P, Philis J, Pantos E, Andritsopoulos G (1984) Absolute vacuum ultraviolet absorption spectra of some gaseous azabenzenes. J Mol Spectrosc 103(2):240–256
187. Innes K, Ross I, Moomaw WR (1988) Electronic states of azabenzenes and azanaphthalenes: a revised and extended critical review. J Mol Spectrosc 132(2):492–544
188. Fridh C, Åsbrink L, Jonsson B, Lindholm E (1972) Rydberg series in small molecules: XVIII. Photoelectron, UV, mass, and electron impact spectra of s-tetrazine. Int J Mass Spectrom 9(5):485–497
189. Weber J, Malsch K, Hohlneicher G (2001) Excited electronic states of p-benzoquinone. Chem Phys 264(3):275–318
190. Hunt H, Simpson W (1953) Spectra of simple amides in the vacuum ultraviolet1. J Am Chem Soc 75(18):4540–4543
191. Nielsen EB, Schellman JA (1967) The absorption spectra of simple amides and peptides. J Phys Chem C 71(7):2297–2304
192. Silva-Junior MR, Sauer SP, Schreiber M, Thiel W (2010) Basis set effects on coupled cluster benchmarks of electronically excited states: CC3, CCSDR(3) and CC2. Mol Phys 108:453–465
193. Schreiber M, Silva-Junior MR, Sauer SPA, Thiel W (2008) Benchmarks for electronically excited states: CASPT2, CC2, CCSD, and CC3. J Chem Phys 128(13):134,110
194. Serrano-Andrés L, Merchán M, Nebot-Gil I, Lindh R, Roos BO (1993) Towards an accurate molecular orbital theory for excited states: ethene, butadiene, and hexatriene. J Chem Phys 98(4):3151–3162
195. Khokhlov D, Belov A (2021) Toward an accurate Ab initio description of low-lying singlet excited states of polyenes. J Chem Theory Comput 17:4301–4315
196. Dunning TH Jr (1989) Gaussian basis sets for use in correlated molecular calculations. I. The atoms boron through neon and hydrogen. J Chem Phys 90(2):1007–1023
197. Lechner MH, Izsák R, Nooijen M, Neese F (2021) A perturbative approach to multireference equation-of-motion coupled cluster. Mol Phys 20:e1939185

Publisher's Note Springer Nature remains neutral with regard to jurisdictional claims in published maps and institutional affiliations.

Authors and Affiliations

Yangyang Song[1] · Yang Guo[1] · Yibo Lei[2] · Ning Zhang[3] · Wenjian Liu[1]

✉ Wenjian Liu
liuwj@sdu.edu.cn

1 Qingdao Institute for Theoretical and Computational Sciences, Institute of Frontier and Interdisciplinary Science, Shandong University, Qingdao 266237, Shandong, China

2 Key Laboratory of Synthetic and Natural Functional Molecule of the Ministry of Education, College of Chemistry and Materials Science, Shaanxi key Laboratory of Physico-Inorganic Chemistry, Northwest University, Xi'an 710127, Shaanxi, China

3 Beijing National Laboratory for Molecular Sciences, Institute of Theoretical and Computational Chemistry, College of Chemistry and Molecular Engineering, Peking University, Beijing 100871, China

Topics in Current Chemistry (2022) 380:4
https://doi.org/10.1007/s41061-021-00359-1

REVIEW

Ensemble Density Functional Theory of Neutral and Charged Excitations

Exact Formulations, Standard Approximations, and Open Questions

Filip Cernatic[1] · Bruno Senjean[2] · Vincent Robert[1] · Emmanuel Fromager[1]

Received: 1 September 2021 / Accepted: 2 November 2021 / Published online: 26 November 2021
© The Author(s), under exclusive licence to Springer Nature Switzerland AG 2021

Abstract
Recent progress in the field of (time-independent) ensemble density-functional theory (DFT) for excited states are reviewed. Both Gross–Oliveira–Kohn (GOK) and N-centered ensemble formalisms, which are mathematically very similar and allow for an in-principle-exact description of neutral and charged electronic excitations, respectively, are discussed. Key exact results, for example, the equivalence between the infamous derivative discontinuity problem and the description of weight dependencies in the ensemble exchange-correlation density functional, are highlighted. The variational evaluation of orbital-dependent ensemble Hartree-exchange (Hx) energies is discussed in detail. We show in passing that state-averaging individual exact Hx energies can lead to severe (although solvable) v-representability issues. Finally, we explore the possibility of using the concept of density-driven correlation, which has been introduced recently and does not exist in regular ground-state DFT, for improving state-of-the-art correlation density-functional approximations for ensembles. The present review reflects the efforts of a growing community to turn ensemble DFT into a rigorous and reliable low-cost computational method for excited states. We hope that, in the near future, this contribution will stimulate new formal and practical developments in the field.

Keywords Density-functional theory · Excited states · Many-body ensembles

Chapter 8 was originally published as Cernatic, F., Senjean, B., Robert, V. & Fromager, E. Topics in Current Chemistry (2022) 380: 4. https://doi.org/10.1007/s41061-021-00359-1.

✉ Emmanuel Fromager
fromagere@unistra.fr

Extended author information available on the last page of the article

1 Introduction

Kohn–Sham density-functional theory (KS-DFT) [1] has become over the last two decades the method of choice for modeling the electronic structure of molecules and materials. This success originates from the relatively low computational cost of the method and its relatively good accuracy in the description of ground-state properties such as equilibrium structures and activation barriers. KS-DFT is an in-principle-exact ground-state theory. As such, it cannot be used straightforwardly for calculating excited-state properties. The formal beauty of KS-DFT lies in its universal description of electronic ground states. Indeed, in KS-DFT, all the quantum many-electron effects are encoded into a system-independent exchange-correlation (xc) density functional $E_{xc}[n]$ which, if it were known, would allow us to compute the exact ground-state energy of any electronic system, simply by solving self-consistent one-electron equations, instead of the many-electron Schrödinger equation. This universality is somehow lost, at least partially, when turning to the excited states. An obvious reason is that there are different types of electron excitations. First, we should distinguish charged excitations, where the number of electrons in the system is modified, from neutral excitations, which occur at a fixed electron number [2]. In the context of DFT, these two processes are usually approached very differently. In the case of charged excitations, one traditionally refers to the extension of DFT to fractional electron numbers [3–9]. Its implementation at the simplest (semi-) local xc density functional level of approximation usually yields too small fundamental gaps for solids [10]. This can be related to the discontinuities that the density functional derivative of the xc energy should in principle exhibit, when crossing an integer electron number, but that are completely absent from standard approximations. This is the reason why hybrid functionals (where a fraction of orbital-dependent exchange energy is combined with density functionals) [8, 11–13] or even more involved frequency-dependent post KS-DFT approaches like *GW* [14–21] are usually employed to improve the description of fundamental gaps, which implies a substantial increase in computational cost. If we now turn to neutral excitation processes, the most popular approach is (linear response) time-dependent-DFT (TD-DFT) [22, 23]. In TD-DFT, the excitation energies are determined by searching for the poles of the KS linear response function. The success of TD-DFT lies in the fact that in many (but not all) cases the (rather crude) adiabatic approximation performs relatively well with moderate computational cost. In the latter approximation, the TD density-functional xc energy (it is in fact an action [24, 25], to be more precise) is evaluated, in the time range $t_0 \leq t \leq t_1$, from the regular (time-independent) ground-state xc functional and the density $n(t) \equiv n(\mathbf{r}, t)$ at time t as follows, $\mathcal{A}_{xc}[n] \approx \int_{t_0}^{t_1} \mathrm{d}t\, E_{xc}[n(t)]$ [23–26]. Nevertheless, the description of charge-transfer excitations becomes problematic when semi-local functionals are employed [23, 27–29]. Moreover, multiple excitations are completely absent from standard TD-DFT spectra, precisely because of the adiabatic approximation [23, 29, 30]. Let us finally mention that TD-DFT can be seen as a single-reference post-DFT method, because it relies on the single-configuration KS ground-state wave function. This can be problematic, for example, when the system under study has near-degenerate low-lying states, such as in the vicinity of a conical intersection, for example [23].

The various limitations (in terms of computational cost, accuracy, or physics) of the above-mentioned frequency-dependent post-DFT approaches explain why, in recent years, time-independent formulations of DFT for excited states have attracted increasing attention. For charged excitations, the extended Koopman's theorem turns out to be an appealing alternative [31, 32]. DFT for fractional electron numbers has also been further developed in the last decade (see, for example, Refs. [33–40]). It has also been argued that, in the calculation of charged excitations, restricting to integer electron numbers is a completely valid alternative [41, 42]. For neutral excitations, various state-specific approaches have been explored at the formal [43, 44] but also practical level. In the latter case, orbital optimizations must be performed under proper constraints in order to avoid variational collapses. This leads to various variational computational schemes such as the Δ self-consistent field (ΔSCF) approach [45–47], the maximum overlap method (MOM) [48–52], orthogonally-constrained DFT [53–56], and constricted DFT [57–61]. Interestingly, when a system has a Coulombic one-electron external local potential, which is the case for any real molecule, an excited state can be identified directly from its density [62–65]. This fundamental property can be used for constructing an in-principle-exact DFT for individual excited states. The practical implementation of such a theory is not straightforward, in particular because density functionals must be defined also for non-Coulombic densities, so that functional derivatives can be evaluated. Another strategy, which is the main topic of this review, is ensemble DFT (eDFT). The ensemble formalism is often referred to in DFT for mathematical purposes like, for example, extending the domain of definition of density functionals [66] or describing strict degeneracies [67, 68]. It has probably been underestimated as a potential alternative to standard TD methods for the practical calculation of (charged or neutral) excitations [69–75]. A clear advantage of eDFT over TD approaches is that its computational cost is essentially that of a standard KS-DFT calculation. The only difference is that, in an ensemble, orbitals can be fractionally occupied. Moreover, like in TD-DFT, regular ground-state xc functionals can be recycled in eDFT [71]. Note that, unlike in thermal DFT [76–78], the fractional orbital occupation numbers are actually known before the eDFT calculation starts. They are determined by the ensemble weights that the user (arbitrarily) assigns to the M ($M = 1, 2, \ldots$) lowest excited states that she/he wants to study. In (say canonical) thermal DFT [78], the ensemble weights are determined not only from the temperature (which is arbitrarily fixed by the user) but also from the (to-be-calculated) KS orbital energies. Another important feature of eDFT is that it can, in principle, describe any kind of excitation, including the double excitations [79, 80] that standard approximate TD-DFT misses. The eDFT formalism for neutral excitations is often referred to as Gross–Oliveira–Kohn DFT (GOK-DFT) because it relies on the GOK variational principle [81, 82]. A similar formalism, referred to as N-centered eDFT, has been derived recently by Senjean and Fromager [83] for the description of charged excitations. The present review aims at highlighting recent progress in eDFT, with a particular focus on the exact theory and the development of approximations from first principles.

The chapter is organized as follows. An introduction to GOK-DFT and N-centered eDFT is given in the following section on "Unified Ensemble DFT Formalism for Neutral and Charged Excitations". Even though the two theories describe completely different physical processes, their mathematical formulations are very similar, as highlighted in the section. We also explain how individual energy levels (which give access to excited-state properties) can be extracted, in principle exactly, from these theories. Then, in "Equivalence Between Weight Derivatives and xc Derivative Discontinuities", we discuss the equivalence between the xc derivative discontinuity, which is a fundamental concept in DFT, and the ensemble weight derivative of the xc density functional, which is central in eDFT. Strategies for developing weight-dependent xc density-functional approximations (DFAs) for ensembles, which is the most challenging task in eDFT, are then reviewed. Key concepts will be illustrated with the prototypical asymmetric Hubbard dimer model [84, 85]. The rigorous construction of hybrid functionals for ensembles is discussed in "The Exact Hartree-Exchange Dilemma in eDFT". We reveal that using state-averaged exact exchange energies, which is common in computational eDFT studies [75], can lead to severe (but solvable) v-representability issues. Finally, we discuss, in "Individual Correlations within Ensembles: An Exact Construction", the concept of ensemble density-driven correlation, which was introduced recently by Gould and Pittalis [86], and how it could be used in the design of correlation DFAs for ensembles. We round off the review with "Conclusions and Perspectives". Detailed derivations of some key equations are provided in the appendices.

2 Unified Ensemble DFT Formalism for Neutral and Charged Excitations

In this chapter, we are interested in the evaluation of neutral ($E_I^N - E_0^N$) and charged ($E_0^{N\pm1} - E_0^N$) excitation energies, where the Ith lowest energy E_I^M of the M-electron system under study is in principle obtained by solving the following Schrödinger equation,

$$\hat{H}|\Psi_I^M\rangle = E_I^M|\Psi_I^M\rangle, \tag{1}$$

where

$$\hat{H} = \hat{T} + \hat{W}_{\text{ee}} + \hat{V}_{\text{ext}} \tag{2}$$

is the electronic Hamiltonian within the Born–Oppenheimer approximation and

$$\begin{aligned}\hat{T} &\equiv -\sum_{i=1}^{M}\frac{1}{2}\nabla^2_{\mathbf{r}_i}\\ \hat{W}_{\text{ee}} &\equiv \sum_{i=1}^{M}\sum_{j>i}^{M}\frac{1}{|\mathbf{r}_i - \mathbf{r}_j|}\times\\ \hat{V}_{\text{ext}} &\equiv \sum_{i=1}^{M} v_{\text{ext}}(\mathbf{r}_i)\times\end{aligned} \tag{3}$$

are the M-electron kinetic energy, Coulomb repulsion, and local (i.e., multiplicative) external potential operators, respectively. Both neutral and charged excitations can be described within a unified eDFT formalism. The calculations will differ simply in the type of excited states (charged or neutral) that are included into the ensemble. On the one hand, the GOK-DFT formalism [82] will be employed for neutral excitations while, for charged excitations, we will use the more recent N-centered eDFT formalism [83]. In this section, we derive key equations for each theory, and we show how individual excited-state properties (energy levels and densities) can be extracted, in principle exactly, from the KS ensemble. Real algebra will be used throughout this work. For the sake of clarity, derivations will be detailed only for ensembles consisting of non-degenerate states. The theory obviously applies to more general cases [82, 87].

2.1 DFT of Neutral Excitations

GOK-DFT was formulated at the end of the 1980s by Gross, Oliveira, and Kohn [81, 82, 88] and is a generalization of the equiensemble DFT of Theophilou [89]. In contrast to standard DFT, which is a ground-state theory, GOK-DFT can describe both ground and (neutral) excited states. In this context, the ensemble density is used as a basic variable (in place of the ground-state density).

2.1.1 GOK Ensembles

Before deriving the main equations of GOK-DFT, let us introduce the exact ensemble theory. We start with the ensemble GOK energy expression [81]

$$E^{\mathbf{w}} = \sum_I \mathtt{w}_I E_I, \tag{4}$$

which is simply a state-averaged energy where $\mathbf{w} = (\mathtt{w}_1, \mathtt{w}_2, \ldots)$ denotes the collection of ensemble weights that are assigned to the excited states, and $E_I \equiv E_I^N$ are the energies of the N-electron ground ($I = 0$) and excited ($I > 0$) states $|\Psi_I^N\rangle$. We assumed in Eq. (4) that the full set of weights (which includes the weight $\mathtt{w}_0$ assigned to the ground state) is normalized, i.e., $\mathtt{w}_0 = 1 - \sum_{I>0} \mathtt{w}_I$, so that

$$\begin{aligned} E^{\mathbf{w}} &= \left(1 - \sum_{I>0} \mathtt{w}_I\right) E_0 + \sum_{I>0} \mathtt{w}_I E_I \\ &= E_0 + \sum_{I>0} \mathtt{w}_I \left(E_I - E_0\right). \end{aligned} \tag{5}$$

For ordered weights $\mathtt{w}_I \geq \mathtt{w}_{I+1} \geq 0$, with $I \geq 0$, the following (so-called GOK) variational principle holds [81],

$$E^{\mathbf{w}} \leq \sum_I \mathtt{w}_I \langle \tilde{\Psi}_I | \hat{H} | \tilde{\Psi}_I \rangle, \tag{6}$$

where $\{\tilde{\Psi}_I\}$ is a trial set of orthonormal N-electron wave functions. Note that the lower bound $E^{\mathbf{w}}$, which is the exact ensemble energy, is not an observable. It is just an (artificial) auxiliary quantity from which properties of interest, such as the excitation energies, can be extracted. Since it varies linearly with the ensemble weights, the extraction of individual energy levels is actually trivial. Indeed, combining the following two relations [see Eq. (5)],

$$\frac{\partial E^{\mathbf{w}}}{\partial \mathtt{w}_I} = E_I - E_0, \tag{7}$$

and

$$\begin{aligned} E_K \underset{K\geq 0}{=} & E_0 + \sum_{I>0} \delta_{IK}(E_I - E_0) \\ = & E^{\mathbf{w}} + \sum_{I>0} (\delta_{IK} - \mathtt{w}_I)(E_I - E_0), \end{aligned} \tag{8}$$

leads to

$$E_K = E^{\mathbf{w}} + \sum_{I>0} (\delta_{IK} - \mathtt{w}_I) \frac{\partial E^{\mathbf{w}}}{\partial \mathtt{w}_I}. \tag{9}$$

Despite its simplicity, the above expression has not been used until very recently for extracting excited-state energies from a GOK-DFT calculation [90, 91]. Further details will be given in the next section.

2.1.2 DFT of GOK Ensembles

In GOK-DFT, the ensemble energy is obtained variationally as follows [82],

$$E^{\mathbf{w}} = \min_{n \to N} \left\{ F^{\mathbf{w}}[n] + \int d\mathbf{r}\, v_{\text{ext}}(\mathbf{r}) n(\mathbf{r}) \right\}, \tag{10}$$

where the minimization is restricted to N-electron densities, i.e., $\int d\mathbf{r}\, n(\mathbf{r}) = N$, and the universal GOK density functional

$$F^{\mathbf{w}}[n] := \sum_I \mathtt{w}_I \left\langle \Psi_I^{\mathbf{w}}[n] \middle| \hat{T} + \hat{W}_{\text{ee}} \middle| \Psi_I^{\mathbf{w}}[n] \right\rangle, \tag{11}$$

which is evaluated from the density-functional eigenfunctions $\{\Psi_I^{\mathbf{w}}[n]\}$ that fulfill the density constraint $\sum_I \mathtt{w}_I n_{\Psi_I^{\mathbf{w}}[n]}(\mathbf{r}) = n(\mathbf{r})$, is the analog for GOK ensembles of the universal Hohenberg–Kohn functional. Its construction relies on a potential-ensemble-density map that is established for a given and fixed set $\mathbf{w}$ of ensemble weight values. Therefore, the universality of the functional implies, as in ground-state DFT, that it does not depend on the local external potential. However, it does not mean that it is ensemble-independent and therefore applicable to any excited state. As discussed in further detail below, encoding ensemble dependencies into density functionals is probably the most challenging task in eDFT.

In the standard KS formulation of GOK-DFT [82], the GOK functional is split into non-interacting kinetic and Hartree-xc (Hxc) ensemble energy contributions, by analogy with regular KS-DFT:

$$F^{\mathbf{w}}[n] = T_{\mathrm{s}}^{\mathbf{w}}[n] + E_{\mathrm{Hxc}}^{\mathbf{w}}[n]. \tag{12}$$

The non-interacting ensemble kinetic energy functional can be expressed more explicitly as follows within the constrained-search formalism [92],

$$T_{\mathrm{s}}^{\mathbf{w}}[n] = \min_{\hat{\gamma}^{\mathbf{w}} \to n} \left\{ \mathrm{Tr}\left[\hat{\gamma}^{\mathbf{w}} \hat{T}\right] \right\} \tag{13}$$

$$\equiv \sum_I \mathtt{w}_I \langle \Phi_I^{\mathbf{w}}[n] | \hat{T} | \Phi_I^{\mathbf{w}}[n] \rangle, \tag{14}$$

where Tr denotes the trace, $\hat{\gamma}^{\mathbf{w}} = \sum_I \mathtt{w}_I |\Phi_I\rangle\langle\Phi_I|$ is a trial ensemble density matrix operator that fulfills the density constraint $n_{\hat{\gamma}^{\mathbf{w}}}(\mathbf{r}) \equiv \mathrm{Tr}[\hat{\gamma}^{\mathbf{w}} \hat{n}(\mathbf{r})] = \sum_I \mathtt{w}_I n_{\Phi_I}(\mathbf{r}) = n(\mathbf{r})$, and $\hat{n}(\mathbf{r}) \equiv \sum_{i=1}^{N} \delta(\mathbf{r} - \mathbf{r}_i)$ is the electron density operator at position $\mathbf{r}$. Combining Eqs. (10), (12) and (13) leads to the final GOK-DFT variational energy expression

$$\begin{aligned} E^{\mathbf{w}} &= \min_{\{\varphi_p\}} \left\{ \mathrm{Tr}\left[\hat{\gamma}^{\mathbf{w}} \left(\hat{T} + \hat{V}_{\mathrm{ext}}\right)\right] + E_{\mathrm{Hxc}}^{\mathbf{w}}\left[n_{\hat{\gamma}^{\mathbf{w}}}\right] \right\} \\ &= \mathrm{Tr}\left[\hat{\gamma}_{\mathrm{KS}}^{\mathbf{w}} \left(\hat{T} + \hat{V}_{\mathrm{ext}}\right)\right] + E_{\mathrm{Hxc}}^{\mathbf{w}}\left[n_{\hat{\gamma}_{\mathrm{KS}}^{\mathbf{w}}}\right], \end{aligned} \tag{15}$$

where the minimization can be restricted to single-configuration wave functions (determinants or configuration state functions), hence the minimization over orbitals $\{\varphi_p\}$ on the first line of Eq. (15). The minimizing KS orbitals $\{\varphi_p^{\mathbf{w}}\}$, from which the KS wave functions $\{\Phi_I^{\mathbf{w}}[n^{\mathbf{w}}] \equiv \Phi_I^{\mathbf{w}}\}$ in the minimizing density matrix operator $\hat{\gamma}_{\mathrm{KS}}^{\mathbf{w}}$ are constructed, fulfill the following self-consistent GOK-DFT equations,

$$\left(-\frac{\nabla_{\mathbf{r}}^2}{2} + v_{\mathrm{ext}}(\mathbf{r}) + v_{\mathrm{Hxc}}^{\mathbf{w}}[n^{\mathbf{w}}](\mathbf{r}) \right) \varphi_p^{\mathbf{w}}(\mathbf{r}) = \varepsilon_p^{\mathbf{w}} \varphi_p^{\mathbf{w}}(\mathbf{r}), \tag{16}$$

where

$$v_{\mathrm{Hxc}}^{\mathbf{w}}[n](\mathbf{r}) = \frac{\delta E_{\mathrm{Hxc}}^{\mathbf{w}}[n]}{\delta n(\mathbf{r})} \tag{17}$$

is the ensemble Hxc density-functional potential. In the exact theory, the ensemble KS orbitals reproduce the exact (interacting) ensemble density, i.e.,

$$\sum_I \mathtt{w}_I n_{\Phi_I^{\mathbf{w}}}(\mathbf{r}) = \sum_I \mathtt{w}_I n_{\Psi_I}(\mathbf{r}) = n^{\mathbf{w}}(\mathbf{r}), \tag{18}$$

where the individual KS densities read as

$$n_{\Phi_I^{\mathbf{w}}}(\mathbf{r}) = \sum_p n_p^I |\varphi_p^{\mathbf{w}}(\mathbf{r})|^2, \tag{19}$$

and n_p^I is the (weight-independent) occupation number of the orbital $\varphi_p^{\mathbf{w}}$ in the single-configuration wave function $\Phi_I^{\mathbf{w}}$.

Let us now focus on the ensemble Hxc density functional. By analogy with regular KS-DFT, it can be decomposed into Hx and correlation energy contributions: $E_{\rm Hxc}^{\mathbf{w}}[n] = E_{\rm Hx}^{\mathbf{w}}[n] + E_{\rm c}^{\mathbf{w}}[n]$. In the original formulation of GOK-DFT [82], the Hx functional is further decomposed as follows,

$$E_{\rm Hx}^{\mathbf{w}}[n] = E_{\rm H}[n] + E_{\rm x}^{\mathbf{w}}[n], \tag{20}$$

where

$$E_{\rm H}[n] = \frac{1}{2}\int d\mathbf{r} \int d\mathbf{r}' \frac{n(\mathbf{r})n(\mathbf{r}')}{|\mathbf{r}-\mathbf{r}'|} \tag{21}$$

is the standard weight-independent Hartree functional, and

$$E_{\rm x}^{\mathbf{w}}[n] = \sum_I \mathtt{w}_I \langle \Phi_I^{\mathbf{w}}[n] | \hat{W}_{\rm ee} | \Phi_I^{\mathbf{w}}[n] \rangle - E_{\rm H}[n] \tag{22}$$

is the exact (complementary and weight-dependent) ensemble exchange functional. Note that $\Phi_I^{\mathbf{w}}[n]$, which describes one of the configurations included into the ensemble, may not be a pure Slater determinant [93]. Other (weight-dependent) definitions for the ensemble Hartree energy, where the explicit dependence on the ensemble density n is lost, have been explored [94]. In the most intuitive one, the ensemble Hartree energy is evaluated as the weighted sum of the individual KS Hartree energies:

$$E_{\rm H}^{\mathbf{w}}[n] := \sum_I \mathtt{w}_I E_{\rm H}\left[n_{\Phi_I^{\mathbf{w}}[n]}\right]. \tag{23}$$

For the sake of generality, we will keep in the following both Hartree and exchange energies into a single functional $E_{\rm Hx}^{\mathbf{w}}[n]$ which is defined as

$$E_{\rm Hx}^{\mathbf{w}}[n] = \sum_I \mathtt{w}_I \langle \Phi_I^{\mathbf{w}}[n] | \hat{W}_{\rm ee} | \Phi_I^{\mathbf{w}}[n] \rangle. \tag{24}$$

The remaining weight-dependent correlation energy can then be expressed as follows, according to Eqs. (11) and (14),

$$\begin{aligned} E_{\rm c}^{\mathbf{w}}[n] &= F^{\mathbf{w}}[n] - T_{\rm s}^{\mathbf{w}}[n] - E_{\rm Hx}^{\mathbf{w}}[n] \\ &= \sum_I \mathtt{w}_I \left(\left\langle \Psi_I^{\mathbf{w}}[n] \middle| \hat{T} + \hat{W}_{\rm ee} \middle| \Psi_I^{\mathbf{w}}[n] \right\rangle - \left\langle \Phi_I^{\mathbf{w}}[n] \middle| \hat{T} + \hat{W}_{\rm ee} \middle| \Phi_I^{\mathbf{w}}[n] \right\rangle \right), \end{aligned} \tag{25}$$

where the non-interacting KS $\{\Phi_I^{\mathbf{w}}[n]\}$ and interacting $\{\Psi_I^{\mathbf{w}}[n]\}$ wave functions, which both reproduce the (weight-independent here) trial ensemble density n, whatever the ensemble weight values $\mathbf{w}$, are in principle weight-dependent [95, 96].

Interestingly, the interacting density-functional wave functions lose their weight dependence when the trial density n matches the exact physical ensemble density $n^{\mathbf{w}}$, i.e., $\Psi_I^{\mathbf{w}}[n^{\mathbf{w}}] = \Psi_I \equiv \Psi_I^N$. However, as shown in the section "Weight Dependence of the KS Wave Functions in GOK-DFT", the KS wave functions remain weight-dependent, even in this special case.

As readily seen from Eqs. (12), (15) and (16), the only (but crucial) difference between regular ground-state KS-DFT and GOK-DFT is the weight dependence in the ensemble density-functional Hxc energy and potential. The computational cost should essentially be the same in both approaches. The challenge lies in the proper description of the weight-dependent ensemble Hxc density functional. Different approximations have been considered, such as the use of (weight-independent) regular ground-state functionals [78, 97], or the use of an ensemble exact-exchange energy [93, 98, 99] with or without approximate weight-dependent correlation functionals [79, 91, 99]. Note that the expected linearity-in-weight of the ensemble energy is not always reproduced in (approximate) practical GOK-DFT calculations [97]. As a result, different weights can give different excitation energies, which is a serious issue. This led to different computation strategies, such as trying to find an optimal value for the weights [100], using Boltzmann weights instead [78], restricting to equiensembles [79, 91], or considering the ground-state $\mathbf{w} = 0$ limit of the theory, like in the direct ensemble correction (DEC) scheme [87, 101]. A linear interpolation method has also been proposed [97, 102].

Designing weight-dependent ensemble DFAs that systematically reduce the curvature in weight of the ensemble energy, while providing at the same time accurate excitation energies, is an important and challenging task. Recent progress in this matter will be discussed extensively below.

2.1.3 Extraction of Individual State Properties

In the section "DFT of GOK Ensembles", we have shown that both exact ensemble energy and density can be calculated, in principle exactly, within GOK-DFT. At this point, we should stress that the KS and true physical densities are not expected to match individually, even though they both reproduce the same ensemble density [see Eq. (18)]. This subtle point will be discussed in more detail in the section "Weight Dependence of the KS Wave Functions in GOK-DFT". Nevertheless, in complete analogy with Eq. (9), the exact individual densities can be extracted from the ensemble density as follows [103]:

$$n_{\Psi_J}(\mathbf{r}) = n^{\mathbf{w}}(\mathbf{r}) + \sum_{I>0} \left(\delta_{IJ} - \mathtt{w}_I\right) \frac{\partial n^{\mathbf{w}}(\mathbf{r})}{\partial \mathtt{w}_I}, \tag{26}$$

which, by inserting the expression in Eq. (18), leads to the key result [103]

$$n_{\Psi_J}(\mathbf{r}) = n_{\Phi_J^{\mathbf{w}}}(\mathbf{r}) + \sum_{I>0} \sum_{K\geq 0} \left(\delta_{IJ} - \mathtt{w}_I\right) \mathtt{w}_K \frac{\partial n_{\Phi_K^{\mathbf{w}}}(\mathbf{r})}{\partial \mathtt{w}_I}, \tag{27}$$

where, according to Eq. (19), the weight derivative of the individual KS densities

$$\frac{\partial n_{\Phi_K^{\mathbf{w}}}(\mathbf{r})}{\partial \mathrm{w}_I} = 2\sum_p n_p^K \varphi_p^{\mathbf{w}}(\mathbf{r}) \frac{\partial \varphi_p^{\mathbf{w}}(\mathbf{r})}{\partial \mathrm{w}_I} \tag{28}$$

can be evaluated from the (static) linear response of the KS orbitals. This can be done, in practice, by solving, for example, an ensemble coupled-perturbed equation [73, 103].

Turning to the excitation energies, we obtain, from the variational GOK-DFT ensemble energy expression and the Hellmann–Feynman theorem, the following expression, where the derivatives of the minimizing (and therefore stationary) KS wave functions do not contribute,

$$\begin{aligned}\frac{\partial E^{\mathbf{w}}}{\partial \mathrm{w}_I} &= \mathrm{Tr}\left[\Delta\hat{\gamma}^{\mathbf{w}}_{\mathrm{KS},I}\left(\hat{T}+\hat{V}_{\mathrm{ext}}\right)\right] + \left.\frac{\partial E^{\mathbf{w}}_{\mathrm{Hxc}}[n]}{\partial \mathrm{w}_I}\right|_{n=n_{\hat{\gamma}^{\mathbf{w}}_{\mathrm{KS}}}} \\ &+ \int d\mathbf{r}\, \frac{\delta E^{\mathbf{w}}_{\mathrm{Hxc}}[n_{\hat{\gamma}^{\mathbf{w}}_{\mathrm{KS}}}]}{\delta n(\mathbf{r})} \mathrm{Tr}\left[\Delta\hat{\gamma}^{\mathbf{w}}_{\mathrm{KS},I}\hat{n}(\mathbf{r})\right],\end{aligned} \tag{29}$$

with $\Delta\hat{\gamma}^{\mathbf{w}}_{\mathrm{KS},I} = |\Phi^{\mathbf{w}}_I\rangle\langle\Phi^{\mathbf{w}}_I| - |\Phi^{\mathbf{w}}_0\rangle\langle\Phi^{\mathbf{w}}_0|$. This expression can be further simplified as follows [90]:

$$\frac{\partial E^{\mathbf{w}}}{\partial \mathrm{w}_I} = E_I - E_0 = \mathcal{E}^{\mathbf{w}}_I - \mathcal{E}^{\mathbf{w}}_0 + \left.\frac{\partial E^{\mathbf{w}}_{\mathrm{Hxc}}[n]}{\partial \mathrm{w}_I}\right|_{n=n_{\hat{\gamma}^{\mathbf{w}}_{\mathrm{KS}}}}, \tag{30}$$

where $\mathcal{E}^{\mathbf{w}}_I$ denotes the Ith (weight-dependent) KS energy which is obtained by summing up the energies $\{\varepsilon^{\mathbf{w}}_p\}$ of the KS orbitals that are occupied in $\Phi^{\mathbf{w}}_I$. Hence, the excitation energies can all be determined, in principle exactly, from a single GOK-DFT calculation.

As shown by Deur and Fromager [90], individual energy levels can also be extracted (from the KS ensemble) and written in a compact form. For that purpose, we will use the exact expression of Eq. (9) where we see, in the light of Eq. (30), that it is convenient to express the total ensemble energy [first term on the right-hand side of Eq. (9)] in terms of total KS energies. Levy and Zahariev (LZ) made such a suggestion in the context of regular ground-state DFT [104]. For that purpose, they introduced a shift in the Hxc potential that can be trivially generalized to GOK ensembles as follows [90]:

$$\frac{\delta E^{\mathbf{w}}_{\mathrm{Hxc}}[n]}{\delta n(\mathbf{r})} \rightarrow \overline{v}^{\mathbf{w}}_{\mathrm{Hxc}}[n](\mathbf{r}) = \frac{\delta E^{\mathbf{w}}_{\mathrm{Hxc}}[n]}{\delta n(\mathbf{r})} + \frac{E^{\mathbf{w}}_{\mathrm{Hxc}}[n] - \int d\mathbf{r}\, \frac{\delta E^{\mathbf{w}}_{\mathrm{Hxc}}[n]}{\delta n(\mathbf{r})} n(\mathbf{r})}{\int d\mathbf{r}\, n(\mathbf{r})}. \tag{31}$$

Note that, if the exact LZ-shifted Hxc potential were known, we would be able to evaluate exact ensemble density-functional Hxc energies as follows:

$$E^{\mathbf{w}}_{\mathrm{Hxc}}[n] = \int d\mathbf{r}\, \overline{v}^{\mathbf{w}}_{\mathrm{Hxc}}[n](\mathbf{r}) n(\mathbf{r}). \tag{32}$$

Once the LZ shift has been applied to the ensemble Hxc potential, the (total N-electron) KS energies will be modified as follows:

$$\mathcal{E}_I^{\mathbf{w}} \rightarrow \overline{\mathcal{E}}_I^{\mathbf{w}} = \mathcal{E}_I^{\mathbf{w}} + E_{\mathrm{Hxc}}^{\mathbf{w}}[n_{\hat{\gamma}_{\mathrm{KS}}^{\mathbf{w}}}] - \int d\mathbf{r}\, \frac{\delta E_{\mathrm{Hxc}}^{\mathbf{w}}[n_{\hat{\gamma}_{\mathrm{KS}}^{\mathbf{w}}}]}{\delta n(\mathbf{r})} n_{\hat{\gamma}_{\mathrm{KS}}^{\mathbf{w}}}(\mathbf{r}), \tag{33}$$

and the true ensemble energy will simply read as a weighted sum of (LZ-shifted) KS energies,

$$E^{\mathbf{w}} = \left(1 - \sum_{I>0} \mathtt{w}_I\right) \overline{\mathcal{E}}_0^{\mathbf{w}} + \sum_{I>0} \mathtt{w}_I \overline{\mathcal{E}}_I^{\mathbf{w}}. \tag{34}$$

Note that the KS excitation energies are not affected by the shift,

$$\overline{\mathcal{E}}_I^{\mathbf{w}} - \overline{\mathcal{E}}_0^{\mathbf{w}} = \mathcal{E}_I^{\mathbf{w}} - \mathcal{E}_0^{\mathbf{w}}. \tag{35}$$

Thus, by combining Eqs. (9), (30), (34) and (35), we recover the exact expression of Ref. [90] for ground- and excited-state energy levels:

$$E_K = \overline{\mathcal{E}}_K^{\mathbf{w}} + \sum_{I>0} \left(\delta_{IK} - \mathtt{w}_I\right) \left.\frac{\partial E_{\mathrm{Hxc}}^{\mathbf{w}}[n]}{\partial \mathtt{w}_I}\right|_{n=n_{\hat{\gamma}_{\mathrm{KS}}^{\mathbf{w}}}}. \tag{36}$$

As readily seen from Eq. (36), applying the LZ shift is not sufficient for reaching an exact energy level. The ensemble weight derivatives of the Hxc density functional are also needed for that purpose.

Finally, it is instructive to consider the general expression of Eq. (36) in the ground-state $\mathbf{w} = 0$ limit of the theory, which gives

$$E_I = \overline{\mathcal{E}}_I^{\mathbf{w}=0} + (1 - \delta_{I0}) \left.\frac{\partial E_{\mathrm{Hxc}}^{\mathbf{w}}[n_{\Psi_0}]}{\partial \mathtt{w}_I}\right|_{\mathbf{w}=0}, \tag{37}$$

where n_{Ψ_0} is the exact ground-state density. As readily seen from Eq. (37), as we start from a pure $I = 0$ ground-state theory (we recover the energy LZ expression in this case [104]), the inclusion of a given $I > 0$ excited state into the ensemble induces an additional shift in the Hxc potential, which corresponds to the weight derivative $\partial E_{\mathrm{Hxc}}^{\mathbf{w}}[n_{\Psi_0}]/\partial \mathtt{w}_I\big|_{\mathbf{w}=0}$ and can be interpreted as a derivative discontinuity, as shown in Ref. [105] and discussed extensively in the section "Equivalence Between Weight Derivatives and xc DerivativeDiscontinuities", in the context of charged excitations.

2.2 DFT of Charged Excitations: *N*-Centered Ensemble Formalism

A recent adaptation of GOK-DFT to charged excitations, referred to as N-centered eDFT [83], is introduced here.

2.2.1 *N*-Centered Ensembles

The N-centered ensemble [83] can be seen as the "grand canonical" ground-state version of GOK ensembles. It is constructed from the M-electron ground states, where the three possible values of the integer $M \in \{N-1, N, N+1\}$ are, like the corresponding ensemble density (see below), centered in N, hence the name "N-centered". The exact N-centered ensemble energy is defined as follows [83],

$$E_0^{\xi} = \xi_- E_0^{N-1} + \xi_+ E_0^{N+1} + \left(1 - \xi_- \frac{N-1}{N} - \xi_+ \frac{N+1}{N}\right) E_0^N, \tag{38}$$

where the two N-centered ensemble weights ξ_- and ξ_+, which describe the removal/addition of an electron from/to the N-electron system, respectively, are collected in

$$\xi \equiv \left(\xi_-, \xi_+\right). \tag{39}$$

Similarly, the N-centered ensemble density reads as

$$n_0^{\xi}(\mathbf{r}) = \xi_- n_{\Psi_0^{N-1}}(\mathbf{r}) + \xi_+ n_{\Psi_0^{N+1}}(\mathbf{r}) + \left(1 - \xi_- \frac{N-1}{N} - \xi_+ \frac{N+1}{N}\right) n_{\Psi_0^N}(\mathbf{r}). \tag{40}$$

Designed by analogy with GOK-DFT (which describes neutral excitations), the N-centered ensemble density integrates to the central integer number N of electrons:

$$\int d\mathbf{r}\, n_0^{\xi}(\mathbf{r}) = N. \tag{41}$$

In other words, even though we describe charged excitation processes, the number of electrons remains fixed and equal to the integer N, whatever the value of the ensemble weights ξ. This major difference with the conventional DFT for fractional electron numbers [3] has important implications that will be discussed extensively in the section "Equivalence Between Weight Derivatives and xc DerivativeDiscontinuities".

In this context, the ensemble energy can be determined variationally, as a direct consequence of the conventional Rayleigh–Ritz variational principle for a fixed number of electrons, i.e.,

$$\begin{aligned} E_0^{\xi} \leq\ & \xi_- \left\langle \tilde{\Psi}^{N-1} \middle| \hat{H} \middle| \tilde{\Psi}^{N-1} \right\rangle + \xi_+ \left\langle \tilde{\Psi}^{N+1} \middle| \hat{H} \middle| \tilde{\Psi}^{N+1} \right\rangle \\ & + \left(1 - \xi_- \frac{N-1}{N} - \xi_+ \frac{N+1}{N}\right) \left\langle \tilde{\Psi}^{N} \middle| \hat{H} \middle| \tilde{\Psi}^{N} \right\rangle, \end{aligned} \tag{42}$$

where $\{\tilde{\Psi}^M\}$ are trial M-electron normalized wave functions, provided that the (so-called convexity) conditions $\xi_- \geq 0, \xi_+ \geq 0$, and $\xi_-(N-1) + \xi_+(N+1) \leq N$ are fulfilled. Like in GOK-DFT, the ensemble energy E_0^{ξ} varies linearly with the ensemble weights. As a result, charged excitation energies can be extracted through differentiation with respect to the N-centered ensemble weights. For example, since

$$\frac{\partial E_0^{\xi}}{\partial \xi_{\pm}} = E_0^{N\pm 1} - \left(\frac{N \pm 1}{N}\right) E_0^{N}, \tag{43}$$

the exact fundamental gap can be determined as follows:

$$\frac{\partial E_0^{\xi}}{\partial \xi_-} + \frac{\partial E_0^{\xi}}{\partial \xi_+} = E_0^{N-1} + E_0^{N+1} - 2E_0^{N} = E_{\text{gap}}^{\text{fund}}. \tag{44}$$

We can also extract the individual cationic, anionic, and neutral energies, respectively, as follows:

$$E_0^{N-1} = \frac{N-1}{N}\left(E_0^{\xi} - \xi_+ \frac{\partial E_0^{\xi}}{\partial \xi_+} + \left(\frac{N}{N-1} - \xi_-\right)\frac{\partial E_0^{\xi}}{\partial \xi_-}\right), \tag{45}$$

$$E_0^{N+1} = \frac{N+1}{N}\left(E_0^{\xi} - \xi_- \frac{\partial E_0^{\xi}}{\partial \xi_-} + \left(\frac{N}{N+1} - \xi_+\right)\frac{\partial E_0^{\xi}}{\partial \xi_+}\right), \tag{46}$$

and

$$E_0^{N} = E_0^{\xi} - \xi_- \frac{\partial E_0^{\xi}}{\partial \xi_-} - \xi_+ \frac{\partial E_0^{\xi}}{\partial \xi_+}. \tag{47}$$

Equations (45)–(47) will be used in the following for deriving exact ionization potential and electron affinity theorems.

2.2.2 DFT of *N*-Centered Ensembles

In complete analogy with GOK-DFT, the *N*-centered ensemble energy can be determined variationally as follows:

$$E_0^{\xi} = \min_{n \to N} \left\{ F^{\xi}[n] + \int \mathrm{d}\mathbf{r}\, v_{\text{ext}}(\mathbf{r}) n(\mathbf{r}) \right\}, \tag{48}$$

where, in the KS formulation of the theory [83], the universal *N*-centered ensemble functional reads as

$$F^{\xi}[n] = T_s^{\xi}[n] + E_{\text{Hxc}}^{\xi}[n]. \tag{49}$$

The non-interacting kinetic energy functional

$$T_s^{\xi}[n] = \min_{\hat{\gamma}^{\xi} \to n} \left\{ \mathrm{Tr}\left[\hat{\gamma}^{\xi} \hat{T}\right] \right\} \tag{50}$$

is now determined through a minimization over *N*-centered density matrix operators

$$\hat{\gamma}^{\xi} \equiv \xi_{-}|\Phi^{N-1}\rangle\langle\Phi^{N-1}| + \xi_{+}|\Phi^{N+1}\rangle\langle\Phi^{N+1}| + \left(1 - \xi_{-}\frac{N-1}{N} - \xi_{+}\frac{N+1}{N}\right)|\Phi^{N}\rangle\langle\Phi^{N}| \tag{51}$$

that fulfill the density constraint $n_{\hat{\gamma}^{\xi}}(\mathbf{r}) = \mathrm{Tr}\left[\hat{\gamma}^{\xi}\hat{n}(\mathbf{r})\right] = n(\mathbf{r})$. Combining Eqs. (48), (49) and (50) leads to the final ensemble energy expression,

$$\begin{aligned} E_0^{\xi} &= \min_{\{\varphi_p\}}\left\{\mathrm{Tr}\left[\hat{\gamma}^{\xi}\left(\hat{T} + \hat{V}_{\mathrm{ext}}\right)\right] + E_{\mathrm{Hxc}}^{\xi}\left[n_{\hat{\gamma}^{\xi}}\right]\right\} \\ &= \mathrm{Tr}\left[\hat{\gamma}_{\mathrm{KS}}^{\xi}\left(\hat{T} + \hat{V}_{\mathrm{ext}}\right)\right] + E_{\mathrm{Hxc}}^{\xi}\left[n_{\hat{\gamma}_{\mathrm{KS}}^{\xi}}\right], \end{aligned} \tag{52}$$

which is mathematically identical to its analog in GOK-DFT [see Eq. (15)], even though the physics it describes is completely different. The orbitals $\{\varphi_p^{\xi}\}$, from which the minimizing single-configuration KS wave functions $\left\{\Phi_0^{M,\xi}\right\}$ in $\hat{\gamma}_{\mathrm{KS}}^{\xi}$ are constructed, fulfill self-consistent KS equations that are similar to those of regular (N-electron ground-state) KS-DFT:

$$\left(-\frac{\nabla_{\mathbf{r}}^2}{2} + v_{\mathrm{ext}}(\mathbf{r}) + v_{\mathrm{Hxc}}^{\xi}[n^{\xi}](\mathbf{r})\right)\varphi_p^{\xi}(\mathbf{r}) = \varepsilon_p^{\xi}\varphi_p^{\xi}(\mathbf{r}). \tag{53}$$

The only difference is that the N-centered ensemble Hxc potential $v_{\mathrm{Hxc}}^{\xi}[n](\mathbf{r}) = \delta E_{\mathrm{Hxc}}^{\xi}[n]/\delta n(\mathbf{r})$ is now employed. In the exact theory, the ensemble KS orbitals are expected to reproduce the interacting N-centered ensemble density, i.e.,

$$n_0^{\xi}(\mathbf{r}) = n_{\hat{\gamma}_{\mathrm{KS}}^{\xi}}(\mathbf{r}) \tag{54}$$

$$= \xi_{-}n_{\Phi_0^{N-1,\xi}}(\mathbf{r}) + \xi_{+}n_{\Phi_0^{N+1,\xi}}(\mathbf{r}) + \left(1 - \xi_{-}\frac{N-1}{N} - \xi_{+}\frac{N+1}{N}\right)n_{\Phi_0^{N,\xi}}(\mathbf{r}), \tag{55}$$

or, equivalently [83],

$$n_0^{\xi}(\mathbf{r}) = \left(1 + \frac{\xi_{-} - \xi_{+}}{N}\right)\sum_{p=1}^{N}|\varphi_p^{\xi}(\mathbf{r})|^2 - \xi_{-}|\varphi_N^{\xi}(\mathbf{r})|^2 + \xi_{+}|\varphi_{N+1}^{\xi}(\mathbf{r})|^2. \tag{56}$$

Turning to the N-centered ensemble Hxc density functional, it can be decomposed as $E_{\mathrm{Hxc}}^{\xi}[n] = E_{\mathrm{Hx}}^{\xi}[n] + E_{\mathrm{c}}^{\xi}[n]$, where, by analogy with GOK-DFT, the exact Hx energy is expressed in terms of the N-centered ensemble density-functional KS wave functions as follows:

$$
\begin{aligned}
E_{\mathrm{Hx}}^{\xi}[n] = {} & \xi_{-}\left\langle \Phi_0^{N-1,\xi}[n] \middle| \hat{W}_{\mathrm{ee}} \middle| \Phi_0^{N-1,\xi}[n] \right\rangle \\
& + \xi_{+}\left\langle \Phi_0^{N+1,\xi}[n] \middle| \hat{W}_{\mathrm{ee}} \middle| \Phi_0^{N+1,\xi}[n] \right\rangle \\
& + \left(1 - \xi_{-}\frac{N-1}{N} - \xi_{+}\frac{N+1}{N}\right)\left\langle \Phi_0^{N,\xi}[n] \middle| \hat{W}_{\mathrm{ee}} \middle| \Phi_0^{N,\xi}[n] \right\rangle,
\end{aligned}
\tag{57}
$$

and the complementary correlation functional reads as

$$
\begin{aligned}
E_{\mathrm{c}}^{\xi}[n] &= F^{\xi}[n] - T_{\mathrm{s}}^{\xi}[n] - E_{\mathrm{Hx}}^{\xi}[n] \\
&= \xi_{-}\left(\left\langle \hat{T} + \hat{W}_{\mathrm{ee}} \right\rangle_{\Psi_0^{N-1,\xi}[n]} - \left\langle \hat{T} + \hat{W}_{\mathrm{ee}} \right\rangle_{\Phi_0^{N-1,\xi}[n]} \right) \\
&\quad + \xi_{+}\left(\left\langle \hat{T} + \hat{W}_{\mathrm{ee}} \right\rangle_{\Psi_0^{N+1,\xi}[n]} - \left\langle \hat{T} + \hat{W}_{\mathrm{ee}} \right\rangle_{\Phi_0^{N+1,\xi}[n]} \right) \\
&\quad + \left(1 - \xi_{-}\frac{N-1}{N} - \xi_{+}\frac{N+1}{N}\right) \\
&\quad \times \left[\left\langle \hat{T} + \hat{W}_{\mathrm{ee}} \right\rangle_{\Psi_0^{N,\xi}[n]} - \left\langle \hat{T} + \hat{W}_{\mathrm{ee}} \right\rangle_{\Phi_0^{N,\xi}[n]} \right],
\end{aligned}
\tag{58}
$$

where $\{\Psi_0^{M,\xi}[n]\}$ denotes the interacting density-functional N-centered ensemble.

When comparison is made with the section "DFT of GOK Ensembles", it becomes clear that N-centered and GOK eDFTs are essentially the same theory (differing only in the definition of the ensemble). From that point of view, we now have a unified eDFT for charged and neutral electronic excitations. As a result, N-centered eDFT can benefit from progress made in GOK-DFT, and vice versa.

2.2.3 Exact Ionization Potential and Electron Affinity Theorems

We have shown above that neutral, anionic, and cationic ground-state energies can be extracted exactly from the N-centered ensemble energy [see Eqs. (45), (46), and (47)]. We can now use the variational density-functional expression of Eq. (52) to obtain expressions for the fundamental gap, the ionization potential (IP), and the electron affinity (EA). Note that these quantities are derived traditionally in the context of DFT for fractional electron numbers [3] (see "Equivalence Between Weight Derivatives and xc DerivativeDiscontinuities" for a detailed comparison). According to the Hellmann–Feynman theorem, we can express the weight derivatives of the ensemble energy as follows:

$$
\begin{aligned}
\frac{\partial E_0^{\xi}}{\partial \xi_{\pm}} = {} & \mathrm{Tr}\left[\Delta_{\pm}\hat{\gamma}_{\mathrm{KS}}^{\xi}\left(\hat{T} + \hat{V}_{\mathrm{ext}}\right)\right] + \left.\frac{\partial E_{\mathrm{Hxc}}^{\xi}[n]}{\partial \xi_{\pm}}\right|_{n=n_{\hat{\gamma}_{\mathrm{KS}}^{\xi}}} \\
& + \int \mathrm{d}\mathbf{r} \frac{\delta E_{\mathrm{Hxc}}^{\xi}[n_{\hat{\gamma}_{\mathrm{KS}}^{\xi}}]}{\delta n(\mathbf{r})} \mathrm{Tr}\left[\Delta_{\pm}\hat{\gamma}_{\mathrm{KS}}^{\xi}\hat{n}(\mathbf{r})\right],
\end{aligned}
\tag{59}
$$

where $\Delta_{\pm}\hat{\gamma}_{\mathrm{KS}}^{\xi} = |\Phi_0^{N\pm1,\xi}\rangle\langle\Phi_0^{N\pm1,\xi}| - \frac{N\pm1}{N}|\Phi_0^{N,\xi}\rangle\langle\Phi_0^{N,\xi}|$. Since the single-configuration M-electron KS wave functions $\Phi_0^{M,\xi}$ are constructed from orbitals that fulfill the KS

Eq. (53), the above energy derivative can be rewritten in terms of the KS orbital energies as [83]

$$\frac{\partial E_0^\xi}{\partial \xi_\pm} = \pm\frac{1}{N}\sum_{p=1}^{N}\left(\varepsilon^\xi_{N+\frac{1}{2}\pm\frac{1}{2}} - \varepsilon_p^\xi\right) + \left.\frac{\partial E^\xi_{\rm Hxc}[n]}{\partial \xi_\pm}\right|_{n=n_{\hat{\gamma}^\xi_{\rm KS}}}. \tag{60}$$

By plugging Eq. (60) into Eq. (44), we immediately obtain the following exact expression for the fundamental gap:

$$E_{\rm gap}^{\rm fund} = \varepsilon^\xi_{N+1} - \varepsilon^\xi_N + \left.\left(\frac{\partial E^\xi_{\rm Hxc}[n]}{\partial \xi_+} + \frac{\partial E^\xi_{\rm Hxc}[n]}{\partial \xi_-}\right)\right|_{n=n_{\hat{\gamma}^\xi_{\rm KS}}}. \tag{61}$$

If we now apply the LZ shift-in-potential procedure [104], by analogy with GOK-DFT (see "Extraction of Individual State Properties"), i.e.,

$$\frac{\delta E^\xi_{\rm Hxc}[n]}{\delta n(\mathbf{r})} \rightarrow \overline{v}^\xi_{\rm Hxc}[n](\mathbf{r}) = \frac{\delta E^\xi_{\rm Hxc}[n]}{\delta n(\mathbf{r})} + \frac{E^\xi_{\rm Hxc}[n] - \int d\mathbf{r}\,\frac{\delta E^\xi_{\rm Hxc}[n]}{\delta n(\mathbf{r})}n(\mathbf{r})}{\int d\mathbf{r}\, n(\mathbf{r})}, \tag{62}$$

we can express both the ensemble energy and its derivatives in terms of the LZ-shifted KS orbital energies $\overline{\varepsilon}^\xi_p$, thus leading to the following compact expressions for the ensemble and individual energies [83], respectively:

$$E_0^\xi = \left(1 + \frac{\xi_- - \xi_+}{N}\right)\sum_{p=1}^{N}\overline{\varepsilon}^\xi_p - \xi_-\overline{\varepsilon}^\xi_N + \xi_+\overline{\varepsilon}^\xi_{N+1}, \tag{63}$$

$$\begin{aligned} E_0^{N-1} = \sum_{p=1}^{N-1}\overline{\varepsilon}^\xi_p + \left(1 - \frac{(N-1)\xi_-}{N}\right)\left.\frac{\partial E^\xi_{\rm Hxc}[n]}{\partial \xi_-}\right|_{n=n_{\hat{\gamma}^\xi_{\rm KS}}} \\ - \frac{(N-1)\xi_+}{N}\left.\frac{\partial E^\xi_{\rm Hxc}[n]}{\partial \xi_+}\right|_{n=n_{\hat{\gamma}^\xi_{\rm KS}}}, \end{aligned} \tag{64}$$

$$\begin{aligned} E_0^{N+1} = \sum_{p=1}^{N+1}\overline{\varepsilon}^\xi_p + \left(1 - \frac{(N+1)\xi_+}{N}\right)\left.\frac{\partial E^\xi_{\rm Hxc}[n]}{\partial \xi_+}\right|_{n=n_{\hat{\gamma}^\xi_{\rm KS}}} \\ - \frac{(N+1)\xi_-}{N}\left.\frac{\partial E^\xi_{\rm Hxc}[n]}{\partial \xi_-}\right|_{n=n_{\hat{\gamma}^\xi_{\rm KS}}}, \end{aligned} \tag{65}$$

and

$$E_0^N = \sum_{p=1}^{N} \overline{\varepsilon}_p^{\xi} - \xi_- \left.\frac{\partial E_{\text{Hxc}}^{\xi}[n]}{\partial \xi_-}\right|_{n=n_{\hat{\gamma}_{\text{KS}}^{\xi}}} - \xi_+ \left.\frac{\partial E_{\text{Hxc}}^{\xi}[n]}{\partial \xi_+}\right|_{n=n_{\hat{\gamma}_{\text{KS}}^{\xi}}}. \tag{66}$$

By subtraction, we immediately obtain in-principle-exact IP and EA theorems:

$$\begin{aligned} I_0^N &= E_0^{N-1} - E_0^N \\ &= -\overline{\varepsilon}_N^{\xi} + \left(1 + \frac{\xi_-}{N}\right) \left.\frac{\partial E_{\text{Hxc}}^{\xi}[n]}{\partial \xi_-}\right|_{n=n_{\hat{\gamma}_{\text{KS}}^{\xi}}} + \frac{\xi_+}{N} \left.\frac{\partial E_{\text{Hxc}}^{\xi}[n]}{\partial \xi_+}\right|_{n=n_{\hat{\gamma}_{\text{KS}}^{\xi}}}, \end{aligned} \tag{67}$$

and

$$\begin{aligned} A_0^N &= E_0^N - E_0^{N+1} \\ &= -\overline{\varepsilon}_{N+1}^{\xi} - \left(1 - \frac{\xi_+}{N}\right) \left.\frac{\partial E_{\text{Hxc}}^{\xi}[n]}{\partial \xi_+}\right|_{n=n_{\hat{\gamma}_{\text{KS}}^{\xi}}} + \frac{\xi_-}{N} \left.\frac{\partial E_{\text{Hxc}}^{\xi}[n]}{\partial \xi_-}\right|_{n=n_{\hat{\gamma}_{\text{KS}}^{\xi}}}. \end{aligned} \tag{68}$$

Interestingly, in the regular ground-state N-electron limit (i.e., when $\xi = 0$), the LZ expression [104] is recovered for the IP,

$$I_0^N = -\overline{\varepsilon}_N^{\xi=0} + \left.\frac{\partial E_{\text{Hxc}}^{\xi}[n_{\Psi_0}]}{\partial \xi_-}\right|_{\xi=0}, \tag{69}$$

where the asymptotic value of the LZ-shifted Hxc potential away from the system [see Ref. [104] and Eq. (132)] can now be expressed explicitly, within the N-centered ensemble formalism, as $\partial E_{\text{Hxc}}^{\xi}[n_{\Psi_0}]/\partial \xi_-\big|_{\xi=0}$. Similarly, we obtain the following expression for the EA:

$$A_0^N = -\overline{\varepsilon}_{N+1}^{\xi=0} - \left.\frac{\partial E_{\text{Hxc}}^{\xi}[n_{\Psi_0}]}{\partial \xi_+}\right|_{\xi=0}. \tag{70}$$

As readily seen from the above expressions, neutral and charged systems cannot be described with the same (LZ-shifted) Hxc potential. As shown in "Equivalence Between Weight Derivatives and xc DerivativeDiscontinuities", the additional ensemble weight derivative correction [second term on the right-hand side of Eqs. (69) and (70)] is actually connected to the concept of derivative discontinuity, which manifests in conventional DFT for fractional electron numbers, when crossing an integer [106].

3 Equivalence Between Weight Derivatives and xc Derivative Discontinuities

The concept of derivative discontinuity originally appeared in the context of DFT for fractional electron numbers [3], which is the conventional theoretical framework for the description of charged excitations. The (xc functional) derivative discontinuities play a crucial role in the evaluation of fundamental gaps [106]. More specifically, they correct the bare KS gap, which is only an approximation to the true interacting gap. It is well known that standard (semi)-local DFAs do not contain such discontinuities, which explains why post-DFT methods based on Green functions, for example, are preferred for the computation of accurate gaps [14–21]. Their substantially higher computational cost is a motivation for exploring simpler (frequency-independent) strategies. The recently proposed *N*-centered ensemble formalism [83, 107], as introduced in "DFT of Charged Excitations: The N-Centered Ensemble Formalism", is (among others [34, 36, 42, 108–112]) promising in this respect.

From a more fundamental point of view, it is important to clarify the similarities and differences between *N*-centered eDFT and the standard formulation of DFT for charged excitations, which is often referred to as Perdew–Parr–Levy–Balduz (PPLB) DFT [3]. More specifically, we should explain what the derivative discontinuity, which is central in PPLB, becomes when switching to the *N*-centered formalism. This is the purpose of this section.

After a brief review of the PPLB formalism and its implications, we will show, on the basis of Ref. [113], that derivative discontinuities exist also in *N*-centered eDFT and that they are directly connected to the ensemble weight derivatives of the xc functional, like in GOK-DFT [105]. Finally, we will explain why these discontinuities can essentially be removed from the theory, unlike in PPLB, and discuss the practical implications.

3.1 Review of the Regular PPLB Approach to Charged Excitations

3.1.1 Ensemble Formalism for Open Systems

The key idea in PPLB is to describe electron ionization or affinity processes through a continuous variation of the electron number, hence the need for an extension of DFT to fractional electron numbers. For that purpose, the energy of an artificial (zero-temperature) grand-canonical-type ensemble, which should not be confused with physical finite-temperature grand-canonical ensembles of statistical physics [114], is constructed as follows:

$$\mathcal{G}(\mu) = \min_{M} \left\{ E_0^M - \mu M \right\}, \tag{71}$$

where we minimize over integer numbers M of electrons and E_0^M denotes the exact M-electron ground-state energy of the system. In this formalism, the number of electrons in the system can be arbitrarily fixed by tuning the chemical potential μ. For example, if the following inequalities are fulfilled,

$$E_0^{N-1} - \mu(N-1) > E_0^N - \mu N < E_0^{N+1} - \mu(N+1), \tag{72}$$

or, equivalently,

$$-I_0^N < \mu < -A_0^N, \tag{73}$$

then the system contains an integer number N of electrons [it is assumed that the N-electron fundamental gap $E_g^N = I_0^N - A_0^N$ is positive, so that Eq. (73) can be fulfilled]. In the special case where one of the inequalities becomes a strict equality, say

$$E_0^{N-1} - \mu(N-1) = E_0^N - \mu N, \tag{74}$$

which means that the chemical potential is exactly equal to minus the N-electron ionization potential,

$$\mu = E_0^N - E_0^{N-1} = -I_0^N, \tag{75}$$

the N- and $(N-1)$-electron solutions are degenerate (grand-canonical energy wise). Therefore, they can be mixed as follows,

$$\mathcal{G}(\mu) \overset{\mu=-I_0^N}{=} (1-\alpha)\Big(E_0^{N-1} - \mu(N-1)\Big) + \alpha\Big(E_0^N - \mu N\Big) \tag{76}$$

$$= \Big((1-\alpha)E_0^{N-1} + \alpha E_0^N\Big) - \mu\Big(N-1+\alpha\Big) \tag{77}$$

$$\equiv E_0^{\mathcal{N}} - \mu\mathcal{N}, \tag{78}$$

where $0 \leq \alpha \leq 1$, thus allowing for a continuous variation of the electron number $\mathcal{N}$ (which now becomes fractional) from $N-1$ to N:

$$\mathcal{N} \equiv N - 1 + \alpha. \tag{79}$$

This is the central idea in PPLB for describing the ionization of an N-electron system. Ionizing the $(N+1)$-electron system gives access to the N-electron affinity. Interestingly, we recover from Eqs. (77), (78), and (79) the well-known piecewise linearity of the energy with respect to the electron number [3]:

$$E_0^{\mathcal{N}} \equiv (1-\alpha)E_0^{N-1} + \alpha E_0^N = (N-\mathcal{N})E_0^{N-1} + (\mathcal{N}-N+1)E_0^N. \tag{80}$$

In order to establish a clearer connection between the PPLB and N-centered formalisms, we follow the approach of Kraisler and Kronik [34], where the ensemble weight α is used as a variable, in place of the electron number $\mathcal{N}$. Therefore, in PPLB, the ensemble energy reads as

$$E^{\alpha} = (1-\alpha)E_0^{N-1} + \alpha E_0^N \tag{81}$$

$$\begin{aligned} &= (1-\alpha)\left\langle \Psi_0^{N-1}\middle|\hat{T}+\hat{W}_{\mathrm{ee}}\middle|\Psi_0^{N-1}\right\rangle + \alpha\left\langle \Psi_0^{N}\middle|\hat{T}+\hat{W}_{\mathrm{ee}}\middle|\Psi_0^{N}\right\rangle \\ &+ \int d\mathbf{r}\, v_{\mathrm{ext}}(\mathbf{r}) n_0^{\alpha}(\mathbf{r}), \end{aligned} \tag{82}$$

where

$$n_0^{\alpha}(\mathbf{r}) = (1-\alpha)n_{\Psi_0^{N-1}}(\mathbf{r}) + \alpha n_{\Psi_0^{N}}(\mathbf{r}) \tag{83}$$

is the exact ground-state ensemble density. Note that, if we introduce the ensemble density matrix operator

$$\hat{\Gamma}_0^{\alpha} = (1-\alpha)|\Psi_0^{N-1}\rangle\langle\Psi_0^{N-1}| + \alpha|\Psi_0^{N}\rangle\langle\Psi_0^{N}|, \tag{84}$$

the ensemble energy and density can be expressed in a compact way as follows:

$$E^{\alpha} = \mathrm{Tr}\left[\hat{\Gamma}_0^{\alpha}\hat{H}\right] \tag{85}$$

and

$$n_0^{\alpha}(\mathbf{r}) = \mathrm{Tr}\left[\hat{\Gamma}_0^{\alpha}\hat{n}(\mathbf{r})\right], \tag{86}$$

respectively.

3.1.2 DFT for Fractional Electron Numbers

On the basis of the "grand-canonical" ensemble formalism introduced in the previous section, we can extend the domain of definition of the universal Hohenberg–Kohn functional $F[n]$ to densities n that integrate to fractional electron numbers, i.e.,

$$\int d\mathbf{r}\, n(\mathbf{r}) = N - 1 + \alpha, \tag{87}$$

as follows,

$$F[n] = (1-\alpha)\left\langle \hat{T}+\hat{W}_{\mathrm{ee}}\right\rangle_{\Psi_0^{N-1}[n]} + \alpha\left\langle \hat{T}+\hat{W}_{\mathrm{ee}}\right\rangle_{\Psi_0^{N}[n]}, \tag{88}$$

where the ground-state density-functional wave functions fulfill the density constraint

$$(1-\alpha)n_{\Psi_0^{N-1}[n]}(\mathbf{r}) + \alpha n_{\Psi_0^{N}[n]}(\mathbf{r}) = n(\mathbf{r}). \tag{89}$$

From now on we will take α in the range

$$0 < \alpha \leq 1, \tag{90}$$

so that the integer electron number case systematically corresponds to $\alpha = 1$. Therefore, in the present density-functional PPLB ensemble, the N-electron state will always contribute (even infinitesimally), and

$$N = \lceil \int \mathrm{d}\mathbf{r}\, n(\mathbf{r}) \rceil. \tag{91}$$

At this point it is essential to realize that, unlike in N-centered eDFT, the ensemble weight α is not an independent variable. Indeed, according to Eqs. (87) and (91), it is an explicit functional of the density:

$$\alpha \equiv \alpha[n] = \int \mathrm{d}\mathbf{r}\, n(\mathbf{r}) - \lceil \int \mathrm{d}\mathbf{r}\, n(\mathbf{r}) \rceil + 1. \tag{92}$$

Therefore, in PPLB, the ensemble is determined fully from the density. The latter remains, as in regular DFT for integer electron numbers, the sole basic variable in the theory. Following Levy and Lieb [66, 92, 115], the extended universal functional of Eq. (88) can be expressed in a compact way as follows:

$$F[n] = \min_{\hat{\gamma}^\alpha \rightarrow n} \mathrm{Tr}\left[\hat{\gamma}^\alpha \left(\hat{T} + \hat{W}_{\mathrm{ee}}\right)\right], \tag{93}$$

where we minimize over grand-canonical ensemble density matrix operators

$$\hat{\gamma}^\alpha \equiv (1-\alpha)|\Psi^{N-1}\rangle\langle\Psi^{N-1}| + \alpha|\Psi^{N}\rangle\langle\Psi^{N}| \tag{94}$$

that fulfill the following density constraint:

$$\mathrm{Tr}\left[\hat{\gamma}^\alpha \hat{n}(\mathbf{r})\right] = n_{\hat{\gamma}^\alpha}(\mathbf{r}) = (1-\alpha) n_{\Psi^{N-1}}(\mathbf{r}) + \alpha n_{\Psi^{N}}(\mathbf{r}) = n(\mathbf{r}). \tag{95}$$

3.1.3 Kohn–Sham PPLB

The commonly used KS formulation of PPLB is recovered when introducing the non-interacting kinetic energy functional

$$T_{\mathrm{s}}[n] = \min_{\hat{\gamma}^\alpha \rightarrow n} \mathrm{Tr}\left[\hat{\gamma}^\alpha \hat{T}\right] \tag{96}$$

and the in-principle-exact decomposition

$$F[n] = T_{\mathrm{s}}[n] + E_{\mathrm{Hxc}}[n], \tag{97}$$

where the Hxc functional now applies to fractional electron numbers. Let us stress that, unlike in N-centered eDFT, the Hxc functional has no ensemble weight dependence because the weight is determined from the density n. Any dependence in α is incorporated into the functional through the density. This is a major difference with N-centered eDFT where the ensemble weight and the density are independent variables, like in GOK-DFT. This subtle point will be central later on when comparing the two theories.

According to the variational principle, the exact ensemble energy can be determined, for a given and *fixed* value of α, as follows:

$$E^\alpha = \min_{n \to N-1+\alpha} \left\{ F[n] + \int d\mathbf{r}\, v_{\text{ext}}(\mathbf{r}) n(\mathbf{r}) \right\}, \tag{98}$$

where we minimize over densities that integrate to the desired number $N - 1 + \alpha$ of electrons. According to Eqs. (96) and (97), the ensemble energy can be rewritten as

$$\begin{aligned} E^\alpha &= \min_{n \to N-1+\alpha} \left\{ \min_{\hat{\gamma}^\alpha \to n} \left\{ \text{Tr}\left[\hat{\gamma}^\alpha \left(\hat{T} + \hat{V}_{\text{ext}}\right)\right] + E_{\text{Hxc}}[n_{\hat{\gamma}^\alpha}] \right\} \right\} \\ &= \min_{\hat{\gamma}^\alpha} \left\{ \text{Tr}\left[\hat{\gamma}^\alpha \left(\hat{T} + \hat{V}_{\text{ext}}\right)\right] + E_{\text{Hxc}}[n_{\hat{\gamma}^\alpha}] \right\} \\ &\equiv \text{Tr}\left[\hat{\gamma}^\alpha_{\text{KS}} \left(\hat{T} + \hat{V}_{\text{ext}}\right)\right] + E_{\text{Hxc}}[n_{\hat{\gamma}^\alpha_{\text{KS}}}], \end{aligned} \tag{99}$$

where the minimizing KS density matrix operator

$$\hat{\gamma}^\alpha_{\text{KS}} = (1-\alpha)|\Phi_0^{N-1,\alpha}\rangle\langle\Phi_0^{N-1,\alpha}| + \alpha|\Phi_0^{N,\alpha}\rangle\langle\Phi_0^{N,\alpha}| \tag{100}$$

reproduces the exact ensemble density of Eq. (83):

$$n_{\hat{\gamma}^\alpha_{\text{KS}}}(\mathbf{r}) = \text{Tr}\left[\hat{\gamma}^\alpha \hat{n}(\mathbf{r})\right] = n_0^\alpha(\mathbf{r}). \tag{101}$$

The orbitals from which $\Phi_0^{N-1,\alpha}$ and $\Phi_0^{N,\alpha}$ are constructed fulfill self-consistent KS equations,

$$\left(-\frac{1}{2}\nabla^2_{\mathbf{r}} + v_{\text{ext}}(\mathbf{r}) + \frac{\delta E_{\text{Hxc}}[n_{\hat{\gamma}^\alpha_{\text{KS}}}]}{\delta n(\mathbf{r})} \right) \varphi_i^\alpha(\mathbf{r}) = \varepsilon_i^\alpha \varphi_i^\alpha(\mathbf{r}), \tag{102}$$

where, as readily seen from the following ensemble density expression,

$$\begin{aligned} n_{\hat{\gamma}^\alpha_{\text{KS}}}(\mathbf{r}) &= (1-\alpha)\sum_{i=1}^{N-1} |\varphi_i^\alpha(\mathbf{r})|^2 + \alpha \sum_{i=1}^{N} |\varphi_i^\alpha(\mathbf{r})|^2 \\ &= \sum_{i=1}^{N-1} |\varphi_i^\alpha(\mathbf{r})|^2 + \alpha|\varphi_N^\alpha(\mathbf{r})|^2, \end{aligned} \tag{103}$$

the highest occupied molecular orbital (HOMO) [i.e., φ_N^α] is fractionally occupied. This is the main difference from conventional DFT calculations for integer electron numbers.

3.1.4 Janak's Theorem and Its Implications

Once the ensemble energy E^α has been determined (variationally), we can evaluate the IP, which is the quantity we are interested in, by differentiation with respect to the ensemble weight α [see Eq. (81)], i.e.,

$$\frac{\mathrm{d}E^\alpha}{\mathrm{d}\alpha} = -I_0^N, \tag{104}$$

which, according to the Hellmann–Feynman theorem and Eqs. (99), (100), and (102), can be written more explicitly as follows,

$$\begin{aligned}\frac{dE^\alpha}{d\alpha} &= \left\langle \Phi_0^{N,\alpha}\middle|\hat{T}+\hat{V}_{\text{ext}}\middle|\Phi_0^{N,\alpha}\right\rangle - \left\langle \Phi_0^{N-1,\alpha}\middle|\hat{T}+\hat{V}_{\text{ext}}\middle|\Phi_0^{N-1,\alpha}\right\rangle \\ &+ \int \mathrm{d}\mathbf{r}\frac{\delta E_{\mathrm{Hxc}}[n_{\hat{\gamma}^\alpha_{\mathrm{KS}}}]}{\delta n(\mathbf{r})}\left(n_{\Phi_0^{N,\alpha}}(\mathbf{r}) - n_{\Phi_0^{N-1,\alpha}}(\mathbf{r})\right)\end{aligned} \tag{105}$$

$$\begin{aligned}&\equiv \sum_{i=1}^{N}\varepsilon_i^\alpha - \sum_{i=1}^{N-1}\varepsilon_i^\alpha \\ &= \varepsilon_N^\alpha,\end{aligned} \tag{106}$$

thus leading to the famous Janak's theorem [116]:

$$I_0^N = -\varepsilon_N^\alpha, \forall\alpha \in]0,1]. \tag{107}$$

As readily seen from Eq. (107), the energy ε_N^α of the KS HOMO does not vary with the fraction $\alpha > 0$ of electron that is introduced into the $(N-1)$-electron system. Therefore, it matches the N-electron KS HOMO energy that we denote simply ε_N^N:

$$\varepsilon_N^\alpha = \varepsilon_N^{\alpha=1} \equiv \varepsilon_N^N = -I_0^N. \tag{108}$$

At this point, it is important to mention that, unlike in N-centered eDFT [see Eq. (69)], there is no ensemble weight derivative of the Hxc functional involved in Janak's theorem. Such a quantity does not exist in PPLB, simply because the ensemble weight α and the density n cannot vary independently. However, while the number of electrons is held constant artificially in the N-centered formalism, this is not the case in PPLB. Indeed, variations in α induce a change in density [see the third contribution on the right-hand side of Eq. (105)] that does not integrate to zero:

$$1 = \int \mathrm{d}\mathbf{r}\left(n_{\Phi_0^{N,\alpha}}(\mathbf{r}) - n_{\Phi_0^{N-1,\alpha}}(\mathbf{r})\right) \neq 0. \tag{109}$$

Therefore, it is crucial, when evaluating the functional derivative of the Hxc energy $\delta E_{\mathrm{Hxc}}[n_{\hat{\gamma}^\alpha_{\mathrm{KS}}}]/\delta n(\mathbf{r})$ (i.e., the Hxc potential), to consider variations of the density $\delta n(\mathbf{r})$ that do not integrate to zero. This is unnecessary in N-centered eDFT. In PPLB, however, proper modeling of the xc potential is essential for describing charged excitations. This is illustrated clearly by the fact that the exact xc potential exhibits derivative discontinuities when crossing an integer electron number, as discussed further in section "Connection Between PPLB and N-Centered Pictures".

Let us finally discuss the unicity of the xc potential. We recall that, in the present review, the external potential is simply the (Coulomb) nuclear potential of the molecule under study. It is fixed and it vanishes away from the system:

$$v_{\mathrm{ext}}(\mathbf{r}) \underset{|\mathbf{r}|\rightarrow+\infty}{\rightarrow} 0, \tag{110}$$

which we simply denote $v_{\mathrm{ext}}(\infty) = 0$ in the following. As readily seen from Eq. (107), when describing a continuous variation of the electron number $\mathcal{N}$ in the range $N-1 < \mathcal{N} < N$, the KS potential becomes truly unique, no longer up to a constant. This can be related to the unicity of the chemical potential, which allows for fractional electron numbers, as discussed previously in the interacting case [see Eq. (75)]. As a result, the xc potential is truly unique. More precisely, as illustrated in Appendix 1 for a one-dimensional (1D) system, Janak's theorem implies that [117]

$$\left.\frac{\delta E_{\mathrm{xc}}[n_{\hat{\gamma}^{\alpha}_{\mathrm{KS}}}]}{\delta n(\mathbf{r})}\right|_{|\mathbf{r}|\rightarrow+\infty} \equiv v^{\alpha}_{\mathrm{xc}}(\infty) = 0. \tag{111}$$

3.1.5 Fundamental Gap Problem

According to Janak's theorem, the fundamental gap can be evaluated in PPLB, in principle exactly, from the HOMO energies as follows:

$$E_g^N = I_0^N - I_0^{N+1} = \varepsilon^{N+1}_{N+1} - \varepsilon^N_N. \tag{112}$$

What is truly challenging in practice, in particular in solids [10], is the extraction of this gap from a single N-electron calculation. Indeed, the HOMO energy ε^{N+1}_{N+1} of the $(N+1)$-electron system has no reason to match the lowest unoccupied molecular orbital (LUMO) energy ε^N_{N+1} of the N-electron system, simply because the infinitesimal addition of an electron to the latter system will affect the density [see Eq. (299)] and, consequently, the xc potential. The impact of an electron addition on the xc potential will be scrutinized in "Connection Between PPLB and N-Centered Pictures", in the context of N-centered eDFT. If we denote

$$\Delta^N_{\mathrm{xc}} = \varepsilon^{N+1}_{N+1} - \varepsilon^N_{N+1} \tag{113}$$

the deviation in energy between the above-mentioned HOMO and LUMO, we recover the usual expression [106]

$$E_g^N = \varepsilon^N_{N+1} - \varepsilon^N_N + \Delta^N_{\mathrm{xc}}, \tag{114}$$

where Δ^N_{xc} can now be interpreted as the difference in gap between the physical and KS systems. As readily seen from the key Eq. (61) of N-centered eDFT, that we take in the regular N-electron ground-state DFT limit (i.e., $\xi_+ = \xi_- = 0$), Δ^N_{xc} is indeed a nonzero correction to the KS gap that can be expressed more explicitly as follows,

$$\Delta^N_{\mathrm{xc}} = \left.\frac{\partial E^{(\xi_-,0)}_{\mathrm{xc}}[n_{\Psi^N_0}]}{\partial \xi_-}\right|_{\xi_-=0} + \left.\frac{\partial E^{(0,\xi_+)}_{\mathrm{xc}}[n_{\Psi^N_0}]}{\partial \xi_+}\right|_{\xi_+=0}. \tag{115}$$

Note that we used in Eq. (115) the in-principle-exact decomposition

$$E_{\rm Hxc}^{(\xi_-,\xi_+)}[n] = E_{\rm H}[n] + E_{\rm xc}^{(\xi_-,\xi_+)}[n], \tag{116}$$

where the regular (weight-independent) Hartree functional is employed. The practical disadvantage of such a decomposition will be discussed extensively below in "The Exact Hartree-Exchange Dilemma in eDFT". We focus here on the exact theory.

In the language of N-centered eDFT, $\Delta_{\rm xc}^N$ describes the variation in ensemble weights (while holding the ensemble density fixed and equal to the N-electron ground-state density $n_{\Psi_0^N}$) of the N-centered ensemble xc energy due to the infinitesimal removal/addition of an electron from/to the N-electron system. Evidently, standard (local or semi-local) DFAs do not incorporate such a weight dependence because they were not designed for N-centered eDFT calculations (we recall that the concept of N-centered ensemble has been proposed quite recently [83, 107, 113]). Therefore, when such DFAs are used, the physical gap is systematically approximated by the (also approximate) KS one. Note that the resulting underestimation of the fundamental gap is highly problematic, for example, when computing transport properties [118, 119]. The interpretation that is given in PPLB for $\Delta_{\rm xc}^N$ is completely different. The latter actually originates from the discontinuity that the xc potential (which is the functional derivative of the xc energy) exhibits when crossing an integer electron number, hence the name derivative discontinuity. In the language of PPLB, $\Delta_{\rm xc}^N$ is not described at all when (semi-) local xc functionals are employed, simply because the latter do not incorporate functional derivative discontinuities. The connection between these two very different interpretations will be made in "Connection Between PPLB and N-Centered Picture".

3.1.6 Exchange-Only Derivative Discontinuity

Let us finish the previous discussion with a detailed comment on the use of (orbital-dependent) exact exchange energies, which is often recommended for improving the description of fundamental gaps [10]. From the perspective of N-centered eDFT, using an exact exchange energy (or a fraction of it) is a way to incorporate weight dependencies into the ensemble exchange density functional. Indeed, according to Eq. (116), the exact exchange-only derivative discontinuity can be rewritten as

$$\Delta_{\rm x}^N = \left.\frac{\partial E_{\rm x}^{(\xi,\xi)}[n_{\Psi_0^N}]}{\partial \xi}\right|_{\xi=0} = \left.\frac{\partial E_{\rm Hx}^{(\xi,\xi)}[n_{\Psi_0^N}]}{\partial \xi}\right|_{\xi=0}, \tag{117}$$

or, equivalently,

$$\begin{aligned}\Delta_{\mathrm{x}}^{N} =& \left.\frac{\partial E_{\mathrm{Hx}}^{(\xi,\xi)}[n_0^{\xi,\xi}]}{\partial \xi}\right|_{\xi=0} \\ &- \left.\frac{\partial E_{\mathrm{Hx}}^{(\xi,\xi)}[n_0^{\xi,\alpha}]}{\partial \alpha}\right|_{\alpha=\xi=0} - \left.\frac{\partial E_{\mathrm{Hx}}^{(\xi,\xi)}[n_0^{\alpha,\xi}]}{\partial \alpha}\right|_{\alpha=\xi=0},\end{aligned} \tag{118}$$

where we have introduced the double-weight N-centered ensemble KS density

$$n_0^{\alpha,\xi}(\mathbf{r}) := (1-2\alpha)n_{\Phi_0^{N,(\xi,\xi)}}(\mathbf{r}) + \alpha n_{\Phi_0^{N-1,(\xi,\xi)}}(\mathbf{r}) + \alpha n_{\Phi_0^{N+1,(\xi,\xi)}}(\mathbf{r}), \tag{119}$$

which reduces to $n_{\Psi_0^N}$ when $\alpha = \xi = 0$. Thus, we can remove all the contributions involving the derivative of the ensemble density [see the second line of Eq. (118)] that are erroneously introduced by the first term on the right-hand side of Eq. (118). We recall that, in the evaluation of Δ_{x}^{N}, we must differentiate with respect to the weight for a fixed density, as readily seen from Eq. (117). At first sight, Eq. (118) is uselessly complicated, when compared with Eq. (117), but it will actually enable us to obtain simpler expressions. This trick has been introduced in the context of GOK-DFT [91, 103]. First, we need to realize that, according to the exact expression of the N-centered ensemble Hx functional in Eq. (57),

$$\begin{aligned}E_{\mathrm{Hx}}^{(\xi,\xi)}[n_0^{\xi,\alpha}] =& (1-2\xi)\left\langle \hat{W}_{\mathrm{ee}}\right\rangle_{\Phi_0^{N,(\alpha,\alpha)}} + \xi\left\langle \hat{W}_{\mathrm{ee}}\right\rangle_{\Phi_0^{N-1,(\alpha,\alpha)}} \\ &+ \xi\left\langle \hat{W}_{\mathrm{ee}}\right\rangle_{\Phi_0^{N+1,(\alpha,\alpha)}}.\end{aligned} \tag{120}$$

Moreover, we have

$$\frac{\partial E_{\mathrm{Hx}}^{(\xi,\xi)}[n_0^{\alpha,\xi}]}{\partial \alpha} = \int \mathrm{d}\mathbf{r}\, \frac{\delta E_{\mathrm{Hx}}^{(\xi,\xi)}[n_0^{\alpha,\xi}]}{\delta n(\mathbf{r})}\frac{\partial n_0^{\alpha,\xi}(\mathbf{r})}{\partial \alpha}, \tag{121}$$

where, in the $\alpha = \xi = 0$ limit, the derivative of the density

$$\begin{aligned}\left.\frac{\partial n_0^{\alpha,\xi}(\mathbf{r})}{\partial \alpha}\right|_{\alpha=\xi=0} &= n_{\Phi_0^{N+1}}(\mathbf{r}) + n_{\Phi_0^{N-1}}(\mathbf{r}) - 2n_{\Phi_0^{N}}(\mathbf{r}) \\ &= |\varphi_{N+1}^{N}(\mathbf{r})|^2 - |\varphi_{N}^{N}(\mathbf{r})|^2\end{aligned} \tag{122}$$

can be evaluated from the regular N-electron KS frontier orbitals. By combining Eqs. (120)–(122) we finally obtain a simple (orbital-dependent) expression for the exchange-only derivative discontinuity:

$$\begin{aligned}\Delta_{\mathrm{x}}^{N} =& \left\langle \hat{W}_{\mathrm{ee}}\right\rangle_{\Phi_0^{N+1}} + \left\langle \hat{W}_{\mathrm{ee}}\right\rangle_{\Phi_0^{N-1}} - 2\left\langle \hat{W}_{\mathrm{ee}}\right\rangle_{\Phi_0^{N}} \\ &- \int \mathrm{d}\mathbf{r}\, \frac{\delta E_{\mathrm{Hx}}[n_{\Psi_0}]}{\delta n(\mathbf{r})}\left(|\varphi_{N+1}^{N}(\mathbf{r})|^2 - |\varphi_{N}^{N}(\mathbf{r})|^2\right).\end{aligned} \tag{123}$$

Adding this correction to the exact KS gap leads to the following approximate fundamental gap expression,

$$E_g^N \approx \varepsilon_{N+1}^N - \varepsilon_N^N + \Delta_x^N = \langle \hat{H} \rangle_{\Phi_0^{N+1}} + \langle \hat{H} \rangle_{\Phi_0^{N-1}} - 2\langle \hat{H} \rangle_{\Phi_0^N} + \int d\mathbf{r} \frac{\delta E_c[n_{\Psi_0}]}{\delta n(\mathbf{r})} \left(|\varphi_{N+1}^N(\mathbf{r})|^2 - |\varphi_N^N(\mathbf{r})|^2 \right), \tag{124}$$

where the physical interacting wave functions have been replaced by the KS ones, and correlation is introduced only through the correlation potential.

Note that a consistent implementation of Eq. (124) on the basis of Eq. (117) would in principle require using optimized effective potentials (OEPs) [120, 121]. Indeed, in the present formulation of N-centered eDFT, the KS orbitals are expected to be generated from a local (i.e., multiplicative) xc potential [see Eq. (53)], unlike in Hartree–Fock (HF)-based methods where the exchange potential is nonlocal. In practice, we would have to consider a trial local potential v and determine the corresponding KS orbitals,

$$\{\varphi_i \equiv \varphi_i[v]\} \longleftarrow \quad \left(-\frac{1}{2}\nabla_{\mathbf{r}}^2 + v(\mathbf{r})\right)\varphi_i(\mathbf{r}) = \varepsilon_i \varphi_i(\mathbf{r}), \tag{125}$$

thus ensuring that the KS wave functions in Eq. (120) are eigenfunctions of a noninteracting Hamiltonian, like in the exact theory. The ensemble energy would then be minimized with respect to v rather than the orbitals (hence the name OEP given to the method). Obviously, such a procedure induces a substantial increase in computational complexity, even though simplifications can be made in the optimization process [122]. Alternative variational evaluations of orbital-dependent ensemble exchange energies exist [120]. Their practical advantages and drawbacks will be discussed in detail below.

3.2 Connection Between PPLB and *N*-Centered Pictures

Crossing an integer electron number, which is a key concept in PPLB, can be described in the context of N-centered eDFT by considering so-called left and right N-centered ensembles [107]. These ensembles are recovered when $\xi_+ = 0$ (electron removal only) and $\xi_- = 0$ (electron addition only), respectively. In the following, we will use the following shorthand notations,

$$\text{left } N\text{-centered ensemble:} \quad (\xi_-, 0) \overset{\text{notation}}{\equiv} \xi_-, \tag{126}$$

$$\text{right } N\text{-centered ensemble:} \quad (0, \xi_+) \overset{\text{notation}}{\equiv} \xi_+, \tag{127}$$

for convenience. For example, the right N-centered ensemble Hxc functional and KS orbital energies will be denoted as $E_{\text{Hxc}}^{\xi_+}[n] \equiv E_{\text{Hxc}}^{(0,\xi_+)}[n]$ and $\varepsilon_i^{\xi_+} \equiv \varepsilon_i^{(0,\xi_+)}$, respectively. The exact left and right N-centered ensemble densities read as [see Eq. (40)]

$$n^{\xi_-}(\mathbf{r}) \equiv \left(1 - \frac{(N-1)\xi_-}{N}\right) n_{\Psi_0^N}(\mathbf{r}) + \xi_- n_{\Psi_0^{N-1}}(\mathbf{r}), \tag{128}$$

and

$$n^{\xi_+}(\mathbf{r}) \equiv \left(1 - \frac{(N+1)\xi_+}{N}\right) n_{\Psi_0^N}(\mathbf{r}) + \xi_+ n_{\Psi_0^{N+1}}(\mathbf{r}), \tag{129}$$

respectively. Note that, with these notations, we have the following equivalence relation,

$$\xi_- = 0 \Leftrightarrow \xi_+ = 0, \tag{130}$$

as readily seen from Eqs. (128) and (129). When Eq. (130) is fulfilled, the system is in its pure N-electron ground state, which means, in the language of PPLB, that it contains exactly the integer number N of electrons.

A clearer connection between the two theories can be established by comparing the two limits $\xi_+ \to 0^+$ (which describes the infinitesimal addition of an electron to the N-electron system) and $\xi_+ = 0$ (or, equivalently, $\xi_- = 0$). For that purpose, we first need to realize that, by analogy with PPLB (see Appendix 1 for the proof in the simpler 1D case), the exact IP/EA theorems of N-centered eDFT in Eqs. (67) and (68) can be alternatively written as follows [113]:

$$A_0^N = I_0^{N+1} \overset{\xi_+>0}{=} -\varepsilon_{N+1}^{\xi_+} + v_{\rm xc}^{\xi_+}(\infty) \tag{131}$$

and

$$I_0^N \overset{\xi_-\geq 0}{=} -\varepsilon_N^{\xi_-} + v_{\rm xc}^{\xi_-}(\infty), \tag{132}$$

where we recall that $v_{\rm xc}^{\xi_\pm}(\mathbf{r}) \equiv \delta E_{\rm xc}^{\xi_\pm}[n]/\delta n(\mathbf{r})\Big|_{n=n^{\xi_\pm}}$. Thus, from the explicit expression of the LZ shift [see the second term on the right-hand side of Eq. (62)], we obtain the following exact expressions for the asymptotic values of the right and left N-centered ensemble xc potentials, respectively:

$$\begin{aligned} v_{\rm xc}^{\xi_+}(\infty) \overset{\xi_+>0}{=} & \left(\frac{\xi_+}{N} - 1\right) \left.\frac{\partial E_{\rm xc}^{\xi_+}[n]}{\partial \xi_+}\right|_{n=n^{\xi_+}} \\ & - \frac{1}{N}\left(E_{\rm Hxc}^{\xi_+}[n^{\xi_+}] - \int d\mathbf{r}\, v_{\rm Hxc}^{\xi_+}(\mathbf{r}) n^{\xi_+}(\mathbf{r})\right) \end{aligned} \tag{133}$$

and

$$\begin{aligned} v_{\rm xc}^{\xi_-}(\infty) \overset{\xi_-\geq 0}{=} & \left(\frac{\xi_-}{N} + 1\right) \left.\frac{\partial E_{\rm xc}^{\xi_-}[n]}{\partial \xi_-}\right|_{n=n^{\xi_-}} \\ & - \frac{1}{N}\left(E_{\rm Hxc}^{\xi_-}[n^{\xi_-}] - \int d\mathbf{r}\, v_{\rm Hxc}^{\xi_-}(\mathbf{r}) n^{\xi_-}(\mathbf{r})\right). \end{aligned} \tag{134}$$

We recall that the decomposition of Eq. (116) is employed for a direct comparison with PPLB. Let us now consider the $\xi_+ \rightarrow 0^+$ and $\xi_- = 0$ limits in Eqs. (133) and (134), respectively. Since

$$n^{\xi_+ \rightarrow 0^+}(\mathbf{r}) = n^{\xi_- = 0}(\mathbf{r}) = n_{\Psi_0^N}(\mathbf{r}), \tag{135}$$

$$v_{\mathrm{H}}^{\xi_+ \rightarrow 0^+}(\mathbf{r}) = v_{\mathrm{H}}^{\xi_- = 0}(\mathbf{r}), \tag{136}$$

$$E_{\mathrm{Hxc}}^{\xi_+}[n^{\xi_+}]\Big|_{\xi_+ \rightarrow 0^+} = E_{\mathrm{Hxc}}^{\xi_-}[n^{\xi_-}]\Big|_{\xi_- = 0} = E_{\mathrm{Hxc}}[n_{\Psi_0^N}], \tag{137}$$

it comes, by subtraction,

$$\int \frac{d\mathbf{r}}{N}\left(v_{\mathrm{xc}}^{\xi_+ \rightarrow 0^+}(\mathbf{r}) - v_{\mathrm{xc}}^{\xi_+ = 0}(\mathbf{r})\right) n_{\Psi_0^N}(\mathbf{r}) = v_{\mathrm{xc}}^{\xi_+ \rightarrow 0^+}(\infty) - v_{\mathrm{xc}}^{\xi_+ = 0}(\infty) + \Delta_{\mathrm{xc}}^N, \tag{138}$$

or, equivalently,

$$\Delta_{\mathrm{xc}}^N = \int \frac{d\mathbf{r}}{N}\left[\left(v_{\mathrm{xc}}^{\xi_+ \rightarrow 0^+}(\mathbf{r}) - v_{\mathrm{xc}}^{\xi_+ \rightarrow 0^+}(\infty)\right) - \left(v_{\mathrm{xc}}^{\xi_+ = 0}(\mathbf{r}) - v_{\mathrm{xc}}^{\xi_+ = 0}(\infty)\right)\right] n_{\Psi_0^N}(\mathbf{r}), \tag{139}$$

where we used Eq. (115) and the relation $v_{\mathrm{xc}}^{\xi_- = 0}(\mathbf{r}) = v_{\mathrm{xc}}^{\xi_+ = 0}(\mathbf{r})$, according to Eq. (130). Note that, as readily seen from Eq. (139), Δ_{xc}^N is insensitive to constant shifts in the xc potential, as expected from Eq. (114).

The connection that is made explicit in Eq. (139) between the N-centered ensemble weight derivative Δ_{xc}^N of the xc density-functional energy [see Eq. (115)] and the xc potential is an important result that was highlighted very recently in Ref. [113]. It proves that weight derivatives and derivative discontinuities are equivalent, thus extending to charged excitations what was already known for neutral excitations [105]. Indeed, if we systematically choose (but, unlike in PPLB, we do not have to in the N-centered formalism) the xc potential that asymptotically goes to zero, i.e.,

$$v_{\mathrm{xc}}^{\xi_\pm}(\infty) = 0, \quad \xi_\pm \geq 0, \tag{140}$$

then we recover what looks like a Janak's theorem [see Eqs. (131) and (132)] and, according to Eq. (138),

$$\int d\mathbf{r}\left(v_{\mathrm{xc}}^{\xi_+ \rightarrow 0^+}(\mathbf{r}) - v_{\mathrm{xc}}^{\xi_+ = 0}(\mathbf{r})\right) n_{\Psi_0^N}(\mathbf{r}) = N\Delta_{\mathrm{xc}}^N \neq 0. \tag{141}$$

It then becomes clear that, in the region of the system under study (i.e., where the density $n_{\Psi_0^N}(\mathbf{r})$ is nonzero), the xc potentials obtained in the $\xi_+ \rightarrow 0^+$ and $\xi_+ = 0$ limits, respectively, cannot match. In order to fulfill the arbitrary constraint of Eq. (140), while still reproducing for $\xi_+ > 0$ the correct density in all regions of space [which includes a proper description of the density's asymptotic behavior (see Appendix 1)], the xc potential must be shifted in the region the system, thus ensuring that the ground-state density $n_{\Psi_0^N}(\mathbf{r})$ is also correctly reproduced in that region. This has been nicely illustrated in Ref. [113] for an atom in 1D. Thus, we recover a well-known result of PPLB: When an electron is infinitesimally added (i.e., $\xi_+ \rightarrow 0^+$) to a system with an integer number of electrons ($\xi_+ = 0$ case), the xc potential exhibits a jump (in the region of the system) which, according to Eqs. (114) and (141), corresponds exactly to the deviation in fundamental gap of the true system from the KS system.

3.3 Suppression of the Derivative Discontinuity

The fundamental gap expression of Eq. (61), which has been derived within the N-centered eDFT formalism, may intrigue PPLB practitioners. Indeed, it makes it possible to describe charged excitations, in principle exactly, without invoking explicitly the concept of derivative discontinuity. Instead, all our attention should be focused on the weight dependence of the N-centered ensemble xc density functional. Note that, despite this major difference between N-centered eDFT and PPLB, the xc potential exhibits derivative discontinuities in both theories, as we saw above. One may argue that modeling weight dependencies in ensemble xc density functionals is actually easier than modeling functional derivative discontinuities. Nevertheless, as discussed in further detail in "The Exact Hartree-Exchange Dilemma in eDFT" and "Individual Correlations Within Ensembles: An Exact Construction", designing weight-dependent exchange and correlation DFAs from first principles raises several fundamental questions, to which, up to now, no definitive answers have been given.

From a conceptual point of view, the fact that we no longer need to put efforts into the explicit description of derivative discontinuities, once we have moved from the standard PPLB picture to the N-centered one, can be interpreted as follows. Unlike in PPLB, the constraint in Eq. (140) is arbitrary because the KS potential remains unique up to a constant when charged excitations occur in N-centered eDFT, by construction. If, for simplicity, we keep this constraint for $\xi_+ = 0$, i.e., we set $\tilde{v}_{\rm xc}^{\xi_+=0}(\mathbf{r}) \equiv v_{\rm xc}^{\xi_+=0}(\mathbf{r})$ so that $\tilde{v}_{\rm xc}^{\xi_+=0}(\infty) = 0$, which is likely to be fulfilled in a practical N-electron DFT calculation, it can be relaxed as follows, when $\xi_+ \rightarrow 0^+$,

$$
v_{\rm xc}^{\xi_+\rightarrow 0^+}(\mathbf{r}) \rightarrow \tilde{v}_{\rm xc}^{\xi_+\rightarrow 0^+}(\mathbf{r}) = v_{\rm xc}^{\xi_+\rightarrow 0^+}(\mathbf{r}) - \Delta_{\rm xc}^N, \tag{142}
$$

thus leading to $\tilde{v}_{\rm xc}^{\xi_+\rightarrow 0^+}(\infty) = -\Delta_{\rm xc}^N$. We stress that $\tilde{v}_{\rm xc}^{\xi_+}$ is as exact as $v_{\rm xc}^{\xi_+}$. However, according to Eq. (139), which also holds for the new (shifted) potential $\tilde{v}_{\rm xc}^{\xi_+}(\mathbf{r})$, the relation in Eq. (141) now reads as

$$
\int d\mathbf{r} \left(\tilde{v}_{\rm xc}^{\xi_+\rightarrow 0^+}(\mathbf{r}) - \tilde{v}_{\rm xc}^{\xi_+=0}(\mathbf{r}) \right) n_{\Psi_0^N}(\mathbf{r}) = 0. \tag{143}
$$

In other words, via the shifting procedure of Eq. (142), we can simply move the derivative discontinuity away from the system, i.e., in regions where the density is essentially equal to zero. Consequently, with this change of paradigm, the absence of derivative discontinuities in standard semi-local DFAs should no longer be considered an issue. The ability of the local density approximation (LDA) to reproduce relatively accurate LZ-shifted KS orbital energies, as shown in a 1D atomic model [113], is actually encouraging since the latter are central in the evaluation of both the IP and the EA [see Eqs. (69) and (70)]. On the other hand, the resulting charged excitation energies are rather poor because weight dependencies are completely absent from standard LDA [113]. We hope that, in the near future, (much) more efforts will be put into the design of weight-dependent DFAs. Recent developments based on uniform electron gas models [79, 91] are a first and important step in this direction.

4 The Exact Hartree-Exchange Dilemma in eDFT

We have shown in "Equivalence Between Weight Derivatives and xc DerivativeDiscontinuities" that the infamous derivative discontinuity problem, which appears in DFT when describing electronic excitations, can be bypassed, in principle exactly, via a proper modeling of the ensemble weight dependence in the xc density functional. We focus in this section on the design of weight-dependent exchange DFAs. Despite several (albeit scarce) attempts [79, 100, 123, 124], it is still unclear how weight dependencies can be introduced into standard (semi-) local exchange functionals in a general and systematically improvable way. The use of orbital-dependent exchange functionals seems much more promising in this respect [71, 91, 120].

Combining (a fraction of) orbital-dependent HF-like exchange energies with semi-local DFAs has been a key ingredient in the success of regular ground-state DFT in chemistry. This procedure finds its rigorous foundation in the generalized KS theory of Seidl et al. [125]. As we will see in the following, its extension to ensembles is nontrivial because different formulations that have pros and cons are possible. The resulting dilemma is nicely summarized by the title of a recent paper by Gould and Kronik [120]: *Ensemble generalized Kohn–Sham theory: The good, the bad, and the ugly*. Their discussion of the current situation will serve as a guideline for this section. Following Lieb [66], we will show how an in-principle-exact (OEP-free) hybrid eDFT approach can be derived simply by exploiting the concavity (in potential) of the state-averaged HF energy. For simplicity, we will focus on GOK ensembles but the discussion applies to other eDFTs like, for example, PPLB or N-centered eDFT (see "Equivalence Between Weight Derivatives and xc Derivative Discontinuities").

4.1 Extending the HF Method to Ensembles

The reason why extending generalized KS-DFT [125] to ensembles leads to a dilemma has actually nothing to do with DFT. It is more a wave function theory problem that arises at the HF level of approximation. Therefore, for clarity, we will first discuss the extension of HF theory to ensembles.

4.1.1 Ensemble Density Matrix Functional Approach

We start with a brief review of the procedure usually followed by DFT practitioners for extending HF to (GOK in the present case) ensembles. In the regular scheme, the ensemble HF energy is evaluated variationally by inserting the ensemble (spin-summed one-electron reduced) density matrix (eDM) into the ground-state DM-functional HF energy. For this reason, we refer to the approach as eDMHF. The corresponding potential-functional ensemble energy can be expressed as follows:

$$E^{\mathbf{w}}_{\mathrm{eDMHF}}[v] \equiv \min_{\{\Phi_I\}} \left\{ \sum_I \mathtt{w}_I \left\langle \Phi_I \middle| \hat{T} + \hat{V} \middle| \Phi_I \right\rangle + \mathcal{E}_{\mathrm{Hx}}\left[\sum_I \mathtt{w}_I \mathbf{D}^{\Phi_I} \right] \right\}, \tag{144}$$

where $\hat{V} = \int \mathrm{d}\mathbf{r}\, v(\mathbf{r})\hat{n}(\mathbf{r})$ is a local potential operator (in practice it would correspond to the nuclear potential). The eDM is evaluated from the trial orthonormal set $\{\Phi_I\}$ of single-configuration wave functions (i.e., Slater determinants or configuration state functions). In an arbitrary orthonormal orbital basis $\{\varphi_p\}$, the eDM reads in second quantization as

$$\sum_I \mathtt{w}_I \mathbf{D}^{\Phi_I} \equiv \left\{ \sum_I \mathtt{w}_I D^{\Phi_I}_{pq} \right\} = \left\{ \sum_I \mathtt{w}_I \sum_{\tau=\uparrow,\downarrow} \left\langle \hat{a}^\dagger_{p\tau} \hat{a}_{q\tau} \right\rangle_{\Phi_I} \right\}. \tag{145}$$

In this context, the ensemble Hx energy is evaluated as follows,

$$\mathcal{E}_{\mathrm{Hx}}\left[\sum_I \mathtt{w}_I \mathbf{D}^{\Phi_I} \right] \equiv W^{\mathrm{HF}}\left[\sum_I \mathtt{w}_I \gamma^{\Phi_I} \right], \tag{146}$$

where the ground-state HF interaction functional reads as

$$W^{\mathrm{HF}}[\gamma] = \frac{1}{2} \int \mathrm{d}\mathbf{r} \int \mathrm{d}\mathbf{r}' \frac{\gamma(\mathbf{r},\mathbf{r})\gamma(\mathbf{r}',\mathbf{r}') - \frac{1}{2}\gamma^2(\mathbf{r},\mathbf{r}')}{|\mathbf{r}-\mathbf{r}'|}, \tag{147}$$

and

$$\gamma^{\Phi_I}(\mathbf{r},\mathbf{r}') = \sum_{pq} \varphi_p(\mathbf{r})\varphi_q(\mathbf{r}') D^{\Phi_I}_{pq}. \tag{148}$$

As shown in Appendix 2, the orbitals $\left\{ \overline{\varphi}^{\mathbf{w}}_p \right\}$, from which the minimizing wave functions $\left\{ \overline{\Phi}^{\mathbf{w}}_I \right\}$ in Eq. (144) are constructed, fulfill the following stationarity condition:

$$\left(\theta^{\mathbf{w}}_p - \theta^{\mathbf{w}}_q \right) f^{\mathbf{w}}_{qp} = 0. \tag{149}$$

The (possibly fractional) occupation number $\theta^{\mathbf{w}}_p$ of the orbital $\overline{\varphi}^{\mathbf{w}}_p$ within the ensemble is determined from the ensemble weights and the (fixed) integer occupation numbers n^I_p of $\overline{\varphi}^{\mathbf{w}}_p$ in each $\overline{\Phi}^{\mathbf{w}}_I$ as follows,

$$\theta_p^{\mathbf{w}} = \sum_I \mathtt{w}_I n_p^I. \tag{150}$$

The ensemble Fock matrix elements $f_{rs}^{\mathbf{w}} \equiv f_{rs}(\mathbf{D}^{\mathbf{w}})$ in Eq. (149) are functionals of the eDM $\mathbf{D}^{\mathbf{w}} \equiv \left\{D_{nl}^{\mathbf{w}} = \delta_{nl}\theta_l^{\mathbf{w}}\right\}$:

$$f_{rs}(\mathbf{D}) = h_{rs} + \sum_{nl} \Big(\langle rn|sl\rangle - \frac{1}{2}\langle rn|ls\rangle \Big) D_{nl}, \tag{151}$$

where $h_{rs} \equiv \left\langle \overline{\varphi}_r^{\mathbf{w}} \middle| \hat{h} \middle| \overline{\varphi}_s^{\mathbf{w}} \right\rangle$ [with $\hat{h} \equiv -\frac{1}{2}\nabla_{\mathbf{r}}^2 + v(\mathbf{r})$] and

$$\langle rn|sl\rangle \equiv \int \mathrm{d}\mathbf{r} \int \mathrm{d}\mathbf{r}' \overline{\varphi}_r^{\mathbf{w}}(\mathbf{r}) \overline{\varphi}_n^{\mathbf{w}}(\mathbf{r}') \overline{\varphi}_s^{\mathbf{w}}(\mathbf{r}) \overline{\varphi}_l^{\mathbf{w}}(\mathbf{r}')/|\mathbf{r} - \mathbf{r}'| \tag{152}$$

are regular one- and two-electron integrals, respectively.

By analogy with the complete active space SCF (CASSCF) method [126], we can distinguish the doubly occupied (so-called inactive) $\overline{\varphi}_i^{\mathbf{w}}, \overline{\varphi}_j^{\mathbf{w}}$ orbitals from the partially occupied (so-called active) $\overline{\varphi}_u^{\mathbf{w}}, \overline{\varphi}_v^{\mathbf{w}}$ and unoccupied (virtual) $\overline{\varphi}_a^{\mathbf{w}}, \overline{\varphi}_b^{\mathbf{w}}$ orbitals. Thus, the stationarity condition of Eq. (149) can be detailed as follows:

$$f_{ai}^{\mathbf{w}} = f_{au}^{\mathbf{w}} = f_{iu}^{\mathbf{w}} = 0 \tag{153}$$

and

$$(\theta_u^{\mathbf{w}} - \theta_v^{\mathbf{w}}) f_{uv}^{\mathbf{w}} = 0. \tag{154}$$

Since $\theta_i^{\mathbf{w}} = \theta_j^{\mathbf{w}} = 2$ and $\theta_a^{\mathbf{w}} = \theta_b^{\mathbf{w}} = 0$, there is no specific condition for the inactive–inactive and virtual–virtual blocks of the Fock matrix, like in a regular ground-state HF calculation. Therefore, we can freely rotate the orbitals within the inactive and virtual orbital subspaces. As readily seen from Eq. (154), this statement holds also for active orbital subspaces in which the orbitals have the same fractional occupation ($\theta_u^{\mathbf{w}} = \theta_v^{\mathbf{w}}$). However, if $\theta_u^{\mathbf{w}} \neq \theta_v^{\mathbf{w}}$, then $f_{uv}^{\mathbf{w}} = 0$. In conclusion, an optimal set of orbitals can be determined by diagonalizing the ensemble Fock matrix, i.e., by solving the (self-consistent) eigenvalue equation

$$\hat{f}^{\mathbf{w}} \overline{\varphi}_p^{\mathbf{w}}(\mathbf{r}) = \overline{\varepsilon}_p^{\mathbf{w}} \overline{\varphi}_p^{\mathbf{w}}(\mathbf{r}). \tag{155}$$

The fact that standard SCF routines can be recycled trivially in this context is the main reason why ensemble HF and, more generally, hybrid eDFT calculations are performed this way. However, as discussed further in the following, the eDMHF energy is unphysical in many ways. For example, by construction, it varies quadratically with the ensemble weights [see Eqs. (146) and (147)] while the true physical ensemble energy is expected to vary linearly.

4.1.2 Ghost Interaction Errors

The most severe issue with the eDMHF energy expression of Eq. (144) is that it incorporates unphysical interactions between the states of the ensemble. These are known as ghost interactions (GIs) [127]. The error, which is inherent to the eDMHF method, originates from the fact that, at the HF level of approximation, the ground-state interaction energy is simply a quadratic functional of the density matrix. More explicitly, we have

$$W^{\mathrm{HF}}\left[\sum_I \mathtt{w}_I \gamma^{\Phi_I}\right] = \frac{1}{2}\sum_{IJ} \mathtt{w}_I \mathtt{w}_J \int d\mathbf{r} \int d\mathbf{r}' \frac{1}{|\mathbf{r}-\mathbf{r}'|} \times \left(\gamma^{\Phi_I}(\mathbf{r},\mathbf{r})\gamma^{\Phi_J}(\mathbf{r}',\mathbf{r}') - \frac{1}{2}\gamma^{\Phi_I}(\mathbf{r},\mathbf{r}')\gamma^{\Phi_J}(\mathbf{r},\mathbf{r}')\right), \tag{156}$$

where, as readily seen, GI terms arise from all "$I \neq J$" pairs. Even though GI corrections can be applied on top of the converged eDMHF energies [91, 128], the procedure is not variational, thus making the evaluation of energy derivatives [and therefore, according to Eq. (9), of excited-state properties] less straightforward. Let us stress that, in the original formulation of GOK-DFT [82], the ensemble Hartree energy, which is evaluated from the standard (ground-state) Hartree functional [see Eq. (20)], includes GI errors [see the first term on the right-hand side of Eq. (156)]. In the exact theory, the latter are supposed to be removed by the weight-dependent ensemble exchange functional. It is not necessarily the case in practice when, for example, standard (weight-independent) local or semi-local DFAs are employed [97, 128].

For wave function theory practitioners, using eDMHF with ad hoc GI corrections would probably seem uselessly complicated. Indeed, substituting the (GI-free) weighted sum of individual (single-configuration) interaction energies, which are evaluated from the individual density matrices, for the HF interaction eDM functional of Eq. (144) looks, at least at first sight, like a simple and straightforward solution to the problem:

$$\mathcal{E}_{\mathrm{Hx}}\left[\sum_I \mathtt{w}_I \mathbf{D}^{\Phi_I}\right] \rightarrow \sum_I \mathtt{w}_I \langle \hat{W}_{\mathrm{ee}} \rangle_{\Phi_I} \rightarrow \sum_I \mathtt{w}_I \mathcal{E}^I_{\mathrm{Hx}}\left[\mathbf{D}^{\Phi_I}\right] = \sum_I \mathtt{w}_I \left(E_{\mathrm{H}}[n_{\Phi_I}] + \mathcal{E}^I_{\mathrm{x}}\left[\mathbf{D}^{\Phi_I}\right]\right). \tag{157}$$

Note that, in Eq. (157), we assumed that individual interaction energies can be written as functionals of the individual (one-electron reduced) density matrices, for simplicity. The variational evaluation of state-averaged interaction energies is discussed in detail in the following section on that basis. Such a simplification is always valid for single Slater determinants. For more general multideterminant (single configuration though) wave functions, the simplification in Eq. (157) might be used as an (additional) approximation. Alternatively, one may evaluate exactly multideterminant interaction energies from the individual two-electron reduced density matrices, by analogy with the state-averaged CASSCF (SA-CASSCF) method [126]. The latter approach is not described further in the present review. Both (approximate anyway) strategies lead ultimately to an in-principle-exact eDFT, once a proper complementary correlation ensemble density functional has been introduced (see "Exact Self-Consistent eDFT Based on SAHF").

As discussed in the next section, the reason why the state-averaging of interaction energies is not as popular as one would expect is that, as we switch from eDMHF to the state-averaged energy paradigm, the orbital optimization can no longer be performed with standard SCF routines. Additional implementation work is needed in this case [71]. We stress that this statement holds even when individual one-electron reduced density matrix-functional interaction energies are employed [see Eq. (157)], as assumed in the rest of the review.

4.1.3 State-Averaged HF Approach

The paradigm on the right-hand side of Eq. (157) can be seen as an adaptation of the SA-CASSCF method [126] to single-configuration wave functions. Whereas in SA-CASSCF, (correlated) multiconfigurational wave functions are employed, in the present case we simply restrict the energy minimization to sets of (uncorrelated) single-configuration wave functions. The resulting (GI-free) total ensemble energy, which is obtained from the eDMHF energy expression and the substitution in Eq. (157), will be referred to as state-averaged HF (SAHF) energy in the following. It reads as follows:

$$
\begin{aligned}
E^{\mathbf{w}}_{\mathrm{SAHF}}[v] &\equiv \min_{\{\Phi_I\}}\left\{\sum_I \mathtt{w}_I\left\langle\Phi_I\middle|\hat{T}+\hat{W}_{\mathrm{ee}}+\hat{V}\middle|\Phi_I\right\rangle\right\}\\
&= \min_{\{\Phi_I\}}\left\{\sum_I \mathtt{w}_I\left(\left\langle\Phi_I\middle|\hat{T}\middle|\Phi_I\right\rangle + E_{\mathrm{H}}[n_{\Phi_I}] + \mathcal{E}^I_{\mathrm{x}}\left[\mathbf{D}^{\Phi_I}\right] + \int d\mathbf{r}\, v(\mathbf{r}) n_{\Phi_I}(\mathbf{r})\right)\right\},
\end{aligned}
\tag{158}
$$

where, as already mentioned after Eq. (157), we assume for simplicity that interaction energies can be evaluated from the individual (one-electron reduced) density matrices. Let us denote $\left\{\tilde{\Phi}^{\mathbf{w}}_I\right\}$ (with a tilde symbol) the minimizing single-configuration wave functions so that they can be distinguished clearly from the eDMHF wave functions. As further discussed in Appendix 3, these minimizing SAHF wave functions are all constructed from the same set of orthonormal molecular orbitals, which can be optimized variationally through orbital rotations. The minimizing orbitals $\left\{\tilde{\varphi}^{\mathbf{w}}_p\right\}$ fulfill the following stationarity conditions [see Appendix 3],

$$
\left(\theta^{\mathbf{w}}_p-\theta^{\mathbf{w}}_q\right)\left\langle\tilde{\varphi}^{\mathbf{w}}_p\middle|\hat{h}\middle|\tilde{\varphi}^{\mathbf{w}}_q\right\rangle + \sum_I \mathtt{w}_I\left(n^I_p-n^I_q\right)\left\langle\tilde{\varphi}^{\mathbf{w}}_p\middle|\hat{v}^{\mathbf{w}}_{\mathrm{Hx},I}\middle|\tilde{\varphi}^{\mathbf{w}}_q\right\rangle = 0, \tag{159}
$$

where the individual (non-local) density-matrix-functional Hx operators read as

$$
\hat{v}^{\mathbf{w}}_{\mathrm{Hx},I} = \hat{v}_{\mathrm{H}}[n_{\tilde{\Phi}^{\mathbf{w}}_I}] + \hat{v}^I_{\mathrm{x}}\left[\mathbf{D}^{\tilde{\Phi}^{\mathbf{w}}_I}\right], \tag{160}
$$

$\hat{v}_{\rm H}[n] \equiv v_{\rm H}[n](\mathbf{r})\times = \delta E_{\rm H}[n]/\delta n(\mathbf{r})\times$ being the standard local (multiplicative) density-functional Hartree potential operator and

$$\left\langle \varphi_r \middle| \hat{v}_{\rm x}^I[\mathbf{D}] \middle| \varphi_s \right\rangle \equiv \frac{\partial \mathcal{E}_{\rm x}^I[\mathbf{D}]}{\partial D_{rs}}. \tag{161}$$

The major difference between SAHF and eDMHF lies in the fact that, as we now employ individual density matrices separately in the evaluation of the ensemble Hx energy [see Eq. (157)], differentiating with respect to any variational orbital rotation parameter κ_{pq} generates individual Hx potentials,

$$\begin{aligned} &\frac{\partial}{\partial \kappa_{pq}}\left(\mathcal{E}_{\rm Hx}\left[\sum_I \mathtt{w}_I \mathbf{D}^{\Phi_I}\right]\right) = \sum_{rs}\left(\sum_I \mathtt{w}_I \frac{\partial D_{rs}^{\Phi_I}}{\partial \kappa_{pq}}\right)\frac{\partial \mathcal{E}_{\rm Hx}[\mathbf{D}]}{\partial D_{rs}}\Bigg|_{\mathbf{D}=\sum_I \mathtt{w}_I \mathbf{D}^{\Phi_I}} \\ &\longrightarrow \frac{\partial}{\partial \kappa_{pq}}\left(\sum_I \mathtt{w}_I \mathcal{E}_{\rm Hx}^I\left[\mathbf{D}^{\Phi_I}\right]\right) = \sum_{rs}\sum_I \mathtt{w}_I \frac{\partial D_{rs}^{\Phi_I}}{\partial \kappa_{pq}}\frac{\partial \mathcal{E}_{\rm Hx}^I[\mathbf{D}]}{\partial D_{rs}}\Bigg|_{\mathbf{D}=\mathbf{D}^{\Phi_I}}. \end{aligned} \tag{162}$$

In eDMHF, the same (ensemble) Hx operator is recovered systematically by differentiation [see the left-hand side of Eq. (162)] and, from the stationarity condition and the definition in Eq. (150) of the fractional orbital occupation numbers, a unique to-be-diagonalized Fock operator can be extracted. This does not happen in SAHF, as seen readily from the second term on the left-hand side of Eq. (159) or, equivalently, on the right-hand side of Eq. (162). Still, we can introduce an orbital-dependent ensemble Fock operator [71, 75, 120],

$$\hat{\mathcal{F}}_p^{\mathbf{w}} := \hat{h} + \frac{1}{\theta_p^{\mathbf{w}}}\sum_I \mathtt{w}_I n_p^I \, \hat{v}_{{\rm Hx},I}^{\mathbf{w}}, \tag{163}$$

so that the stationarity condition of Eq. (159) can be rewritten as follows,

$$\theta_p^{\mathbf{w}}\left\langle \tilde{\varphi}_p^{\mathbf{w}} \middle| \hat{\mathcal{F}}_p^{\mathbf{w}} \middle| \tilde{\varphi}_q^{\mathbf{w}} \right\rangle - \theta_q^{\mathbf{w}}\left\langle \tilde{\varphi}_p^{\mathbf{w}} \middle| \hat{\mathcal{F}}_q^{\mathbf{w}} \middle| \tilde{\varphi}_q^{\mathbf{w}} \right\rangle = 0. \tag{164}$$

Let us stress that, unlike in GOK-DFT (see "DFT of GOK Ensembles") or eDMHF, the minimizing orbitals are a priori not eigenfunctions of their associated Fock operator. Indeed, if we consider two fractionally occupied orbitals $\tilde{\varphi}_u^{\mathbf{w}}$ and $\tilde{\varphi}_v^{\mathbf{w}}$, and assume that $\left\langle \tilde{\varphi}_u^{\mathbf{w}} \middle| \hat{\mathcal{F}}_u^{\mathbf{w}} \middle| \tilde{\varphi}_v^{\mathbf{w}} \right\rangle = 0$, Eq. (164) would immediately imply that $\left\langle \tilde{\varphi}_u^{\mathbf{w}} \middle| \hat{\mathcal{F}}_v^{\mathbf{w}} \middle| \tilde{\varphi}_v^{\mathbf{w}} \right\rangle = 0$, which is unlikely due to the orbital dependence of the Fock operator [see Eq. (163)]. As a result, the self-consistent SAHF equations will have the following general structure,

$$\theta_p^{\mathbf{w}}\hat{\mathcal{F}}_p^{\mathbf{w}}\tilde{\varphi}_p^{\mathbf{w}}(\mathbf{r}) = \sum_q \tilde{\varepsilon}_{qp}^{\mathbf{w}}\,\tilde{\varphi}_q^{\mathbf{w}}(\mathbf{r}), \tag{165}$$

or, more explicitly [see Eq. (163)],

$$\left(-\frac{1}{2}\nabla^2_{\mathbf{r}} + v(\mathbf{r}) + \frac{1}{\theta^{\mathbf{w}}_p}\sum_I \mathtt{w}_I n^I_p \hat{v}^{\mathbf{w}}_{\mathrm{Hx},I}\right)\tilde{\varphi}^{\mathbf{w}}_p(\mathbf{r}) = \frac{1}{\theta^{\mathbf{w}}_p}\sum_q \tilde{\varepsilon}^{\mathbf{w}}_{qp}\,\tilde{\varphi}^{\mathbf{w}}_q(\mathbf{r}), \tag{166}$$

where off-diagonal one-electron energy couplings $\left\{\tilde{\varepsilon}^{\mathbf{w}}_{qp}\right\}_{p\neq q}$ cannot be removed. In practice, Eq. (165) can be solved with the coupling operator technique [75, 129], which consists of diagonalizing $(\theta^{\mathbf{w}}_p\hat{\mathcal{F}}^{\mathbf{w}}_p - \theta^{\mathbf{w}}_q\hat{\mathcal{F}}^{\mathbf{w}}_q)/(\theta^{\mathbf{w}}_p - \theta^{\mathbf{w}}_q)$ repeatedly until convergence is reached. In the latter case, the matrix $\tilde{\boldsymbol{\varepsilon}}^{\mathbf{w}} \equiv \left\{\tilde{\varepsilon}^{\mathbf{w}}_{qp}\right\}$ becomes Hermitian, as a consequence of Eqs. (164) and (165), and the hermiticity of the Fock operators:

$$\tilde{\varepsilon}^{\mathbf{w}}_{qp} = \left\langle \tilde{\varphi}^{\mathbf{w}}_q \middle| \theta^{\mathbf{w}}_p \hat{\mathcal{F}}^{\mathbf{w}}_p \middle| \tilde{\varphi}^{\mathbf{w}}_p \right\rangle = \left\langle \tilde{\varphi}^{\mathbf{w}}_p \middle| \theta^{\mathbf{w}}_p \hat{\mathcal{F}}^{\mathbf{w}}_p \middle| \tilde{\varphi}^{\mathbf{w}}_q \right\rangle = \left\langle \tilde{\varphi}^{\mathbf{w}}_p \middle| \theta^{\mathbf{w}}_q \hat{\mathcal{F}}^{\mathbf{w}}_q \middle| \tilde{\varphi}^{\mathbf{w}}_q \right\rangle = \tilde{\varepsilon}^{\mathbf{w}}_{pq}. \tag{167}$$

4.1.4 eDMHF Versus SAHF

Let us summarize what we have learned from the previous subsections. While the orbital optimization in eDMHF is relatively straightforward, because standard SCF routines can be recycled in this context, it is more involved in SAHF because the not to-be-diagonalized Fock operator emerges from the stationarity condition. On the other hand, the commonly used eDMHF energy expression suffers from severe GI errors, while SAHF is completely GI-free. The latter point is probably the strongest argument for promoting SAHF over eDMHF. Note that both schemes would be good starting points for turning the recently formulated ensemble reduced density matrix functional theory (w-RDMFT) [130] into a practical method for the computation of low-lying excited states. In the following, we will show how eDMHF and SAHF can be merged rigorously with eDFT.

4.2 Concavity of Approximate Energies and Lieb Maximization

In order to derive in-principle-exact hybrid eDFT schemes, where (a fraction of) orbital-dependent exchange energies are combined with ensemble density functionals, we need to "exactify" the eDMHF and SAHF approximations reviewed previously. In a DFT perspective, an approximation becomes exact when it reproduces the exact density of the system under study. This is how KS-DFT transforms an approximate non-interacting problem into an exact one. In the present case, we want to extend the Hohenberg–Kohn theorem to the more advanced eDMHF and SAHF approximations. For that purpose, convex analysis [66, 131, 132] turns out to be a powerful mathematical tool because, as we will see, it allows for the derivation of several exact eDFTs within the same (unified) formalism.

For the sake of generality, we will express the various approximate ensemble energies discussed previously as follows:

$$\mathscr{E}^{\mathbf{w}}_{\mathrm{approx.}}[v] = \min_{\boldsymbol{\kappa}}\left\{\mathscr{F}^{\mathbf{w}}_{\mathrm{approx.}}(\boldsymbol{\kappa}) + \int d\mathbf{r}\, v(\mathbf{r})n^{\mathbf{w}}(\boldsymbol{\kappa},\mathbf{r})\right\}, \tag{168}$$

where $\boldsymbol{\kappa}$ denotes the collection of variational parameters (in the present case, the latter will be orbital rotation parameters). Each approximation (non-interacting (KS), eDMHF, or SAHF) is characterized by a specific potential-independent function of $\boldsymbol{\kappa}$:

$$\mathscr{F}^{\mathbf{w}}_{\text{approx.}}(\boldsymbol{\kappa}) \stackrel{\text{KS}}{\equiv} \sum_I \mathtt{w}_I \langle \hat{T} \rangle_{\Phi_I(\boldsymbol{\kappa})}, \tag{169}$$

$$\mathscr{F}^{\mathbf{w}}_{\text{approx.}}(\boldsymbol{\kappa}) \stackrel{\text{eDMHF}}{\equiv} \sum_I \mathtt{w}_I \langle \hat{T} \rangle_{\Phi_I(\boldsymbol{\kappa})} + \mathcal{E}_{\text{Hx}}\left[\sum_I \mathtt{w}_I \mathbf{D}^{\Phi_I(\boldsymbol{\kappa})}\right], \tag{170}$$

$$\mathscr{F}^{\mathbf{w}}_{\text{approx.}}(\boldsymbol{\kappa}) \stackrel{\text{SAHF}}{\equiv} \sum_I \mathtt{w}_I \left[\langle \hat{T} \rangle_{\Phi_I(\boldsymbol{\kappa})} + E_{\text{H}}[n_{\Phi_I(\boldsymbol{\kappa})}] + \mathcal{E}^I_{\text{x}}\left[\mathbf{D}^{\Phi_I(\boldsymbol{\kappa})}\right]\right], \tag{171}$$

where single-configuration ground- and excited-state wave functions $\{\Phi_I(\boldsymbol{\kappa})\}$ are employed. The potential-dependent contribution to the energy expression of Eq. (168) is determined from the ensemble density $n^{\mathbf{w}}(\boldsymbol{\kappa}, \mathbf{r}) = \sum_I \mathtt{w}_I n_{\Phi_I(\boldsymbol{\kappa})}(\mathbf{r})$.

From a mathematical point of view, the fact that the approximate ensemble energies are evaluated variationally has important implications. Even though they are approximate, these energies still share a fundamental property with the exact ensemble energy, namely the concavity with respect to the local potential v. Indeed, for two potentials v_a and v_b, α in the range $0 \leq \alpha \leq 1$, and any set $\boldsymbol{\kappa}$ of variational parameters, we have

$$\begin{aligned}
&\mathscr{F}^{\mathbf{w}}_{\text{approx.}}(\boldsymbol{\kappa}) + \int d\mathbf{r}\, \big((1-\alpha)v_a(\mathbf{r}) + \alpha v_b(\mathbf{r})\big) n^{\mathbf{w}}(\boldsymbol{\kappa}, \mathbf{r}) \\
&= (1-\alpha)\left[\mathscr{F}^{\mathbf{w}}_{\text{approx.}}(\boldsymbol{\kappa}) + \int d\mathbf{r}\, v_a(\mathbf{r}) n^{\mathbf{w}}(\boldsymbol{\kappa}, \mathbf{r})\right] \\
&\quad + \alpha\left[\mathscr{F}^{\mathbf{w}}_{\text{approx.}}(\boldsymbol{\kappa}) + \int d\mathbf{r}\, v_b(\mathbf{r}) n^{\mathbf{w}}(\boldsymbol{\kappa}, \mathbf{r})\right] \\
&\geq (1-\alpha)\mathscr{E}^{\mathbf{w}}_{\text{approx.}}[v_a] + \alpha \mathscr{E}^{\mathbf{w}}_{\text{approx.}}[v_b],
\end{aligned} \tag{172}$$

thus leading to

$$\mathscr{E}^{\mathbf{w}}_{\text{approx.}}[(1-\alpha)v_a + \alpha v_b] \geq (1-\alpha)\mathscr{E}^{\mathbf{w}}_{\text{approx.}}[v_a] + \alpha \mathscr{E}^{\mathbf{w}}_{\text{approx.}}[v_b]. \tag{173}$$

Following Lieb [66], we can now construct (thanks to this concavity property) an approximation to the universal GOK density functional as follows:

$$F^{\mathbf{w}}[n] \approx F^{\mathbf{w}}_{\text{approx.}}[n] = \max_v \left\{ \mathscr{E}^{\mathbf{w}}_{\text{approx.}}[v] - \int d\mathbf{r}\, v(\mathbf{r}) n(\mathbf{r}) \right\}, \tag{174}$$

where we assume, for simplicity, that a maximum is reached. An even more rigorous definition (from a mathematical point of view [66]) would actually be obtained by using a "sup" instead of a "max" [and a "inf" instead of a "min", in Eq. (168)]. The maximizing potential $v^{\mathbf{w}}_{\text{approx.}}[n]$ in Eq. (174) fulfills the following stationarity condition:

$$\left.\frac{\delta \mathscr{E}^{\mathbf{w}}_{\text{approx.}}[v]}{\delta v(\mathbf{r})}\right|_{v=v^{\mathbf{w}}_{\text{approx.}}[n]} = n(\mathbf{r}). \tag{175}$$

We conclude from the Hellmann–Feynman theorem that, when the approximate ensemble energy of Eq. (168) is calculated with $v = v^{\mathbf{w}}_{\text{approx.}}[n]$, the minimizing single-configuration wave functions (which are determined from the minimizing $\boldsymbol{\kappa}$) reproduce the desired ensemble density n. Thus, we automatically extend the Hohenberg–Kohn theorem to eDMHF and SAHF ensembles. If we choose for n the true physical ensemble density of a given system, both approximations become exact density wise, because they reproduce the correct density. Exact ensemble energies can then be recovered from the approximate ensembles once a complementary ensemble (x)c density functional has been introduced. This final step will be discussed in "Exact Self-Consistent eDFT Based on SAHF".

Let us finally focus on the OEP- and GI-free SAHF approximation. The corresponding universal density functional reads more explicitly as [see Eqs. (168), (171), and (174)]

$$\begin{aligned} F^{\mathbf{w}}_{\text{SAHF}}[n] &= \mathscr{E}^{\mathbf{w}}_{\text{SAHF}}[v^{\mathbf{w}}_{\text{SAHF}}[n]] - \int d\mathbf{r}\, v^{\mathbf{w}}_{\text{SAHF}}[n](\mathbf{r})n(\mathbf{r}) \\ &\equiv \sum_I \mathtt{w}_I \left\langle \tilde{\Phi}^{\mathbf{w}}_I[n] \middle| \hat{T} + \hat{W}_{\text{ee}} \middle| \tilde{\Phi}^{\mathbf{w}}_I[n] \right\rangle, \end{aligned} \tag{176}$$

where $v^{\mathbf{w}}_{\text{SAHF}}[n]$ denotes the stationary (maximizing) density-functional potential of Eq. (175) in the particular case of the SAHF approximation. By analogy with the constrained-search formalism of Levy [92], the SAHF functional can be rewritten as follows:

$$F^{\mathbf{w}}_{\text{SAHF}}[n] = \min_{\{\Phi_I\} \overset{\mathbf{w}}{\to} n} \left\{ \sum_I \mathtt{w}_I \left\langle \Phi_I \middle| \hat{T} + \hat{W}_{\text{ee}} \middle| \Phi_I \right\rangle \right\}, \tag{177}$$

where the density constraint $\left\{\Phi_I\right\} \overset{\mathbf{w}}{\to} n$ imposed on the single-configuration wave functions $\left\{\Phi_I\right\}$ reads as $\sum_I \mathtt{w}_I n_{\Phi_I}(\mathbf{r}) = n(\mathbf{r})$. Note that, as illustrated in "Insights from the Hubbard Dimer Model", some densities may not be v-representable by a single SAHF ensemble. In this case, a more general Levy–Lieb-like [66] functional, where an ensemble of ensembles is considered, should be employed:

$$F^{\mathbf{w}}_{\text{SAHF}}[n] = \min_{\left\{\left\{\Phi^{(i)}_I\right\}, \alpha^{(i)}\right\} \overset{\mathbf{w}}{\to} n} \sum_i \alpha^{(i)} \left\{ \sum_I \mathtt{w}_I \left\langle \Phi^{(i)}_I \middle| \hat{T} + \hat{W}_{\text{ee}} \middle| \Phi^{(i)}_I \right\rangle \right\}, \tag{178}$$

where the density constraint reads as

$$\sum_i \alpha^{(i)} \left(\sum_I \mathtt{w}_I n_{\Phi_I^{(i)}}(\mathbf{r}) \right) = \sum_I \mathtt{w}_I \left(\sum_i \alpha^{(i)} n_{\Phi_I^{(i)}}(\mathbf{r}) \right) = n(\mathbf{r}), \tag{179}$$

with

$$\sum_i \alpha^{(i)} = 1. \tag{180}$$

We stress that, in the above more general definition, the additional ensemble weights $\{\alpha^{(i)}\}$ are not given, unlike the GOK ensemble weights $\mathbf{w}$. They are determined from the density constraint of Eq. (179). An interesting feature of the constrained-search formalism is that it applies to densities that might be ensemble N-representable but not SAHF v-representable, i.e., densities that cannot be generated from a given potential v according to Eq. (168). For clarity, we will use the simpler density functional expression of Eq. (177) [rather than the one in Eq. (178)] in the section "Exact Self-Consistent eDFT Based on SAHF".

4.3 Insights from the Hubbard Dimer Model

In order to compare SAHF with eDMHF, both methods are applied in this section to the (two-electron) Hubbard dimer model [84, 85, 96]. Despite its simplicity, it is nontrivial and has become in recent years the model of choice for analyzing and understanding failures of DFT or TD-DFT, but also for exploring new concepts [99, 101, 133–136]. In this model, the ab initio Hamiltonian is simplified as follows:

$$\begin{aligned} \hat{T} \rightarrow \hat{\mathcal{T}} &= -t \sum_{\tau=\uparrow\downarrow} (\hat{c}^\dagger_{0\tau}\hat{c}_{1\tau} + \hat{c}^\dagger_{1\tau}\hat{c}_{0\tau}), \hat{W}_{\text{ee}} \rightarrow \hat{U} = U \sum_{i=0}^{1} \hat{n}_{i\uparrow}\hat{n}_{i\downarrow}, \\ \hat{V} &\rightarrow \Delta v(\hat{n}_1 - \hat{n}_0)/2, \hat{n}_{i\tau} = \hat{c}^\dagger_{i\tau}\hat{c}_{i\tau}, \end{aligned} \tag{181}$$

where operators are written in second quantization and $\hat{n}_i = \sum_{\tau=\uparrow\downarrow} \hat{n}_{i\tau}$ is the density operator on site i ($i = 0, 1$). Note that the local potential reduces to a single number Δv which controls the asymmetry of the dimer. The density also reduces to a single number $n = n_0$, which is the occupation of site 0, given that $n_1 = 2 - n$.

We consider in the following a two-electron singlet biensemble consisting of the ground state and the singly excited state.

4.3.1 SAHF and eDMHF Energy Expressions

In (restricted) SAHF theory, both ground- and excited-state wave functions are approximated by configuration state functions. Following this view, trial singlet ground-state Φ_0 and first excited-state Φ_1 wave functions of the two-electron Hubbard dimer read as follows, in second quantization,

$$|\Phi_0\rangle = |\sigma_0^2\rangle \equiv \hat{c}^\dagger_{\sigma_0\uparrow}\hat{c}^\dagger_{\sigma_0\downarrow}|\text{vac}\rangle \tag{182}$$

and

$$|\Phi_1\rangle = |\sigma_0\sigma_1\rangle \equiv \frac{1}{\sqrt{2}}\left(\hat{c}^\dagger_{\sigma_1\uparrow}\hat{c}^\dagger_{\sigma_0\downarrow} - \hat{c}^\dagger_{\sigma_1\downarrow}\hat{c}^\dagger_{\sigma_0\uparrow}\right)|\text{vac}\rangle, \tag{183}$$

respectively, where σ_0 and σ_1 stand for the molecular orbitals (MOs) written in the basis of the orthonormal local atomic orbitals (AOs) a and b. In this context, the latter simply correspond to site 0 ($\hat{c}^\dagger_{a\tau} \equiv \hat{c}^\dagger_{0\tau}$) and 1 ($\hat{c}^\dagger_{b\tau} \equiv \hat{c}^\dagger_{1\tau}$), respectively. Bonding σ_0 and anti-bonding σ_1 MOs can be determined through orbital rotation as follows,

$$\begin{aligned} \sigma_0 &= a\cos(\alpha) + b\sin(\alpha), \\ \sigma_1 &= -a\sin(\alpha) + b\cos(\alpha), \end{aligned} \tag{184}$$

where the angle α is the sole variational parameter in the model. In SAHF, the energy that must be minimized, for a fixed ensemble weight $\mathtt{w}$ in the range $0 \leq \mathtt{w} \leq 1/2$, is constructed as follows,

$$E^{\mathtt{w}}_{\text{SAHF}} \equiv (1-\mathtt{w})\langle\hat{H}\rangle_{\Phi_0} + \mathtt{w}\langle\hat{H}\rangle_{\Phi_1}, \tag{185}$$

where

$$\langle\hat{H}\rangle_{\Phi_0} = 2h_{\sigma_0\sigma_0} + J_{\sigma_0\sigma_0} \tag{186}$$

and

$$\langle\hat{H}\rangle_{\Phi_1} = h_{\sigma_0\sigma_0} + h_{\sigma_1\sigma_1} + J_{\sigma_0\sigma_1} + K_{\sigma_0\sigma_1}. \tag{187}$$

We use standard notations for Coulomb and exchange two-electron integrals,

$$J_{ij} = (ii,jj) = \langle ij|ij\rangle, \tag{188}$$

$$K_{ij} = (ij,ji) = \langle ij|ji\rangle, \tag{189}$$

both expressed in the MO basis $\{\sigma_0, \sigma_1\}$ of Eq. (184). At this point, let us stress that the SAHF energy differs substantially from that of a (truncated) configuration interaction calculation. Indeed, in the present case, the weight $\mathtt{w}$ is fixed. Moreover, the configurations Φ_0 and Φ_1 are never coupled explicitly, whether the dimer is symmetric or not. They are just mixed through the ensemble formalism. Note that, in the symmetric $\Delta v = 0$ case, symmetry can in principle be artificially broken, like in (spin) unrestricted calculations. This feature will be discussed in further detail in the next section. For convenience, we denote

$$\theta = \frac{\pi}{4} - \alpha, \tag{190}$$

so that the symmetric solution corresponds to $\theta = 0$. Consequently, the SAHF energy expression of Eq. (185) reduces to

$$\begin{aligned} E^{\mathrm{w}}_{\mathrm{SAHF}}(\Delta v, \theta) = &-(1-\mathrm{w})[2t\cos(2\theta) + \Delta v \sin(2\theta)] \\ &+ \frac{U}{4}[3 - \mathrm{w} + (3\mathrm{w} - 1)\cos(4\theta)]. \end{aligned} \tag{191}$$

As readily seen from the above expression, the SAHF energy is π-periodic, which means that it is sufficient to vary θ in the range $-\frac{\pi}{2} \leq \theta \leq \frac{\pi}{2}$.

Similarly, from the general expression in Eq. (144), we obtain the following analytical expression for the eDMHF energy:

$$\begin{aligned} E^{\mathrm{w}}_{\mathrm{eDMHF}}(\Delta v, \theta) = &-(1-\mathrm{w})[2t\cos(2\theta) + \Delta v \sin(2\theta)] \\ &+ \frac{U}{4}\left[\mathrm{w}^2 - 2\mathrm{w} + 3 - (1-\mathrm{w})^2\cos(4\theta)\right]. \end{aligned} \tag{192}$$

Note that, according to Eq. (184), the ensemble density (on site 0) varies with the trial angle θ as follows,

$$\begin{aligned} n^{\mathrm{w}}(\theta) &\equiv (1-\mathrm{w})\sum_{\tau=\uparrow\downarrow}\left\langle \hat{c}^{\dagger}_{0\tau}\hat{c}_{0\tau}\right\rangle_{\Phi_0} + \mathrm{w}\sum_{\tau=\uparrow\downarrow}\left\langle \hat{c}^{\dagger}_{0\tau}\hat{c}_{0\tau}\right\rangle_{\Phi_1} \\ &= 1 + (1-\mathrm{w})\sin(2\theta). \end{aligned} \tag{193}$$

4.3.2 Symmetric Case

Let us concentrate on the symmetric dimer, for which $\Delta v = 0$. In this case, the dimer is a prototype for the H_2 molecule. If we denote

$$\rho(\theta) = \cos(2\theta) = 2\cos^2\theta - 1, \tag{194}$$

then the SAHF energy reads as

$$E^{\mathrm{w}}_{\mathrm{SAHF}}(\Delta v = 0, \theta) \equiv E^{\mathrm{w}}_{\mathrm{SAHF}}(\theta) = \mathcal{E}^{\mathrm{w}}_{\mathrm{SAHF}}(\rho(\theta)), \tag{195}$$

where

$$\mathcal{E}^{\mathrm{w}}_{\mathrm{SAHF}}(\rho) = \frac{U}{2}(3\mathrm{w} - 1)\rho^2 - 2t(1-\mathrm{w})\rho + U(1-\mathrm{w}). \tag{196}$$

By taking its first derivative with respect to θ, we obtain the following stationarity condition,

$$\frac{dE^{\mathrm{w}}_{\mathrm{SAHF}}(\theta)}{d\theta} = -2\sin(2\theta)[(U(3\mathrm{w} - 1)\rho(\theta) - 2t(1-\mathrm{w})] = 0. \tag{197}$$

Therefore, $\theta = 0$ is systematically an extremum where the traditional in-phase and out-of-phase linear combinations for σ_0 and σ_1 are recovered. The nature (maximum or minimum) of this stationary point, which is discussed in the following, emerges from a straightforward evaluation of the energy curvature:

$$\left.\frac{d^2E^{\mathrm{w}}_{\mathrm{SAHF}}(\theta)}{d\theta^2}\right|_{\theta=0} = -4[U(3\mathrm{w}-1) - 2t(1-\mathrm{w})]. \tag{198}$$

Remembering that $|\rho(\theta)| \leq 1$, the stationarity condition of Eq. (197) is also fulfilled for two additional (opposite) θ values given by

$$\rho(\theta) = \rho_0 \equiv \frac{2t(1-\mathrm{w})}{U(3\mathrm{w}-1)}, \tag{199}$$

as long as

$$|\rho_0| \leqslant 1. \tag{200}$$

Note that, with this notation, the successive energy derivatives can be expressed as follows,

$$\frac{dE^{\mathrm{w}}_{\mathrm{SAHF}}(\theta)}{d\theta} = -4t(1-\mathrm{w})\sin(2\theta)\left[\frac{\rho(\theta)}{\rho_0} - 1\right] \tag{201}$$

and

$$\left.\frac{d^2E^{\mathrm{w}}_{\mathrm{SAHF}}(\theta)}{d\theta^2}\right|_{\theta=0} = -8t(1-\mathrm{w})\left[\frac{1}{\rho_0} - 1\right]. \tag{202}$$

The symmetric solution $\theta = 0$ will not be the absolute minimum anymore when the above curvature becomes strictly negative, which implies $1/\rho_0 > 1$, as readily seen from Eq. (202). Obviously, this constraint can be fulfilled only if $\mathrm{w} > 1/3$, since $1/\rho_0$ must be strictly positive. This is a necessary but not sufficient condition. More precisely, for weights in the range $1/3 < \mathrm{w} \leq 1/2$, electron correlation should be strong enough such that $\rho_0 < 1$ or, equivalently,

$$\frac{U}{t} > \frac{2(1-\mathrm{w})}{3\mathrm{w}-1}. \tag{203}$$

Interestingly, if we introduce effective weight-dependent hopping $\tilde{t} = t(1-\mathrm{w})$ and on-site interaction $\tilde{U} = U(3\mathrm{w}-1)$ parameters, the condition in Eq. (203) can be rewritten as

$$\frac{2\tilde{t}}{\tilde{U}} < 1, \tag{204}$$

which resembles the usual definition of moderate (up to strong) electron correlation in lattices. Actually, in the commonly used equiensemble case ($\mathrm{w} = 1/2$), the effective ratio matches the physical one $2t/U$.

Finally, when $\mathrm{w} \leq 1/3$, the SAHF energy becomes convex at $\theta = 0$ and, since it can have two additional (say $\theta_+ > 0$ and $\theta_- = -\theta_+$) stationary points at most in the

Table 1 θ values minimizing the state-averaged Hartree–Fock (SAHF) energy of the symmetric Hubbard dimer. See text for further details

	$\mathtt{w} \leqslant 1/3$	$\mathtt{w} > 1/3$
$\lvert\rho_0\rvert > 1$	$\theta = 0$	$\theta = 0$
$\lvert\rho_0\rvert \leqslant 1$	$\theta = 0$	$\theta \neq 0$

range $-\pi/2 \le \theta \le \pi/2$ [see Eqs. (194) and (201)], the symmetric solution has to be the absolute minimum. The different possible scenarios are summarized in Table 1.

Note that a connection can be made with the (spin) unrestricted energy of a symmetric dimer (e.g., the H_2 molecule in the minimal basis) [137]. For sufficiently large bond distances, the restricted solution becomes a saddle point and two unrestricted lower-in-energy solutions emerge. Accordingly, the constraint in Eq. (203) is compatible with a reduction of the t value featuring an increasing bond length. Still, even in the strictly correlated $t \rightarrow 0$ limit, $\theta = 0$ remains the global minimum for weights in the range $0 \le \mathtt{w} \le 1/3$.

In order to illustrate the above discussion, trial SAHF energies are plotted as functions of the rotation angle θ in the top panel of Fig. 1 for the strongly correlated $U/t = 3.5$ dimer and various biensemble weight values. As expected, the symmetric $\theta = 0$ solution gives systematically the lowest ensemble energy as long as $\mathtt{w} \le 1/3$. When $\mathtt{w} > 1/3$, two scenarios can be observed. For example, when $\mathtt{w} = 0.35$, which gives $2(1-\mathtt{w})/(3\mathtt{w}-1) = 26 \gg U/t$, the dimer is not strongly correlated enough to break the symmetry and $\theta = 0$ is still the global minimum. However, for the larger $\mathtt{w} = 0.475$ weight value [$2(1-\mathtt{w})/(3\mathtt{w}-1) = 2.47 < U/t$ in this case] or in the commonly used equiensemble case [$\mathtt{w} = 0.5$ and $2(1-\mathtt{w})/(3\mathtt{w}-1) = 2 < U/t$], the energy has the expected double-well shape, thus leading to two degenerate (non-zero) minima, both corresponding to asymmetric solutions. Interestingly, in such situations, the popular eDMHF approach always favors the symmetric solution, as shown in the bottom panel of Fig. 1.

4.3.3 Single SAHF Ensemble *v*-Representability Issue

We show in Fig. 2 the potential-ensemble-density maps

$$\Delta v \rightarrow n^{\mathtt{w}}(\theta_{\min}(\Delta v)) \tag{205}$$

generated from Eq. (193) and

$$\theta_{\min}(\Delta v) = \arg\min_{\theta}\left\{E^{\mathtt{w}}_{\text{approx.}}(\Delta v, \theta)\right\}, \tag{206}$$

at both eDMHF and SAHF levels of approximation for the fixed $U/t = 3.5$ interaction strength value. Various scenarios are illustrated, in particular those where broken symmetry SAHF solutions are obtained when the dimer is symmetric (see "Symmetric Case"). As we will see, what happens in the symmetric case can play a crucial role in the density-functional description of the asymmetric dimer.

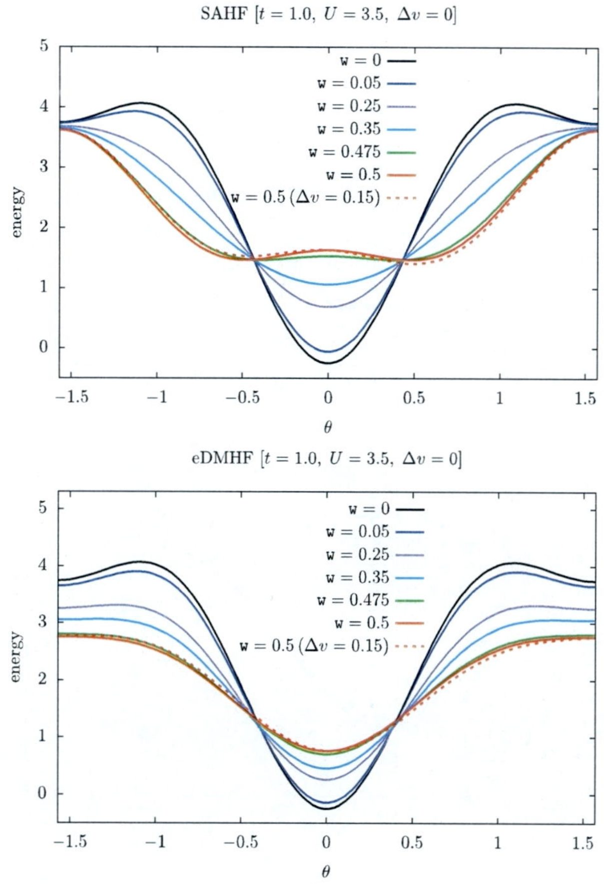

Fig. 1 Trial state-averaged Hartree–Fock (HF) (SAHF) (top panel) and ensemble density matrix HF (eDMHF) (bottom panel) energies of the symmetric Hubbard dimer plotted as functions of the orbital rotation angle θ for $U/t = 3.5$ and various ensemble weight values. In the equiensemble ($\mathtt{w} = 0.5$) case, results are also shown for a slightly asymmetric ($\Delta v/t = +0.15$) dimer, for analysis purposes. In the latter case, a non-degenerate (positive) minimizing angle is recovered at the SAHF level (red dashed curve in the top panel), unlike in the strictly symmetric $\Delta v = 0$ case. See text for further details

In cases where $\mathtt{w} \leq 1/3$, or $\mathtt{w} > 1/3$ and $2(1-\mathtt{w})/(3\mathtt{w}-1) > U/t$, both approximations give smooth density profiles. We note in passing the one-to-one correspondence between potentials and ensemble densities, as expected from the concavity of the eDMHF and SAHF energies (see "Concavity of Approximate Energies and Lieb Maximization"). However, when $\mathtt{w} > 1/3$ and $2(1-\mathtt{w})/(3\mathtt{w}-1) < U/t$, the SAHF density profile exhibits a discontinuity at $\Delta v = 0$, unlike the eDMHF one. This step in density can be interpreted as follows. If $\mathtt{w} > 1/3$ and the constraint of Eq. (203) is fulfilled, as $\Delta v \rightarrow 0^{\pm}$, we will recover the SAHF biensemble solution $\hat{\gamma}_{\pm} \equiv (1-\mathtt{w})|\Phi_0^{\pm}\rangle\langle\Phi_0^{\pm}| + \mathtt{w}|\Phi_1^{\pm}\rangle\langle\Phi_1^{\pm}|$, where $\Phi_I^{\pm} \equiv \Phi_I(\theta_{\pm})$ and $\theta_{\pm}$ are the minimizing angles associated to the broken-symmetry orbitals. Any slight deviation from $\Delta v = 0$ will favor one of these solutions, depending on its sign, as illustrated in the top panel

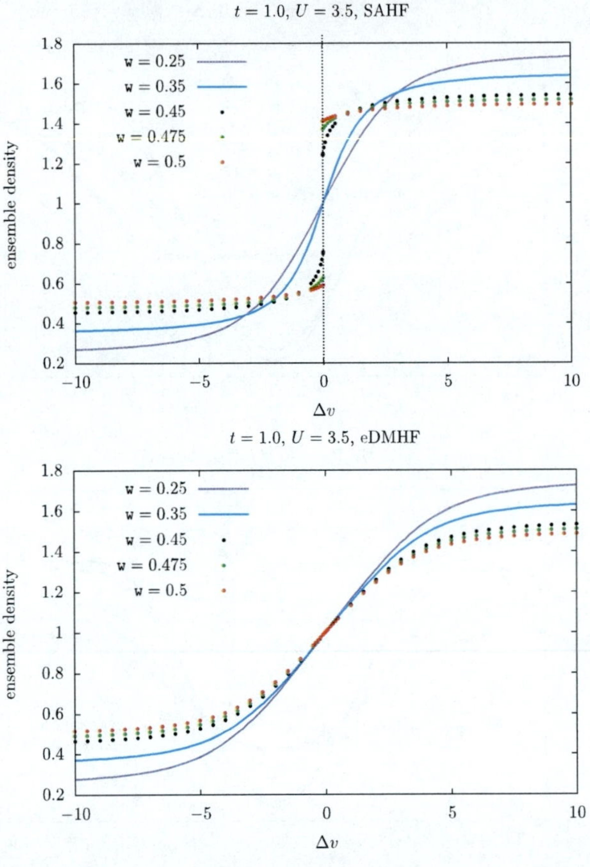

Fig. 2 Potential-ensemble-density maps generated for the Hubbard dimer at the SAHF (top panel) and eDMHF (bottom panel) levels of approximation for various ensemble weight values and $U/t = 3.5$. See text for further details

of Fig. 1 (see the "$\Delta v = +0.15$" curve, which exhibits, in the equiensemble case, a single absolute minimum in the vicinity of θ_+). Note that, in eDMHF, the minimizing angle simply passes through $\theta = 0$ when the potential changes from $\Delta v = 0^-$ to $\Delta v = 0^+$ [see the "$\Delta v = +0.15$" curve in the bottom panel of Fig. 1], and no discontinuity is observed in the density profile. The step in density observed in SAHF covers the density range $n_- \leq n \leq n_+$, where

$$n\pm \equiv 1 + (1 - \mathtt{w})\sin(2\theta_\pm) = 1 \pm (1 - \mathtt{w})\sqrt{1 - \rho_0^2}. \tag{207}$$

In the equiensemble case ($\mathtt{w} = 0.5$), we have $n\pm = 1.0 \pm 0.410326$, as readily seen from the top panel of Fig. 2. We keep many digits for analysis purposes (see the bottom panel of Fig. 3). It is important to stress that none of the densities in the range $n_- < n < n_+$, which includes the a priori simple symmetric $n = 1$ case, can be represented by a single SAHF ensemble. This severe v-representability issue, which

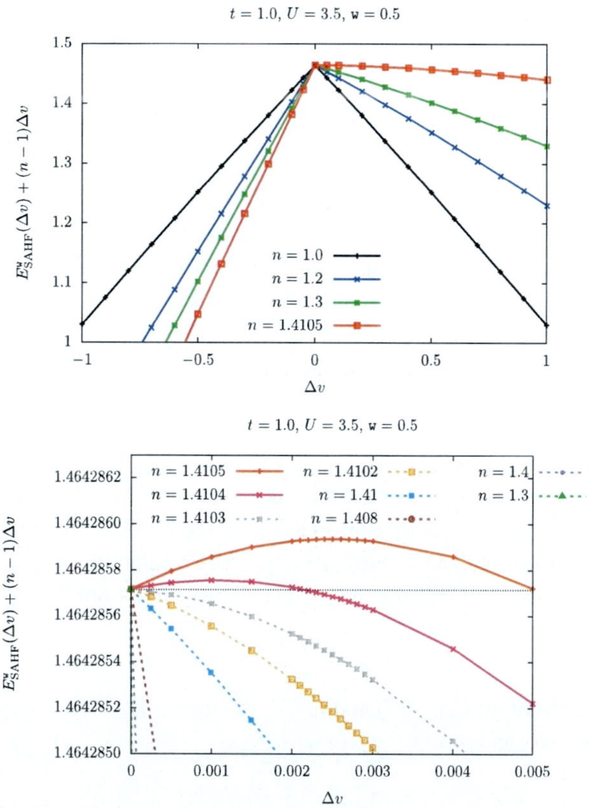

Fig. 3 Top: To-be-maximized Lieb's potential functional (see Refs. [96, 99] and the main text for further details) evaluated at the equal-weight SAHF level of approximation and plotted as a function of Δv for various input densities n. Bottom: Zoom around $\Delta v = 0$ for densities inside and outside the single SAHF-ensemble representability domain. See text for further details

would deserve further investigation at the ab initio level, for example, in the stretched H_2 molecule, might be used as an argument for promoting the eDMHF approach over the SAHF one in practical calculations. A counter-argument is of course the presence of GI errors in eDMHF. At the formal level, the v-representability issue can be solved easily as follows. As illustrated in Fig. 3, Lieb's maximization of Eq. (174) systematically returns $\Delta v = 0$ for input densities in the range $n_- \leq n \leq n_+$. As soon as the input density leaves this interval, a non-zero maximizing potential is obtained [see the bottom panel of Fig. 3]. If we exploit the strict degeneracy of the broken-symmetry solutions, we can write, for $\Delta v = 0$ [see Eq. (181)],

$$
\begin{aligned}
E^{\mathtt{w}}_{\mathrm{SAHF}}(\theta_+) &= E^{\mathtt{w}}_{\mathrm{SAHF}}(\theta_-) \\
&= (1-\alpha)E^{\mathtt{w}}_{\mathrm{SAHF}}(\theta_-) + \alpha E^{\mathtt{w}}_{\mathrm{SAHF}}(\theta_+) \\
&= \mathrm{Tr}\left[\hat{\gamma}(\alpha)\left(\hat{\mathcal{T}} + \hat{U}\right)\right],
\end{aligned}
\tag{208}
$$

where $0 \leq \alpha \leq 1$ and

$$\hat{\gamma}(\alpha) := (1-\alpha)\hat{\gamma}_{-} + \alpha\hat{\gamma}_{+} \tag{209}$$

is the convex combination of the two degenerate SAHF ensemble density matrix operators. The density (on site 0) of the resulting "ensemble of ensembles" reads as

$$n(\alpha) = \mathrm{Tr}\left[\hat{\gamma}(\alpha)\hat{n}_0\right] = (1-\alpha)n_{-} + \alpha n_{+}, \tag{210}$$

and, as readily seen, it can vary continuously from n_{-} to n_{+}. The generalization of this approach to the ab initio theory is provided in Eqs. (178) and (179).

4.4 Exact Self-Consistent eDFT Based on SAHF

Let us continue with the general ab initio SAHF-based formulation of eDFT that we left at the end of section "Concavity of Approximate Energies and Lieb Maximization". As already mentioned, the SAHF universal density functional $F^{\mathbf{w}}_{\mathrm{SAHF}}[n]$ that is defined in Eq. (177) is an approximation to the universal GOK functional $F^{\mathbf{w}}[n]$. As readily seen from Eq. (177), the former misses all correlation effects. These effects can actually be introduced into the theory as a density-functional complement:

$$\tilde{E}^{\mathbf{w}}_{\mathrm{c}}[n] := F^{\mathbf{w}}[n] - F^{\mathbf{w}}_{\mathrm{SAHF}}[n]. \tag{211}$$

As a result, according to Eqs. (10) and (177), the exact ensemble energy can be calculated variationally as follows:

$$\begin{aligned} E^{\mathbf{w}} = \min_{n}\Bigg\{ & \min_{\{\Phi_I\}\stackrel{\mathbf{w}}{\rightarrow} n}\left\{\sum_I \mathtt{w}_I\left\langle\Phi_I\middle|\hat{T}+\hat{W}_{\mathrm{ee}}\middle|\Phi_I\right\rangle\right\} + \tilde{E}^{\mathbf{w}}_{\mathrm{c}}[n] \\ & + \int \mathrm{d}\mathbf{r}\, v_{\mathrm{ext}}(\mathbf{r})n(\mathbf{r})\Bigg\}, \end{aligned} \tag{212}$$

or, equivalently,

$$E^{\mathbf{w}} = \min_{n}\left\{\min_{\{\Phi_I\}\stackrel{\mathbf{w}}{\rightarrow} n}\left\{\sum_I \mathtt{w}_I\left\langle\Phi_I\middle|\hat{H}\middle|\Phi_I\right\rangle + \tilde{E}^{\mathbf{w}}_{\mathrm{c}}\left[\sum_I \mathtt{w}_I n_{\Phi_I}\right]\right\}\right\}, \tag{213}$$

thus leading to the final expression

$$E^{\mathbf{w}} = \min_{\{\Phi_I\}}\left\{\sum_I \mathtt{w}_I\left\langle\Phi_I\middle|\hat{H}\middle|\Phi_I\right\rangle + \tilde{E}^{\mathbf{w}}_{\mathrm{c}}\left[\sum_I \mathtt{w}_I n_{\Phi_I}\right]\right\}. \tag{214}$$

By differentiating the density-functional correlation energy with respect to any (orbital rotation) variational parameter κ_{pq} [see Appendix 3],

$$\frac{\partial}{\partial \kappa_{pq}}\left(\tilde{E}_{\mathrm{c}}^{\mathbf{w}}\left[\sum_I \mathrm{w}_I n_{\Phi_I(\kappa)}\right]\right) = \int \mathrm{d}\mathbf{r}\, \left.\frac{\delta \tilde{E}_{\mathrm{c}}^{\mathbf{w}}[n]}{\delta n(\mathbf{r})}\right|_{n=\sum_I \mathrm{w}_I n_{\Phi_I(\kappa)}} \times \frac{\partial}{\partial \kappa_{pq}}\left(\sum_I \mathrm{w}_I n_{\Phi_I(\kappa)}(\mathbf{r})\right), \tag{215}$$

we realize that the minimizing orbitals in Eq. (214), from which the single-configuration wave functions that reproduce the exact ensemble density $n^{\mathbf{w}}$ are constructed (we denote them $\tilde{\Phi}_I^{\mathbf{w}}$ for convenience), fulfill SAHF-like self-consistent equations [see Eq. (166)] where the density-functional correlation potential $\delta \tilde{E}_{\mathrm{c}}^{\mathbf{w}}[n]/\delta n(\mathbf{r})\big|_{n=\sum_I \mathrm{w}_I n_{\tilde{\Phi}_I^{\mathbf{w}}}}$ is simply added to the local external one.

In practice, the quantities of interest are usually the excitation energies and, more generally, the ground- and excited-state energy levels (which are needed, for example, for geometry optimizations). According to Eqs. (9) and (214), and the Hellmann–Feynman theorem, the latter can be evaluated, in principle exactly, as follows:

$$E_I = \langle \hat{H} \rangle_{\tilde{\Phi}_I^{\mathbf{w}}} + \int \mathrm{d}\mathbf{r}\, \tilde{\mathcal{V}}_{\mathrm{c}}^{\mathbf{w}}(\mathbf{r}) n_{\tilde{\Phi}_I^{\mathbf{w}}}(\mathbf{r}) + \sum_{J>0} (\delta_{IJ} - \mathrm{w}_J) \left.\frac{\partial \tilde{E}_{\mathrm{c}}^{\mathbf{w}}[n]}{\partial \mathrm{w}_J}\right|_{n=n^{\mathbf{w}}}, \tag{216}$$

where the following relation has been used,

$$\sum_{J>0} (\delta_{IJ} - \mathrm{w}_J)\left(n_{\tilde{\Phi}_J^{\mathbf{w}}}(\mathbf{r}) - n_{\tilde{\Phi}_0^{\mathbf{w}}}(\mathbf{r})\right) = n_{\tilde{\Phi}_I^{\mathbf{w}}}(\mathbf{r}) - n^{\mathbf{w}}(\mathbf{r}), \tag{217}$$

and $\tilde{\mathcal{V}}_{\mathrm{c}}^{\mathbf{w}} \equiv \tilde{\mathcal{V}}_{\mathrm{c}}^{\mathbf{w}}[n^{\mathbf{w}}]$ is the LZ-shifted [91, 104] ensemble correlation-only density-functional potential:

$$\tilde{\mathcal{V}}_{\mathrm{c}}^{\mathbf{w}}[n](\mathbf{r}) = \frac{\delta \tilde{E}_{\mathrm{c}}^{\mathbf{w}}[n]}{\delta n(\mathbf{r})} + \frac{\tilde{E}_{\mathrm{c}}^{\mathbf{w}}[n] - \int \mathrm{d}\mathbf{r}\, \frac{\delta \tilde{E}_{\mathrm{c}}^{\mathbf{w}}[n]}{\delta n(\mathbf{r})} n(\mathbf{r})}{\int \mathrm{d}\mathbf{r}\, n(\mathbf{r})}. \tag{218}$$

Interestingly, an expression similar to that of Eq. (216) has been derived in the context of GOK-DFT [91, 103] (see also "Individual Correlations Within Ensembles: An Exact Construction") where, unlike in the present case, a local ensemble exchange potential was used.

4.5 Connection with Practical Hybrid eDFT Calculations

By analogy with conventional (ground-state) hybrid functionals, we can modify the SAHF-based functional of Eq. (176) as follows, in order to combine rigorously a fraction λ of SAHF exchange energy with a weighted sum of (approximate) individual density-functional exchange energies, thus leading to another (Hx-only) approximation to the GOK functional:

$$F^{\mathbf{w}}[n] \approx F^{\mathbf{w},\lambda}_{\text{hybrid}}[n] = \max_{v} \left\{ \mathscr{E}^{\mathbf{w},\lambda}_{\text{hybrid}}[v] - \int d\mathbf{r}\, v(\mathbf{r})n(\mathbf{r}) \right\}, \tag{219}$$

where

$$\mathscr{E}^{\mathbf{w},\lambda}_{\text{hybrid}}[v] = \min_{\{\Phi_I\}} \left\{ \sum_I \mathtt{w}_I \left[\left\langle \hat{T} + \lambda \hat{W}_{\text{ee}} + \hat{V} \right\rangle_{\Phi_I} + (1-\lambda) E_{\text{Hx}}[n_{\Phi_I}] \right] \right\}, \tag{220}$$

and $E_{\text{Hx}}[n]$ is the regular ground-state Hx density functional of KS-DFT. The above approximate ensemble energy can be computed from SAHF routines simply by scaling the individual non-local exchange potentials and then adding to each of them the complementary fraction $(1-\lambda)$ of individual local density-functional exchange potential. As readily seen from Eq. (220), in the present scheme, the total ensemble Hx energy remains GI-free, even in the limiting $\lambda = 0$ case. On that basis, an exact hybrid eDFT can be derived along the lines of "Exact self-consistent eDFT based on SAHF" by considering the following in-principle-exact decomposition of the universal GOK functional (where correlation is described with an ensemble density functional):

$$F^{\mathbf{w}}[n] = F^{\mathbf{w},\lambda}_{\text{hybrid}}[n] + (1-\lambda)\Delta\tilde{E}^{\mathbf{w},\lambda}_{\text{x}}[n] + \tilde{E}^{\mathbf{w},\lambda}_{\text{c}}[n]. \tag{221}$$

A density-functional correction $\Delta\tilde{E}^{\mathbf{w},\lambda}_{\text{x}}[n]$ to the ensemble exchange energy must in principle be introduced, since each individual exchange energy $E_{\text{x}}[n_{\Phi_I}]$ is evaluated, for both ground- and excited-state densities, as a ground-state exchange energy. This correction is usually neglected in practical calculations [75]. This observation would actually hold also for approximate ensemble correlation energies that are constructed from the regular ground-state correlation functional $E_{\text{c}}[n]$ of KS-DFT (see below and "State-of-the-Art Ensemble Correlation DFAs and Beyond"). Note finally the λ-dependence of both $\Delta\tilde{E}^{\mathbf{w},\lambda}_{\text{x}}[n]$ and $\tilde{E}^{\mathbf{w},\lambda}_{\text{c}}[n]$ functionals in Eq. (221). It originates from the fact that the ensemble xc energy is now evaluated from the (SAHF-like) λ -dependent single-configuration wave functions that reproduce the desired density n.

Finally, as discussed in further detail in "Individual Correlations Within Ensembles: An Exact Construction", it is quite common to construct weight-dependent ensemble density-functional correlation energies by recycling the ground-state correlation functional as follows [71],

$$\tilde{E}^{\mathbf{w},\lambda}_{\text{c}}\left[\sum_I \mathtt{w}_I n_{\Phi_I}\right] \approx \sum_I \mathtt{w}_I E_{\text{c}}[n_{\Phi_I}]. \tag{222}$$

This standard approximation, which is referred to as *ground-state individual correlations* (GS-ic) scheme in the following, can also be made formally exact within the present formalism. Indeed, once we have introduced the following (approximate) potential-functional energy

$$\mathcal{E}^{\mathbf{w},\lambda}_{\text{GS-ic}}[v] = \min_{\{\Phi_I\}} \left\{ \sum_I \mathtt{w}_I \Big(\langle \hat{T} + \lambda \hat{W}_{\text{ee}} + \hat{V} \rangle_{\Phi_I} + (1-\lambda) E_{\text{Hx}}[n_{\Phi_I}] + E_{\text{c}}[n_{\Phi_I}] \Big) \right\}, \tag{223}$$

and the subsequent density functional

$$F^{\mathbf{w},\lambda}_{\text{GS-ic}}[n] = \max_{v} \left\{ \mathcal{E}^{\mathbf{w},\lambda}_{\text{GS-ic}}[v] - \int \mathbf{dr}\, v(\mathbf{r}) n(\mathbf{r}) \right\}, \tag{224}$$

we only need to consider the alternative (but still exact) partitioning of the GOK functional

$$F^{\mathbf{w}}[n] = F^{\mathbf{w},\lambda}_{\text{GS-ic}}[n] + (1-\lambda)\Delta\check{E}^{\mathbf{w},\lambda}_{\text{x}}[n] + \Delta\check{E}^{\mathbf{w},\lambda}_{\text{c}}[n], \tag{225}$$

where complementary density-functional corrections to both exchange and (GS-ic) correlation energies have been introduced. Note that, in practice, these corrections are simply neglected [75]. While SAHF is expected to provide, through its orbital dependence, a proper description of the ensemble exchange energy, modeling the correlation density-functional correction $\Delta\check{E}^{\mathbf{w},\lambda}_{\text{c}}[n]$ remains a necessary and challenging task that has attracted too little attention until now. For that purpose, it is essential to have a deeper understanding of how individual correlation energies are connected to the ensemble one. This is the main focus of the next section.

5 Individual Correlations within Ensembles: An Exact Construction

While the previous section was dedicated to the description of orbital- and weight-dependent ensemble Hx energies, this last section deals with correlation effects in many-body ensembles. For convenience, we continue focusing on GOK ensembles but the discussion applies to other types of ensembles like, for example, N-centered [83] or thermal ones [76–78, 136]. We will work within the original GOK-DFT formalism [82], where a local multiplicative ensemble-density-functional Hxc potential is employed, but the discussion holds also when orbital-dependent exchange energies are employed (see "The Exact Hartree-Exchange Dilemma in eDFT").

5.1 State-of-the-Art Ensemble Correlation DFAs and Beyond

To the best of our knowledge, very few works have addressed the construction of weight-dependent ensemble correlation DFAs from first principles. We can essentially distinguish two different general strategies. In the first and most straightforward, which was introduced in Eq. (222) and which we referred to as GS-ic, the (weight-independent) ground-state correlation functional is recycled as follows,

$$E_{\rm c}^{\bf w}[n^{\bf w}] \overset{\text{GS-ic}}{\approx} \sum_I {\tt w}_I E_{\rm c}[n_{\Phi_I^{\bf w}}], \tag{226}$$

where, in the exact theory, the KS wave functions $\{\Phi_I^{\bf w}\}$ are expected to reproduce the true ensemble density $n^{\bf w}$.

More recently [79, 91], Loos and coworkers explored another path. They designed a first generation of weight-dependent ensemble LDA (eLDA) correlation functionals where the regular ground-state LDA functional $E_{\rm c}^{\rm LDA}[n] = \int {\rm d}{\bf r}\, n({\bf r})\epsilon_{\rm c}(n({\bf r}))$, which is based on the infinite uniform electron gas (UEG) model, is combined with the density-functional correlation excitation energies of a finite UEG (hence the acronym *f*LDA used below) as follows:

$$E_{\rm c}^{\bf w}[n] \overset{\text{eLDA}}{\approx} E_{\rm c}^{\rm LDA}[n] + \sum_{I>0} {\tt w}_I \left(\mathcal{E}_{{\rm c},I}^{f{\rm LDA}}[n] - \mathcal{E}_{{\rm c},I=0}^{f{\rm LDA}}[n] \right). \tag{227}$$

The individual correlation functional $\mathcal{E}_{{\rm c},I}^{f{\rm LDA}}[n] = \int {\rm d}{\bf r}\, n({\bf r})\epsilon_{{\rm c},I}^{f}(n({\bf r}))$ is constructed from the Ith state of the finite UEG:

$$N_f\, \epsilon_{{\rm c},I}^{f}(n) \equiv \left\langle \Psi_I(n) \middle| \hat{T} + \hat{W}_{\rm ee} \middle| \Psi_I(n) \right\rangle - \left\langle \Phi_I(n) \middle| \hat{T} + \hat{W}_{\rm ee} \middle| \Phi_I(n) \right\rangle, \tag{228}$$

where N_f is the number of electrons in the finite gas. If N_f is fixed, a parameterization of the correlation energy per particle $\epsilon_{{\rm c},I}^{f}(n)$ as a function of the uniform density n is obtained by varying the volume of the gas [91]. Note that, in a uniform system, there is no need to introduce weight dependencies into the interacting and non-interacting (ground- or excited-state) density-functional wave functions, unlike in the general definition of Eq. (25). Indeed, all the eigenstates of the (interacting or non-interacting) uniform gas have the same (uniform) density n, which then becomes the ensemble density of the gas, whatever the value of the ensemble weights:

$$\sum_I {\tt w}_I n_{\Psi_I(n)} = \sum_I {\tt w}_I n_{\Phi_I(n)} = n \sum_I {\tt w}_I = n. \tag{229}$$

Note also that, while the finite UEG allows for the incorporation of weight dependencies into the correlation functional, the use of a regular LDA correlation functional reduces finite-size errors. Refinements are possible, for example, by including a dependence in the Fermi hole curvature [138].

The strategies depicted in Eqs. (226) and (227) miss various correlation effects that we briefly review below. More insight will be given in the next subsections. Let us start with the GS-ic approximation. From the exact expression,

$$E_{\rm c}[n_{\Phi_I^{\bf w}}] = \left\langle \hat{T} + \hat{W}_{\rm ee} \right\rangle_{\Psi_0[n_{\Phi_I^{\bf w}}]} - \left\langle \hat{T} + \hat{W}_{\rm ee} \right\rangle_{\Phi_0[n_{\Phi_I^{\bf w}}]}, \tag{230}$$

where $\Psi_0[n]$ and $\Phi_0[n]$ are the interacting and KS non-interacting ground-state density-functional wave functions of regular KS-DFT, respectively, we immediately identify two sources of errors. The first is related to the fact that, as already mentioned in "Extraction of Individual State Properties", the individual KS density $n_{\Phi_I^{\bf w}}$ does not necessarily match the interacting individual one n_{Ψ_I}. This subtle point was highlighted recently by Gould and Pittalis [86, 139]. It induces what the authors referred to as density-driven (DD) correlation effects. Even if the true individual densities $\left\{n_{\Psi_I}\right\}$ (which can be extracted in principle exactly from the KS ensemble, as shown in Eq. (27) and Ref. [103]) were inserted into the expression of Eq. (230), we would still not recover the correct individual excited-state correlation energies simply because $\Psi_0[n_{\Psi_I}]$ will always be a ground-state wave function, even when n_{Ψ_I} is an excited-state density. The missing energy contribution is connected to the concept of state-driven (SD) correlation [86]. Interestingly, eLDA describes (approximately) SD correlations, as readily seen from Eq. (228). However, it completely misses DD correlations, simply because KS and interacting (ground- or excited-state) wave functions have the same density in a uniform system.

Even though the physical meaning of DD and SD correlations is rather clear, it is less obvious how their contributions to the total exact ensemble correlation energy should be defined mathematically [86, 103, 139, 140]. Addressing this fundamental question is of primary importance for the design of more accurate and systematically improvable ensemble correlation DFAs, which is probably the most challenging task in GOK-DFT. Up to now, we have discussed the concept of DD and SD correlations in the light of the GS-ic approximation [see Eqs. (226) and (230)]. We may actually wonder if a proper definition can be (or should be) given without referring explicitly to the GS correlation functional of KS-DFT. Indeed, the latter appears naturally in GOK-DFT only in the limiting $\mathbf{w} = 0$ case. Gould and Pittalis [86], and then Fromager [103], recently addressed this SD/DD ensemble correlation energy decomposition issue from that perspective. A detailed and complemented review of the two approaches is presented in the following.

5.2 Weight Dependence of the KS Wave Functions in GOK-DFT

Before proceeding with the extraction of individual correlation energies from the GOK-DFT ensemble energy, which is convenient for deriving in-principle-exact SD/DD decompositions [103], we would like to highlight the importance of weight dependencies in the KS wave functions. It might be surprising at first sight because the true ground and excited states of the system under study are of course weight-independent. We explain below, with a simple argument, why this cannot be the case in the KS ensemble.

Since the KS and true ensemble densities match for any set of weights $\mathbf{w}$, their derivatives with respect to the weights also match. Therefore, if we consider the ground-state $\mathbf{w} = 0$ limit of GOK-DFT, it comes

$$\frac{\partial}{\partial \mathtt{w}_J}\left(\sum_I \mathtt{w}_I n_{\Psi_I}(\mathbf{r})\right)\Bigg|_{\mathbf{w}=0} \overset{J>0}{=} \frac{\partial}{\partial \mathtt{w}_J}\left(\sum_I \mathtt{w}_I n_{\Phi_I^{\mathbf{w}}}(\mathbf{r})\right)\Bigg|_{\mathbf{w}=0}, \tag{231}$$

or, equivalently,

$$n_{\Psi_J}(\mathbf{r}) - n_{\Psi_0}(\mathbf{r}) = n_{\Phi_J}(\mathbf{r}) - n_{\Phi_0}(\mathbf{r}) + \left.\frac{\partial n_{\Phi_0^{\mathbf{w}}}(\mathbf{r})}{\partial \mathrm{w}_J}\right|_{\mathbf{w}=0}, \tag{232}$$

where $\{\Phi_I\}$ denote here the ground- and excited-state KS wave functions generated from a regular ground-state KS-DFT calculation. In KS-DFT, the density constraint applies to the ground state only, i.e., $n_{\Phi_0}(\mathbf{r}) = n_{\Psi_0}(\mathbf{r})$, not to the excited states. Thus, we obtain the exact individual excited-state density expression, which can be recovered from Eq. (27) when $\mathbf{w} = 0$,

$$n_{\Psi_J}(\mathbf{r}) - n_{\Phi_J}(\mathbf{r}) \overset{J>0}{=} \left.\frac{\partial n_{\Phi_0^{\mathbf{w}}}(\mathbf{r})}{\partial \mathrm{w}_J}\right|_{\mathbf{w}=0} \neq 0, \tag{233}$$

where we readily see that, in GOK-DFT, the KS wave functions (the ground-state one in the present case) are necessarily weight-dependent. This feature is central to the design of DD correlation energies [103], as discussed further in the following. We refer to Eq. (275) for an illustrative example (based on the prototypical Hubbard dimer) of weight-dependent KS ground-state density.

5.3 Extraction of Individual Correlation Energies

In this section we revisit the derivation of the individual energy levels in Eq. (36) in order to construct individual correlation energies within the ensemble under study. For that purpose, we start from the exact relation between individual and ensemble energies in Eq. (9), and the variational GOK-DFT ensemble energy expression of Eq. (15), thus leading to, according to the Hellmann–Feynman theorem,

$$\begin{aligned} E_J = {} & \sum_{I\geq 0} \mathrm{w}_I \left\langle \Phi_I^{\mathbf{w}} \middle| \hat{T} + \hat{V}_{\mathrm{ext}} \middle| \Phi_I^{\mathbf{w}} \right\rangle + E_{\mathrm{Hxc}}^{\mathbf{w}}[n^{\mathbf{w}}] \\ & + \sum_{I>0} (\delta_{IJ} - \mathrm{w}_I) \left[\left\langle \Phi_I^{\mathbf{w}} \middle| \hat{T} + \hat{V}_{\mathrm{ext}} \middle| \Phi_I^{\mathbf{w}} \right\rangle - \left\langle \Phi_0^{\mathbf{w}} \middle| \hat{T} + \hat{V}_{\mathrm{ext}} \middle| \Phi_0^{\mathbf{w}} \right\rangle \right] \\ & + \sum_{I>0} (\delta_{IJ} - \mathrm{w}_I) \left[\frac{\partial E_{\mathrm{Hxc}}^{\mathbf{w}}[n^{\mathbf{w}}]}{\partial \mathrm{w}_I} - \left.\frac{\partial E_{\mathrm{Hxc}}^{\xi}[n^{\xi,\mathbf{w}}]}{\partial \mathrm{w}_I}\right|_{\xi=\mathbf{w}} \right], \end{aligned} \tag{234}$$

or, equivalently,

$$\begin{aligned} E_J = {} & \left\langle \Phi_J^{\mathbf{w}} \middle| \hat{T} + \hat{V}_{\mathrm{ext}} \middle| \Phi_J^{\mathbf{w}} \right\rangle + E_{\mathrm{Hxc}}^{\mathbf{w}}[n^{\mathbf{w}}] \\ & + \sum_{I>0} (\delta_{IJ} - \mathrm{w}_I) \left[\frac{\partial E_{\mathrm{Hxc}}^{\mathbf{w}}[n^{\mathbf{w}}]}{\partial \mathrm{w}_I} - \left.\frac{\partial E_{\mathrm{Hxc}}^{\xi}[n^{\xi,\mathbf{w}}]}{\partial \mathrm{w}_I}\right|_{\xi=\mathbf{w}} \right], \end{aligned} \tag{235}$$

where, in analogy with Eq. (119), the following double-weight ensemble KS density has been introduced:

$$n^{\xi,\mathbf{w}}(\mathbf{r}) = \sum_{I\geq 0} \xi_I n_{\Phi_I^{\mathbf{w}}}(\mathbf{r}). \tag{236}$$

The last contribution (that is subtracted) on the right-hand side of Eq. (235) originates from the Hellmann–Feynman theorem. In other words, derivatives of the KS wave functions (and, therefore, of their densities) do not contribute to the derivatives of the total ensemble energy, because the latter is variational.

As shown in Refs. [91, 103], the Hx contribution to the individual Jth energy level reduces to the expectation value of the two-electron repulsion operator evaluated for the Jth KS state, as one would guess. Indeed, once we have realized that, for given weight values ξ, the ensemble KS potential that reproduces $n^{\xi,\mathbf{w}}$ is simply the one that reproduces the true ensemble density $n^{\mathbf{w}}$, we deduce from Eq. (24) that

$$E^{\xi}_{\mathrm{Hx}}[n^{\xi,\mathbf{w}}] = \sum_{K\geq 0} \xi_K \left\langle \Phi_K^{\mathbf{w}} \middle| \hat{W}_{\mathrm{ee}} \middle| \Phi_K^{\mathbf{w}} \right\rangle, \tag{237}$$

and, consequently,

$$\left.\frac{\partial E^{\xi}_{\mathrm{Hx}}[n^{\xi,\mathbf{w}}]}{\partial \mathtt{w}_I}\right|_{\xi=\mathbf{w}} = \sum_{K\geq 0} \mathtt{w}_K \frac{\partial \left\langle \Phi_K^{\mathbf{w}} \middle| \hat{W}_{\mathrm{ee}} \middle| \Phi_K^{\mathbf{w}} \right\rangle}{\partial \mathtt{w}_I}. \tag{238}$$

As a result, since $E^{\mathbf{w}}_{\mathrm{Hx}}[n^{\mathbf{w}}] = E^{\mathbf{w}}_{\mathrm{Hx}}[n^{\mathbf{w},\mathbf{w}}]$, it comes

$$\frac{\partial E^{\mathbf{w}}_{\mathrm{Hx}}[n^{\mathbf{w}}]}{\partial \mathtt{w}_I} - \left.\frac{\partial E^{\xi}_{\mathrm{Hx}}[n^{\xi,\mathbf{w}}]}{\partial \mathtt{w}_I}\right|_{\xi=\mathbf{w}} = \left\langle \Phi_I^{\mathbf{w}} \middle| \hat{W}_{\mathrm{ee}} \middle| \Phi_I^{\mathbf{w}} \right\rangle - \left\langle \Phi_0^{\mathbf{w}} \middle| \hat{W}_{\mathrm{ee}} \middle| \Phi_0^{\mathbf{w}} \right\rangle, \tag{239}$$

thus leading to the expected result:

$$\begin{aligned} &E^{\mathbf{w}}_{\mathrm{Hx}}[n^{\mathbf{w}}] + \sum_{I>0} (\delta_{IJ} - \mathtt{w}_I) \left[\frac{\partial E^{\mathbf{w}}_{\mathrm{Hx}}[n^{\mathbf{w}}]}{\partial \mathtt{w}_I} - \left.\frac{\partial E^{\xi}_{\mathrm{Hx}}[n^{\xi,\mathbf{w}}]}{\partial \mathtt{w}_I}\right|_{\xi=\mathbf{w}} \right] \\ &= \left\langle \Phi_J^{\mathbf{w}} \middle| \hat{W}_{\mathrm{ee}} \middle| \Phi_J^{\mathbf{w}} \right\rangle. \end{aligned} \tag{240}$$

We conclude from Eq. (235) that the energy levels can be evaluated exactly within GOK-DFT as follows:

$$E_J = \left\langle \Phi_J^{\mathbf{w}} \middle| \hat{H} \middle| \Phi_J^{\mathbf{w}} \right\rangle + E^{\mathbf{w}}_{\mathrm{c},J}[n^{\mathbf{w}}], \tag{241}$$

where the individual correlation energy of the Jth state is determined from the ensemble correlation density functional as

$$E^{\mathbf{w}}_{\mathrm{c},J}[n^{\mathbf{w}}] = E^{\mathbf{w}}_{\mathrm{c}}[n^{\mathbf{w}}] + \sum_{I>0} (\delta_{IJ} - \mathtt{w}_I) \left[\frac{\partial E^{\mathbf{w}}_{\mathrm{c}}[n^{\mathbf{w}}]}{\partial \mathtt{w}_I} - \left.\frac{\partial E^{\xi}_{\mathrm{c}}[n^{\xi,\mathbf{w}}]}{\partial \mathtt{w}_I}\right|_{\xi=\mathbf{w}} \right]. \tag{242}$$

In the following section, we will see how the concept of DD correlation emerges from Eq. (242), once it has been rewritten more explicitly in terms of individual densities.

5.4 Individual Correlations Versus Individual Components

According to the definition of the ensemble correlation functional in GOK-DFT [see Eq. (25)], the exact ensemble correlation energy can be decomposed as follows,

$$E_{\rm c}^{\rm w}[n^{\rm w}] = \sum_{J\geq 0} {\rm w}_J \mathcal{E}_{{\rm c},J}^{\rm w}[n^{\rm w}], \tag{243}$$

where the individual components read as

$$\mathcal{E}_{{\rm c},J}^{\rm w}[n^{\rm w}] = \left\langle \Psi_J \middle| \hat{T} + \hat{W}_{\rm ee} \middle| \Psi_J \right\rangle - \left\langle \Phi_J^{\rm w} \middle| \hat{T} + \hat{W}_{\rm ee} \middle| \Phi_J^{\rm w} \right\rangle. \tag{244}$$

Let us stress that these components do not match the individual correlation energies of Eq. (242). Indeed, unlike the latter [see Eq. (241)], they do not give access to the exact individual energy levels,

$$\begin{aligned} \left\langle \Phi_J^{\rm w} \middle| \hat{H} \middle| \Phi_J^{\rm w} \right\rangle + \mathcal{E}_{{\rm c},J}^{\rm w}[n^{\rm w}] &= \left\langle \Psi_J \middle| \hat{T} + \hat{W}_{\rm ee} \middle| \Psi_J \right\rangle + \int d\mathbf{r}\, v_{\rm ext}(\mathbf{r}) n_{\Phi_J^{\rm w}}(\mathbf{r}) \\ &\neq E_J, \end{aligned} \tag{245}$$

simply because the KS density $n_{\Phi_J^{\rm w}}$ does not match, in general, the true physical density n_{Ψ_J}. The concept of DD correlation, which was introduced recently by Gould and Pittalis [86], originates from this observation. The important property that the true individual correlation energies share with the individual correlation components is that both of them can be used to construct the total ensemble correlation energy, i.e.,

$$E_{\rm c}^{\rm w}[n^{\rm w}] = \sum_{J\geq 0} {\rm w}_J E_{{\rm c},J}^{\rm w}[n^{\rm w}]. \tag{246}$$

The above expression can be deduced from Eq. (242) and the fact that, for any $\{\Delta_I\}_{I>0}$,

$$\begin{aligned} \sum_{J\geq 0} {\rm w}_J \left(\sum_{I>0} (\delta_{IJ} - {\rm w}_I)\Delta_I \right) &= \sum_{I>0}\sum_{J\geq 0} \delta_{IJ}{\rm w}_J\Delta_I - \left(\sum_{J\geq 0} {\rm w}_J \right) \sum_{I>0} {\rm w}_I \Delta_I \\ &= \sum_{I>0} {\rm w}_I\Delta_I - \sum_{I>0} {\rm w}_I\Delta_I \\ &= 0. \end{aligned} \tag{247}$$

From now on we will substitute the decomposition of Eq. (246) for the more conventional one of Eq. (243). As shown in the following, with this change of paradigm, DD-type correlation energy contributions will naturally emerge from the derivation of a more explicit expression. Unlike in Ref. [86], the approach of Ref. [103], which is reviewed in the next section, does not require additional (state-specific) KS systems, thus avoiding formal issues such as the non-uniqueness of KS potentials for individual excited states or v-representability issues [103].

5.5 Density-Driven Ensemble Correlation Energy Expression

Let us now derive a more explicit expression for the ensemble correlation energy, on the basis of Eq. (246). We start with a simplification of the true individual correlation energy expression of Eq. (242), where the standard decomposition into components [see Eq. (243)] of the ensemble correlation energy will be employed. On the one hand, we will have

$$\frac{\partial E_{\mathrm{c}}^{\mathbf{w}}[n^{\mathbf{w}}]}{\partial \mathtt{w}_I} = \mathcal{E}_{\mathrm{c},I}^{\mathbf{w}}[n^{\mathbf{w}}] - \mathcal{E}_{\mathrm{c},0}^{\mathbf{w}}[n^{\mathbf{w}}] + \sum_{K\geq 0} \mathtt{w}_K \frac{\partial \mathcal{E}_{\mathrm{c},K}^{\mathbf{w}}[n^{\mathbf{w}}]}{\partial \mathtt{w}_I}, \tag{248}$$

where, according to Eq. (244),

$$\begin{aligned} \frac{\partial \mathcal{E}_{\mathrm{c},K}^{\mathbf{w}}[n^{\mathbf{w}}]}{\partial \mathtt{w}_I} &= -\frac{\partial}{\partial \mathtt{w}_I}\left[\left\langle \Phi_K^{\mathbf{w}} \middle| \hat{T} + \hat{W}_{\mathrm{ee}} \middle| \Phi_K^{\mathbf{w}} \right\rangle\right] \\ &= -2\left\langle \Phi_K^{\mathbf{w}} \middle| \hat{T} + \hat{W}_{\mathrm{ee}} \middle| \frac{\partial \Phi_K^{\mathbf{w}}}{\partial \mathtt{w}_I} \right\rangle. \end{aligned} \tag{249}$$

As readily seen from Eq. (249), the weight derivatives of the individual correlation components are evaluated solely from the KS wave functions and their (static) linear response to variations in the ensemble weights. The true interacting wave functions are not involved since, unlike the KS wave functions, they do not vary with the ensemble weights [see the comment that follows Eq. (25), and Eq. (244)]. Combining Eqs. (243), (248), and (249) leads to the following expression for the first two contributions in Eq. (242) to the true individual correlation energy:

$$\begin{aligned} &E_{\mathrm{c}}^{\mathbf{w}}[n^{\mathbf{w}}] + \sum_{I>0} (\delta_{IJ} - \mathtt{w}_I) \frac{\partial E_{\mathrm{c}}^{\mathbf{w}}[n^{\mathbf{w}}]}{\partial \mathtt{w}_I} \\ &= \mathcal{E}_{\mathrm{c},J}^{\mathbf{w}}[n^{\mathbf{w}}] - 2\sum_{I>0}\sum_{K\geq 0} (\delta_{IJ} - \mathtt{w}_I)\mathtt{w}_K \left\langle \Phi_K^{\mathbf{w}} \middle| \hat{T} + \hat{W}_{\mathrm{ee}} \middle| \frac{\partial \Phi_K^{\mathbf{w}}}{\partial \mathtt{w}_I} \right\rangle. \end{aligned} \tag{250}$$

On the other hand, according to Eq. (236),

$$\left.\frac{\partial E_{\mathrm{c}}^{\xi}[n^{\xi,\mathbf{w}}]}{\partial \mathtt{w}_I}\right|_{\xi=\mathbf{w}} = \int d\mathbf{r} \frac{\delta E_{\mathrm{c}}^{\mathbf{w}}[n^{\mathbf{w}}]}{\delta n(\mathbf{r})} \sum_{K\geq 0} \mathtt{w}_K \frac{\partial n_{\Phi_K^{\mathbf{w}}}(\mathbf{r})}{\partial \mathtt{w}_I}, \tag{251}$$

thus leading to [see Eq. (27)]

$$-\sum_{I>0} (\delta_{IJ} - \mathtt{w}_I) \left.\frac{\partial E_{\mathrm{c}}^{\xi}[n^{\xi,\mathbf{w}}]}{\partial \mathtt{w}_I}\right|_{\xi=\mathbf{w}} = \int d\mathbf{r} \frac{\delta E_{\mathrm{c}}^{\mathbf{w}}[n^{\mathbf{w}}]}{\delta n(\mathbf{r})} \left(n_{\Phi_J^{\mathbf{w}}}(\mathbf{r}) - n_{\Psi_J}(\mathbf{r})\right). \tag{252}$$

Finally, by combining Eqs. (242), (250), and (252), we recover the expression of Ref. [103] for the deviation of the true Jth individual correlation energy from the component $\mathcal{E}^{\mathbf{w}}_{c,J}[n^{\mathbf{w}}]$,

$$E^{\mathbf{w}}_{c,J}[n^{\mathbf{w}}] - \mathcal{E}^{\mathbf{w}}_{c,J}[n^{\mathbf{w}}] = -2\sum_{I>0}\sum_{K\geq 0}(\delta_{IJ} - \mathtt{w}_I)\mathtt{w}_K\left\langle \Phi^{\mathbf{w}}_K \middle| \hat{T} + \hat{W}_{\mathrm{ee}} \middle| \frac{\partial \Phi^{\mathbf{w}}_K}{\partial \mathtt{w}_I} \right\rangle + \int d\mathbf{r}\, \frac{\delta E^{\mathbf{w}}_{c}[n^{\mathbf{w}}]}{\delta n(\mathbf{r})}\left(n_{\Phi^{\mathbf{w}}_J}(\mathbf{r}) - n_{\Psi_J}(\mathbf{r})\right), \tag{253}$$

thus leading [see Eq. (244)] to the following exact expression for individual correlation energies:

$$\begin{aligned} E^{\mathbf{w}}_{c,J}[n^{\mathbf{w}}] = {}& \left\langle \Psi_J \middle| \hat{T} + \hat{W}_{\mathrm{ee}} \middle| \Psi_J \right\rangle - \left\langle \Phi^{\mathbf{w}}_J \middle| \hat{T} + \hat{W}_{\mathrm{ee}} \middle| \Phi^{\mathbf{w}}_J \right\rangle \\ & - 2\sum_{I>0}\sum_{K\geq 0}(\delta_{IJ} - \mathtt{w}_I)\mathtt{w}_K\left\langle \Phi^{\mathbf{w}}_K \middle| \hat{T} + \hat{W}_{\mathrm{ee}} \middle| \frac{\partial \Phi^{\mathbf{w}}_K}{\partial \mathtt{w}_I} \right\rangle \\ & + \int d\mathbf{r}\, \frac{\delta E^{\mathbf{w}}_{c}[n^{\mathbf{w}}]}{\delta n(\mathbf{r})}\left(n_{\Phi^{\mathbf{w}}_J}(\mathbf{r}) - n_{\Psi_J}(\mathbf{r})\right). \end{aligned} \tag{254}$$

The above expression is a key result of Ref. [103] which, as we will see, allows us to explore in-principle-exact SD/DD correlation energy decompositions.

Let us now analyze the different contributions on the right-hand side of Eq. (254). While, on the first line, the bare Jth correlation energy component is recovered, the additional terms on the second and third lines ensure that the external potential energy is evaluated with the correct true density (see Eqs. (241) and (245), and the supplementary material of Ref. [103]). Interestingly, in the summation (in K) over all the states that belong to the ensemble [see the second line of Eq. (254)], one may separate the contribution of the state under consideration (i.e., the Jth state) from the others, thus defining an individual SD correlation energy:

$$\begin{aligned} E^{\mathbf{w},\mathrm{SD}}_{c,J}[n^{\mathbf{w}}] = {}& \left\langle \Psi_J \middle| \hat{T} + \hat{W}_{\mathrm{ee}} \middle| \Psi_J \right\rangle - \left\langle \Phi^{\mathbf{w}}_J \middle| \hat{T} + \hat{W}_{\mathrm{ee}} \middle| \Phi^{\mathbf{w}}_J \right\rangle \\ & - 2\mathtt{w}_J\sum_{I>0}(\delta_{IJ} - \mathtt{w}_I)\left\langle \Phi^{\mathbf{w}}_J \middle| \hat{T} + \hat{W}_{\mathrm{ee}} \middle| \frac{\partial \Phi^{\mathbf{w}}_J}{\partial \mathtt{w}_I} \right\rangle. \end{aligned} \tag{255}$$

The above definition, which was denoted $\overline{\mathrm{SD}}$ in Ref. [103] (the "overline" notation is dropped in the present work, for simplicity), differs substantially from the definition of Gould and Pittalis [86]. In the latter, an additional state-specific KS wave function, which is expected to reproduce the true individual density of the state under consideration, is introduced. In this case, the name "state-driven" means that the correlation energy is evaluated from interacting and non-interacting wave functions which share the same density. Here, no additional KS wave function is introduced, which is obviously appealing from a computational point of view. One possible criticism about the definition in Eq. (255) is its arbitrariness. Indeed, we may opt for a more density-based definition, in the spirit of what Gould and Pittalis proposed, by introducing, for example, the following auxiliary wave functions:

$$\overline{\Phi}_J^{\mathbf{w}} = \Phi_J^{\mathbf{w}} + \sum_{I>0}\sum_{K\geq 0}\sqrt{|\delta_{IJ} - \mathtt{w}_I|\mathtt{w}_K}\left(\operatorname{sgn}(\delta_{IJ} - \mathtt{w}_I)\,\Phi_K^{\mathbf{w}} + \frac{\partial \Phi_K^{\mathbf{w}}}{\partial \mathtt{w}_I}\right). \tag{256}$$

Note that, in the ground-state $\mathbf{w} = 0$ limit, $\overline{\Phi}_0^{\mathbf{w}}$ reduces to the conventional KS wave function $\Phi_0^{\mathbf{w}=0}$ of KS-DFT. What might be interesting in the (somehow artificial) construction of the above individual auxiliary KS states is the possibility it gives to recover, like in the Gould-Pittalis approach [86], all the KS contributions (to the individual correlation energy) that appear on the first two lines of Eq. (254) from a single expectation value, thus generating, on the other hand, (several) additional terms that should ultimately be removed:

$$\begin{aligned}
&\left\langle \overline{\Phi}_J^{\mathbf{w}} \middle| \hat{T} + \hat{W}_{\text{ee}} \middle| \overline{\Phi}_J^{\mathbf{w}} \right\rangle \\
&= \left\langle \Phi_J^{\mathbf{w}} \middle| \hat{T} + \hat{W}_{\text{ee}} \middle| \Phi_J^{\mathbf{w}} \right\rangle + 2\sum_{I>0}\sum_{K\geq 0}(\delta_{IJ} - \mathtt{w}_I)\mathtt{w}_K\left\langle \Phi_K^{\mathbf{w}} \middle| \hat{T} + \hat{W}_{\text{ee}} \middle| \frac{\partial \Phi_K^{\mathbf{w}}}{\partial \mathtt{w}_I} \right\rangle \\
&+ \ldots
\end{aligned} \tag{257}$$

Moreover, according to Eq. (27), we recover (among other terms) the correct physical density:

$$\begin{aligned}
\left\langle \overline{\Phi}_J^{\mathbf{w}} |\hat{n}(\mathbf{r})| \overline{\Phi}_J^{\mathbf{w}} \right\rangle &= n_{\Phi_J^{\mathbf{w}}}(\mathbf{r}) + \sum_{I>0}\sum_{K\geq 0}(\delta_{IJ} - \mathtt{w}_I)\mathtt{w}_K \frac{\partial n_{\Phi_K^{\mathbf{w}}}(\mathbf{r})}{\partial \mathtt{w}_I} + \ldots \\
&= n_{\Psi_J}(\mathbf{r}) + \ldots
\end{aligned} \tag{258}$$

On that basis, we could argue that the first two lines on the right-hand side of Eq. (254) should be interpreted as a SD correlation energy, while the third line would correspond to the missing DD correlation energy. The issue with such a decomposition is that the individual DD correlation energies would then cancel out in the weighted sum:

$$\begin{aligned}
&\sum_{J\geq 0}\mathtt{w}_J \int d\mathbf{r}\,\frac{\delta E_{\text{c}}^{\mathbf{w}}[n^{\mathbf{w}}]}{\delta n(\mathbf{r})}\left(n_{\Phi_J^{\mathbf{w}}}(\mathbf{r}) - n_{\Psi_J}(\mathbf{r})\right) \\
&= \int d\mathbf{r}\,\frac{\delta E_{\text{c}}^{\mathbf{w}}[n^{\mathbf{w}}]}{\delta n(\mathbf{r})}\left(n^{\mathbf{w}}(\mathbf{r}) - n^{\mathbf{w}}(\mathbf{r})\right) = 0,
\end{aligned} \tag{259}$$

which means that the ensemble DD correlation energy would be zero. As a result, with such an interpretation, the concept of DD correlation would not be of any help in the development of correlation DFAs for ensembles. This is of course not what we want [86]. In this respect, the definition in Eq. (255) is much more appealing. We will stick to this definition from now on. Consequently, the complementary ensemble DD correlation energy will read as [see Eqs. (243), (244), and (255)]

$$E_{\rm c}^{\rm w,DD}[n^{\rm w}] = E_{\rm c}^{\rm w}[n^{\rm w}] - \sum_{J\geq 0} \mathtt{w}_J E_{{\rm c},J}^{\rm w,SD}[n^{\rm w}] \tag{260}$$

$$= 2\sum_{J\geq 0} \mathtt{w}_J^2 \sum_{I>0} (\delta_{IJ} - \mathtt{w}_I) \left\langle \Phi_J^{\rm w} \middle| \hat{T} + \hat{W}_{\rm ee} \middle| \frac{\partial \Phi_J^{\rm w}}{\partial \mathtt{w}_I} \right\rangle. \tag{261}$$

Thus, we recover another key result of Ref. [103]. As readily seen from Eq. (261), the exact evaluation of the DD correlation energy only requires computing the static linear response of the KS wave functions that belong to the ensemble, which is computationally affordable.

Finally, at a more formal level, we note that the ensemble DD correlation energy expression of Eq. (261) is related to the individual components $f_J^{\rm w}[n] = \left\langle \Phi_J^{\rm w}[n] \middle| \hat{T} + \hat{W}_{\rm ee} \middle| \Phi_J^{\rm w}[n] \right\rangle$ of the Hx-only approximation to the universal GOK functional [see Eqs. (12), (14), and (24)]

$$f^{\rm w}[n] := T_{\rm s}^{\rm w}[n] + E_{\rm Hx}^{\rm w}[n] = \sum_K \mathtt{w}_K \left\langle \Phi_K^{\rm w}[n] \middle| \hat{T} + \hat{W}_{\rm ee} \middle| \Phi_K^{\rm w}[n] \right\rangle, \tag{262}$$

as follows,

$$E_{\rm c}^{\rm w,DD}[n^{\rm w}] = \sum_{J\geq 0} \mathtt{w}_J^2 \sum_{I>0} (\delta_{IJ} - \mathtt{w}_I) \frac{\partial f_J^{\rm w}[n^{\rm w}]}{\partial \mathtt{w}_I}. \tag{263}$$

We can even establish a direct connection with the total ensemble functional $f^{\rm w}[n]$, by analogy with Eq. (237). Indeed, since

$$f^{\xi}\left[n^{\xi,\rm w}\right] = \sum_K \xi_K \left\langle \Phi_K^{\rm w} \middle| \hat{T} + \hat{W}_{\rm ee} \middle| \Phi_K^{\rm w} \right\rangle, \tag{264}$$

it comes

$$f_J^{\rm w}\left[n^{\rm w}\right] = f^{\rm w}\left[n^{\rm w}\right] + \sum_{I>0} \left(\delta_{IJ} - \mathtt{w}_I\right) \left. \frac{\partial f^{\xi}\left[n^{\xi,\rm w}\right]}{\partial \xi_I} \right|_{\xi = \rm w}. \tag{265}$$

5.6 Application to the Hubbard Dimer

The importance of DD correlations, which was revealed in Ref. [86], has been confirmed in Ref. [103], in the weakly asymmetric and strongly correlated regime of the two-electron Hubbard dimer. We propose in the following to complete the study of Ref. [103] by exploring all asymmetry and correlation regimes, and comparing exact results with that of standard approximations.

5.6.1 Exact Theory and Approximations

The Hubbard dimer was introduced in the section "Insights from the Hubbard Dimer Model". In this simple model system, exact (two-electron and singlet) biensemble density-functional correlation energies $E_c^{w}(n)$ can be evaluated through Lieb maximizations [96, 99] from the following analytical expressions for the exact potential-functional interacting energies [84, 85, 136]:

$$E_I(\Delta v) = \frac{2U}{3} + \frac{2r}{3}\cos\left(\theta + \frac{2\pi}{3}(I+1)\right), \quad I = 0, 1, \tag{266}$$

where

$$r = \sqrt{3(4t^2 + \Delta v^2) + U^2} \tag{267}$$

and

$$\theta = \frac{1}{3}\arccos\left[\frac{9U(\Delta v^2 - 2t^2) - U^3}{r^3}\right]. \tag{268}$$

Exact ground- and excited-state densities are then obtained from the Hellmann–Feynman theorem [see Eq. (181)],

$$n_{\Psi_I} = 1 - \frac{\partial E_I(\Delta v)}{\partial \Delta v}, \tag{269}$$

and the cubic polynomial equation that the energies fulfill (see the Appendix of Ref. [96]). The resulting biensemble density reads as $n^{w} = (1-w)n_{\Psi_0} + w\, n_{\Psi_1}$. The Hx-only GOK functional introduced in Eq. (262) can be expressed analytically as follows [96],

$$\begin{aligned} f^{\xi}(n) &= T_s^{\xi}(n) + E_{Hx}^{\xi}(n) \\ &= -2t\sqrt{(1-\xi)^2 - (1-n)^2} + \frac{U}{2}\left[1 + \xi - \frac{(3\xi-1)(1-n)^2}{(1-\xi)^2}\right], \end{aligned} \tag{270}$$

so that, as shown in Appendix 4, the exact DD ensemble correlation energy reads explicitly as

$$\begin{aligned} E_c^{w,DD}(n^{w}) &= -w(n^{w}-1)(n_{\Psi_1}-1) \\ &\times \left[\frac{2t}{\sqrt{(1-w)^2 - (1-n^{w})^2}} + \frac{U(1+w)}{(1-w)^2}\right]. \end{aligned} \tag{271}$$

Since the KS excited-state density is always equal to 1 in this model [96], the prefactor $(n_{\Psi_1} - 1)$ matches the deviation in density of the true physical excited state from the KS one:

$$n_{\Psi_1} - 1 = n_{\Psi_1} - n_{\Phi_1^{w}}. \tag{272}$$

Then it becomes clear that $E_c^{w,DD}(n^w)$ is a DD correlation energy. We also see from the expression of Eq. (271) that, in the regular ground-state DFT limit ($w = 0$), this type of correlation disappears.

In the following, we test two common DFAs: a (weight-independent) ground-state density-functional description of the ensemble correlation energy (GS-ec) [96, 97],

$$E_c^w(n^w) \overset{\text{GS-ec}}{\approx} E_c(n^w), \tag{273}$$

where $E_c(n) = E_c^{w=0}(n)$, and the GS-ic approximation introduced in "State-of-the-Art Ensemble Correlation DFAs and Beyond", which, in the present case, gives

$$\begin{aligned} E_c^w(n^w) &\overset{\text{GS-ic}}{\approx} (1-w)E_c(n_{\Phi_0^w}) + wE_c(n_{\Phi_1^w}) \\ &= (1-w)E_c(n_{\Phi_0^w}) + wE_c(n=1). \end{aligned} \tag{274}$$

Note that the KS ground-state density $n_{\Phi_0^w}$ fulfills the constraint $(1-w)n_{\Phi_0^w} + wn_{\Phi_1^w} = n^w$, thus leading to

$$n_{\Phi_0^w} = \frac{n^w - w}{(1-w)} = n_{\Psi_0} + \frac{w(n_{\Psi_1} - 1)}{(1-w)}, \tag{275}$$

where we readily see that, in general, $n_{\Phi_0^w} \neq n_{\Psi_0}$. In the following, the local potential will be fixed. It is then analogous to the external potential of ab initio calculations, hence the notation $\Delta v = \Delta v_{ext}$.

5.6.2 Results and Discussion

Let us first discuss the strictly symmetric ($\Delta v_{ext} = 0$) dimer in which simple analytical expressions can be derived for both exact and approximate ensemble correlation energies. In this special case, ground- and excited-state densities are equal to 1, in both KS and physical interacting systems. Consequently, the ensemble DD correlation energy vanishes [see Eq. (271)]. Total and SD ensemble correlation energies are equal and vary linearly with the ensemble weight [99] as

$$E_c^w(n^w = 1) = 2t(1-w)\left(1 - \sqrt{1 + \frac{U^2}{16t^2}}\right) = (1-w)E_c(n=1), \tag{276}$$

with the positive slope $-E_c(n=1)$. As readily seen, the excited state exhibits no correlation effects in this density regime. Turning to the approximations, GS-ic erroneously assigns a (ground-state) correlation energy to the excited state [see Eq. (274)], thus leading to a total ensemble correlation energy that is wrong and equal to $E_c(n=1)$, like in GS-ec [see Eq. (273)]. In conclusion, in the symmetric case, both GS-ic and GS-ec approximations completely miss the weight dependence of the ensemble correlation energy.

We now discuss the performance of GS-ec and GS-ic in the asymmetric dimer. Results are shown in Fig. 4. The features described in the symmetric case are preserved

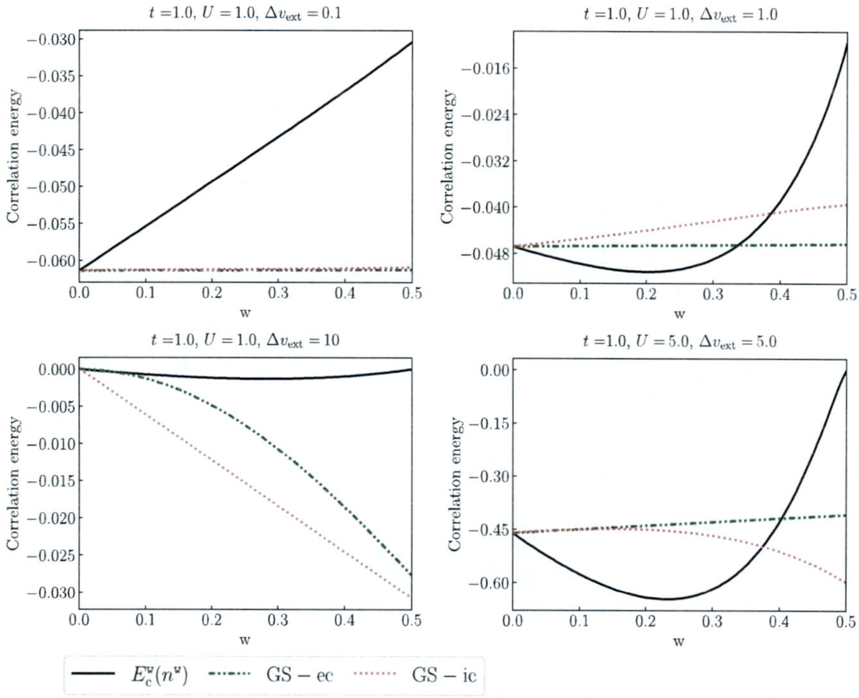

Fig. 4 Exact (solid black lines) and approximate ensemble correlation energies plotted as functions of the biensemble weight for the Hubbard dimer in various correlation and asymmetry regimes. See text for further details

in the weakly asymmetric regime (see the top left panel of Fig. 4). When the asymmetry is more pronounced, both exact and approximate ensemble correlation energies exhibit curvature. By construction, these energies all reduce to the same (ground-state) correlation energy when $\mathtt{w} = 0$. They differ substantially by their slope in the ground-state limit ($\mathtt{w} = 0$). Further insight into GS-ic, for example, is obtained from the following analytical expression:

$$\left.\frac{\partial E_c^{\mathtt{w}}(n^{\mathtt{w}})}{\partial \mathtt{w}}\right|_{\mathtt{w}=0} \overset{\text{GS-ic}}{\approx} E_c(n=1) - E_c(n=n_{\Psi_0}) + \left(n_{\Psi_1} - 1\right) \left.\frac{\partial E_c(n)}{\partial n}\right|_{n=n_{\Psi_0}}, \tag{277}$$

where $E_c(n=1) - E_c(n=n_{\Psi_0}) \leq 0$, as readily seen from Fig. 4 of Ref. [99]. Interestingly, in the strongly asymmetric $\Delta v_{\rm ext}/U \gg 1$ regime, the true ground- (n_{Ψ_0}) and excited-state (n_{Ψ_1}) densities tend to 2 and 1, respectively (see Fig. 1 of Ref. [96]). This is the situation where the slope in weight expressed in Eq. (277) reaches its maximum (in absolute value), thus inducing a large deviation from the exact slope, as shown in the bottom left panel of Fig. 4. At the GS-ec level of approximation, the situation is less critical, at least for small weight values. As readily seen from the following expression [see Eq. (273)],

$$\left.\frac{\partial E_c^{\mathtt{w}}(n^{\mathtt{w}})}{\partial \mathtt{w}}\right|_{\mathtt{w}=0} \overset{\text{GS-ec}}{\approx} \left(n_{\Psi_1} - n_{\Psi_0}\right) \left.\frac{\partial E_c(n)}{\partial n}\right|_{n=n_{\Psi_0}}, \tag{278}$$

when $\Delta v_{\text{ext}} \sim U$, the slope (at $\mathtt{w} = 0$) is relatively small since $n_{\Psi_1} \sim n_{\Psi_0}$ (see Fig. 1 of Ref. [96]). This is in agreement with the right panels of Fig. 4. Note that, in this regime, GS-ic can exhibit positive slopes (see the top right panel of Fig. 4). In this case, the density derivative contribution [second line of Eq. (277)], which is positive [96, 99], is no longer negligible and it (more than) compensates the negative correlation energy difference [first line of Eq. (277)]. When the asymmetry of the dimer is more pronounced (i.e., $\Delta v_{\text{ext}} \gg U$), the GS-ec slope (in weight) remains negligible, as shown in the bottom left panel of Fig. 4. Indeed, in this case, n_{Ψ_0} tends to 2. Moreover, since the ground-state correlation functional expands as follows in the weakly and strongly correlated regimes [99],

$$E_c(n) = E_c^{\mathtt{w}=0}(n) \overset{U/t \ll 1}{\approx} -\frac{U^2\left(1-(1-n)^2\right)^{5/2}}{16t}, \tag{279}$$

and

$$E_c(n) \overset{U/t \gg 1}{\approx} U\left[|n-1| - \frac{1}{2}(1+(n-1)^2)\right], \tag{280}$$

respectively, we immediately see that $\partial E_c(n)/\partial n \approx 0$ when n approaches 2, whether U/t is large or small. Note finally that, as already mentioned, in regimes where the asymmetry is weaker than the correlation, i.e., $\Delta v_{\text{ext}}/t < t/U < 1$ (see the top left panel of Fig. 4), the slopes obtained at $\mathtt{w} = 0$ with GS-ec and GS-ic are identical and relatively weak. This can now be understood from the expressions in Eqs. (277) and (278), and the fact that $\partial E_c(n)/\partial n\big|_{n=1} = 0$ [99], knowing that $n_{\Psi_0} \approx 1$ [96] in this case.

We see in Fig. 4 that the overall weight dependence of the exact ensemble correlation energy differs substantially from that of the GS-ec and GS-ic approximations. It is again instructive to look at the slope at $\mathtt{w} = 0$. It can be expressed exactly as follows:

$$\left.\frac{\partial E_c^{\mathtt{w}}(n^{\mathtt{w}})}{\partial \mathtt{w}}\right|_{\mathtt{w}=0} = \left(n_{\Psi_1} - n_{\Psi_0}\right) \left.\frac{\partial E_c(n)}{\partial n}\right|_{n=n_{\Psi_0}} + \left.\frac{\partial E_c^{\mathtt{w}}(n_{\Psi_0})}{\partial \mathtt{w}}\right|_{\mathtt{w}=0}, \tag{281}$$

where we readily see from Eq. (278) that GS-ec neglects the derivative in weight of the ensemble correlation density functional. As highlighted in Eq. (37) [see also Connection Between PPLB and N-Centered Pictures for a more detailed discussion in the context of charged excitations], the latter contribution is connected to the derivative discontinuity that the xc potential exhibits when an excited state is incorporated into the ensemble. Since, in the weakly correlated regime [99],

$$E_c^{\mathrm{w}}(n) \overset{U/t\ll 1}{\approx} -\frac{U^2\left((1-\mathrm{w})^2-(1-n)^2\right)^{3/2}}{16t(1-\mathrm{w})^2} \times\left[1+\frac{(1-n)^2}{(1-\mathrm{w})^2}\left(3-\frac{4(1-3\mathrm{w})^2}{(1-\mathrm{w})^2}\right)\right], \tag{282}$$

it comes

$$\left.\frac{\partial E_c^{\mathrm{w}}(n_{\Psi_0})}{\partial \mathrm{w}}\right|_{\mathrm{w}=0} \overset{U/t\ll 1}{\approx} \frac{U^2}{16t}\left(n_{\Psi_0}(2-n_{\Psi_0})\right)^{3/2}\left(1-12(n_{\Psi_0}-1)^2\right). \tag{283}$$

Therefore, as long as the ground state does not deviate too much from the symmetric $n_{\Psi_0}=1$ density profile, which is the case when $\Delta v_{\mathrm{ext}} \ll U$, the exact slope is not negligible, and it is positive. This is in agreement with the top left panel of Fig. 4. Interestingly, in this density regime, this feature is preserved when the strength of electron correlation increases (not shown). Indeed, in this case, the ensemble correlation functional reads as [99]

$$E_c^{\mathrm{w}}(n) \overset{U/t\gg 1,\,|n-1|\leq \mathrm{w}}{\approx} -\frac{U}{2}\left[(1-\mathrm{w})-\frac{(3\mathrm{w}-1)(n-1)^2}{(1-\mathrm{w})^2}\right], \tag{284}$$

$$\overset{n=n_{\Psi_0}}{\approx} \frac{U}{2}(\mathrm{w}-1), \tag{285}$$

thus leading to $\partial E_c^{\mathrm{w}}(n_{\Psi_0})/\partial \mathrm{w} \approx U/2$. When the dimer is strongly asymmetric, the ground-state density approaches 2 and, in this case [99],

$$E_c^{\mathrm{w}}(n) \overset{U/t\gg 1,\,\mathrm{w}\leq|n-1|\leq 1-\mathrm{w}}{\approx} U|n-1| -\frac{U}{2}\left[(1+\mathrm{w})-\frac{(3\mathrm{w}-1)(n-1)^2}{(1-\mathrm{w})^2}\right], \tag{286}$$

so that

$$\left.\frac{\partial E_c^{\mathrm{w}}(n_{\Psi_0})}{\partial \mathrm{w}}\right|_{\mathrm{w}=0} \overset{U/t\gg 1}{\approx} -\frac{U}{2}\left(1-(n_{\Psi_0}-1)^2\right), \tag{287}$$

thus leading to $\left.\partial E_c^{\mathrm{w}}(n_{\Psi_0})/\partial \mathrm{w}\right|_{\mathrm{w}=0} \approx 0$. As readily seen from Eq. (283), the same result is obtained in the weakly correlated regime. This is in complete agreement with the bottom left panel of Fig. 4. It also explains why the deviation of GS-ec from the exact result drastically reduces when Δv_{ext} increases for a fixed interaction strength U and relatively small weight values. Finally, in the particular case where $\Delta v_{\mathrm{ext}}=U$, the computed ground-state densities equal $n_{\Psi_0} \approx 1.30$ and $n_{\Psi_0} \approx 1.46$ in the moderately $U/t=1$ and strongly $U/t=5$ correlated regimes, respectively. As expected from Eqs. (283) and (287), the exact slope will be substantial and negative, which agrees with the right panels of Fig. 4.

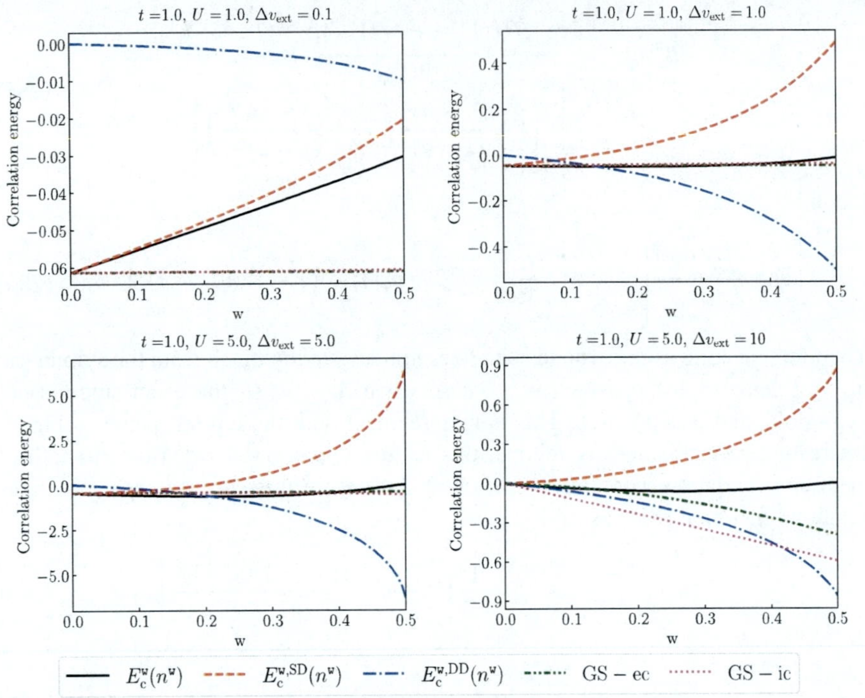

Fig. 5 Exact state-driven (SD)/density-driven (DD) decomposition of the ensemble correlation energy plotted as a function of the biensemble weight w in various asymmetry and correlation regimes. Comparison is made with the approximate GS-ec and GS-ic ensemble correlation energies, for analysis purposes. See text for further details

We now focus on the exact SD/DD decomposition of the ensemble correlation energy. Results are shown in Fig. 5 for various correlation and asymmetry regimes. In the ground-state limit, the slope of the DD ensemble correlation energy reads as [see Eq. (271)]

$$\left.\frac{\partial E_c^{\tt w,DD}(n^{\tt w})}{\partial {\tt w}}\right|_{{\tt w}=0} = -(n_{\Psi_0}-1)(n_{\Psi_1}-1)\left[\frac{2t}{\sqrt{1-(1-n_{\Psi_0})^2}}+U\right]. \tag{288}$$

Interestingly, when $\Delta v_{\rm ext} \approx U$ [$n_{\Psi_0} \approx n_{\Psi_1}$ in this case], the slope is nonzero (and negative), whether the dimer is strongly correlated or not, as seen from the top right and bottom left panels of Fig. 5. In the strongly correlated regime, the DD ensemble correlation energy essentially varies in w as [see Eq. (271)]

$$E_c^{\tt w,DD}(n^{\tt w}) \overset{U/t\gg 1}{\approx} -U(n_{\Psi_1}-1)(n_{\Psi_0}-1+{\tt w}(n_{\Psi_1}-n_{\Psi_0}))\frac{{\tt w}(1+{\tt w})}{(1-{\tt w})^2}, \tag{289}$$

which means that, when approaching the equiensemble ${\tt w}=1/2$ case, it decreases systematically with the ensemble weight (because of the term $(1-{\tt w})^2$ in the denominator), unlike the total ensemble correlation energy [see Eq. (284)]. As long as the dimer remains close to symmetric, which requires that Δv_{ext} reduces as U increases (see Fig. 1 of Ref. [96]), the numerator in Eq. (289) will be small enough such that DD correlations are at most equal to the total correlation energy. This feature is actually observed in the moderately correlated $U/t = 1$ regime (see the top left panel of Fig. 5). However, in asymmetric and strongly correlated regimes where $0 < \Delta v_{\text{ext}} \ll U$ ($n_{\Psi_0} \approx 1$ and $n_{\Psi_1} > 1$ in this case) or $\Delta v_{\text{ext}} \approx U$ (i.e., $n_{\Psi_0} \approx n_{\Psi_1} \approx 1.5$) [96], the numerator is no longer negligible and, consequently, the DD ensemble correlation energy is significantly lower than the total energy (see the bottom left panel of Fig. 5; see also Ref. [103]). In such cases, the complementary SD ensemble correlation energy can be large and positive. This may look unphysical at first sight but, if we return to the definition of Eq. (255), we see that the individual SD correlation energies are not guaranteed to be negative. The reason is that, unlike the total ensemble correlation energy, they are not evaluated variationally. Note finally that, when $\Delta v_{\text{ext}} > U \gg t$ ($n_{\Psi_0} \approx 2$ and $n_{\Psi_1} \approx 1$ in this case), the numerator in Eq. (289) will be relatively small, because of the $(n_{\Psi_1} - 1)$ prefactor, thus reducing the energy difference between total and DD correlations (see the bottom right panel of Fig. 5).

In summary, with the present SD/DD decomposition [see Eqs. (260) and (261)], both SD and DD correlation energies become relatively large (when compared to the total ensemble correlation energy), especially in the commonly used equiensemble case, and they mostly compensate when the Hubbard dimer has a pronounced asymmetry. This is clearly not a favorable situation for the development of DFAs, which was the initial motivation for introducing the SD/DD decomposition [86, 103]. The latter should definitely be implemented for atoms and diatomics, for example, in order to get further insight. In the case of stretched diatomics, the present study of the Hubbard dimer might be enlightening [141]. We should also stress that, in the asymmetric $\Delta v_{\text{ext}} = U$ case, standard GS-ic and GS-ec approximations give ensemble correlation energies that are of the same order of magnitude as the exact one, unlike the SD and DD correlation energies. As briefly mentioned in "State-of-the-Art Ensemble Correlation DFAs and Beyond", exploring alternative SD/DD decompositions that rely explicitly on GS-ic, which is maybe a better starting point, would be relevant in this respect. Work is currently in progress in this direction.

6 Conclusions and Perspectives

Despite the success of TD post-DFT approaches for the description of charged and neutral electronic excitations, various limitations (in terms of accuracy, computational cost, or physics) have motivated the development of time-independent alternatives in recent years. In the present review, we focused on GOK-DFT and N-centered eDFT, which are two flavors of eDFT for neutral and charged excitations, respectively. Their computational cost is essentially that of a standard KS-DFT calculation, because they both rely on self-consistent one-electron KS equations [see Eqs. (16) and (53)]. A major difference is that, in eDFT, the Hxc density functional is ensemble weight-dependent. This weight dependence is central in eDFT. It allows for the in-principle-exact extraction, from the KS ensemble, of individual (ground- and excited-state) energy levels [see Eqs. (36), (64) and (65)] and densities [Eq. (27)]. We have also shown that the infamous derivative discontinuity problem that must be addressed when computing fundamental (or optical) gaps [see Eqs. (30), (61) and (141)] can be bypassed, in principle exactly, via a relocation of the derivative discontinuity away from the system [see Eq. (143)] and a proper modeling of the ensemble weight dependence in the xc density functional. Recent progress in the design of weight-dependent xc DFAs has been reviewed. The pros and cons of using an (orbital-dependent) ensemble density matrix functional exchange energy or state-averaging individual exact exchange energies have been discussed in detail. We reveal in passing that, in the latter case, severe (but solvable) v-representability issues can occur when electron correlation becomes strong. Turning to the design of DFAs for ensemble correlation energies, state-of-the-art strategies have been discussed, in particular the combination of finite and infinite uniform electron gas models as well as the recycling of standard (ground-state) correlation DFAs through state-averaging. In the latter case, further improvements may emerge from the concept of density-driven correlation, which does not exist in ground-state KS-DFT. How to define the corresponding correlation energy mathematically is an open question to which we provided a tentative answer [see Eqs. (261) and (263)]. Test calculations on the Hubbard dimer reveal how difficult it is to have a definition that is both rigorous and useful for the development of approximations. Work is currently in progress in other (briefly discussed) directions. Even though it was not mentioned explicitly in the review, we would like to stress that current formulations of eDFT do not give a direct access to exact response properties such as oscillator strengths, Dyson orbitals, or non-adiabatic couplings. Extending Görling–Levy perturbation theory [142–144] to ensembles might be enlightening in this respect. We recently became aware of such an extension [145] for the computation of excitation energies within the DEC scheme [87, 101], which is an important first step. Nevertheless, a general quasi-degenerate density-functional perturbation theory based on ensembles, where individual energy levels and properties can be evaluated, is still highly desirable. Work is currently in progress in this direction.

In conclusion, we have summarized in the present review recent efforts of a growing community to put eDFT to the front of the scene. We highlighted several formal and practical aspects of the theory that should be investigated further in the near future in order to turn eDFT into a reliable and low-cost computational method for excited states.

Asymptotic Behavior of the xc Potential

Let us consider the simpler one-dimensional (1D) case in which the KS-PPLB equations read as

$$-\frac{1}{2}\frac{d^2\varphi_i^\alpha(x)}{dx^2} + \left(v_{\rm ext}(x) + v_{\rm Hxc}^\alpha(x)\right)\varphi_i^\alpha(x) = \varepsilon_i^\alpha\varphi_i^\alpha(x), \tag{290}$$

thus leading to

$$\frac{d^2\varphi_i^\alpha(x)}{dx^2} \underset{|x|\to+\infty}{=} -2\big(\varepsilon_i^\alpha - v_{\rm xc}^\alpha(\infty)\big)\varphi_i^\alpha(x), \tag{291}$$

where we used the limits $v_{\rm ext}(\infty) = v_{\rm H}^\alpha(\infty) = 0$. Note that $|\varphi_i^\alpha(x)|$ is expected to decay as $|x| \to +\infty$, which implies $-2\big(\varepsilon_i^\alpha - v_{\rm xc}^\alpha(\infty)\big) > 0$. Therefore,

$$\varphi_i^\alpha(x) \underset{|x|\to+\infty}{\sim} e^{-\sqrt{-2(\varepsilon_i^\alpha - v_{\rm xc}^\alpha(\infty))}|x|}, \tag{292}$$

and

$$n_{\hat{\gamma}_{\rm KS}^\alpha}(x) \underset{|x|\to+\infty}{\sim} \alpha|\varphi_N^\alpha(x)|^2 \sim \alpha\, e^{-2\sqrt{-2(\varepsilon_N^\alpha - v_{\rm xc}^\alpha(\infty))}|x|}. \tag{293}$$

In the true interacting system, the N-electron ground-state wave function Ψ_0^N fulfills

$$\begin{aligned}&\left[\sum_{i=1}^N\left(-\frac{1}{2}\frac{\partial^2}{\partial x_i^2} + v_{\rm ext}(x_i)\right) + \sum_{1\le i<j}^N w_{\rm ee}(|x_i - x_j|)\right]\Psi_0^N(x_1,\ldots,x_N)\\ &= E_0^N\Psi_0^N(x_1,\ldots,x_N),\end{aligned} \tag{294}$$

where $w_{\rm ee}(|x_i - x_j|)$ is a well-behaved two-electron repulsion energy in 1D. Let us consider the situation where $|x_1| \to +\infty$ while $x_2,\ldots,x_N$ remain in the region of the system, which corresponds to an ionization process in the ground state. Since $w_{\rm ee}(|x_1 - x_j|) \to 0$, the (to-be-antisymmetrized) wave function and its density can be rewritten as

$$\Psi_0^N(x_1,\ldots,x_N) \underset{|x_1|\to+\infty}{\sim} \varphi^{[N]}(x_1)\,\Psi_0^{N-1}(x_2,\ldots,x_N) \tag{295}$$

and

$$n_{\Psi_0^N}(x_1) \underset{|x_1|\to+\infty}{\sim} |\varphi^{[N]}(x_1)|^2, \tag{296}$$

respectively, where

$$\frac{{\rm d}^2\varphi^{[N]}(x_1)}{{\rm d}x_1^2} \underset{|x_1|\to+\infty}{\sim} -2\big(E_0^N - E_0^{N-1}\big)\varphi^{[N]}(x_1) = 2I_0^N\varphi^{[N]}(x_1), \tag{297}$$

thus leading to the explicit expression

$$\varphi^{[N]}(x) \underset{|x|\to+\infty}{\sim} e^{-\sqrt{2I_0^N}|x|}. \tag{298}$$

From the exact mapping of the ensemble PPLB density onto the KS system, we deduce from Eqs. (296) and (298) that

$$n_{\hat{\gamma}_{\rm KS}^\alpha}(x) \underset{|x|\to+\infty}{\sim} (1-\alpha)e^{-2\sqrt{2I_0^{N-1}}|x|} + \alpha\, e^{-2\sqrt{2I_0^N}|x|} \sim \alpha\, e^{-2\sqrt{2I_0^N}|x|}, \tag{299}$$

where we assumed that $E_g^{N-1} = I_0^{N-1} - I_0^N > 0$. Thus, we conclude from Eq. (293) that

$$I_0^N = -(\varepsilon_N^\alpha - v_{\rm xc}^\alpha(\infty)). \tag{300}$$

Any constant shift in the xc potential $v_{\rm xc}^\alpha(\mathbf{r})$ does not affect the above expression. Since, according to Janak's theorem, $I_0^N = -\varepsilon_N^\alpha$, the constant is imposed in PPLB and

$$v_{\rm xc}^\alpha(\infty) = 0. \tag{301}$$

We now turn to the left and right formulations of N-centered eDFT. We recall the shorthand notations $(\xi_-, 0) \overset{\text{notation}}{\equiv} \xi_-$ and $(0, \xi_+) \overset{\text{notation}}{\equiv} \xi_+$. When $\xi_+ > 0$, the right N-centered ensemble density, which is mapped onto a non-interacting KS ensemble, has the following asymptotic behavior [we just need to substitute $N+1$ for N in Eqs. (293), (296), and (298)],

$$n^{\xi_+}(x) \underset{|x|\to+\infty}{\sim} \xi_+ e^{-2\sqrt{2I_0^{N+1}}|x|} \tag{302}$$

$$\sim \xi_+ e^{-2\sqrt{-2(\varepsilon_{N+1}^{\xi_+} - v_{\rm xc}^{\xi_+}(\infty))}|x|}. \tag{303}$$

Similarly, for $\xi_- \geq 0$, we have

$$n^{\xi_-}(x) \underset{|x|\to+\infty}{\sim} \left(1 - \frac{(N-1)\xi_-}{N}\right) e^{-2\sqrt{2I_0^N}|x|} \tag{304}$$

$$\sim \left(1 - \frac{(N-1)\xi_-}{N}\right) e^{-2\sqrt{-2(\varepsilon_N^{\xi_-} - v_{\rm xc}^{\xi_-}(\infty))}|x|}. \tag{305}$$

Thus, we conclude that

$$A_0^N = I_0^{N+1} \overset{\xi_+>0}{=} -\varepsilon_{N+1}^{\xi_+} + v_{\rm xc}^{\xi_+}(\infty) \tag{306}$$

and

$$I_0^N \overset{\xi_-\geq 0}{=} -\varepsilon_N^{\xi_-} + v_{\rm xc}^{\xi_-}(\infty). \tag{307}$$

Derivation of the eDMHF Equations

For convenience, we use the following exponential parameterization of the single-configuration wave functions [126],

$$|\Phi_I\rangle \equiv |\Phi_I(\boldsymbol{\kappa})\rangle = e^{-\hat{\kappa}}|\overline{\Phi}_I^{\mathbf{w}}\rangle, \tag{308}$$

where $\boldsymbol{\kappa} \equiv \left\{\kappa_{pq}\right\}_{p<q}$ are the variational orbital rotation parameters and $\hat{\kappa}$ is the corresponding real singlet rotation quantum operator. The latter reads as follows in second quantization,

$$\hat{\kappa} = \sum_{p<q} \kappa_{pq}(\hat{E}_{pq} - \hat{E}_{qp}) = -\hat{\kappa}^{\dagger}, \tag{309}$$

where the index p refers to the orbital $\overline{\varphi}_p^{\mathtt{w}}$ and $\hat{E}_{pq} = \sum_{\tau=\uparrow,\downarrow} \hat{a}^{\dagger}_{p\tau}\hat{a}_{q\tau}$. Therefore, the eDMHF energy becomes a function of $\boldsymbol{\kappa}$,

$$E^{\mathbf{w}}_{\mathrm{eDMHF}}(\boldsymbol{\kappa}) = E_{\mathrm{HF}}\left(\mathbf{D}^{\mathbf{w}}(\boldsymbol{\kappa})\right), \tag{310}$$

where $\mathbf{D}^{\mathbf{w}}(\boldsymbol{\kappa}) = \sum_I \mathtt{w}_I \mathbf{D}^{\Phi_I(\boldsymbol{\kappa})}$ is a trial ensemble density matrix, and $E_{\mathrm{HF}}(\mathbf{D})$ is the conventional ground-state HF density matrix functional energy:

$$E_{\mathrm{HF}}(\mathbf{D}) = \sum_{mk} h_{mk}D_{mk} + \frac{1}{2}\sum_{klmn}\left(\langle mn|kl\rangle - \frac{1}{2}\langle mn|lk\rangle\right)D_{mk}D_{nl}. \tag{311}$$

By construction, the minimum is reached when $\boldsymbol{\kappa} = 0$, and we denote $\mathbf{D}^{\mathbf{w}} = \mathbf{D}^{\mathbf{w}}(\boldsymbol{\kappa} = 0)$. Note that

$$D_{pq}^{\overline{\Phi}_I^{\mathtt{w}}} = \left\langle \overline{\Phi}_I^{\mathtt{w}} \middle| \hat{E}_{pq} \middle| \overline{\Phi}_I^{\mathtt{w}} \right\rangle = \delta_{pq} n_p^I, \tag{312}$$

where the occupation number n_p^I is an integer, and

$$D^{\mathbf{w}}_{pq} = \sum_I \mathtt{w}_I D_{pq}^{\overline{\Phi}_I^{\mathtt{w}}} = \delta_{pq}\sum_I \mathtt{w}_I n_p^I = \delta_{pq}\theta_p^{\mathbf{w}}, \tag{313}$$

where $\theta_p^{\mathbf{w}}$ can be fractional. The stationarity condition that is fulfilled by the minimizing eDMHF orbitals can now be written explicitly as follows,

$$\left.\frac{\partial E^{\mathbf{w}}_{\mathrm{eDMHF}}(\boldsymbol{\kappa})}{\partial \kappa_{pq}}\right|_{\boldsymbol{\kappa}=0} = \sum_{rs} \left.\frac{\partial D^{\mathbf{w}}_{rs}(\boldsymbol{\kappa})}{\partial \kappa_{pq}}\right|_{\boldsymbol{\kappa}=0} \left.\frac{\partial E_{\mathrm{HF}}(\mathbf{D})}{\partial D_{rs}}\right|_{\mathbf{D}=\mathbf{D}^{\mathbf{w}}} = 0, \tag{314}$$

where

$$\begin{aligned}\frac{\partial E_{\mathrm{HF}}(\mathbf{D})}{\partial D_{rs}} &= h_{rs} + \frac{1}{2}\sum_{nl}\left(\langle rn|sl\rangle - \frac{1}{2}\langle rn|ls\rangle\right)D_{nl} \\ &\quad + \frac{1}{2}\sum_{mk}\left(\langle mr|ks\rangle - \frac{1}{2}\langle mr|sk\rangle\right)D_{mk} \\ &= h_{rs} + \sum_{nl}\left(\langle rn|sl\rangle - \frac{1}{2}\langle rn|ls\rangle\right)D_{nl} \\ &\equiv f_{rs}(\mathbf{D})\end{aligned} \tag{315}$$

is the conventional density matrix functional Fock operator matrix element, and

$$\begin{aligned}\left.\frac{\partial D^{\mathrm{w}}_{rs}(\boldsymbol{\kappa})}{\partial \kappa_{pq}}\right|_{\boldsymbol{\kappa}=0} &= \sum_I \mathtt{w}_I \langle [\hat{E}_{pq} - \hat{E}_{qp}, \hat{E}_{rs}] \rangle_{\Phi^{\mathrm{w}}_I} \\ &= \sum_I \mathtt{w}_I \left(\delta_{qr}\delta_{ps} n^I_p - \delta_{ps}\delta_{qr} n^I_q - \delta_{pr}\delta_{qs} n^I_q + \delta_{qs}\delta_{pr} n^I_p \right) \\ &= \left(\delta_{qr}\delta_{ps} + \delta_{pr}\delta_{qs}\right) \sum_I \mathtt{w}_I \left(n^I_p - n^I_q \right) \\ &= \left(\delta_{qr}\delta_{ps} + \delta_{pr}\delta_{qs}\right)(\theta^{\mathrm{w}}_p - \theta^{\mathrm{w}}_q),\end{aligned} \tag{316}$$

where we used the relation $[\hat{E}_{pq}, \hat{E}_{rs}] = \delta_{qr}\hat{E}_{ps} - \delta_{ps}\hat{E}_{rq}$ (see Ref. [126]) with Eqs. (312) and (313). If we denote $f^{\mathrm{w}}_{rs} = f_{rs}(\mathbf{D}^{\mathrm{w}})$, Eq. (314) can be written in a compact form as follows,

$$(\theta^{\mathrm{w}}_p - \theta^{\mathrm{w}}_q) \sum_{rs} \left(\delta_{qr}\delta_{ps} + \delta_{pr}\delta_{qs}\right) f^{\mathrm{w}}_{rs} = 0, \tag{317}$$

thus leading to the final result:

$$\left(\theta^{\mathrm{w}}_p - \theta^{\mathrm{w}}_q\right) f^{\mathrm{w}}_{qp} = 0. \tag{318}$$

Derivation of the SAHF Equations

We use the same parameterization as in Appendix 2, i.e.,

$$|\Phi_I\rangle \equiv |\Phi_I(\boldsymbol{\kappa})\rangle = e^{-\hat{\kappa}} |\tilde{\Phi}^{\mathrm{w}}_I\rangle, \tag{319}$$

where the indices $\{p\}$ in creation/annihilation operators (as well as in one- and two-electron integrals) now refer to the minimizing SAHF orbitals $\left\{ \tilde{\varphi}^{\mathrm{w}}_p \right\}$. The to-be-minimized SAHF energy can be expressed as follows,

$$E^{\mathrm{w}}_{\mathrm{SAHF}}(\boldsymbol{\kappa}) = \sum_I \mathtt{w}_I \left(\sum_{rs} h_{rs} D^{\Phi_I(\boldsymbol{\kappa})}_{rs} + E_{\mathrm{H}}[n_{\Phi_I(\boldsymbol{\kappa})}] + \mathcal{E}^I_{\mathrm{x}}\left[\mathbf{D}^{\Phi_I(\boldsymbol{\kappa})}\right] \right), \tag{320}$$

so that the stationarity condition reads as

$$
\begin{aligned}
&\left.\frac{\partial E^{\mathbf{w}}_{\mathrm{SAHF}}(\boldsymbol{\kappa})}{\partial \kappa_{pq}}\right|_{\boldsymbol{\kappa}=0} = 0\\
&= \sum_I \mathtt{w}_I \left[\sum_{rs}\left(h_{rs} + \left[v^{\mathbf{w}}_{\mathrm{x},I}\right]_{rs}\right)\frac{\partial D^{\Phi_I(\boldsymbol{\kappa})}_{rs}}{\partial \kappa_{pq}} + \int \mathrm{d}\mathbf{r}\, v_{\mathrm{H}}[n_{\tilde{\Phi}^{\mathbf{w}}_I}](\mathbf{r})\frac{\partial n_{\Phi_I(\boldsymbol{\kappa})}(\mathbf{r})}{\partial \kappa_{pq}}\right]_{\boldsymbol{\kappa}=0},
\end{aligned}
\tag{321}
$$

where $\left[v^{\mathbf{w}}_{\mathrm{x},I}\right]_{rs} \equiv \partial \mathcal{E}^I_{\mathrm{x}}[\mathbf{D}]/\partial D_{rs}\Big|_{\mathbf{D}=\mathbf{D}^{\tilde{\Phi}^{\mathbf{w}}_I}}$ and $v_{\mathrm{H}}[n](\mathbf{r}) = \delta E_{\mathrm{H}}[n]/\delta n(\mathbf{r})$. Note that the individual densities are recovered from the density matrices as follows,

$$
n_{\Phi_I(\boldsymbol{\kappa})}(\mathbf{r}) = \gamma^{\Phi_I(\boldsymbol{\kappa})}(\mathbf{r},\mathbf{r}) = \sum_{rs} \tilde{\varphi}^{\mathbf{w}}_r(\mathbf{r})\tilde{\varphi}^{\mathbf{w}}_s(\mathbf{r})D^{\Phi_I(\boldsymbol{\kappa})}_{rs}.
\tag{322}
$$

Therefore, if we use the notation

$$
\left\langle \tilde{\varphi}^{\mathbf{w}}_r \middle| \hat{h} + \hat{v}^{\mathbf{w}}_{\mathrm{Hx},I} \middle| \tilde{\varphi}^{\mathbf{w}}_s \right\rangle = h_{rs} + \int \mathrm{d}\mathbf{r}\, \tilde{\varphi}^{\mathbf{w}}_r(\mathbf{r}) v_{\mathrm{H}}[n_{\tilde{\Phi}^{\mathbf{w}}_I}](\mathbf{r})\tilde{\varphi}^{\mathbf{w}}_s(\mathbf{r}) + \left[v^{\mathbf{w}}_{\mathrm{x},I}\right]_{rs},
\tag{323}
$$

Eq. (321) can be rewritten in a compact form as follows,

$$
\sum_I \mathtt{w}_I \sum_{rs} \left\langle \tilde{\varphi}^{\mathbf{w}}_r \middle| \hat{h} + \hat{v}^{\mathbf{w}}_{\mathrm{Hx},I} \middle| \tilde{\varphi}^{\mathbf{w}}_s \right\rangle \left.\frac{\partial D^{\Phi_I(\boldsymbol{\kappa})}_{rs}}{\partial \kappa_{pq}}\right|_{\boldsymbol{\kappa}=0} = 0.
\tag{324}
$$

We conclude from Eq. (316) that

$$
\begin{aligned}
0 &= \sum_I \mathtt{w}_I\left(n^I_p - n^I_q\right)\sum_{rs}\left(\delta_{qr}\delta_{ps} + \delta_{pr}\delta_{qs}\right)\left\langle \tilde{\varphi}^{\mathbf{w}}_r \middle| \hat{h} + \hat{v}^{\mathbf{w}}_{\mathrm{Hx},I} \middle| \tilde{\varphi}^{\mathbf{w}}_s \right\rangle\\
&= 2\sum_I \mathtt{w}_I\left(n^I_p - n^I_q\right)\left\langle \tilde{\varphi}^{\mathbf{w}}_p \middle| \hat{h} + \hat{v}^{\mathbf{w}}_{\mathrm{Hx},I} \middle| \tilde{\varphi}^{\mathbf{w}}_q \right\rangle,
\end{aligned}
\tag{325}
$$

thus leading to the final result:

$$
(\theta^{\mathbf{w}}_p - \theta^{\mathbf{w}}_q)\left\langle \tilde{\varphi}^{\mathbf{w}}_p \middle| \hat{h} \middle| \tilde{\varphi}^{\mathbf{w}}_q \right\rangle + \sum_I \mathtt{w}_I\left(n^I_p - n^I_q\right)\left\langle \tilde{\varphi}^{\mathbf{w}}_p \middle| \hat{v}^{\mathbf{w}}_{\mathrm{Hx},I} \middle| \tilde{\varphi}^{\mathbf{w}}_q \right\rangle = 0.
\tag{326}
$$

Exact DD Ensemble Correlation Energy in the Hubbard Dimer

For convenience, we will use the following exact expression for the ensemble DD correlation energy:

$$
E^{\mathtt{w},\mathrm{DD}}_{\mathrm{c}}(n^{\mathtt{w}}) = -(1-\mathtt{w})^2\mathtt{w}\frac{\partial f^{\mathtt{w}}_0(n^{\mathtt{w}})}{\partial \mathtt{w}} + \mathtt{w}^2(1-\mathtt{w})\frac{\partial f^{\mathtt{w}}_1(n^{\mathtt{w}})}{\partial \mathtt{w}}.
\tag{327}
$$

The individual Hx-only GOK energies are extracted from the ensemble energy,

$$f^{\xi}(n) = -2t\sqrt{(1-\xi)^2 - (1-n)^2} + \frac{U}{2}\left[1 + \xi - \frac{(3\xi - 1)(1-n)^2}{(1-\xi)^2}\right], \quad (328)$$

as follows,

$$f_0^{\mathtt{w}}(n^{\mathtt{w}}) = f^{\mathtt{w}}(n^{\mathtt{w}}) - \mathtt{w}\left.\frac{\partial f^{\xi}\left(n^{\xi,\mathtt{w}}\right)}{\partial \xi}\right|_{\xi=\mathtt{w}}, \quad (329)$$

and

$$f_1^{\mathtt{w}}(n^{\mathtt{w}}) = f^{\mathtt{w}}(n^{\mathtt{w}}) + (1-\mathtt{w})\left.\frac{\partial f^{\xi}\left(n^{\xi,\mathtt{w}}\right)}{\partial \xi}\right|_{\xi=\mathtt{w}}, \quad (330)$$

where

$$n^{\xi,\mathtt{w}} = (1-\xi)n_{\Phi_0^{\mathtt{w}}} + \xi n_{\Phi_1^{\mathtt{w}}}. \quad (331)$$

Since $n_{\Phi_1^{\mathtt{w}}} = 1$ and

$$(1-\mathtt{w})n_{\Phi_0^{\mathtt{w}}} + \mathtt{w}n_{\Phi_1^{\mathtt{w}}} = n^{\mathtt{w}}, \quad (332)$$

or, equivalently,

$$n_{\Phi_0^{\mathtt{w}}} = \frac{n^{\mathtt{w}} - \mathtt{w}}{(1-\mathtt{w})}, \quad (333)$$

it comes

$$n^{\xi,\mathtt{w}} = (1-\xi)\frac{(n^{\mathtt{w}} - \mathtt{w})}{1-\mathtt{w}} + \xi \quad (334)$$

and

$$\left.\frac{\partial n^{\xi,\mathtt{w}}}{\partial \xi}\right|_{\xi=\mathtt{w}} = 1 - \frac{(n^{\mathtt{w}} - \mathtt{w})}{1-\mathtt{w}} = \frac{1-n^{\mathtt{w}}}{1-\mathtt{w}}. \quad (335)$$

From the weight derivative expression

$$\left.\frac{\partial f^{\xi}\left(n^{\xi,\mathtt{w}}\right)}{\partial \xi}\right|_{\xi=\mathtt{w}} = \left.\frac{\partial f^{\xi}(n^{\mathtt{w}})}{\partial \xi}\right|_{\xi=\mathtt{w}} + \left.\frac{\partial n^{\xi,\mathtt{w}}}{\partial \xi}\right|_{\xi=\mathtt{w}} \times \left.\frac{\partial f^{\mathtt{w}}(n)}{\partial n}\right|_{n=n^{\mathtt{w}}}, \quad (336)$$

where

$$\frac{\partial f^{\xi}(n)}{\partial \xi} = \frac{2t(1-\xi)}{\sqrt{(1-\xi)^2-(1-n)^2}} + \frac{U}{2}\left[1 - \frac{(n-1)^2(1+3\xi)}{(1-\xi)^3}\right] \tag{337}$$

and

$$\frac{\partial f^{\mathtt{w}}(n)}{\partial n} = \frac{2t(n-1)}{\sqrt{(1-\mathtt{w})^2-(1-n)^2}} + U\frac{(3\mathtt{w}-1)(1-n)}{(1-\mathtt{w})^2}, \tag{338}$$

thus leading to

$$\begin{aligned}\left.\frac{\partial f^{\xi}\left(n^{\xi,\mathtt{w}}\right)}{\partial \xi}\right|_{\xi=\mathtt{w}} = & \frac{2t(1-\mathtt{w})}{\sqrt{(1-\mathtt{w})^2-(1-n^{\mathtt{w}})^2}} \\ & - \frac{2t(1-n^{\mathtt{w}})^2}{(1-\mathtt{w})\sqrt{(1-\mathtt{w})^2-(1-n^{\mathtt{w}})^2}} \\ & + \frac{U}{2}\left[1 - \frac{(n^{\mathtt{w}}-1)^2(1+3\mathtt{w})}{(1-\mathtt{w})^3}\right] \\ & + U\frac{(3\mathtt{w}-1)(1-n^{\mathtt{w}})^2}{(1-\mathtt{w})^3},\end{aligned} \tag{339}$$

or, equivalently,

$$\left.\frac{\partial f^{\xi}\left(n^{\xi,\mathtt{w}}\right)}{\partial \xi}\right|_{\xi=\mathtt{w}} = \frac{2t\sqrt{(1-\mathtt{w})^2-(1-n^{\mathtt{w}})^2}}{(1-\mathtt{w})} + \frac{U}{2}\left[1 - \frac{3(1-n^{\mathtt{w}})^2}{(1-\mathtt{w})^2}\right], \tag{340}$$

it comes

$$f_0^{\mathtt{w}}(n^{\mathtt{w}}) = -\frac{2t\sqrt{(1-\mathtt{w})^2-(1-n^{\mathtt{w}})^2}}{(1-\mathtt{w})} + \frac{U}{2}\left[1 + \frac{(1-n^{\mathtt{w}})^2}{(1-\mathtt{w})^2}\right] \tag{341}$$

and

$$f_1^{\mathtt{w}}(n^{\mathtt{w}}) = U\left[1 - \frac{(1-n^{\mathtt{w}})^2}{(1-\mathtt{w})^2}\right]. \tag{342}$$

As a result,

$$\frac{\partial f_0^{\mathtt{w}}(n^{\mathtt{w}})}{\partial \mathtt{w}} = \frac{2t(n^{\mathtt{w}}-1)(n_{\Psi_1}-1)}{(1-\mathtt{w})^2\sqrt{(1-\mathtt{w})^2-(1-n^{\mathtt{w}})^2}} + U\frac{(n^{\mathtt{w}}-1)(n_{\Psi_1}-1)}{(1-\mathtt{w})^3} \tag{343}$$

and

$$\frac{\partial f_1^{\mathrm{w}}(n^{\mathrm{w}})}{\partial \mathrm{w}} = -\frac{2U(n^{\mathrm{w}}-1)(n_{\Psi_1}-1)}{(1-\mathrm{w})^3}, \tag{344}$$

which leads, according to Eq. (327), to the final compact expression

$$E_{\mathrm{c}}^{\mathrm{w,DD}}(n^{\mathrm{w}}) = -\mathrm{w}(n^{\mathrm{w}}-1)(n_{\Psi_1}-1) \times \left[\frac{2t}{\sqrt{(1-\mathrm{w})^2-(1-n^{\mathrm{w}})^2}} + \frac{U(1+\mathrm{w})}{(1-\mathrm{w})^2}\right]. \tag{345}$$

Acknowledgements E.F. would like to thank M. Levy, A. Savin, P.-F. Loos, T. Gould, M. J. P. Hodgson, and J. Wetherell for fruitful discussions as well as LabEx CSC (grant no.: ANR-10-LABX-0026-CSC) for funding. E.F. is also grateful to Trygve Helgaker for his introductory lectures on convex analysis and DFT for fractional electron numbers. The authors also thank ANR (CoLab project, grant no.: ANR-19-CE07-0024-02) for funding.

Declarations

Conflict of interest The authors declare that they have no conflict of interest.

References

1. Kohn W, Sham L (1965) Phys Rev A 140:1133. https://doi.org/10.1103/PhysRev.140.A1133
2. Bredas JL (2014) Mater Horizons 1(1):17
3. Perdew JP, Parr RG, Levy M, Balduz JL Jr (1982) Phys Rev Lett 49(23):1691. https://doi.org/10.1103/PhysRevLett.49.1691
4. Cohen AJ, Mori-Sánchez P, Yang W (2011) Chem Rev 112(1):289. https://doi.org/10.1021/cr200107z
5. Mori-Sánchez P, Cohen AJ, Yang W (2008) Phys Rev Lett 100(14):146401
6. Cohen AJ, Mori-Sánchez P, Yang W (2008) Science 321(5890):792
7. Cohen AJ, Mori-Sánchez P, Yang W (2008) Phys Rev B 77(11):115123. https://doi.org/10.1103/PhysRevB.77.115123
8. Stein T, Eisenberg H, Kronik L, Baer R (2010) Phys Rev Lett 105(26):266802
9. Zheng X, Cohen AJ, Mori-Sánchez P, Hu X, Yang W (2011) Phys Rev Lett 107(2):026403. https://doi.org/10.1103/PhysRevLett.107.026403
10. Perdew JP, Yang W, Burke K, Yang Z, Gross EKU, Scheffler M, Scuseria GE, Henderson TM, Zhang IY, Ruzsinszky A et al (2017) Proc Natl Acad Sci USA 114(11):2801
11. Imamura Y, Kobayashi R, Nakai H (2011) J Chem Phys 134(12):124113. https://doi.org/10.1063/1.3569030
12. Atalla V, Zhang IY, Hofmann OT, Ren X, Rinke P, Scheffler M (2016) Phys Rev B 94(3):035140. https://doi.org/10.1103/PhysRevB.94.035140
13. Stein T, Autschbach J, Govind N, Kronik L, Baer R (2012) J Chem Phys Lett 3(24):3740. https://doi.org/10.1021/jz3015937
14. Onida G, Reining L, Rubio A (2002) Rev Mod Phys 74(2):601. https://doi.org/10.1103/RevModPhys.74.601
15. Sottile F, Marsili M, Olevano V, Reining L (2007) Phys Rev B 76(16):161103
16. Bruneval F (2012) J Chem Phys 136(19):194107
17. Bruneval F, Marques MA (2012) J Chem Theory Comput 9(1):324
18. Jiang H (2015) Int J Quantum Chem 115(11):722

19. Pacchioni G (2015) Catal Lett 145(1):80
20. Ou Q, Subotnik JE (2016) J Phys Chem A 120(26):4514
21. Reining L (2018) WIREs Comput Mol Sci 8(3):e1344. https://doi.org/10.1002/wcms.1344
22. Runge E, Gross EK (1984) Phys Rev Lett 52(12):997
23. Casida M, Huix-Rotllant M (2012) Annu Rev Phys Chem 63:287
24. Vignale G (2008) Phys Rev A 77:062511
25. Vignale G (2011) Phys Rev A 83:046501
26. Fromager E, Knecht S, Aa Jensen HJ (2013) J Chem Phys 138:084101
27. Fuks JI, Maitra NT (2014) Phys Chem Chem Phys 16(28):14504. https://doi.org/10.1039/C4CP00118D
28. Dreuw A, Head-Gordon M (2004) J Am Chem Soc 126(12):4007
29. Maitra NT (2021) arXiv:2107.05600
30. Maitra NT, Zhang F, Cave RJ, Burke K (2004) J Chem Phys 120:5932
31. Filatov M, Lee S, Choi CH (2020) J Chem Theory Comput 16(7):4489. https://doi.org/10.1021/acs.jctc.0c00218
32. Filatov M, Lee S, Nakata H, Choi CH (2020) J Phys Chem A 124(38):7795
33. Chan GKL (1999) J Chem Phys 110(10):4710. https://doi.org/10.1063/1.478357
34. Kraisler E, Kronik L (2013) Phys Rev Lett 110(12):126403. https://doi.org/10.1103/PhysRevLett.110.126403
35. Gould T, Toulouse J (2014) Phys Rev A 90(5):050502. https://doi.org/10.1103/PhysRevA.90.050502
36. Kraisler E, Kronik L (2014) J Chem Phys 140(18):18A540
37. Kraisler E, Kronik L (2015) Phys Rev A 91(3):032504
38. Kraisler E, Schmidt T, Kümmel S, Kronik L (2015) J Chem Phys 143(10):104105
39. Li C, Zheng X, Cohen AJ, Mori-Sánchez P, Yang W (2015) Phys Rev Lett 114(5):053001. https://doi.org/10.1103/PhysRevLett.114.053001
40. Li C, Lu J, Yang W (2017) J Chem Phys 146(21):214109. https://doi.org/10.1063/1.4982951
41. Görling A (2015) Phys Rev B 91(24):245120. https://doi.org/10.1103/PhysRevB.91.245120
42. Baerends EJ (2017) Phys Chem Chem Phys 19(24):15639. https://doi.org/10.1039/C7CP02123B
43. Görling A (1999) Phys Rev A 59(5):3359. https://doi.org/10.1103/PhysRevA.59.3359
44. Levy M, Nagy A (1999) Phys Rev Lett 83(21):4361. https://doi.org/10.1103/PhysRevLett.83.4361
45. Ziegler T, Rauk A, Baerends EJ (1977) Theor Chim Acta 43(3):261
46. Gavnholt J, Olsen T, Engelund M, Schiøtz J (2008) Phys Rev B 78(7):075441. https://doi.org/10.1103/PhysRevB.78.075441
47. Kowalczyk T, Yost SR, Voorhis TV (2011) J Chem Phys 134(5):054128. https://doi.org/10.1063/1.3530801
48. Levi G, Ivanov AV, Jónsson H (2020) J Chem Theory Comput 16(11):6968. https://doi.org/10.1021/acs.jctc.0c00597
49. Hait D, Head-Gordon M (2020) J Chem Theory Comput 16(3):1699. https://doi.org/10.1021/acs.jctc.9b01127
50. Carter-Fenk K, Herbert JM (2020) J Chem Theory Comput 16(8):5067. https://doi.org/10.1021/acs.jctc.0c00502
51. Gilbert ATB, Besley NA, Gill PMW (2008) J Phys Chem A 112(50):13164. https://doi.org/10.1021/jp801738f
52. Ivanov AV, Levi G, Jónsson EO, Jónsson H (2021) J Chem Theory Comput https://doi.org/10.1021/acs.jctc.1c00157
53. Evangelista FA, Shushkov P, Tully JC (2013) J Phys Chem A 117(32):7378. https://doi.org/10.1021/jp401323d
54. Ramos P, Pavanello M (2018) J Chem Phys 148(14):144103. https://doi.org/10.1063/1.5018615
55. Roychoudhury S, Sanvito S, O'Regan DD (2020) Sci Rep 10(1):1. https://doi.org/10.1038/s41598-020-65209-4
56. Karpinski N, Ramos P, Pavanello M (2020) Phys Rev A 101(3):032510. https://doi.org/10.1103/PhysRevA.101.032510
57. Ziegler T, Seth M, Krykunov M, Autschbach J, Wang F (2009) J Chem Phys 130(15):154102. https://doi.org/10.1063/1.3114988

58. Cullen J, Krykunov M, Ziegler T (2011) Chem Phys 391(1):11. https://doi.org/10.1016/j.chemphys.2011.05.021
59. Ziegler T, Krykunov M, Cullen J (2012) J Chem Phys 136(12):124107. https://doi.org/10.1063/1.3696967
60. Krykunov M, Ziegler T (2013) J Chem Theory Comput 9(6):2761. https://doi.org/10.1021/ct300891k
61. Park YC, Senn F, Krykunov M, Ziegler T (2016) J Chem Theory Comput 12(11):5438. https://doi.org/10.1021/acs.jctc.6b00333
62. Ayers PW, Levy M, Nagy A (2012) Phys Rev A 85(4):042518. https://doi.org/10.1103/PhysRevA.85.042518
63. Ayers P, Levy M, Nagy A (2015) J Chem Phys 143(19):191101. https://doi.org/10.1063/1.4934963
64. Glushkov V, Levy M (2016) Computation 4(3):28. https://www.mdpi.com/2079-3197/4/3/28
65. Ayers P, Levy M, Nagy Á (2018) Theor Chem Acc 137(11):152. https://doi.org/10.1007/s00214-018-2352-7
66. Lieb EH (1983) Int J Quantum Chem 24(3):243
67. Ullrich CA, Kohn W (2001) Phys Rev Lett 87:093001. https://doi.org/10.1103/PhysRevLett.87.093001
68. Yang W, Zhang Y, Ayers PW (2000) Phys Rev Lett 84:5172. https://doi.org/10.1103/PhysRevLett.84.5172
69. Filatov M, Shaik S (1999) Chem Phys Lett 304(5–6):429. https://doi.org/10.1016/S0009-2614(99)00336-X
70. Kazaryan A, Heuver J, Filatov M (2008) J Phys Chem A 112(50):12980. https://doi.org/10.1021/jp8033837
71. Filatov M (2015) WIREs Comput Mol Sci 5:146. https://doi.org/10.1002/wcms.1209
72. Filatov M, Huix-Rotllant M, Burghardt I (2015) J Chem Phys 142:184104
73. Filatov M, Liu F, Martínez TJ (2017) J Chem Phys 147(3):034113. https://doi.org/10.1063/1.4994542
74. Liu F, Filatov M, Martínez TJ (2021) J Chem Phys 154(10):104108. https://doi.org/10.1063/5.0041389
75. Filatov M, Lee S, Choi CH (2021) J Chem Theory Comput 17(8):5123. https://doi.org/10.1021/acs.jctc.1c00479
76. Pittalis S, Proetto CR, Floris A, Sanna A, Bersier C, Burke K, Gross EKU (2011) Phys Rev Lett 107:163001. https://doi.org/10.1103/PhysRevLett.107.163001
77. Pribram-Jones A, Burke K (2016) Phys Rev B 93:205140. https://doi.org/10.1103/PhysRevB.93.205140
78. Pastorczak E, Gidopoulos NI, Pernal K (2013) Phys Rev A 87(6):062501. https://doi.org/10.1103/PhysRevA.87.062501
79. Marut C, Senjean B, Fromager E, Loos PF (2020) Faraday Discuss 224:402. https://doi.org/10.1039/D0FD00059K
80. Gould T, Kronik L, Pittalis S (2021) Phys Rev A 104:022803. https://doi.org/10.1103/PhysRevA.104.022803
81. Gross EKU, Oliveira LN, Kohn W (1988) Phys Rev A 37:2805. https://doi.org/10.1103/PhysRevA.37.2805
82. Gross EKU, Oliveira LN, Kohn W (1988) Phys Rev A 37:2809. https://doi.org/10.1103/PhysRevA.37.2809
83. Senjean B, Fromager E (2018) Phys Rev A 98(2):022513. https://doi.org/10.1103/PhysRevA.98.022513
84. Carrascal DJ, Ferrer J, Smith JC, Burke K (2015) J Phys Condens Matter 27(39):393001. http://stacks.iop.org/0953-8984/27/i=39/a=393001
85. Carrascal D, Ferrer J, Smith J, Burke K (2016) J Phys Condens Matter 29(1):019501
86. Gould T, Pittalis S (2019) Phys Rev Lett 123(1):016401. https://doi.org/10.1103/PhysRevLett.123.016401
87. Yang Zh, Pribram-Jones A, Burke K, Ullrich CA (2017) Phys Rev Lett 119(3):033003. https://doi.org/10.1103/PhysRevLett.119.033003
88. Oliveira LN, Gross EKU, Kohn W (1988) Phys Rev A 37(8):2821

89. Theophilou AK (1979) J Phys C Solid State Phys 12(24):5419. https://doi.org/10.1088/0022-3719/12/24/013
90. Deur K, Fromager E (2019) J Chem Phys 150(9):094106. https://doi.org/10.1063/1.5084312
91. Loos PF, Fromager E (2020) J Chem Phys 152(21):214101. https://doi.org/10.1063/5.0007388
92. Levy M (1979) Proc Natl Acad Sci USA 76(12):6062
93. Gould T, Pittalis S (2017) Phys Rev Lett 119(24):243001. https://doi.org/10.1103/PhysRevLett.119.243001
94. Gould T, Stefanucci G, Pittalis S (2020) Phys Rev Lett 125(23):233001. https://doi.org/10.1103/PhysRevLett.125.233001
95. Franck O, Fromager E (2014) Mol Phys 112:1684. https://doi.org/10.1080/00268976.2013.858191
96. Deur K, Mazouin L, Fromager E (2017) Phys Rev B 95(3):035120. https://doi.org/10.1103/PhysRevB.95.035120
97. Senjean B, Knecht S, Jensen HJAa, Fromager E (2015) Phys Rev A 92:012518. https://doi.org/10.1103/PhysRevA.92.012518
98. Gould T, Kronik L, Pittalis S (2018) J Chem Phys 148(17):174101
99. Deur K, Mazouin L, Senjean B, Fromager E (2018) Eur Phys J B 91:162. https://doi.org/10.1140/epjb/e2018-90124-7
100. Nagy A (1996) J Phys B At Mol Opt Phys 29(3):389. https://doi.org/10.1088/0953-4075/29/3/007
101. Sagredo F, Burke K (2018) J Chem Phys 149(13):134103. https://doi.org/10.1063/1.5043411
102. Senjean B, Hedegård ED, Alam MM, Knecht S, Fromager E (2016) Mol Phys 114(7–8):968
103. Fromager E (2020) Phys Rev Lett 124(24):243001. https://doi.org/10.1103/PhysRevLett.124.243001
104. Levy M, Zahariev F (2014) Phys Rev Lett 113(11):113002. https://doi.org/10.1103/PhysRevLett.113.113002
105. Levy M (1995) Phys Rev A 52(6):R4313. https://doi.org/10.1103/PhysRevA.52.R4313
106. Perdew JP, Levy M (1983) Phys Rev Lett 51(20):1884. https://doi.org/10.1103/PhysRevLett.51.1884
107. Senjean B, Fromager E (2020) Int J Quantum Chem 120(21):e26190. https://doi.org/10.1002/qua.26190
108. Hodgson MJ, Kraisler E, Schild A, Gross EK (2017) J Phys Chem Lett 8(24):5974. https://doi.org/10.1021/acs.jpclett.7b02615
109. Guandalini A, Rozzi CA, Räsänen E, Pittalis S (2019) Phys Rev B 99:125140. https://doi.org/10.1103/PhysRevB.99.125140
110. Guandalini A, Ruini A, Räsänen E, Rozzi CA, Pittalis S (2021) Phys Rev B 104:085110. https://doi.org/10.1103/PhysRevB.104.085110
111. Rauch Tcv, Marques MAL, Botti S (2020) Phys Rev B 101:245163. https://doi.org/10.1103/PhysRevB.101.245163
112. Rauch Tcv, Marques MAL, Botti S (2020) Phys Rev B 102:119902. https://doi.org/10.1103/PhysRevB.102.119902
113. Hodgson MJP, Wetherell J, Fromager E (2021) Phys Rev A 103(1):012806. https://doi.org/10.1103/PhysRevA.103.012806
114. Baerends EJ (2020) Mol Phys 118(5):e1612955. https://doi.org/10.1080/00268976.2019.1612955
115. Levy M (1982) Phys Rev A 26(3):1200
116. Janak JF (1978) Phys Rev B 18(12):7165. https://doi.org/10.1103/PhysRevB.18.7165
117. Levy M, Perdew JP, Sahni V (1984) Phys Rev A 30:2745. https://doi.org/10.1103/PhysRevA.30.2745
118. Hofmann D, Kümmel S (2012) Phys Rev B 86:201109. https://doi.org/10.1103/PhysRevB.86.201109
119. Koentopp M, Burke K, Evers F (2006) Phys Rev B 73:121403. https://doi.org/10.1103/PhysRevB.73.121403
120. Gould T, Kronik L (2021) J Chem Phys 154(9):094125. https://doi.org/10.1063/5.0040447
121. Kümmel S, Kronik L (2008) Rev Mod Phys 80:3. https://doi.org/10.1103/RevModPhys.80.3
122. Nagy Á (2001) J Phys B At Mol Phys 34(12):2363. https://doi.org/10.1088/0953-4075/34/12/305
123. Paragi G, Gyémánt I, Van Doren VE (2000) Chem Phys Lett 324(5–6):440. https://doi.org/10.1016/S0009-2614(00)00613-8

124. Paragi G, Gyémánt I, VanDoren VE (2001) J Mol Struct (Theochem) 571(1–3):153. https://doi.org/10.1016/S0166-1280(01)00561-9
125. Seidl A, Görling A, Vogl P, Majewski J, Levy M (1996) Phys Rev B 53(7):3764
126. Helgaker T, Jorgensen P, Olsen J (2014) Molecular electronic-structure theory. Wiley, New York. https://doi.org/10.1002/9781119019572
127. Gidopoulos NI, Papaconstantinou PG, Gross EKU (2002) Phys Rev Lett 88:033003. https://doi.org/10.1103/PhysRevLett.88.033003
128. Pastorczak E, Pernal K (2014) J Chem Phys 140:18A514. https://doi.org/10.1063/1.4866998
129. Hirao K, Nakatsuji H (1973) J Chem Phys 59(3):1457. https://doi.org/10.1063/1.1680203
130. Schilling C, Pittalis S (2021) Phys Rev Lett 127:023001. https://doi.org/10.1103/PhysRevLett.127.023001
131. Kvaal S, Ekström U, Teale AM, Helgaker T (2014) J Chem Phys 140(18):18A518. https://doi.org/10.1063/1.4867005
132. Penz M, Laestadius A, Tellgren EI, Ruggenthaler M (2019) Phys Rev Lett 123:037401. https://doi.org/10.1103/PhysRevLett.123.037401
133. Senjean B, Tsuchiizu M, Robert V, Fromager E (2017) Mol Phys 115(1–2):48. https://doi.org/10.1080/00268976.2016.1182224
134. Li C, Requist R, Gross EKU (2018) J Chem Phys 148(8):084110. https://doi.org/10.1063/1.5011663
135. Carrascal DJ, Ferrer J, Maitra N, Burke K (2018) Eur Phys J B 91(7):142. https://doi.org/10.1140/epjb/e2018-90114-9
136. Smith JC, Pribram-Jones A, Burke K (2016) Phys Rev B 93:245131. https://doi.org/10.1103/PhysRevB.93.245131
137. Burton HGA, Marut C, Daas TJ, Gori-Giorgi P, Loos PF (2021) J Chem Phys 155(5):054107. https://doi.org/10.1063/5.0056968
138. Loos PF (2017) J Chem Phys 146(11):114108. https://doi.org/10.1063/1.4978409
139. Gould T, Pittalis S (2020) Aust J Chem 73(8):714. https://doi.org/10.1071/CH19504
140. Gould T (2020) J Phys Chem Lett 11(22):9907. https://doi.org/10.1021/acs.jpclett.0c02894
141. Fromager E (2015) Mol Phys 113(5):419. https://doi.org/10.1080/00268976.2014.993342
142. Görling A, Levy M (1994) Phys Rev A 50(1):196
143. Görling A, Levy M (1995) Int J Quantum Chem 56(S29):93. https://doi.org/10.1002/qua.560560810
144. Ivanov S, Levy M (2002) J Chem Phys 116(16):6924. https://doi.org/10.1063/1.1453952
145. Yang Zh (2021) arXiv:2109.07697

Publisher's Note Springer Nature remains neutral with regard to jurisdictional claims in published maps and institutional affiliations.

Authors and Affiliations

Filip Cernatic[1] · Bruno Senjean[2] · Vincent Robert[1] · Emmanuel Fromager[1]

Filip Cernatic
filip.cernatic@etu.unistra.fr

Bruno Senjean
bruno.senjean@umontpellier.fr

Vincent Robert
vrobert@unistra.fr

1 Laboratoire de Chimie Quantique, Institut de Chimie, CNRS/Université de Strasbourg, 4 rue Blaise Pascal, 67000 Strasbourg, France

2 ICGM, Univ Montpellier, CNRS, ENSCM, Montpellier, France